Evolutionary Computation: Theories, Techniques, and Applications

Evolutionary Computation: Theories, Techniques, and Applications

Editor

Vincent A. Cicirello

Basel • Beijing • Wuhan • Barcelona • Belgrade • Novi Sad • Cluj • Manchester

Editor
Vincent A. Cicirello
Stockton University
Galloway
USA

Editorial Office
MDPI AG
Grosspeteranlage 5
4052 Basel, Switzerland

This is a reprint of articles from the Special Issue published online in the open access journal *Applied Sciences* (ISSN 2076-3417) (available at: https://www.mdpi.com/journal/applsci/special_issues/Evolutionary_Computation).

For citation purposes, cite each article independently as indicated on the article page online and as indicated below:

Lastname, A.A.; Lastname, B.B. Article Title. *Journal Name* **Year**, *Volume Number*, Page Range.

ISBN 978-3-7258-2123-5 (Hbk)
ISBN 978-3-7258-2124-2 (PDF)
doi.org/10.3390/books978-3-7258-2124-2

© 2024 by the authors. Articles in this book are Open Access and distributed under the Creative Commons Attribution (CC BY) license. The book as a whole is distributed by MDPI under the terms and conditions of the Creative Commons Attribution-NonCommercial-NoDerivs (CC BY-NC-ND) license.

Contents

About the Editor vii

Preface ix

Vincent A. Cicirello
Evolutionary Computation: Theories, Techniques, and Applications
Reprinted from: *Appl. Sci.* 2024, *14*, 2542, doi:10.3390/app14062542 1

Vincent A. Cicirello
Cycle Mutation: Evolving Permutations via Cycle Induction
Reprinted from: *Appl. Sci.* 2022, *12*, 5506, doi:10.3390/app12115506 7

Valentín Osuna-Enciso and Elizabeth Guevara-Martínez
A Stigmergy-Based Differential Evolution
Reprinted from: *Appl. Sci.* 2022, *12*, 6093, doi:10.3390/app12126093 33

Alejandro Tapia Córdoba, Pablo Millán Gata and Daniel Gutiérrez Reina
Optimizing the Layout of Run-of-River Powerplants Using Cubic Hermite Splines and Genetic Algorithms
Reprinted from: *Appl. Sci.* 2022, *12*, 8133, doi:10.3390/app12168133 52

Daniel Parra, Alberto Gutiérrez-Gallego, Oscar Garnica, Jose Manuel Velasco, Khaoula Zekri-Nechar, José J. Zamorano-León, et al.
Predicting the Risk of Overweight and Obesity in Madrid—A Binary Classification Approach with Evolutionary Feature Selection
Reprinted from: *Appl. Sci.* 2022, *12*, 8251, doi:10.3390/app12168251 71

Yu-An Fan and Chiu-Kuo Liang
Hybrid Discrete Particle Swarm Optimization Algorithm with Genetic Operators for Target Coverage Problem in Directional Wireless Sensor Networks
Reprinted from: *Appl. Sci.* 2022, *12*, 8503, doi:10.3390/app12178503 99

Shir Li Wang, Sarah Hazwani Adnan, Haidi Ibrahim, Theam Foo Ng and Parvathy Rajendran
A Hybrid of Fully Informed Particle Swarm and Self-Adaptive Differential Evolution for Global Optimization
Reprinted from: *Appl. Sci.* 2022, *12*, 11367, doi:10.3390/app122211367 118

Tsai-Jung Chen, Ying-Ji Hong, Chia-Han Lin and Jing-Yuan Wang
Optimization on Linkage System for Vehicle Wipers by the Method of Differential Evolution
Reprinted from: *Appl. Sci.* 2023, *13*, 332, doi:10.3390/app13010332 135

Bruce Kwong-Bun Tong, Chi Wan Sung and Wing Shing Wong
Random Orthogonal Search with Triangular and Quadratic Distributions (TROS and QROS): Parameterless Algorithms for Global Optimization
Reprinted from: *Appl. Sci.* 2023, *13*, 1391, doi:10.3390/app13031391 154

Nikola Anđelić, Sandi Baressi Šegota, Matko Glučina, and Zlatan Car
Estimation of Interaction Locations in Super Cryogenic Dark Matter Search Detectors Using Genetic Programming-Symbolic Regression Method
Reprinted from: *Appl. Sci.* 2023, *13*, 2059, doi:10.3390/app13042059 178

Weidong Wu, Xiaoyan Sun, Guangyi Man, Shuai Li and Lin Bao
Interactive Multifactorial Evolutionary Optimization Algorithm with Multidimensional Preference Surrogate Models for Personalized Recommendation
Reprinted from: *Appl. Sci.* **2023**, *13*, 2243, doi:10.3390/app13042243 201

Rahul Dubey and Sushil J. Louis
Genetic Algorithms Optimized Adaptive Wireless Network Deployment
Reprinted from: *Appl. Sci.* **2023**, *13*, 4858, doi:10.3390/app13084858 224

Vasiliki Lazari and Athanasios Chassiakos
Multi-Objective Optimization of Electric Vehicle Charging Station Deployment Using Genetic Algorithms
Reprinted from: *Appl. Sci.* **2023**, *13*, 4867, doi:10.3390/app13084867 245

Hamza Reffad and Adel Alti
Semantic-Based Multi-Objective Optimization for QoS and Energy Efficiency in IoT, Fog, and Cloud ERP Using Dynamic Cooperative NSGA-II
Reprinted from: *Appl. Sci.* **2023**, *13*, 5218, doi:10.3390/app13085218 264

About the Editor

Vincent A. Cicirello

Vincent A. Cicirello is a Professor and Chair of Computer Science at Stockton University. He earned his Ph.D. in Robotics from Carnegie Mellon University in 2003, and he earned his M.Sc. in Computer Science and B.Sc. in Computer Science and Mathematics from Drexel University in 1999. He joined the faculty at Stockton University in 2005. His research interests include algorithms, artificial intelligence, evolutionary computation, computational intelligence, open-source research software, and swarm intelligence. He was honored with a Congressional Proclamation from the U.S. House of Representatives in 2022, was elevated to IEEE Senior Member in 2021, was designated an ACM Senior Member in 2011, received a U.S. Patent in 2010, and received the 2005 AAAI Outstanding Paper award. He has published over 60 journal articles and conference papers. He is a member of AAAI (Life Member), ACM (Senior Member), ACM SIGCSE, ACM SIGAI, ACM SIGEVO, CCSC, EAI (Distinguished Member), IEEE (Senior Member), the IEEE Computer Society, the IEEE Computational Intelligence Society, and SIAM.

Preface

Evolutionary computation offers powerful problem-solving methodologies inspired by models of natural genetics and evolutionary processes. Potential applications are wide-ranging and include problems related to combinatorial optimization, numerical optimization, multi-objective optimization, etc., as well as specific applications of these problems in diverse domains, such as engineering, design, medicine, robotics, science, etc. Techniques from evolutionary computation often lend themselves well to parallel and distributed implementations and are often more effective in dealing with challenging problem characteristics such as non-linearity and high dimensionality than alternative approaches. This Special Issue brings together recent advances in the theory and application of evolutionary computation. The published articles span the breadth of evolutionary computation techniques, covering topics including genetic algorithms, genetic programming, differential evolution, particle swarm optimization, other forms of evolutionary algorithms, and hybrids of multiple techniques.

Vincent A. Cicirello
Editor

Editorial

Evolutionary Computation: Theories, Techniques, and Applications

Vincent A. Cicirello

Computer Science, Stockton University, 101 Vera King Farris Dr, Galloway, NJ 08205, USA; cicirelv@stockton.edu

1. Introduction

Evolutionary computation is now nearly 50 years old, originating with the seminal work of John Holland at the University of Michigan in 1975 which introduced the genetic algorithm [1]. Evolutionary computation [2] encompasses a variety of problem-solving methodologies that take inspiration from natural evolutionary and genetic processes. The most well-known form of evolutionary computation is the genetic algorithm [3,4], which evolves a population of solutions to the problem at hand, each represented as a bit-string—the genotype—with a fitness function measuring the fitness of the bit-string within the context of the problem (i.e., mapping a genotype to a phenotype). Evolutionary operators, such as mutation, crossover, and selection, control the simulated evolution over several generations.

There are now many forms of evolutionary computation (a few of which are illustrated in Figure 1) that have developed over the years, including genetic programming [5], evolution strategies [6], differential evolution [7,8], evolutionary programming [9], permutation-based evolutionary algorithms [10], memetic algorithms [11], the estimation of distribution algorithms [12], particle swarm optimization [13], interactive evolutionary algorithms [14], ant colony optimization [15,16], and artificial immune systems [17], among others [18,19]. Among the characteristics of evolutionary algorithms that lead to powerful problem solving is the fact that they lend themselves very well to parallel implementation [20–22], enabling the exploitation of today's multicore and manycore computer architectures. Rich theoretical foundations also exist which are related to convergence properties [23–25], parameter optimization, and control [26], as well as the powerful analytical tools of fitness landscape analysis [27–29], such as fitness–distance correlation [30] and search landscape calculus [31], among others. These theoretical foundations inform the engineering of evolutionary solutions to specific problems. There are also many open-source libraries and toolkits available for evolutionary computation in a variety of programming languages [32–41], making the application of evolutionary algorithms to new problems and domains particularly easy.

Citation: Cicirello, V.A. Evolutionary Computation: Theories, Techniques, and Applications. *Appl. Sci.* **2024**, *14*, 2542. https://doi.org/10.3390/app14062542

Received: 12 March 2024
Accepted: 14 March 2024
Published: 18 March 2024

Copyright: © 2024 by the author. Licensee MDPI, Basel, Switzerland. This article is an open access article distributed under the terms and conditions of the Creative Commons Attribution (CC BY) license (https://creativecommons.org/licenses/by/4.0/).

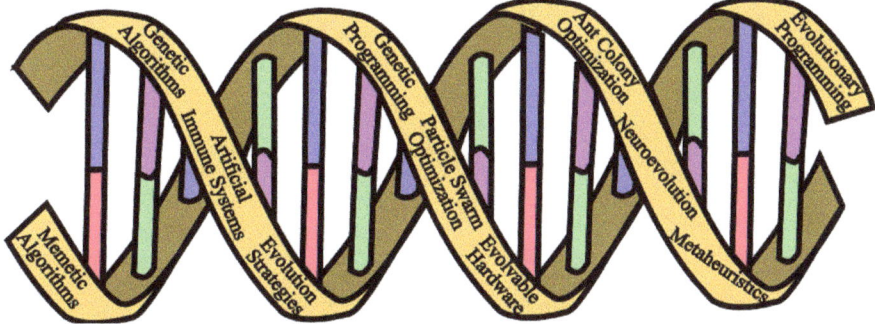

Figure 1. A few of the many forms of evolutionary computation.

Evolutionary computation has been effective in solving problems with a variety of characteristics, and within many application domains, such as multiobjective optimization [42–45], data science [46], machine learning [47–49], classification [50], feature selection [51], neural architecture search [52], neuroevolution [53], bioinformatics [54], scheduling [55], algorithm selection [56], computer vision [57], hardware validation [58], software engineering [59,60], and multi-task optimization [61,62], among many others.

This Special Issue brings together recent advances in the theory and application of evolutionary computation. It includes 13 articles. The authors of the 13 articles represent institutions from 11 different countries, demonstrating the global reach of the topic of evolutionary computation. The published articles span the breadth of evolutionary computation techniques, and cover a variety of applications. The remainder of this Editorial briefly describes the articles included within this Special Issue; and I encourage you to read and explore each.

2. Overview of the Published Articles

This overview of the articles is organized in the order in which the contributions to the Special Issue were published.

Cicirello (contribution 1) presents a new mutation operator for evolutionary algorithms where solutions are represented by permutations. The new mutation operator, cycle mutation, is inspired by cycle crossover. Cycle mutation is designed specifically for assignment and mapping problems (e.g., quadratic assignment, largest common subgraph, etc.) rather than ordering problems like the traveling salesperson. This article includes a fitness landscape analysis exploring the strengths and weaknesses of cycle mutation in terms of permutation features.

Osuna-Enciso and Guevara-Martínez (contribution 2) propose a variation of differential evolution that they call stigmergic differential evolution which can be used for solving continuous optimization problems. Their approach integrates the concept of stigmergy with differential evolution. Stigmergy originated from swarm intelligence, and refers to the indirect communication among members of a swarm that occurs when swarm members manipulate the environment and detect modifications made by others (e.g., the pheromone trail-following behavior of ants, among others).

Córdoba, Gata, and Reina (contribution 3) consider a problem related to energy access in remote, rural areas. Namely, they utilize a $(\mu + \lambda)$-evolutionary algorithm to optimize the design of mini hydropower plants, using cubic Hermite splines to model the terrain in 3D, rather than the more common 2D simplifications.

Parra, et al. (contribution 4) consider the binary classification problem of predicting obesity. In their experiments, they explore utilizing evolutionary computation in feature selection for binary classifier systems. They consider ten different machine learning classifiers, combined with four feature-selection strategies. Two of the feature-selection strategies considered use the classic bit-string-encoded genetic algorithm.

Fan and Liang (contribution 5) consider directional sensor networks and target coverage. In their approach to target coverage, they developed a hybrid of particle swarm optimization and a genetic algorithm. Their experiments demonstrate that the hybrid approach outperforms both particle swarm optimization and the genetic algorithm alone for the problem of maximizing covered targets and minimizing active sensors.

Wang, et al. (contribution 6) developed a hybrid between particle swarm optimization and differential evolution for real-valued function optimization. Their hybrid combines a self-adaptive form of differential evolution with particle swarm optimization, and they experiment with their approach on a variety of function optimization benchmarks.

Chen, et al. (contribution 7) explore the constrained optimization problem of optimizing the linkage system for vehicle wipers. Their aim was to improve steadiness of wipers. They utilize differential evolution to optimize the maximal magnitude of the angular acceleration of the links in the system subject to a set of constraints. They were able to reduce the maximal magnitude of angular acceleration by 10%.

Tong, Sung, and Wong (contribution 8) analyze the performance of a parameter-free evolutionary algorithm known as pure random orthogonal search. They propose improvements to the algorithm involving local search. They performed experiments on a variety of benchmark function optimization problems with a variety of features (e.g., unimodal vs. multi-modal, convex vs. non-convex, separable vs. non-separable).

Anđelić, et al. (contribution 9) approach the problem of searching for candidates for dark matter particles, so-called weakly interacting massive particles, using symbolic regression via genetic programming. Their approach estimates the interaction locations with high accuracy.

Wu, et al. (contribution 10) developed a recommender system utilizing an interactive evolutionary algorithm for making personalized recommendations. In an interactive evolutionary algorithm, human users are directly involved in evaluating the fitness of members of the population. Wu, et al. use a surrogate model in their approach to reduce the number of evaluations required by users.

Dubey and Louis (contribution 11) utilize a $(\mu + \lambda)$-evolutionary algorithm. They developed an approach to deploying a UAV-based ad hoc network to cover an area of interest. UAV motion is controlled by a set of potential fields that are optimized by the $(\mu + \lambda)$-evolutionary algorithm using polynomial mutation and simulated binary crossover.

Lazari and Chassiakos (contribution 12) take on the problem of deploying electric vehicle charging stations. They define it as a multi-objective optimization problem with two cost functions: station deployment costs and user travel costs between areas of demand and the station's location. Their evolutionary algorithm's chromosome representation combines x and y coordinates of candidate charging station locations, using the classic bit-string of genetic algorithms to model whether or not each candidate station is deployed.

Reffad and Alti (contribution 13) use NSGA-II to optimize enterprise resource planning performance. They aimed to optimize average service quality and average energy consumption. They propose an adaptive and dynamic solution within IoT, fog, and cloud environments.

3. Conclusions

This collection of articles spans a variety of forms of evolutionary computation, including genetic algorithms, genetic programming, differential evolution, particle swarm optimization, and evolutionary algorithms more generally, as well as hybrids of multiple forms of evolutionary computation. The evolutionary algorithms represent solutions in several ways, including the common bit-string representation, vectors of reals, and permutations, as well as custom representations. The authors of the articles tackle a very diverse collection of problems of different types and from many application domains. For example, some of the problems considered are discrete optimization problems, while others optimize continuous functions. Although many of the articles focus on optimizing a single objective function, others involve multi-objective optimization. Some of the articles primarily utilize common benchmarking optimization functions and problems, while several others explore a variety of real-world applications, such as optimizing mini hydropower plants, UAV deployment, the deployment of electric vehicle charging stations, target coverage in wireless sensor networks, enterprise resource planning, recommender systems, dark matter detection, and optimizing vehicle wiper linkage systems, among others. The diversity of evolutionary techniques, evolutionary operators, problem features, and applications that are covered within this collection of articles demonstrates the wide reach and applicability of evolutionary computation.

Conflicts of Interest: The author declares no conflicts of interest.

List of Contributions

1. Cicirello, V.A. Cycle Mutation: Evolving Permutations via Cycle Induction. *Appl. Sci.* **2022**, *12*, 5506. https://doi.org/10.3390/app12115506.
2. Osuna-Enciso, V.; Guevara-Martínez, E. A Stigmergy-Based Differential Evolution. *Appl. Sci.* **2022**, *12*, 6093. https://doi.org/10.3390/app12126093.
3. Córdoba, A.T.; Gata, P.M.; Reina, D.G. Optimizing the Layout of Run-of-River Powerplants Using Cubic Hermite Splines and Genetic Algorithms. *Appl. Sci.* **2022**, *12*, 8133. https://doi.org/10.3390/app12168133.
4. Parra, D.; Gutiérrez-Gallego, A.; Garnica, O.; Velasco, J.M.; Zekri-Nechar, K.; Zamorano-León, J.J.; Heras, N.d.l.; Hidalgo, J.I. Predicting the Risk of Overweight and Obesity in Madrid—A Binary Classification Approach with Evolutionary Feature Selection. *Appl. Sci.* **2022**, *12*, 8251. https://doi.org/10.3390/app12168251.
5. Fan, Y.A.; Liang, C.K. Hybrid Discrete Particle Swarm Optimization Algorithm with Genetic Operators for Target Coverage Problem in Directional Wireless Sensor Networks. *Appl. Sci.* **2022**, *12*, 8503. https://doi.org/10.3390/app12178503.
6. Wang, S.L.; Adnan, S.H.; Ibrahim, H.; Ng, T.F.; Rajendran, P. A Hybrid of Fully Informed Particle Swarm and Self-Adaptive Differential Evolution for Global Optimization. *Appl. Sci.* **2022**, *12*, 11367. https://doi.org/10.3390/app122211367.
7. Chen, T.J.; Hong, Y.J.; Lin, C.H.; Wang, J.Y. Optimization on Linkage System for Vehicle Wipers by the Method of Differential Evolution. *Appl. Sci.* **2023**, *13*, 332. https://doi.org/10.3390/app13010332.
8. Tong, B.K.B.; Sung, C.W.; Wong, W.S. Random Orthogonal Search with Triangular and Quadratic Distributions (TROS and QROS): Parameterless Algorithms for Global Optimization. *Appl. Sci.* **2023**, *13*, 1391. https://doi.org/10.3390/app13031391.
9. Anđelić, N.; Baressi Šegota, S.; Glučina, M.; Car, Z. Estimation of Interaction Locations in Super Cryogenic Dark Matter Search Detectors Using Genetic Programming-Symbolic Regression Method. *Appl. Sci.* **2023**, *13*, 2059. https://doi.org/10.3390/app13042059.
10. Wu, W.; Sun, X.; Man, G.; Li, S.; Bao, L. Interactive Multifactorial Evolutionary Optimization Algorithm with Multidimensional Preference Surrogate Models for Personalized Recommendation. *Appl. Sci.* **2023**, *13*, 2243. https://doi.org/10.3390/app13042243.
11. Dubey, R.; Louis, S.J. Genetic Algorithms Optimized Adaptive Wireless Network Deployment. *Appl. Sci.* **2023**, *13*, 4858. https://doi.org/10.3390/app13084858.
12. Lazari, V.; Chassiakos, A. Multi-Objective Optimization of Electric Vehicle Charging Station Deployment Using Genetic Algorithms. *Appl. Sci.* **2023**, *13*, 4867. https://doi.org/10.3390/app13084867.
13. Reffad, H.; Alti, A. Semantic-Based Multi-Objective Optimization for QoS and Energy Efficiency in IoT, Fog, and Cloud ERP Using Dynamic Cooperative NSGA-II. *Appl. Sci.* **2023**, *13*, 5218. https://doi.org/10.3390/app13085218.

References

1. Holland, J.H. *Adaptation in Natural and Artificial Systems: An Introductory Analysis with Applications to Biology, Control, and Artificial Intelligence*; MIT Press: Cambridge, MA, USA, 1992.
2. Eiben, A.; Smith, J. *Introduction to Evolutionary Computing*, 2nd ed.; Springer: Heidelberg, Germany, 2015.
3. Mitchell, M. *An Introduction to Genetic Algorithms*; MIT Press: Cambridge, MA, USA, 1998.
4. Katoch, S.; Chauhan, S.S.; Kumar, V. A review on genetic algorithm: Past, present, and future. *Multimed. Tools Appl.* **2021**, *80*, 8091–8126. [CrossRef]
5. Langdon, W.B.; Poli, R. *Foundations of Genetic Programming*; Springer: Heidelberg, Germany, 2010.
6. Beyer, H.G.; Schwefel, H.P. Evolution strategies—A comprehensive introduction. *Nat. Comput. Int. J.* **2002**, *1*, 3–52. [CrossRef]
7. Das, S.; Suganthan, P.N. Differential Evolution: A Survey of the State-of-the-Art. *IEEE Trans. Evol. Comput.* **2011**, *15*, 4–31. [CrossRef]
8. Bilal; Pant, M.; Zaheer, H.; Garcia-Hernandez, L.; Abraham, A. Differential Evolution: A review of more than two decades of research. *Eng. Appl. Artif. Intell.* **2020**, *90*, 103479. [CrossRef]

9. Yao, X.; Liu, Y.; Lin, G. Evolutionary programming made faster. *IEEE Trans. Evol. Comput.* **1999**, *3*, 82–102. [CrossRef]
10. Cicirello, V.A. A Survey and Analysis of Evolutionary Operators for Permutations. In Proceedings of the 15th International Joint Conference on Computational Intelligence, Rome, Italy, 13–15 November 2023; pp. 288–299. [CrossRef]
11. Osaba, E.; Del Ser, J.; Cotta, C.; Moscato, P. Memetic Computing: Accelerating optimization heuristics with problem-dependent local search methods. *Swarm Evol. Comput.* **2022**, *70*, 101047. [CrossRef]
12. Larrañaga, P.; Bielza, C. Estimation of Distribution Algorithms in Machine Learning: A Survey. *IEEE Trans. Evol. Comput.* **2023**, early access. [CrossRef]
13. Kennedy, J.; Eberhart, R. Particle swarm optimization. In Proceedings of the International Conference on Neural Networks, Perth, WA, Australia, 27 November–1 December 1995; Volume 4, pp. 1942–1948. [CrossRef]
14. Uusitalo, S.; Kantosalo, A.; Salovaara, A.; Takala, T.; Guckelsberger, C. Creative collaboration with interactive evolutionary algorithms: A reflective exploratory design study. *Genet. Program. Evolvable Mach.* **2023**, *25*, 4. [CrossRef]
15. Dorigo, M.; Gambardella, L. Ant colony system: A cooperative learning approach to the traveling salesman problem. *IEEE Trans. Evol. Comput.* **1997**, *1*, 53–66. [CrossRef]
16. Dorigo, M.; Maniezzo, V.; Colorni, A. Ant system: Optimization by a colony of cooperating agents. *IEEE Trans. Syst. Man Cybern. Part B (Cybernetics)* **1996**, *26*, 29–41. [CrossRef]
17. Dasgupta, D. Advances in artificial immune systems. *IEEE Comput. Intell. Mag.* **2006**, *1*, 40–49. [CrossRef]
18. Siarry, P. (Ed.) *Metaheuristics*; Springer Nature: Cham, Switzerland, 2016.
19. Hoos, H.H.; Stützle, T. *Stochastic Local Search: Foundations and Applications*; Morgan Kaufmann: San Francisco, CA, USA, 2005.
20. Harada, T.; Alba, E. Parallel Genetic Algorithms: A Useful Survey. *ACM Comput. Surv.* **2020**, *53*, 86. [CrossRef]
21. Cicirello, V.A. Impact of Random Number Generation on Parallel Genetic Algorithms. In Proceedings of the 31st International Florida Artificial Intelligence Research Society Conference, Melbourne, FL, USA, 21–23 May 2018; AAAI Press: Menlo Park, CA, USA, 2018; pp. 2–7.
22. Luque, G.; Alba, E. *Parallel Genetic Algorithms: Theory and Real World Applications*; Springer: Heidelberg, Germany, 2011.
23. Rudolph, G. Convergence analysis of canonical genetic algorithms. *IEEE Trans. Neural Netw.* **1994**, *5*, 96–101. [CrossRef] [PubMed]
24. Rudolph, G. Convergence of evolutionary algorithms in general search spaces. In Proceedings of the IEEE International Conference on Evolutionary Computation, Nagoya, Japan, 20–22 May 1996; pp. 50–54. [CrossRef]
25. He, J.; Yao, X. Drift analysis and average time complexity of evolutionary algorithms. *Artif. Intell.* **2001**, *127*, 57–85. [CrossRef]
26. Karafotias, G.; Hoogendoorn, M.; Eiben, A.E. Parameter Control in Evolutionary Algorithms: Trends and Challenges. *IEEE Trans. Evol. Comput.* **2015**, *19*, 167–187. [CrossRef]
27. Cicirello, V.A. On Fitness Landscape Analysis of Permutation Problems: From Distance Metrics to Mutation Operator Selection. *Mob. Netw. Appl.* **2023**, *28*, 507–517. [CrossRef]
28. Pimenta, C.G.; de Sá, A.G.C.; Ochoa, G.; Pappa, G.L. Fitness Landscape Analysis of Automated Machine Learning Search Spaces. In Proceedings of the Evolutionary Computation in Combinatorial Optimization: 20th European Conference, EvoCOP 2020, Held as Part of EvoStar 2020, Seville, Spain, 15–17 April 2020; Springer: Cham, Switzerland, 2020; pp. 114–130. [CrossRef]
29. Huang, Y.; Li, W.; Tian, F.; Meng, X. A fitness landscape ruggedness multiobjective differential evolution algorithm with a reinforcement learning strategy. *Appl. Soft Comput.* **2020**, *96*, 106693. [CrossRef]
30. Jones, T.; Forrest, S. Fitness Distance Correlation as a Measure of Problem Difficulty for Genetic Algorithms. In Proceedings of the 6th International Conference on Genetic Algorithms, Pittsburgh, PA, USA, 15–19 July 1995; pp. 184–192.
31. Cicirello, V.A. The Permutation in a Haystack Problem and the Calculus of Search Landscapes. *IEEE Trans. Evol. Comput.* **2016**, *20*, 434–446. [CrossRef]
32. Scott, E.O.; Luke, S. ECJ at 20: Toward a General Metaheuristics Toolkit. In Proceedings of the Genetic and Evolutionary Computation Conference Companion, Prague, Czech Republic, 13–17 July 2019; ACM Press: New York, NY, USA, 2019; pp. 1391–1398. [CrossRef]
33. Cicirello, V.A. Chips-n-Salsa: A Java Library of Customizable, Hybridizable, Iterative, Parallel, Stochastic, and Self-Adaptive Local Search Algorithms. *J. Open Source Softw.* **2020**, *5*, 2448. [CrossRef]
34. Jenetics. Jenetics—Genetic Algorithm, Genetic Programming, Evolutionary Algorithm, and Multi-Objective Optimization. 2024. Available online: https://jenetics.io/ (accessed on 27 January 2024).
35. Bell, I.H. CEGO: C++11 Evolutionary Global Optimization. *J. Open Source Softw.* **2019**, *4*, 1147. [CrossRef]
36. Gijsbers, P.; Vanschoren, J. GAMA: Genetic Automated Machine learning Assistant. *J. Open Source Softw.* **2019**, *4*, 1132. [CrossRef]
37. Detorakis, G.; Burton, A. GAIM: A C++ library for Genetic Algorithms and Island Models. *J. Open Source Softw.* **2019**, *4*, 1839. [CrossRef]
38. de Dios, J.A.M.; Mezura-Montes, E. Metaheuristics: A Julia Package for Single- and Multi-Objective Optimization. *J. Open Source Softw.* **2022**, *7*, 4723. [CrossRef]
39. Izzo, D.; Biscani, F. dcgp: Differentiable Cartesian Genetic Programming made easy. *J. Open Source Softw.* **2020**, *5*, 2290. [CrossRef]
40. Simson, J. LGP: A robust Linear Genetic Programming implementation on the JVM using Kotlin. *J. Open Source Softw.* **2019**, *4*, 1337. [CrossRef]
41. Tarkowski, T. Quilë: C++ genetic algorithms scientific library. *J. Open Source Softw.* **2023**, *8*, 4902. [CrossRef]

42. Deb, K.; Pratap, A.; Agarwal, S.; Meyarivan, T. A fast and elitist multiobjective genetic algorithm: NSGA-II. *IEEE Trans. Evol. Comput.* **2002**, *6*, 182–197. [CrossRef]
43. Liang, J.; Ban, X.; Yu, K.; Qu, B.; Qiao, K.; Yue, C.; Chen, K.; Tan, K.C. A Survey on Evolutionary Constrained Multiobjective Optimization. *IEEE Trans. Evol. Comput.* **2023**, *27*, 201–221. [CrossRef]
44. Tian, Y.; Si, L.; Zhang, X.; Cheng, R.; He, C.; Tan, K.C.; Jin, Y. Evolutionary Large-Scale Multi-Objective Optimization: A Survey. *ACM Comput. Surv.* **2021**, *54*, 174. [CrossRef]
45. Li, M.; Yao, X. Quality Evaluation of Solution Sets in Multiobjective Optimisation: A Survey. *ACM Comput. Surv.* **2019**, *52*, 26. [CrossRef]
46. Sohail, A. Genetic Algorithms in the Fields of Artificial Intelligence and Data Sciences. *Ann. Data Sci.* **2023**, *10*, 1007–1018. [CrossRef]
47. Li, N.; Ma, L.; Yu, G.; Xue, B.; Zhang, M.; Jin, Y. Survey on Evolutionary Deep Learning: Principles, Algorithms, Applications, and Open Issues. *ACM Comput. Surv.* **2023**, *56*, 41. [CrossRef]
48. Telikani, A.; Tahmassebi, A.; Banzhaf, W.; Gandomi, A.H. Evolutionary Machine Learning: A Survey. *ACM Comput. Surv.* **2021**, *54*, 161. [CrossRef]
49. Li, N.; Ma, L.; Xing, T.; Yu, G.; Wang, C.; Wen, Y.; Cheng, S.; Gao, S. Automatic design of machine learning via evolutionary computation: A survey. *Appl. Soft Comput.* **2023**, *143*, 110412. [CrossRef]
50. Espejo, P.G.; Ventura, S.; Herrera, F. A Survey on the Application of Genetic Programming to Classification. *IEEE Trans. Syst. Man, Cybern. Part C (Appl. Rev.)* **2010**, *40*, 121–144. [CrossRef]
51. Xue, B.; Zhang, M.; Browne, W.N.; Yao, X. A Survey on Evolutionary Computation Approaches to Feature Selection. *IEEE Trans. Evol. Comput.* **2016**, *20*, 606–626. [CrossRef]
52. Zhou, X.; Qin, A.K.; Sun, Y.; Tan, K.C. A Survey of Advances in Evolutionary Neural Architecture Search. In Proceedings of the 2021 IEEE Congress on Evolutionary Computation (CEC), Virtually, 28 June–1 July 2021; pp. 950–957. [CrossRef]
53. Papavasileiou, E.; Cornelis, J.; Jansen, B. A Systematic Literature Review of the Successors of "NeuroEvolution of Augmenting Topologies". *Evol. Comput.* **2021**, *29*, 1–73. [CrossRef]
54. Fogel, G.B.; Corne, D.W. (Eds.) *Evolutionary Computation in Bioinformatics*; Morgan Kaufmann: San Francisco, CA, USA, 2003.
55. Zhang, F.; Mei, Y.; Nguyen, S.; Zhang, M. Survey on Genetic Programming and Machine Learning Techniques for Heuristic Design in Job Shop Scheduling. *IEEE Trans. Evol. Comput.* **2023**, *28*, 147–167. [CrossRef]
56. Kerschke, P.; Hoos, H.H.; Neumann, F.; Trautmann, H. Automated Algorithm Selection: Survey and Perspectives. *Evol. Comput.* **2019**, *27*, 3–45. [CrossRef]
57. Bi, Y.; Xue, B.; Mesejo, P.; Cagnoni, S.; Zhang, M. A Survey on Evolutionary Computation for Computer Vision and Image Analysis: Past, Present, and Future Trends. *IEEE Trans. Evol. Comput.* **2023**, *27*, 5–25. [CrossRef]
58. Jayasena, A.; Mishra, P. Directed Test Generation for Hardware Validation: A Survey. *ACM Comput. Surv.* **2024**, *56*, 132. [CrossRef]
59. Sobania, D.; Schweim, D.; Rothlauf, F. A Comprehensive Survey on Program Synthesis with Evolutionary Algorithms. *IEEE Trans. Evol. Comput.* **2023**, *27*, 82–97. [CrossRef]
60. Arcuri, A.; Galeotti, J.P.; Marculescu, B.; Zhang, M. EvoMaster: A Search-Based System Test Generation Tool. *J. Open Source Softw.* **2021**, *6*, 2153. [CrossRef]
61. Tan, Z.; Luo, L.; Zhong, J. Knowledge transfer in evolutionary multi-task optimization: A survey. *Appl. Soft Comput.* **2023**, *138*, 110182. [CrossRef]
62. Zhao, H.; Ning, X.; Liu, X.; Wang, C.; Liu, J. What makes evolutionary multi-task optimization better: A comprehensive survey. *Appl. Soft Comput.* **2023**, *145*, 110545. [CrossRef]

Disclaimer/Publisher's Note: The statements, opinions and data contained in all publications are solely those of the individual author(s) and contributor(s) and not of MDPI and/or the editor(s). MDPI and/or the editor(s) disclaim responsibility for any injury to people or property resulting from any ideas, methods, instructions or products referred to in the content.

Article

Cycle Mutation: Evolving Permutations via Cycle Induction

Vincent A. Cicirello

Computer Science, Stockton University, 101 Vera King Farris Dr, Galloway, NJ 08205, USA; cicirelv@stockton.edu

Abstract: Evolutionary algorithms solve problems by simulating the evolution of a population of candidate solutions. We focus on evolving permutations for ordering problems such as the traveling salesperson problem (TSP), as well as assignment problems such as the quadratic assignment problem (QAP) and largest common subgraph (LCS). We propose cycle mutation, a new mutation operator whose inspiration is the well-known cycle crossover operator, and the concept of a permutation cycle. We use fitness landscape analysis to explore the problem characteristics for which cycle mutation works best. As a prerequisite, we develop new permutation distance measures: cycle distance, k-cycle distance, and cycle edit distance. The fitness landscape analysis predicts that cycle mutation is better suited for assignment and mapping problems than it is for ordering problems. We experimentally validate these findings showing cycle mutation's strengths on problems such as QAP and LCS, and its limitations on problems such as the TSP, while also showing that it is less prone to local optima than commonly used alternatives. We integrate cycle mutation into the open source Chips-n-Salsa library, and the new distance metrics into the open source JavaPermutationTools library.

Keywords: combinatorial optimization; evolutionary algorithms; fitness distance correlation; fitness landscape analysis; genetic algorithms; mutation; permutation cycles; permutation distance

PACS: 02.70.-c; 07.05.Mh; 89.20.Ff

MSC: 05A05; 05C60; 68T05; 68T20; 68W20; 68W50; 90C27; 90C59

Citation: Cicirello, V.A. Cycle Mutation: Evolving Permutations via Cycle Induction. Appl. Sci. 2022, 12, 5506. https://doi.org/10.3390/app12115506

Academic Editor: Giancarlo Mauri

Received: 6 May 2022
Accepted: 27 May 2022
Published: 29 May 2022

Publisher's Note: MDPI stays neutral with regard to jurisdictional claims in published maps and institutional affiliations.

Copyright: © 2022 by the author. Licensee MDPI, Basel, Switzerland. This article is an open access article distributed under the terms and conditions of the Creative Commons Attribution (CC BY) license (https://creativecommons.org/licenses/by/4.0/).

1. Introduction

In an evolutionary algorithm (EA), a problem is solved through the simulated evolution of a population that evolves over many generations. There are many types of EA that mostly differ in the types of problems they solve and in how solutions are represented. For example, a genetic algorithm (GA) [1], the original EA, usually represents a candidate solution to an optimization problem with a vector of bits. Evolution strategies (ESs) [2] focus specifically on real-valued function optimization, utilizing a vector of floating-point values to represent each candidate solution. Genetic programming [3] is an approach to automatic inductive programming, and evolves a population of programs, each of which is typically represented with a tree structure.

This paper focuses on EAs that encode solutions with permutations, often referred to as a permutation-based GA [4–7], while others prefer the more general term EA [8,9] to avoid confusion with a binary-encoded GA. Solutions to some problems are more naturally represented with a permutation than with another representation. The classic example is the traveling salesperson problem (TSP), where a solution is a tour of the cities, and thus can be represented in a straightforward way as a permutation of indexes into a list of cities.

One challenge with designing a permutation EA is deciding which genetic operators to use. This is less of an issue with the classic bit-vector GA or a real-valued ES because it is possible to mutate bits independently of the rest of a bit-vector in a GA, and real-valued alleles in an ES can likewise be mutated independently from the vector as a whole, such as with Gaussian mutation [10] or Cauchy mutation [11]. Within a permutation-based EA, mutation cannot change individual elements independent of the rest of the permutation. For example, in the permutation [2, 4, 0, 3, 1] of the first five integers, we cannot mutate one value in isolation or it leads to an invalid permutation. Crossover cannot naively exchange parts of parents. For example, if one parent is as above, and the other is [4, 2, 1, 0, 3], exchanging the second and third elements between the parents leads to two invalid permutations, [2, 2, 1, 3, 1] and [4, 4, 0, 0, 3]. Thus, for permutations, mutation and crossover must consider the overall structure to ensure a valid encoding.

As a consequence, many mutation and crossover operators exist for permutations. The suitability of each depends upon the characteristics of the permutation that most significantly impacts solution fitness for the problem at hand, such as absolute element positions, relative element positions, or element precedences [12,13]. Only relative positions matter to fitness of a TSP solution. For example, if cities i and j are adjacent in the permutation, then the solution includes the edge (i, j) regardless of where the pair of cities appears. The absolute element positions are most important for other problems, such as the largest common subgraph (LCS). The LCS is an NP-hard [14] optimization problem involving finding a one-to-one mapping between the vertex sets of a pair of graphs to maximize the number of edges of the common subgraph implied by the mapping. As a permutation problem, one holds the vertexes of one graph in a fixed order, and a permutation of the vertexes of the other graph represents a mapping. The absolute index of a vertex in the permutation therefore corresponds to the vertex it is mapped to in the other graph.

Mutation for permutations is most often one of swap mutation, insertion mutation, reversal mutation, or scramble mutation [15]. There are many permutation crossover operators that focus on maintaining different characteristics of the parents, including order crossover [16], non-wrapping order crossover [17], uniform order-based crossover [18], partially matched crossover [19], uniform partially matched crossover [20], precedence preservative crossover [21], edge assembly crossover [22,23], and cycle crossover (CX) [24].

The central aim of this paper is development of a mutation operator for permutations that (a) is characterized by small random perturbations on average, (b) has a large neighborhood size, and (c) is tunable. These properties enable focused search with improved handling of local optima. No such mutation operators currently exist for permutations. For bit-vectors, the standard bit-flip mutation satisfies all of these properties. For example, the bit-flip mutation rate M is usually set to a low value for a small average number of bits flipped, but can be tuned for problems where greater mutation is beneficial, and as long as M is non-zero, the neighborhood includes all other bit-vectors. Similarly, Gaussian mutation in an ES satisfies all of these properties, with a tunable parameter, the standard deviation of the Gaussian. All existing permutation mutation operators have a fixed neighborhood size, which in most cases is small, that depends only on permutation length.

To achieve this aim, we present a new mutation operator called cycle mutation, which relies on the concept of a permutation cycle and is inspired by CX. Rather than operating on two parents as CX does, cycle mutation instead mutates a single member of the population. Thus, it is also applicable to non-population metaheuristics such as simulated annealing (SA) [25–27]. We develop two variations of cycle mutation, Cycle($kmax$) and Cycle(α), offering two ways of addressing locally optimal solutions.

To formally demonstrate that cycle mutation achieves the target properties of large neighborhood but small average changes, we conduct a fitness landscape analysis of cycle mutation, and other permutation mutation operators for three NP-hard optimization problems [14], the TSP, the LCS, and the quadratic assignment problem (QAP). The fitness landscape analysis predicts that cycle mutation likely performs well for assignment and mapping problems, such as LCS and QAP, where absolute positions directly affect fitness, but that it may be less well-suited to relative ordering problems such as the TSP. We use fitness distance correlation (FDC) [28], which requires a measure of distance between solutions that corresponds to the operator under analysis. Appropriate measures of distance exist for the operators to which we compare. However, no existing permutation distance functions are suitable for the new cycle mutation. Therefore, we introduce three new permutation distance measures: cycle distance, k-cycle distance, and cycle edit distance.

To validate the new cycle mutation, we experiment with the LCS, QAP, and TSP, comparing cycle mutation with commonly used permutation mutation operators within a $(1+1)$-EA as well as within SA. The results support the predictions of the fitness landscape analysis, demonstrating the efficacy of cycle mutation for assignment problems such as LCS and QAP, while also showing that cycle mutation is inferior to alternatives for the TSP where relative element positions are more important than absolute locations. The experiments also show that the Cycle(α) variation is especially effective at escaping local optima.

A secondary aim of this research is to enable reproducibility [29], as well as to advance the state of practice. Therefore, we integrate our Java implementation of cycle mutation into the open source Chips-n-Salsa library [30], and cycle distance, k-cycle distance, and cycle edit distance into the open source JavaPermutationTools library [31]. Chips-n-Salsa is a library for stochastic and adaptive local search as well as EAs. JavaPermutationTools is a library for computation on permutations and sequences, with a focus on measures of distance. We also disseminate the source code for the fitness landscape analysis and experiments, as well as the raw and processed experiment data, on GitHub (https://github.com/cicirello/cycle-mutation-experiments, accessed on 26 May 2022).

We begin by introducing necessary background in Section 2. We proceed with our methods in Section 3, including deriving the new cycle mutation and distance metrics, as well as performing the fitness landscape analysis. Results are presented in Section 4. We wrap up with a discussion and conclusions in Section 5 including discussing insights into the situations where the new cycle mutation is likely to excel.

2. Background

This section provides background on common mutation operators for permutations (Section 2.1); permutation cycles (Section 2.2), which is the theoretical basis for the new cycle mutation; the CX operator (Section 2.3) on which the new cycle mutation is based; and NP-hard combinatorial optimization problems (Section 2.4) used later in this article.

2.1. Permutation Mutation Operators

We later compare cycle mutation to the most common permutation mutation operators, including the following. Swap picks two different elements uniformly at random and exchanges their locations within the permutation. All other elements remain in their current positions. Insertion picks an element uniformly at random, removes it from the permutation, and reinserts it at a different position chosen uniformly at random. This has the effect of shifting all elements between the removal and insertion points. Reversal (also known as inversion) reverses the order of a subsequence of the permutation, where the end points are chosen uniformly at random. Scramble (also known as shuffle) randomizes the order of a subsequence of the permutation, where the end points of the subsequence are chosen uniformly at random. Scramble is the most disruptive of these operators.

Prior fitness landscape analyses (e.g., [32]) show that swap strongly dominates when absolute positions are most important to fitness; reversal is best for relative positions with undirected edges, followed by insertion and swap, but reversal performs poorly with directed edges; and insertion is best when element precedences most greatly affect fitness.

2.2. Permutation Cycles

Cycle mutation relies upon the concept of a permutation cycle [33]. Align two permutations such that corresponding positions are vertically adjacent, such as

$$\begin{aligned} p_1 &= [0,1,2,3,4,5,6,7,8,9] \\ p_2 &= [2,3,0,5,6,7,8,9,4,1]. \end{aligned} \quad (1)$$

Consider a directed graph with one vertex for each element. In this example, the hypothetical graph has 10 vertexes. Corresponding positions define directed edges. For example, p_1 and p_2 have 0 and 2 at the beginning, respectively, which implies an edge from vertex 0 to vertex 2. Thus, the directed edges of the graph induced by p_1 and p_2 are

$$\{(0,2),(1,3),(2,0),(3,5),(4,6),(5,7),(6,8),(7,9),(8,4),(9,1)\}. \tag{2}$$

A permutation cycle is a cycle in this graph. Thus, in this example, there are three permutation cycles, consisting of the following sets of vertexes:

$$\{0,2\},\{1,3,5,7,9\},\{4,6,8\}. \tag{3}$$

2.3. Cycle Crossover (CX)

The CX [24] operator creates two children from two parents as follows. It first selects an index into one parent uniformly at random. It computes the permutation cycle for the pair of parents that includes the chosen element. Child c_1 receives the positions of the elements that are in the cycle from p_2 and the positions of the other elements from p_1. Likewise, child c_2 receives the positions of the elements that are in the cycle from p_1 and the positions of the others from p_2. The runtime to apply CX is $\Theta(n)$, where n is the permutation length.

Consider an example where the parents are the permutations of Equation (1), which consist of three permutation cycles (see Equation (3)). The result of CX depends upon the random starting element. If element 0 or 2 begins the cycle, then the children are:

$$\begin{aligned} c_1 &= [2,1,0,3,4,5,6,7,8,9] \\ c_2 &= [0,3,2,5,6,7,8,9,4,1]. \end{aligned} \tag{4}$$

If one of the elements $\{1,3,5,7,9\}$ begins the cycle, then the children are:

$$\begin{aligned} c_1 &= [0,3,2,5,4,7,6,9,8,1] \\ c_2 &= [2,1,0,3,6,5,8,7,4,9]. \end{aligned} \tag{5}$$

Otherwise, if one of the elements $\{4,6,8\}$ begins the cycle, then the children are:

$$\begin{aligned} c_1 &= [0,1,2,3,6,5,8,7,4,9] \\ c_2 &= [2,3,0,5,4,7,6,9,8,1]. \end{aligned} \tag{6}$$

Since the starting element is chosen uniformly at random, larger cycles are chosen with greater probability. In this example, the probability of generating the first pair of children above is 0.2, while the probability of generating the second pair of children is 0.5, and the probability of generating the last pair of children is 0.3.

One characteristic of CX is that every element of each child has its absolute position within the permutation from one or the other of the two parents. In this way, it is particularly well suited to permutation problems where absolute position has greatest effect on fitness, since the children are inheriting absolute element position from the parents.

2.4. Test Problems

We now provide background on the TSP, QAP, and LCS, which are NP-hard combinatorial optimization problems used in the fitness landscape analysis and experiments.

2.4.1. TSP

In the TSP, a salesperson must complete a tour of n cities to minimize total cost, usually distance traveled. The cities are vertexes of a completely connected graph. A tour is a simple cycle that includes all n vertexes. The NP-complete decision variant of the TSP asks whether a tour exists with cost at most C; and the NP-hard optimization problem seeks the minimum cost tour [14]. The TSP is widely studied and is perhaps the most common combinatorial optimization problem in experimental studies. It has been used in machine learning [34,35], ant colony optimization [36,37], GA [19,24,38,39], other forms of EA [22,23,40], and other metaheuristics [41–45]. There are variations of the problem, such as the asymmetric TSP (ATSP), where the cost of using an edge differs depending upon the direction of travel along the edge [46,47]. Some variations include problem domain characteristics, such as in delivery route planning [43] and wireless sensor networks [48].

2.4.2. QAP

The formal definition of the QAP is as follows. We are given an n by n cost matrix C, and an m by m distance matrix D, such that $m \geq n$. The NP-complete decision version of the problem [14] asks the question, for a given bound B: Does there exist a one-to-one function $f : \{0, 1, \ldots, n-1\} \to \{0, 1, \ldots, m-1\}$, such that:

$$\sum_{i=0}^{n-1} \sum_{\substack{j=0 \\ j \neq i}}^{n-1} C_{i,j} D_{f(i),f(j)} \leq B? \tag{7}$$

The NP-hard optimization version of the QAP is to find the one-to-one function $f : \{0, 1, \ldots, n-1\} \to \{0, 1, \ldots, m-1\}$ that minimizes

$$\sum_{i=0}^{n-1} \sum_{\substack{j=0 \\ j \neq i}}^{n-1} C_{i,j} D_{f(i),f(j)}. \tag{8}$$

Most authors consider the case when $n = m$. This problem is naturally represented with permutations. When $n = m$, a solution (i.e., the one-to-one function f) is simply represented as a permutation of the integers $\{0, 1, \ldots, n-1\}$. In the more general case, a solution is the first n integers in a permutation of the integers $\{0, 1, \ldots, m-1\}$. The QAP is an especially challenging NP-hard optimization problem, where you most often find experimental studies utilizing what seems to be rather small instances (e.g., $n < 50$). A wide variety of EA, metaheuristics, and heuristic approaches have been proposed for the QAP [49–55].

2.4.3. LCS

The LCS problem [14] is closely related to the subgraph isomorphism problem. In the LCS problem, we are given graphs $G_1 = (V_1, E_1)$ and $G_2 = (V_2, E_2)$, with vertex sets V_1 and V_2, and edge sets E_1 and E_2. In the optimization variant of the problem, we must find the graph $G_3 = (V_3, E_3)$, such that G_3 is isomorphic to a subgraph of G_1 and G_3 is isomorphic to a subgraph of G_2, and such that the cardinality of edge set $|E_3|$ is maximized. This problem is NP-hard. The NP-complete decision variant of the problem asks whether there exists such a graph G_3 such that $|E_3| \geq K$ for some threshold K.

If $|V_1| = |V_2|$, then a solution is represented by a permutation of $\{0, 1, \ldots, |V_1| - 1\}$, and more generally the solution is represented by the first $\min(|V_1|, |V_2|)$ integers of a permutation of $\{0, 1, \ldots, \max(|V_1|, |V_2|) - 1\}$. Without loss of generality, assume that $|V_1| \leq |V_2|$. For permutation p of integers $\{0, 1, \ldots, |V_2| - 1\}$, the number of edges $|E_3|$ in the common subgraph implied by such a permutation is then computed as

$$|E_3| = \sum_{(u,v) \in E_1} \begin{cases} 1 & \text{if } (p(u), p(v)) \in E_2 \\ 0 & \text{if } (p(u), p(v)) \notin E_2, \end{cases} \tag{9}$$

where $p(u)$ means the element in position u of permutation p. This is most efficiently implemented if the smaller graph, G_1, is represented with adjacency lists or by a simple list of ordered pairs as the edge set; and if the larger graph G_2 is represented with an adjacency matrix to enable constant time checks for existence of edges.

Prior approaches to the LCS problem include GA [20], hill-climbing [56], and heuristic approaches [57–59]. There are applications of the LCS problem in computer-aided engineering [56], software engineering [59], protein molecule comparisons [58], integrated circuit design [57,60], natural language processing [61], and cybersecurity [62], among others.

3. Methods

To achieve the objective of a tunable mutation operator with a large neighborhood but small average changes, we derive two forms of the new cycle mutation in Section 3.1.

To gain an understanding of the topology of fitness landscapes associated with cycle mutation, we use a variety of fitness landscape analysis tools. This requires measures of permutation distance corresponding to the mutation operators. Ideally, these should be edit distances where the mutation operator is the edit operation. Edit distance is defined as the minimum cost of the edit operations necessary to transform one structure into the other, and originated within the context of string distance [63,64]. There are many permutation distance measures available in the literature [12,13,31,65–68], including several edit distances. However, none of these are suitable for characterizing the distance between permutations within the context of cycle mutation. Therefore, in Section 3.2, we derive new measures of permutation distance: cycle distance, cycle edit distance, and k-cycle distance.

We then proceed with the fitness landscape analysis in Section 3.3. We derive the diameter of the fitness landscapes for both forms of cycle mutation, as well as for other common permutation mutation operators. The diameter of a fitness landscape is the distance between the two furthest points, where distance is the minimum number of applications of the operators to transform one point to the other. Thus, diameter directly relates to neighborhood size, where larger neighborhood corresponds to smaller diameter, providing a means to quantify our objective of a mutation with large neighborhood. The fitness landscape analysis then utilizes FDC, which is the Pearson correlation coefficient between the fitness of solutions, and the distance to the nearest optimal solution [28]. The FDC analysis uses the TSP, LCS, and QAP problems, so that we can explore the behavior of the cycle mutation operator on a variety of permutation problems. To directly address the objective of mutation with small average changes, the fitness landscape analysis also uses search landscape calculus [32], which examines the average local rate of fitness change.

3.1. Cycle Mutation

We present two variations of cycle mutation. Both forms mutate permutation p by inducing a permutation cycle of length k. The primary difference between the two versions is in how k is chosen. We provide notation and operations shared by the two variations in Section 3.1.1, followed by the presentations of the two versions, Cycle($kmax$) in Section 3.1.2 and Cycle(α) in Section 3.1.3, and finally a comparison of the asymptotic runtime of the new cycle mutation with commonly used mutation operators in Section 3.1.4.

3.1.1. Shared Notation and Operations

Let p be a permutation of length n, such that $p(i)$ is the element at index i. Assume that indexes into p are 0-based (i.e., valid indexes are $\{0, 1, \ldots, n-1\}$), as are array indexes. Algorithm 1 shows pseudocode for the core operation of both variations of cycle mutation. Namely, it induces a cycle in the given permutation p from an array of indexes into p.

Algorithm 1 CreateCycle(p, indexes)

1: temp $\leftarrow p(\text{indexes}[0])$
2: **for** $i = 1$ **to** $n - 1$ **do**
3: $\quad p(\text{indexes}[i-1]) \leftarrow p(\text{indexes}[i])$
4: $p(\text{indexes}[n-1]) \leftarrow$ temp

As an example of its behavior, consider the following permutation:

$$p = [2, 6, 0, 5, 3, 8, 7, 9, 4, 1]. \tag{10}$$

Now consider the following array of indexes into p:

$$\text{indexes} = [3, 7, 1, 4]. \tag{11}$$

Executing CreateCycle(p, indexes) will produce the permutation p' such that $p'(3) = p(7)$, $p'(7) = p(1)$, $p'(1) = p(4)$, and $p'(4) = p(3)$, resulting in the following:

$$p' = [2, 3, 0, 9, 5, 8, 7, 6, 4, 1]. \tag{12}$$

The runtime of CreateCycle is linear in the induced cycle length.

One of the steps of cycle mutation requires sampling k random indexes into permutation p without replacement. Algorithm 2 shows our sampling approach, which uses one of three algorithms depending on the value of k relative to the permutation length n.

Algorithm 2 Sample(n, k)

1: **if** $k \geq \frac{n}{2}$ **then**
2:     **return** ReservoirSample(n, k)
3: **else if** $k \geq \sqrt{n}$ **then**
4:     **return** PoolSample(n, k)
5: **else**
6:     **return** InsertionSample(n, k)

When $k \geq \frac{n}{2}$, we use Vitter's reservoir sampling algorithm [69] (line 2 of Algorithm 2), which has a runtime of $O(n)$ and utilizes $O(n - k)$ random numbers. When $\sqrt{n} \leq k < \frac{n}{2}$, we use Ernvall and Nevalainen's sampling algorithm [70], which we refer to as PoolSample in line 4 of Algorithm 2, and which also has a runtime of $O(n)$, but requires $O(k)$ random numbers. Asymptotic runtime of both of these options is $O(n)$, but since random number generation is a costly operation with very significant impact on the runtime of an EA [71], our approach chooses the sampling algorithm that requires fewer random numbers.

When $k < \sqrt{n}$, we use what we believe is a brand new sampling algorithm: insertion sampling (Algorithm 2, line 6). Insertion sampling's runtime is $O(k^2)$, and requires $O(k)$ random numbers. Since runtime increases quadratically in k, insertion sampling is only a good choice when k is very small relative to n. Pseudocode for insertion sampling is in Algorithm 3. To ensure a duplicate-free result, it maintains a sorted list of the integers selected thus far, and inserts into that list in a way similar to insertion sort. The Rand(a, b) in Algorithm 3 is a uniform random variable over the interval $[a, b]$, inclusive.

Algorithm 3 InsertionSample(n, k)

1: result $\leftarrow$ a new array of length k
2: **for** $i = 0$ **to** $k - 1$ **do**
3:     $v \leftarrow$ Rand($0, n - i - 1$)
4:     $j \leftarrow k - i$
5:     **while** $j < k$ **and** $v \geq$ result[j] **do**
6:         $v \leftarrow v + 1$
7:         result[$j - 1$] $\leftarrow$ result[j]
8:         $j \leftarrow j + 1$
9:     result[$j - 1$] $\leftarrow v$
10: **return** result

The composite sampling algorithm of Algorithm 2 runs in $O(\min(n, k^2))$ time and requires $O(\min(k, n - k))$ random numbers. We integrated this composite sampling algorithm, the new insertion sampling algorithm, as well as our implementations of reservoir sampling and pool sampling into an open source Java library $\rho\mu$ (https://github.com/cicirello/rho-mu, accessed on 26 May 2022), independent of the application to EA of this article.

3.1.2. Cycle(kmax)

In the first version of cycle mutation (Algorithm 4), called Cycle($kmax$), the induced cycle length k is uniformly random from the interval $[2, kmax]$, such that $kmax \geq 2$.

Algorithm 4 CycleMutation($p, kmax$)

1: $k \leftarrow \text{Rand}(2, kmax)$
2: indexes $\leftarrow \text{Sample}(n, k)$
3: Shuffle(indexes)
4: CreateCycle(p, indexes)

It first selects the cycle length k in $O(1)$ time. Next, k indexes are sampled uniformly at random without replacement (line 2) with a call to Sample(n, k) of Algorithm 2. Since $k \leq kmax$, the worst-case runtime of that step is $O(\min(n, kmax^2))$. This array of indexes is randomized (line 3), with a worst-case cost of $O(kmax)$. Finally, a cycle is induced from the randomized list of k indexes with a call to the CreateCycle of Algorithm 1 in line 4, which also costs $O(kmax)$ time in the worst case. Thus, the worst-case runtime is $O(\min(n, kmax^2))$, since the most costly step is sampling the indexes. The average case is also $O(\min(n, kmax^2))$ since the average cycle length $\bar{k} = \frac{kmax+2}{2}$ is proportional to $kmax$.

Cycle($kmax$) limits the induced permutation cycle to a predetermined maximum length, with all cycle lengths up to $kmax$ equally likely. This enables tuning the size of the local neighborhood, with lower $kmax$ leading to a smaller neighborhood and higher $kmax$ creating a larger neighborhood. Thus, increasing $kmax$ may lead to fewer local optima in the fitness landscape, but would also lead to a more disruptive and less focused search.

3.1.3. Cycle(α)

The second version of cycle mutation, called Cycle(α), maximizes the size of the local neighborhood while retaining the local focus of a small cycle length. Cycle(α) allows any possible cycle length $k \in [2, n]$, but chooses a smaller cycle length with higher probability than a greater cycle length. Specifically, the probability of choosing cycle length k is proportional to α^{k-2}. Thus, the probability $P(k)$ of choosing cycle length k is

$$P(k) = \frac{\alpha^{k-2}}{\sum_{i=2}^{n} \alpha^{i-2}} = \frac{\alpha^{k-2}(1-\alpha)}{1-\alpha^{n-1}}. \tag{13}$$

The α is a parameter of the operator such that $0 < \alpha < 1$. The nearer α is to 0, the more probabilistic weight is placed upon lower values of k relative to higher values.

From this, we can derive a mathematical transformation from uniform random number $U \in [0.0, 1.0)$ to corresponding cycle length k. Choose k according to the following:

$$k = \begin{cases} 2 & \text{if } 0 \leq U < P(2) \\ j & \text{if } \sum_{i=2}^{j-1} P(i) \leq U < \sum_{i=2}^{j} P(i). \end{cases} \tag{14}$$

This is equivalent to

$$k = \underset{j \in [2,n]}{\arg\min} \left\{ \sum_{i=2}^{j} P(i) \right\} \text{ subject to } \left\{ \sum_{i=2}^{j} P(i) > U \right\}. \tag{15}$$

Substituting Equation (13) into the constraint and simplifying arrives at

$$k = \underset{j \in [2,n]}{\mathrm{argmin}} \left\{ \sum_{i=2}^{j} P(i) \right\} \text{ subject to } \left\{ \frac{1 - \alpha^{j-1}}{1 - \alpha^{n-1}} > U \right\}. \tag{16}$$

Solve the constraint for j to derive

$$k = \underset{j \in [2,n]}{\mathrm{argmin}} \left\{ \sum_{i=2}^{j} P(i) \right\} \text{ subject to } \left\{ j > \frac{\log(1 - (1 - \alpha^{n-1})U)}{\log(\alpha)} + 1 \right\}. \tag{17}$$

Since the summation inside the argmin increases as j increases, this is equivalent to

$$k = \underset{j \in [2,n]}{\min} \{j\} \text{ subject to } \left\{ j > \frac{\log(1 - (1 - \alpha^{n-1})U)}{\log(\alpha)} + 1 \right\}. \tag{18}$$

Finally, we compute k from U via

$$k = 2 + \left\lceil \frac{\log(1 - (1 - \alpha^{n-1})U)}{\log(\alpha)} \right\rceil. \tag{19}$$

Since α is a parameter that does not change during the run, and since the permutation length n is likewise fixed based on the problem instance we are solving, the $(1 - \alpha^{n-1})$ and the $\log(\alpha)$ are constants that can be computed a single time at the start of the run.

Algorithm 5 shows pseudocode of Cycle(α). The worst-case runtime is $O(n)$, which occurs when the random k equals n, leading lines 3–4 to cost $O(n)$. The worst case is a rare occurrence. Since mutation is applied a very large number of times during an EA, it is more meaningful to examine the average runtime of mutation. To determine the average runtime, we must first compute the expected cycle length $E[k]$ as follows:

$$\begin{aligned} E[k] &= \sum_{k=2}^{n} k P(k) \\ &= \frac{(1-\alpha)}{1-\alpha^{n-1}} \sum_{k=2}^{n} k \alpha^{k-2} \\ &= \frac{2 - \alpha + n\alpha^n - n\alpha^{n-1} - \alpha^{n-1}}{(1-\alpha)(1-\alpha^{n-1})} \\ &\leq \lim_{n \to \infty} \frac{2 - \alpha + n\alpha^n - n\alpha^{n-1} - \alpha^{n-1}}{(1-\alpha)(1-\alpha^{n-1})} \\ &= \frac{2-\alpha}{1-\alpha}. \end{aligned} \tag{20}$$

The expected cycle lengths for Cycle(0.25), Cycle(0.5), and Cycle(0.75) are $E[k] \leq 2\frac{1}{3}$, $E[k] \leq 3$, and $E[k] \leq 5$, respectively. The average runtime of Cycle(α) is therefore $O(\min(n, (\frac{2-\alpha}{1-\alpha})^2))$ due to the call to Sample(n, k) in line 2 of Algorithm 5.

Algorithm 5 CycleMutation(p, α)

1: $k \leftarrow 2 + \left\lceil \frac{\log(1-(1-\alpha^{n-1})U)}{\log(\alpha)} \right\rceil$
2: indexes $\leftarrow$ Sample(n, k)
3: Shuffle(indexes)
4: CreateCycle(p, indexes)

3.1.4. Asymptotic Runtime Summary

Table 1 summarizes the asymptotic runtime of cycle mutation and common mutation operators. Swap mutation is a constant time operation, while the worst-case runtime of insertion, reversal, scramble, and Cycle(α) is linear in the permutation length n. The worst case for Cycle($kmax$) is between these extremes, depending upon $kmax$.

Since mutation is computed a very large number of times during an EA, average runtime is more meaningful. The average runtime of insertion, reversal, and scramble is $O(n)$. The average runtime of Cycle(α) depends upon α. However, other than values of α very near 1.0, the average runtime is essentially a constant. Although the runtime of swap is constant, and that of cycle mutation is very nearly constant depending on α or $kmax$, they are not strictly superior to the linear time operators for all problems. Some problem characteristics may lead to superior performance with fewer applications of one of the linear time mutation operators than if the constant time swap was instead used.

Table 1. Asymptotic runtime of cycle mutation and common permutation mutation operators.

Mutation Operator	Worst Case	Average Case
Cycle($kmax$)	$O(\min(n, kmax^2))$	$O(\min(n, kmax^2))$
Cycle(α)	$O(n)$	$O(\min(n, (\frac{2-\alpha}{1-\alpha})^2))$
Swap	$O(1)$	$O(1)$
Insertion	$O(n)$	$O(n)$
Reversal	$O(n)$	$O(n)$
Scramble	$O(n)$	$O(n)$

3.2. New Measures of Permutation Distance

We now present three new measures of permutation distance: cycle distance (Section 3.2.1), cycle edit distance (Section 3.2.2), and k-cycle distance (Section 3.2.3).

3.2.1. Cycle Distance

As a first step toward a distance measure appropriate for use in analyzing fitness landscapes associated with the Cycle(α) form of cycle mutation, we present cycle distance. Cycle(α) mutates a permutation by inducing a cycle whose length is only limited by the permutation length itself. Therefore, cycle distance is the count of the number of non-singleton cycles. We can define the cycle distance between permutations p_1 and p_2 as

$$\delta(p_1, p_2) = \text{CycleCount}(p_1, p_2) - \text{FixedPointCount}(p_1, p_2), \tag{21}$$

where CycleCount is the number of permutation cycles, and FixedPointCount is the number of singleton cycles or fixed points, which is a cycle of length one (i.e., a point where both permutations contain the same element). We compute cycle distance in $O(n)$ time.

Cycle distance is a semi-metric, since it satisfies all of the metric properties except for the triangle inequality. First, it obviously satisfies non-negativity since $\delta(p_1, p_2)$ clearly cannot be negative since there cannot be a negative number of non-singleton permutation cycles. Second, it is obvious that $\delta(p_1, p_2) = \delta(p_2, p_1)$, and is thus symmetric. Third, it satisfies the identity of indiscernibles as follows. When $p_1 = p_2 = p$, we have

$$\delta(p, p) = \text{CycleCount}(p, p) - \text{FixedPointCount}(p, p) = n - n = 0, \tag{22}$$

since the cycle count is n (i.e., n singleton cycles), and thus the number of fixed points is also n. In the other direction, we have

$$\delta(p_1, p_2) = 0 \implies \text{CycleCount}(p_1, p_2) = \text{FixedPointCount}(p_1, p_2) \implies p_1 = p_2, \tag{23}$$

since the only way that every cycle is a fixed point is if p_1 and p_2 are identical.

To demonstrate the violation of the triangle inequality, consider the permutations:

$$p_1 = [0, 1, 2, 3, 4, 5, 6, 7, 8, 9]$$
$$p_2 = [1, 0, 3, 2, 5, 4, 7, 6, 9, 8] \qquad (24)$$
$$p_3 = [0, 3, 2, 5, 4, 7, 6, 9, 8, 1].$$

Observe $\delta(p_1, p_2) = 5$ since every consecutive pair of elements in p_1 is swapped in p_2, creating five cycles of length two. All even-numbered elements are fixed points for p_1 and p_3, and all odd-numbered elements form a single cycle. That is, keep the even elements fixed and cycle the odd elements to the left within p_1 to obtain p_3. Thus, $\delta(p_1, p_3) = 1$. Finally, inspect p_3 and p_2 to note that if we cycle all of the elements of p_3 one position to the right, we obtain p_2. Therefore, $\delta(p_3, p_2) = 1$. Thus, since $\delta(p_1, p_2) > \delta(p_1, p_3) + \delta(p_3, p_2)$, cycle distance does not satisfy the triangle inequality, and is only a semi-metric.

The diameter of the space of permutations S_n of length n given a measure of distance δ is the maximal distance between points in that space. Define the diameter $D(n, \delta)$ as

$$D(n, \delta) = \max_{p_1, p_2 \in S_n} \{\delta(p_1, p_2)\}. \qquad (25)$$

Since each non-singleton cycle contributes one to cycle distance, independent of cycle length, the maximum case is when the number of non-singleton cycles is maximized. The smallest non-singleton cycle is length two. The maximum cycle distance therefore occurs when there are $[n/2]$ cycles of length two, leading to a diameter of

$$D(n, \delta) = \left[\frac{n}{2}\right]. \qquad (26)$$

3.2.2. Cycle Edit Distance

Although cycle distance relates to the Cycle(α) operator, it is not actually an edit distance with Cycle(α) as the edit operation. However, we can utilize it to define such an edit distance. To define cycle edit distance as the minimum number of induced permutation cycles necessary to transform p_1 into p_2, we can formally define cycle edit distance as

$$\delta_e(p_1, p_2) = \begin{cases} 0 & \text{if } p_1 = p_2 \\ 1 & \text{if } \delta(p_1, p_2) = 1 \\ 2 & \text{if } \delta(p_1, p_2) > 1, \end{cases} \qquad (27)$$

where $\delta(p_1, p_2)$ is cycle distance (Equation (21)). If there is exactly one non-singleton cycle, we can trivially transform it to all fixed points by cycling the elements of that cycle. If there are two or more non-singleton cycles, there exists a cycle of the union of the elements of those cycles that will merge all of them into a single larger cycle, possibly producing some fixed points. See Equation (24) for an example. Thus, one cycle edit operation merges all non-singleton cycles, and a second cycle edit transforms it into all fixed points. We can compute cycle edit distance in $O(n)$ time since we can compute cycle distance in $O(n)$ time.

Cycle edit distance satisfies all of the metric properties as follows. From Equation (27), it is trivial to confirm non-negativity, symmetry, and the identity of indiscernibles. Without loss of generality, assume that p_1, p_2, and p_3 are all different. Thus, $\delta_e(p_1, p_3) + \delta_e(p_3, p_2) \geq 1 + 1 \geq 2$, implying $\delta_e(p_1, p_2) \leq \delta_e(p_1, p_3) + \delta_e(p_3, p_2)$ since $\delta_e(p_1, p_2) \leq 2$ by definition. Thus, cycle edit distance satisfies the triangle inequality, and it is therefore a metric.

Multiple non-singleton cycles can only exist if permutation length $n \geq 4$, and permutations of length one must be identical. Thus, the diameter of cycle edit distance is

$$D(n, \delta_e) = \begin{cases} 0 & \text{if } n \leq 1 \\ 1 & \text{if } 1 < n \leq 3 \\ 2 & \text{if } 3 < n. \end{cases} \qquad (28)$$

3.2.3. K-Cycle Distance

Cycle distance and cycle edit distance assume that arbitrary length cycles can be induced in a single operation, which is needed for Cycle(α) since it does not restrict the induced cycle length. However, this is not suitable when characterizing fitness landscapes for the Cycle(*kmax*) mutation operator, which does limit the cycle length to *kmax*.

We now define k-cycle distance, which limits operations to inducing cycles of lengths $\{2, 3, \ldots, k\}$. The k-cycle distance is not an edit distance. Instead, it is a weighted sum over the cycles of the pair of permutations, where the weight of each cycle is the minimum number of induced cycles of length at most k necessary to transform the cycle to all fixed points. That is, k-cycle distance does not consider operations that span multiple cycles.

A cycle of length $c \leq k$ can be transformed to c fixed points by inducing a cycle of length c. If the cycle length $c > k$, a sequence of cycles can derive c fixed points. For k-cycle distance, iteratively induce cycles of length k until the remaining cycle length is at most k, completing the transformation with one final cycle operation. Each intermediate cycle creates $k - 1$ fixed points. Consider the following example beginning with permutations:

$$p_1 = [0, 1, 2, 3, 4, 5, 6, 7, 8, 9] \\ p_2 = [2, 3, 1, 5, 6, 7, 8, 9, 4, 0]. \quad (29)$$

There are two non-singleton permutation cycles in this example with element sets:

$$\{1, 3, 5, 7, 9, 0, 2\}, \{4, 6, 8\}. \quad (30)$$

If we are computing 3-cycle distance, then we consider cycle mutations of lengths 2 and 3. The cycle with elements $\{4, 6, 8\}$ can thus be transformed into all fixed points with a single cycle mutation since its length is less than or equal to 3 to obtain

$$p_1 = [0, 1, 2, 3, 4, 5, 6, 7, 8, 9] \\ p_2 = [2, 3, 1, 5, 4, 7, 6, 9, 8, 0]. \quad (31)$$

The longer cycle $\{1, 3, 5, 7, 9, 0, 2\}$ requires a sequence of three operations. The first two produce $k - 1 = 2$ fixed points each with one final cycle mutation for the remaining three elements. First, cycle elements 3, 5, and 7, resulting in fixed points for elements 3 and 5:

$$p_1 = [0, 1, 2, 3, 4, 5, 6, 7, 8, 9] \\ p_2 = [2, 7, 1, 3, 4, 5, 6, 9, 8, 0]. \quad (32)$$

A cycle of $\{1, 7, 9, 0, 2\}$ remains. Cycle elements 7, 9, and 0, creating fixed points 7 and 9:

$$p_1 = [0, 1, 2, 3, 4, 5, 6, 7, 8, 9] \\ p_2 = [2, 0, 1, 3, 4, 5, 6, 7, 8, 9]. \quad (33)$$

A cycle of $\{1, 0, 2\}$ remains, which transforms into fixed points with one final cycle mutation:

$$p_1 = [0, 1, 2, 3, 4, 5, 6, 7, 8, 9] \\ p_2 = [0, 1, 2, 3, 4, 5, 6, 7, 8, 9]. \quad (34)$$

With this example in mind, we now derive the k-cycle distance, assuming $k \geq 2$. First, given a cycle of length c, the fewest cycle edits of length at most k necessary to transform it to c fixed points is $\lceil (c-1)/(k-1) \rceil$. To compute k-cycle distance, sum this over all cycles, CycleSet(p_1, p_2), of the permutations p_1 and p_2. Therefore, define k-cycle distance by

$$\delta_k(p_1, p_2) = \sum_{\substack{\text{cycle} \in \text{CycleSet}(p_1, p_2) \\ c = |\text{cycle}|}} \left\lceil \frac{c-1}{k-1} \right\rceil. \quad (35)$$

The k-cycle distance can be computed in $O(n)$ time. It is a metric when $k \leq 4$, but it is only a semi-metric for $k \geq 5$. Independent of k, it trivially satisfies non-negativity since the expression within the sum of Equation (35) is never negative; and since CycleSet(p_1, p_2) computes the same set of cycles as CycleSet(p_2, p_1), it is also trivially symmetric. It satisfies the identity of indiscernibles as follows. If $p_1 = p_2 = p$, then there are n singleton cycles, and since the expression within the sum of Equation (35) is 0 when $c = 1$, we have $\delta_k(p, p) = 0$. In the other direction, if $\delta_k(p_1, p_2) = 0$, then all elements in the summation must be 0, and that only occurs with fixed points. Thus, $\delta_k(p_1, p_2) = 0 \implies p_1 = p_2$.

The triangle inequality is violated if $k \geq 5$. Consider this case with $k = 6$:

$$\begin{aligned} p_1 &= [0, 1, 2, 3, 4, 5] \\ p_2 &= [1, 0, 3, 2, 5, 4] \\ p_3 &= [0, 3, 2, 5, 4, 1]. \end{aligned} \quad (36)$$

The 6-cycle distance allows cycle operations of length up to six. Note that $\delta_6(p_1, p_2) = 3$ since there are three cycles of length two; $\delta_6(p_1, p_3) = 1$ since the even-numbered elements are fixed points, and the odd-numbered elements form a single cycle of length three; and $\delta_6(p_3, p_2) = 1$, with a single cycle of length six (i.e., cycle the elements of p_3 one position to the right to obtain p_2). Thus, $\delta_6(p_1, p_3) + \delta_6(p_3, p_2) = 1 + 1 = 2 < \delta_6(p_1, p_2)$. Therefore, 6-cycle distance is only a semi-metric, as is k-cycle distance for any other $k \geq 6$.

We can produce a similar example for the case of $k = 5$ as follows:

$$\begin{aligned} p_1 &= [0, 1, 2, 3, 4, 5] \\ p_2 &= [1, 0, 3, 2, 5, 4] \\ p_3 &= [0, 3, 2, 5, 1, 4]. \end{aligned} \quad (37)$$

As in the previous example, $\delta_5(p_1, p_2) = 3$. Note that $\delta_5(p_1, p_3) = 1$ since $\{0, 2\}$ are fixed points, and there is a single cycle of the elements $\{1, 3, 5, 4\}$; and $\delta_5(p_3, p_2) = 1$, with a fixed point for element 4, and a single cycle of the remaining five elements. In this example, $\delta_5(p_1, p_3) + \delta_5(p_3, p_2) < \delta_5(p_1, p_2)$. Thus, 5-cycle distance is also only a semi-metric.

In general, as many as k non-singleton cycles can be merged into a single larger cycle by cycling k elements, where the cycle edit includes at least one element of each of the merged cycles. This is also true when $k < 5$, but for $k < 5$ the resulting merged cycle is larger than k, requiring multiple cycle edits to transform into all fixed points.

When $k \leq 4$, the k-cycle distance satisfies the triangle inequality, and is a metric; but when $k \geq 5$, it is only a semi-metric. Indeed, when $k \leq 4$, the k-cycle distance is equivalent to an edit distance with cycles of length up to k as the edit operations. For example, 2-cycle distance is equivalent to an existing distance metric known as interchange distance, which is an edit distance with swap as its edit operation, which is a metric.

The diameter of the space of permutations for k-cycle distance depends upon the permutation length n in relation to k. When k is high, distance is maximized by maximizing the number of permutation cycles, which occurs when there are $\lfloor n/2 \rfloor$ cycles of length 2. When k is low, distance is maximized when there is a single cycle of length n, which requires $\lceil (n-1)/(k-1) \rceil$ cycle operations to transform to n fixed points. Thus, the diameter is

$$D(n, \delta_k) = \max\left\{ \left\lfloor \frac{n}{2} \right\rfloor, \left\lceil \frac{n-1}{k-1} \right\rceil \right\}. \quad (38)$$

3.3. Fitness Landscape Analysis

The fitness landscape analysis includes calculation of landscape diameters in Section 3.3.1, FDC analysis in Section 3.3.2, and an analysis of the search landscape calculus in Section 3.3.3. We synthesize the fitness landscape analysis findings in Section 3.3.4.

When an edit distance is required, such as to compute FDC and the diameter, we use cycle edit distance for Cycle(α), and k-cycle distance for Cycle($kmax$). Swap's edit distance is interchange distance, the minimum number of swaps to transform p_1 into p_2:

$$\delta_i(p_1, p_2) = n - \text{CycleCount}(p_1, p_2), \tag{39}$$

where CycleCount(p_1, p_2) is the number of permutation cycles. The edit distance for insertion mutation is known as reinsertion distance, the minimum number of insertion mutations needed to transform p_1 into p_2. It is efficiently computed as [32]:

$$\delta_r(p_1, p_2) = n - |\text{LongestCommonSubsequence}(p_1, p_2)|, \tag{40}$$

where LongestCommonSubsequence(p_1, p_2) is the longest common subsequence. It is not feasible to utilize an edit distance to analyze reversal mutation landscapes, because computing reversal edit distance is NP-hard [72]. Instead we utilize cyclic edge distance [68], which interprets a permutation as a cyclic sequence of edges (e.g., 1 following 2 is treated as an edge between 1 and 2) and counts the edges in p_1 that are not in p_2. For scramble mutation, a trivial edit distance from a permutation to itself is 0, and to any other permutation is 1, since any permutation is reachable by some shuffling of any other.

3.3.1. Fitness Landscape Diameter

Table 2 compares the diameters of fitness landscapes for the various mutation operators. Except where noted, we use the metrics identified above. The diameter of both swap and insertion landscapes is $n - 1$. The maximum distance for swap occurs for a single n element cycle. The maximum case for insertion occurs when one permutation is the reverse of the other. For reversal mutation, we use the exact diameter for a reversal edit distance rather than relying on the surrogate distance measure identified earlier. The diameter of a reversal landscape is $n - 1$, proven by Bafna and Pevzner [73], the proof of which is well beyond the scope of this article. The diameter of a scramble landscape is simply 1 since every permutation is reachable from any other via a single scramble operation.

Table 2. Fitness landscape diameter for cycle mutation and common permutation mutation operators.

Mutation Operator	Diameter
Cycle($kmax$), $kmax \geq 5$	$\approx 2n/kmax$
Cycle($kmax$), $kmax \leq 4$	$\max\{[n/2], [(n-1)/(kmax-1)]\}$
Cycle(α)	2
Swap	$n - 1$
Insertion	$n - 1$
Reversal	$n - 1$
Scramble	1

The diameter of a Cycle(α) landscape is 2, which is the diameter of the space of permutations for cycle edit distance (Section 3.2.2). When $kmax \leq 4$, the diameter of Cycle($kmax$) is $\max\{[n/2], [(n-1)/(kmax-1)]\}$, which is the maximum k-cycle distance (Section 3.2.3).

We previously saw that k-cycle distance does not satisfy the triangle inequality if $kmax \geq 5$. Although computing a Cycle($kmax$) edit distance for $kmax \geq 5$ is too costly for our purpose, it is straightforward to compute its diameter. Recall the examples illustrating that k-cycle distance violates the triangle inequality, and specifically that two cycle edits of length k can transform a set of cycles with a total of k elements into k fixed points. Thus, the maximum case of $[n/2]$ cycles of length 2 can be transformed to n fixed points with approximately $2n/kmax$ applications of the Cycle($kmax$) operator when $kmax \geq 5$.

3.3.2. Fitness Distance Correlation

We now compute FDC for small instances of the TSP, LCS, and QAP. To compute FDC for a problem instance, you need all optimal solutions to that instance. This necessitates utilizing an instance small enough to feasibly and reliably determine all optimal solutions. The FDC calculations in this section use a permutation length $n = 10$ for this reason, which means a 10-city TSP, an LCS with graphs of 10 vertexes, and a QAP with 10 by 10 cost and distance matrices. The EA experiments later in the paper use larger problem instances to experimentally compare the performance of the different mutation operators.

We use a TSP instance with 10 cities arranged equidistantly around a circle of radius 10, and Euclidean distance for the edge costs. In this way, the optimal solutions are known a priori. There are 20 optimal permutations, all representing the same TSP tour following the cities around the circle, including ten possible starting cities and two travel directions.

We generate a QAP instance such that a random permutation p of length 10 is the known optimal solution. Let $A = [(1, 90), (2, 89), \ldots, (90, 1)]$. Shuffle this array of ordered pairs. The 90 non-diagonal elements of the 10 by 10 cost matrix C are populated row by row using the first element of the tuples in A. Let (u, v) be one of these tuples. If $C[i][j]$ is set to u, then $D[p(i)][p(j)]$ in the distance matrix D is set to v.

We generate an LCS instance consisting of a pair of isomorphic graphs. In this way, we know that the LCS is either of those graphs or any of the other automorphisms of the graph. We use a strongly regular graph in the analysis. Strongly regular graphs are especially challenging for algorithms for the LCS and other related problems. Specifically, we use the Petersen graph [74], shown in Figure 1, which has 120 automorphisms, and thus 120 optimal vertex mappings. Such instances are hard because many vertexes of a strongly regular graph look the same locally to a solver, which can be simultaneously attracted toward different distinct solutions in a space plagued by complex local optima.

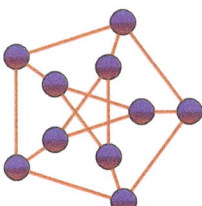

Figure 1. The Petersen graph, a strongly regular graph.

Although guidance on interpreting FDC varies, an $r \leq -0.15$ is commonly considered an easier fitness landscape since it implies that fitness increases as distance to an optimal solution decreases, while $r \geq 0.15$ is likely a deceptive landscape since fitness increases as you move away from optimal solutions, and $-0.15 < r < 0.15$ is considered difficult since there is very little correlation between fitness and distance to an optimal solution [28].

For each combination of problem and mutation operator, we compute FDC exactly, calculating Pearson correlation over all $n! = 3,628,800$ permutations of length $n = 10$. Table 3 summarizes the results. Without loss of generality, for TSP and QAP we compute FDC from the cost function (to minimize) rather than a fitness function (to maximize), flipping the sign of the FDC with positive FDC, implying more straightforward problem solving. Since LCS is a maximization problem, the sign of the FDC is interpreted normally.

Table 3. FDC for combinations of mutation operator and problem.

Mutation Operator	TSP	QAP	LCS
Cycle(α)	−0.0569	0.0213	−0.0278
Cycle(5)	0.1801	0.1339	−0.5342
Cycle(4)	0.1667	0.1737	−0.3984
Cycle(3)	0.2482	0.2210	−0.6180
Swap	0.3318	0.2245	−0.6355
Insertion	0.5277	0.0305	−0.3547
Reversal	0.8459	0.0189	−0.0350
Scramble	0.0117	0.0048	−0.0340

For the TSP, there is very strong FDC for reversal mutation, and lesser but still strong FDC for insertion mutation. The FDC analysis predicts that these will perform better than the others. Swap has the next-highest FDC. Although not as high, all cases of Cycle(*kmax*) exhibit $r > 0.15$. Given the correlation strength of reversal and insertion landscapes, we expect them to dominate the others, but swap and Cycle(*kmax*) may also perform well.

The strength of the correlations are not nearly as high for the QAP, and only three of the mutation operators have an $r \geq 0.15$, including swap mutation and Cycle(*kmax*) with a *kmax* of either 3 or 4. This may be a problem where cycle mutation is better suited.

Because LCS is a maximization problem, FDC should be interpreted in the ordinary sense with FDC computed from fitness where higher fitness implies better solutions. Thus, negative FDC implies easier problem solving. In this case, there is very strong FDC for swap mutation, Cycle(3) mutation, and Cycle(5) mutation. Both Cycle(4) and insertion mutation also have $r \leq -0.15$, so should also progressively lead toward the solution.

The FDC for scramble mutation is very near 0 for all three problems. Scramble is thus unlikely to perform well. The FDC for Cycle(α) is also very near 0 for all three problems. However, we will see that FDC misses an important behavioral property of this operator.

3.3.3. Search Landscape Calculus

Let $\eta(p)$ be the neighbors of p (i.e., solutions reachable with one mutation) and $f(p)$ is fitness. Search landscape calculus [32] defines the average local rate of fitness change:

$$\Delta[f](p) = \frac{1}{|\eta(p)|} \sum_{p' \in \eta(p)} |f(p) - f(p')|. \tag{41}$$

Search landscape calculus then defines $\Delta[f]$ as the average of $\Delta[f](p)$ over all p. It is infeasible to directly compute $\Delta[f]$ for the true fitness function f. Therefore, the search landscape calculus focuses on topological properties that influence fitness, such as absolute positions, relative positions, and element precedences for permutation problems, replacing f by a distance function δ relevant to the context of a problem feature. Thus, define $\Delta[\delta](p)$:

$$\Delta[\delta](p) = \frac{1}{|\eta(p)|} \sum_{p' \in \eta(p)} \delta(p, p'), \tag{42}$$

and from this derive $\Delta[\delta]$ as the average of $\Delta[\delta](p)$ over all points p. The distance δ must correspond to the topological property under analysis, and its maximum must be proportional to the permutation length n (i.e., $\max\{\delta(p_1, p_2)\} \in \Theta(n)$).

We focus on two topological features, absolute element positions and edges, and thus require corresponding distance functions. We use exact match distance [67] (denoted as δ_{em}), which is the count of the number of positions containing different elements, as a measure of how different two permutations are within the context of absolute positioning; and we use cyclic edge distance [68] (denoted as δ_{ce}) as a measure of how different two permutations are within the context of relative positions. Both meet the requirement that the maximum distance is proportional to n. It is exactly n in both cases.

Table 4 summarizes $\Delta[\delta_{em}]$ and $\Delta[\delta_{ce}]$ for both forms of cycle mutation, and the other mutation operators considered. The rates of change of the topological properties of swap, insertion, reversal, and scramble were determined elsewhere [32]. The $\Delta[\delta_{em}]$ is the average number of elements whose absolute positions are changed by a single application of the operator. For cycle mutation, this is the average length of the induced cycle, determined earlier in the article—Cycle(α) in Section 3.1.3 and Cycle($kmax$) in Section 3.1.2. To compute $\Delta[\delta_{ce}]$ we need the average number of edges changed by the operator. For cycle mutation, this depends upon the cycle length, and it also depends upon whether any cycle elements are adjacent. As the permutation length n increases, the probability of adjacent cycle elements decreases. For sufficiently large n, the probability of adjacent cycle elements approaches 0, in which case the number of edges replaced by cycle mutation is, on average, twice the length of the cycle. Thus, $\Delta[\delta_{ce}] = 2\Delta[\delta_{em}]$ for both forms of cycle mutation.

Table 4. Average rates of change of fitness landscape topological properties.

Mutation Operator	$\Delta[\delta_{em}]$	$\Delta[\delta_{ce}]$
Cycle(α)	$(2-\alpha)/(1-\alpha)$	$(4-2\alpha)/(1-\alpha)$
Cycle($kmax$)	$(kmax+2)/2$	$kmax+2$
Swap	2	4
Insertion	$(n+4)/3$	3
Reversal	$[(n+1)/3, (n+4)/3]$	2
Scramble	$(n+1)/3$	$(n+1)/3$

In the search landscape calculus, when $\lim_{n\to\infty}\Delta[\delta] = \infty$, the fitness landscape exhibits very large local changes in fitness, often due to many deep local optima that are difficult to escape. Thus, we expect poor performance from insertion, reversal, and scramble on absolute-positioning problems since $\Delta[\delta_{em}]$ grows with n, as is the case for scramble when relative positions are more important. When $\lim_{n\to\infty}\Delta[\delta] = C$, for a non-zero finite constant C, the search landscape calculus suggests that the landscape is smooth locally. Due to constant $\Delta[\delta_{em}]$, swap and cycle mutation should perform better than the others for problems such as LCS and QAP where absolute positions more greatly impact fitness. All operators except for scramble have constant $\Delta[\delta_{ce}]$, and are thus potentially relevant to problems such as the TSP where edges influence fitness more than absolute element positions.

For cycle mutation, it should be noted that its topological characteristics depend upon the specific parameter settings. For example, higher values of $kmax$ likely lead to the same sort of disruption inherent in scramble mutation, as would values of α very near 1.0.

Previously, we saw that FDC suggested that Cycle(α) would likely perform poorly for all three problems due to FDC very near 0; whereas the search landscape calculus suggests that it may be relevant for all three problems. We now reconcile the discrepancy between these two fitness landscape analysis tools. The number of permutations within one step of a given permutation with respect to Cycle(α) is enormous, leading to extremely low variation in distance and then subsequently to near 0 FDC. However, a large proportion of the Cycle(α) neighbors are very low probability events. Thus, most of the time it behaves more similar to Cycle($kmax$) with a very low $kmax$, but with the ability to make larger jumps.

3.3.4. Summary of Fitness Landscape Analysis Findings

Due to strong FDC, we expect reversal mutation to dominate the others for the TSP, finding lower-cost solutions with the same or fewer evaluations. Insertion should likely be the next best since it also has strong FDC, and a low constant rate of fitness change for edge-focused problems. Swap and cycle mutation, if configured to emphasize smaller cycles, may also be relevant for such problems. Due to strong FDC for QAP and LCS, and low constant rates of change of exact match distance, we anticipate swap and both forms of cycle mutation will find lower cost solutions for absolute-positioning focused problems.

4. Results

We experimentally compare the two forms of cycle mutation with commonly encountered permutation mutation operators. Our experiments are implemented in Java. Our test machine is a Windows 10 PC, with an AMD A10-5700 3.4 GHz CPU, and 8 GB memory. The code is compiled with OpenJDK 17.0.2 for a Java 11 target, and runs on an OpenJDK 64-bit Server VM Temurin-17.0.2+8. The code is open source, licensed via the GPL 3.0, and available on GitHub at https://github.com/cicirello/cycle-mutation-experiments (accessed on 26 May 2022), and includes the code to analyze the experiment data and to generate the figures.

In the experiments, we apply each of the mutation operators within both a $(1+1)$-EA as well as in an SA. In a $(\mu + \lambda)$-EA [15], the population size is μ, λ offspring are created in each generation, and the best μ individuals from the combination of parents and offspring survive into the next generation. The $(1+1)$-EA is commonly used in experimental studies. We employ it here since it removes the impact of choice of selection operator and crossover operator, and also eliminates many parameters such as population size, mutation rate, etc. Additionally, it supports a more direct comparison with the non-population approach of SA. For the SA, we use the parameter-free Self-Tuning Lam Annealing [75] that adaptively adjusts the temperature parameter based on problem-solving feedback.

We consider three cases of Cycle(k_{max}) mutation, including Cycle(3), Cycle(4), and Cycle(5); and three cases of Cycle(α) mutation, including Cycle(0.25), Cycle(0.5), and Cycle(0.75). All results are 50 run averages. We use run lengths $\{10^2, 10^3, 10^4, 10^5, 10^6, 10^7\}$ in number of evaluations. We test significance with the Wilcoxon rank sum test. The results on the TSP, QAP, and LCS are presented in Section 4.1, Section 4.2 and Section 4.3, respectively.

4.1. TSP Results

Each of the 50 TSP instances consists of 100 cities. An instance is defined by a random distance matrix, such that the distance between cities is a uniformly random integer from the interval $[1, 1000]$. The results are shown in Figure 2 with number of evaluations on the horizontal axis at log scale, and average solution cost over 50 runs on the vertical axis.

Consistent with the extremely strong FDC that we earlier observed for reversal mutation for the TSP, we find that reversal is clearly dominant on the 100-city TSP instances within both the EA (Figure 2a) and SA (Figure 2b), finding lower-cost solutions with fewer evaluations. All comparisons to reversal mutation are extremely statistically significant (e.g., Wilcoxon rank sum test p-values very near 0) except for the shortest 100 evaluation runs. For SA, the second-best mutation operator is insertion (results also statistically significant). We earlier saw that insertion had the second strongest FDC for the TSP. Swap and the various cases of cycle mutation all find solutions of approximately equivalent cost within the SA, especially for very long run lengths.

The EA comparison (Figure 2a) is a bit more interesting. Although for mid-length runs, insertion is second-best at statistically significant levels, the various cases of cycle mutation surpass insertion mutation for the longest runs. Insertion mutation exhibits a premature convergence effect, converging to a significantly suboptimal solution 10^5 evaluations into the EA runs, as does swap. However, cycle mutation continues to find lower-cost solutions, overtaking insertion mutation. The convergence effect in the EA is due to the smaller neighborhoods of swap and insertion, leading to greater impact of local optima. Cycle mutation has a larger neighborhood so better avoids this, continuing to show progress. In fact, even though we saw near-zero FDC for Cycle(α), all cases of Cycle(α) continue to make progress, although converging at a slower rate than reversal.

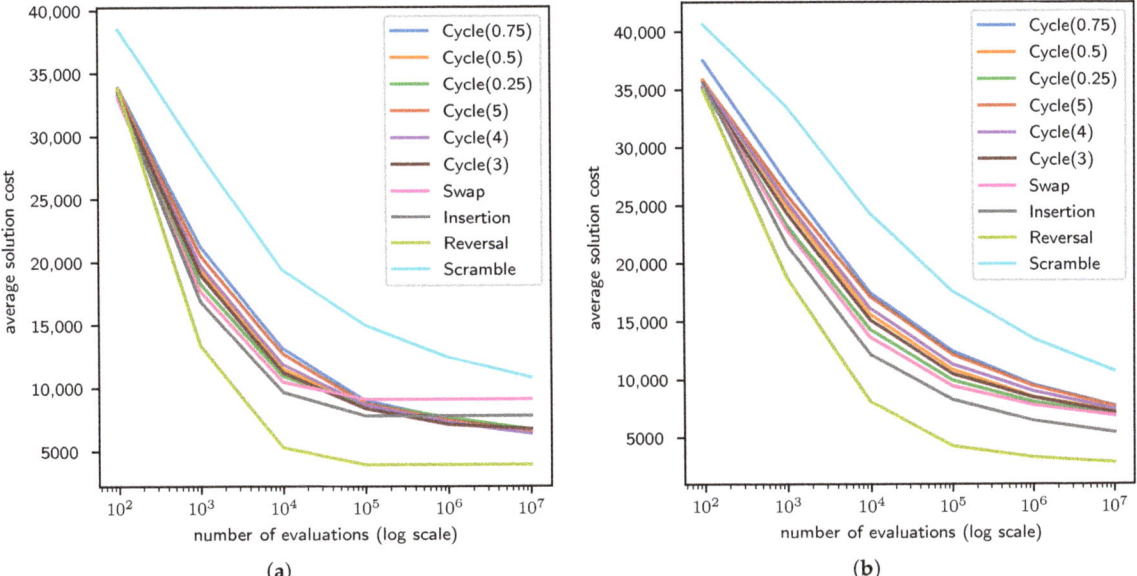

Figure 2. Results on the TSP for (**a**) $(1+1)$-EA, and (**b**) SA.

4.2. QAP Results

Each of the 50 QAP instances consists of a 50 by 50 cost matrix and a 50 by 50 distance matrix, with integer costs and distances generated uniformly at random from [1, 50].

The results are visualized in Figure 3. We earlier saw that FDC suggested that swap, Cycle(3), and Cycle(4) would likely perform better than the others, while the search landscape calculus suggested that swap and all forms of cycle mutation are appropriate for the QAP. Consistent with the fitness landscape analysis, scramble, reversal, and insertion all perform poorly for QAP within both the EA and SA. For the longest SA runs (Figure 3b), there is little difference in solution cost among swap and the various cases of cycle mutation. For runs of 10^4 to 10^5 SA evaluations, swap and Cycle(0.25) find solutions with lower values of the cost function than the others at statistically significant levels.

The EA results are especially interesting (Figure 3a). Swap suffers from a premature convergence effect at 10^4 evaluations, while all cycle mutation cases continue to make substantial progress, minimizing the cost function. Furthermore, Cycle(0.25) and Cycle(3)'s rates of convergence are equivalent to swap up to that point. The superior convergence effect is attributed to cycle mutation's larger neighborhood, especially in the case of Cycle(α). We do not see the same behavior with SA (Figure 3b) because SA has a built-in way of handling local optima, allowing swap to continue to improve solution quality despite swap's much smaller local neighborhood.

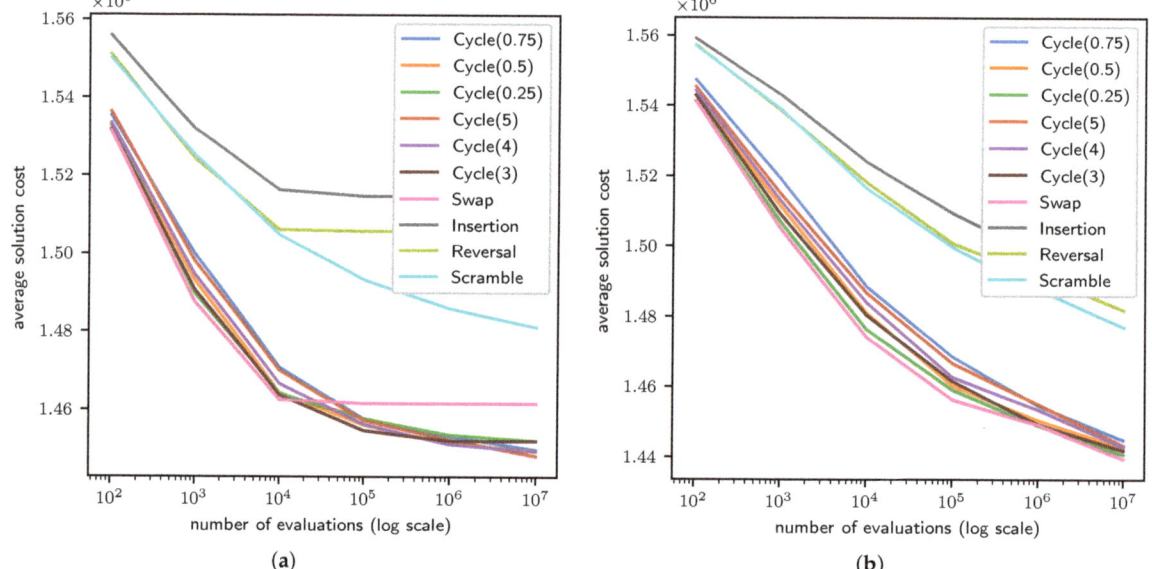

Figure 3. Results on the QAP for (**a**) $(1+1)$-EA, and (**b**) SA.

4.3. LCS Results

Consider two sets of experiments with the LCS problem, one with random graphs and the other with strongly regular graphs. In the random graph case, we have 50 instances, each consisting of a pair of randomly generated isomorphic graphs. We use pairs of isomorphic graphs so that each problem instance has a known optimal solution. That is, the largest common subgraph is simply the graph itself. Each graph has 50 vertexes and edge density 0.5, which means that the probability of each possible edge is 0.5. Thus, each graph has 50 vertexes, and the expected number of edges is $0.5 \times 50 \times 49/2 = 612.5$. The second graph for each instance is formed by relabeling the vertexes randomly.

We use generalized Petersen graph $G(25,2)$ for the strongly regular case. Generalized Petersen graph [76] $G(n,k)$ has $2n$ vertexes $V = \{u_0, u_1, \ldots, u_{n-1}, v_0, v_1, \ldots, v_{n-1}\}$, and $3n$ edges $E = \{(u_i, v_i) \mid 0 \leq i < n\} \bigcup \{(u_i, u_{i+1 \bmod n}) \mid 0 \leq i < n\} \bigcup \{(v_i, v_{i+k \bmod n}) \mid 0 \leq i < n\}$. The original Petersen graph (Figure 1) is $G(5,2)$. Figure 4 shows a generalized Petersen graph $G(25,2)$ that has 50 vertexes and 75 edges. We again average over 50 runs, but in this case, each instance consists of graph $G(25,2)$ and a second graph isomorphic to it that is formed by randomly relabeling the vertexes.

LCS is a maximization problem; however, for consistency we transform it to a minimization problem. Since each instance is a pair of isomorphic graphs, the LCS has $|E|$ edges, where E is the edge set. Thus, redefine LCS to minimize the cost of permutation p: $C(p) = |E| - |E'|$, where E' is the edge set of the subgraph implied by vertex mapping p.

The results are in Figure 5 (random graphs) and Figure 6 (strongly regular graphs). For SA and random graphs (Figure 5b), scramble, insertion, and reversal all perform very poorly compared with the others. Cycle(5) and Cycle(0.75) are inferior to the other cycle mutation cases for the longest runs at statistically significant levels. Within an EA for random graphs (Figure 5a), the behavior is similar to that of the QAP. Swap exhibits a premature convergence effect, while cycle mutation finds increasingly lower cost solutions. Cycle(3) also prematurely converges at 10^6 evaluations, while cycle mutation configured with a larger neighborhood continues to improve solution cost beyond that point. Cycle(0.25) is best at statistically significant levels for the 10^7 evaluation runs.

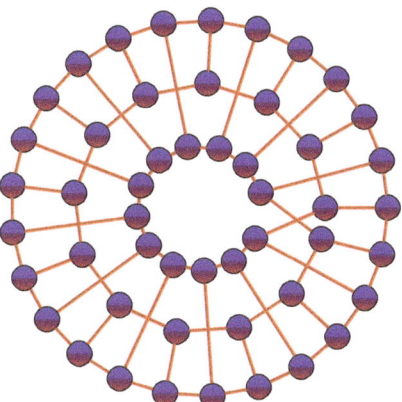

Figure 4. The generalized Petersen graph, $G(25, 2)$, a strongly regular graph.

These results are consistent with the fitness landscape analysis. First, FDC suggested that swap, Cycle(3), and Cycle(5) were best suited to the problem, followed by Cycle(4) and insertion. The FDC analysis was likely misled with respect to insertion mutation due to the small size of the instance used in the FDC analysis, which is likely why we find insertion performing poorly on the larger instance. This is consistent with the search landscape calculus analysis which suggests that swap and cycle mutation are both well suited to the problem, and that the other operators, including insertion, are not. The Cycle(0.25) is most effective within longer runs of the EA due to the combination of small average change (average cycle length of 2.33) and very large neighborhood size. Within SA, cycle mutation is neither better nor worse than swap, converging at the same rate.

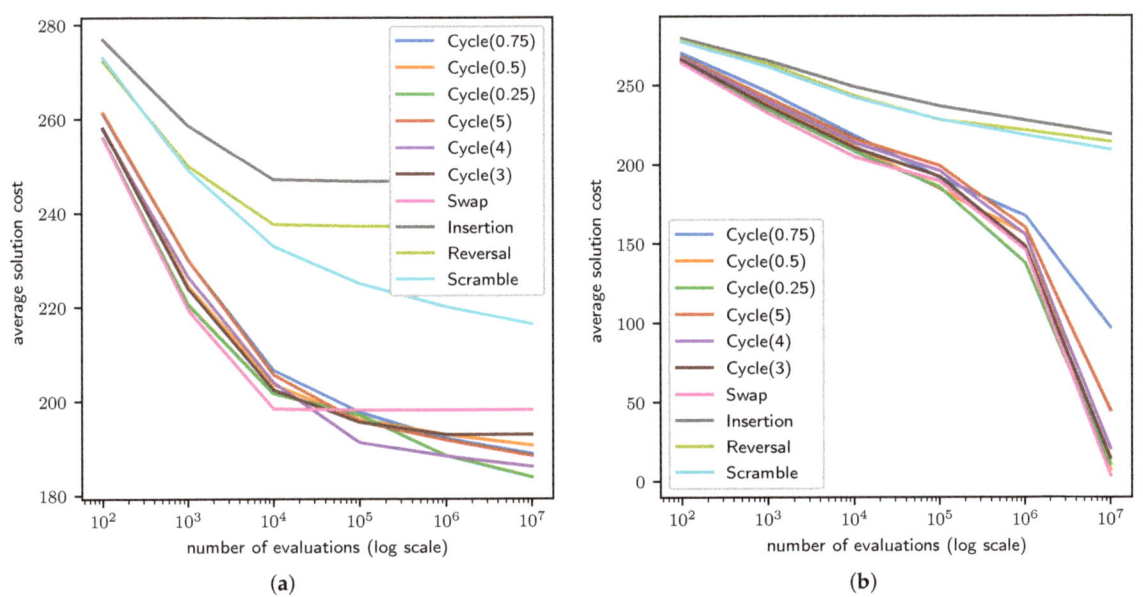

Figure 5. Results on the LCS problem with random graphs for (**a**) $(1+1)$-EA, and (**b**) SA.

Observe that the strongly regular graph results (Figure 6) are similar to those of random graphs, but are more pronounced. For example, in an EA (Figure 6a), swap and Cycle(3) prematurely converge, while Cycle(4) and Cycle(5) exhibit superior convergence effect, outperforming all others at statistically significant levels from 10^4 evaluations onward, but differences between them are not significant.

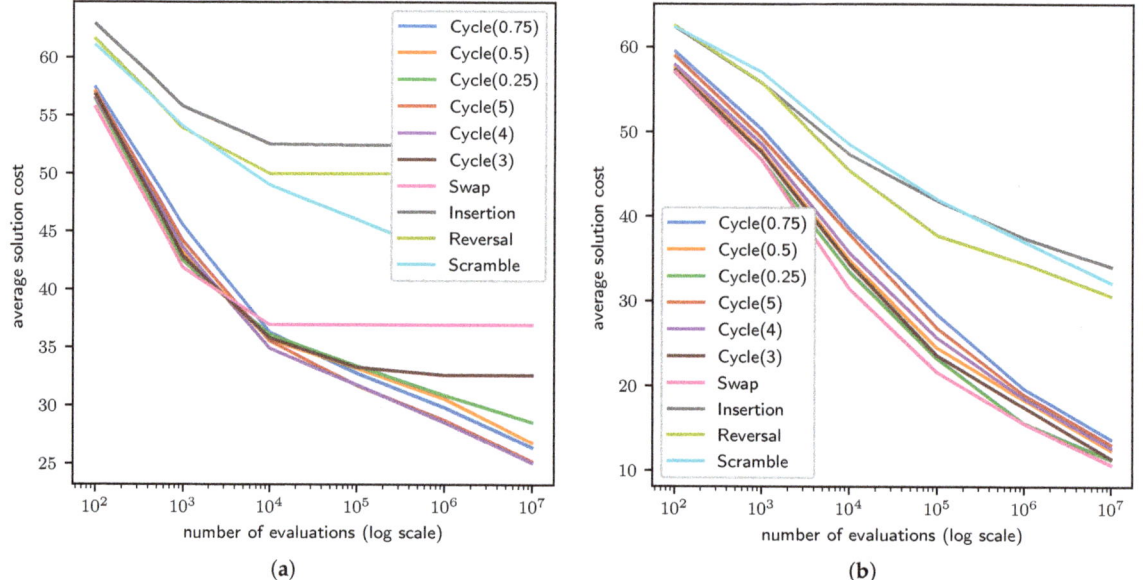

Figure 6. Results on the LCS problem with strongly regular graphs for (**a**) $(1+1)$-EA, and (**b**) SA.

5. Discussion and Conclusions

In this paper, we proposed a new mutation operator for use by EA and other related algorithms, such as SA, when evolving permutations. This new operator is called cycle mutation, and includes two variations: Cycle(α) and Cycle($kmax$). Cycle mutation induces a random permutation cycle. The difference between the two operators is in how the cycle length is chosen. The Cycle($kmax$) operator has a small maximum cycle length, while Cycle(α) does not impose any maximum cycle length. Thus, Cycle(α) has a significantly larger neighborhood size, while retaining a low average cycle length.

The runtime of cycle mutation in the worst case is linear, similar to insertion, reversal, and scramble mutation; but unlike those operators, cycle mutation's average runtime is constant. Therefore, the computational time to use cycle mutation within an EA is little more than that of the constant time swap. Helping to achieve this efficient runtime, cycle mutation relies on a new algorithm for sampling k elements from an n element set, called insertion sampling, which is faster than existing alternatives for low k.

The fitness landscape analysis showed that cycle mutation is well suited to permutation problems such as the QAP and LCS, where absolute positions more greatly impact fitness than relative ordering. The fitness landscape analysis also showed that cycle mutation may be relevant to relative ordering problems such as the TSP, provided cycle length is kept low. While undertaking the fitness landscape analysis, we developed three new measures of permutation distance: cycle distance, k-cycle distance, and cycle edit distance.

Validating the fitness landscape analysis, cycle mutation experimentally outperformed the others for QAP and LCS within an EA, especially for long runs, finding lower-cost solutions with fewer evaluations. Furthermore, while swap suffers from a premature convergence effect due to small neighborhood, cycle mutation continues to make optimization progress even for the longest runs; and the Cycle(α) form of cycle mutation is especially good at managing local optima, due to its very large neighborhood size enabling it to make large jumps when necessary. Thus, Cycle(α) exhibits superior convergence effect.

Cycle mutation does have limitations. Cycle mutation does not show any advantage within SA, although its rate of convergence is similar to that of swap in SA so it may be worth considering for SA nonetheless. Additionally, within an EA, Cycle($kmax$) may exhibit a premature convergence effect for some problems if $kmax$ is set too low, such as what we saw with Cycle(3) for the LCS. However, Cycle(α) overcomes this limitation.

Our Java implementations of cycle mutation are integrated into an open source library, Chips-n-Salsa [30], and our Java implementations of cycle distance, k-cycle distance, and cycle edit distance are integrated into the open source JavaPermutationTools [31] library. By disseminating the implementations in open source libraries, we hope to contribute not only to the research literature, but also to the state of practice. Additionally, all of the code to reproduce the experiments, and to analyze the results, is available in a GitHub repository, https://github.com/cicirello/cycle-mutation-experiments (accessed on 26 May 2022).

Funding: This research received no external funding.

Institutional Review Board Statement: Not applicable.

Informed Consent Statement: Not applicable.

Data Availability Statement: All experiment data (raw and post-processed) are available on GitHub: https://github.com/cicirello/cycle-mutation-experiments (accessed on 26 May 2022), which also includes all source code of our experiments, as well as instructions for compiling and running the experiments.

Conflicts of Interest: The author declares no conflict of interest.

Abbreviations

The following abbreviations are used in this manuscript:

CX	Cycle crossover
EA	Evolutionary algorithm
ES	Evolution strategies
FDC	Fitness distance correlation
GA	Genetic algorithm
LCS	Largest common subgraph
QAP	Quadratic assignment problem
SA	Simulated annealing
TSP	Traveling salesperson problem

References

1. Mitchell, M. *An Introduction to Genetic Algorithms*; MIT Press: Cambridge, MA, USA, 1998.
2. Beyer, H. *The Theory of Evolution Strategies*; Natural Computing Series; Springer: Berlin/Heidelberg, Germany, 2013.
3. Langdon, W.; Poli, R. *Foundations of Genetic Programming*; Springer: Berlin/Heidelberg, Germany, 2013.
4. Cuéllar, M.; Gómez-Torrecillas, J.; Lobillo, F.; Navarro, G. Genetic algorithms with permutation-based representation for computing the distance of linear codes. *Swarm Evol. Comput.* **2021**, *60*, 100797. [CrossRef]
5. Koohestani, B. A crossover operator for improving the efficiency of permutation-based genetic algorithms. *Expert Syst. Appl.* **2020**, *151*, 113381. [CrossRef]
6. Shabash, B.; Wiese, K.C. PEvoSAT: A Novel Permutation Based Genetic Algorithm for Solving the Boolean Satisfiability Problem. In Proceedings of the 15th Annual Conference on Genetic and Evolutionary Computation, Amsterdam, The Netherlands, 6–10 July 2013; Association for Computing Machinery: New York, NY, USA, 2013; pp. 861–868. [CrossRef]
7. Kalita, Z.; Datta, D.; Palubeckis, G. Bi-objective corridor allocation problem using a permutation-based genetic algorithm hybridized with a local search technique. *Soft Comput.* **2019**, *23*, 961–986. [CrossRef]
8. Shakya, S.; Lee, B.S.; Di Cairano-Gilfedder, C.; Owusu, G. Spares parts optimization for legacy telecom networks using a permutation-based evolutionary algorithm. In Proceedings of the IEEE Congress on Evolutionary Computation (CEC), Donostia, Spain, 5–8 June 2017; pp. 1742–1748. [CrossRef]
9. Mironovich, V.; Buzdalov, M.; Vyatkin, V. Evaluation of Permutation-Based Mutation Operators on the Problem of Automatic Connection Matching in Closed-Loop Control System. In *Recent Advances in Soft Computing and Cybernetics*; Matoušek, R., Kůdela, J., Eds.; Springer International Publishing: Berlin/Heidelberg, Germany, 2021; pp. 41–51. [CrossRef]

10. Hinterding, R. Gaussian mutation and self-adaption for numeric genetic algorithms. In Proceedings of the IEEE International Conference on Evolutionary Computation, Perth, WA, Australia, 29 November–1 December 1995; Volume 1, pp. 384–389. [CrossRef]
11. Szu, H.; Hartley, R. Nonconvex optimization by fast simulated annealing. *Proc. IEEE* **1987**, *75*, 1538–1540. [CrossRef]
12. Campos, V.; Laguna, M.; Marti, R. Context-Independent Scatter and Tabu Search for Permutation Problems. *INFORMS J. Comput.* **2005**, *17*, 111–122. [CrossRef]
13. Cicirello, V.A. Classification of Permutation Distance Metrics for Fitness Landscape Analysis. In Proceedings of the 11th International Conference on Bio-Inspired Information and Communication Technologies, Pittsburgh, PA, USA, 13–14 March 2019; Springer: New York, NY, USA, 2019; pp. 81–97._7. [CrossRef]
14. Garey, M.R.; Johnson, D.S. *Computers and Intractability: A Guide to the Theory of NP-Completeness*; W. H. Freeman & Co.: New York, NY, USA, 1979.
15. Eiben, A.E.; Smith, J.E. *Introduction to Evolutionary Computing*, 2nd ed.; Springer: Berlin/Heidelberg, Germany, 2015.
16. Davis, L. Applying Adaptive Algorithms to Epistatic Domains. In Proceedings of the International Joint Conference on Artificial Intelligence, San Francisco, CA, USA, 18–23 August 1985; pp. 162–164.
17. Cicirello, V.A. Non-Wrapping Order Crossover: An Order Preserving Crossover Operator that Respects Absolute Position. In Proceedings of the Genetic and Evolutionary Computation Conference, Seattle, WA, USA, 8–12 July 2006; ACM Press: New York, NY, USA, 2006; Volume 2, pp. 1125–1131. [CrossRef]
18. Syswerda, G. Schedule Optimization using Genetic Algorithms. In *Handbook of Genetic Algorithms*; Davis, L., Ed.; Van Nostrand Reinhold: New York, NY, USA, 1991.
19. Goldberg, D.E.; Lingle, R. Alleles, Loci, and the Traveling Salesman Problem. In Proceedings of the 1st International Conference on Genetic Algorithms, Sheffield, UK, 12–14 September 1995; Lawrence Erlbaum Associates, Inc.: Mahwah, NJ, USA, 1985; pp. 154–159.
20. Cicirello, V.A.; Smith, S.F. Modeling GA Performance for Control Parameter Optimization. In *Proceedings of the Genetic and Evolutionary Computation Conference*; Morgan Kaufmann Publishers: San Francisco, CA, USA, 2000; pp. 235–242.
21. Bierwirth, C.; Mattfeld, D.; Kopfer, H. On permutation representations for scheduling problems. In *Proceedings of the International Conference on Parallel Problem Solving from Nature*; Springer: Berlin/Heidelberg, Germany, 1996; pp. 310–318.
22. Nagata, Y.; Kobayashi, S. Edge Assembly Crossover: A High-Power Genetic Algorithm for the Travelling Salesman Problem. In Proceedings of the International Conference on Genetic Algorithms, East Lansing, MI, USA, 19–23 July 1997; pp. 450–457.
23. Watson, J.P.; Ross, C.; Eisele, V.; Denton, J.; Bins, J.; Guerra, C.; Whitley, L.D.; Howe, A.E. The Traveling Salesrep Problem, Edge Assembly Crossover, and 2-opt. In Proceedings of the International Conference on Parallel Problem Solving from Nature, Amsterdam, The Netherlands, 27–30 September 1998; Springer: Berlin/Heidelberg, Germany, 1998; pp. 823–834.
24. Oliver, I.M.; Smith, D.J.; Holland, J.R.C. A study of permutation crossover operators on the traveling salesman problem. In Proceedings of the 2nd International Conference on Genetic Algorithms, Cambridge, MA, USA, 1 July 1987; Lawrence Erlbaum Associates, Inc.: Mahwah, NJ, USA, 1987; pp. 224–230.
25. Delahaye, D.; Chaimatanan, S.; Mongeau, M. Simulated Annealing: From Basics to Applications. In *Handbook of Metaheuristics*; Gendreau, M., Potvin, J.Y., Eds.; Springer: Berlin/Heidelberg, Germany, 2019; pp. 1–35._1. [CrossRef]
26. Laarhoven, P.J.M.; Aarts, E.H.L. *Simulated Annealing: Theory and Applications*; Kluwer Academic Publishers: Norwell, MA, USA, 1987.
27. Kirkpatrick, S.; Gelatt, C.D.; Vecchi, M.P. Optimization by Simulated Annealing. *Science* **1983**, *220*, 671–680. [CrossRef]
28. Jones, T.; Forrest, S. Fitness Distance Correlation as a Measure of Problem Difficulty for Genetic Algorithms. In Proceedings of the 6th International Conference on Genetic Algorithms, Pittsburgh, PA, USA, 15–19 July 1995; Morgan Kaufmann: San Francisco, CA, USA, 1995; pp. 184–192.
29. National Academies of Sciences, Engineering, and Medicine. *Reproducibility and Replicability in Science*; The National Academies Press: Washington, DC, USA, 2019. [CrossRef]
30. Cicirello, V.A. Chips-n-Salsa: A Java Library of Customizable, Hybridizable, Iterative, Parallel, Stochastic, and Self-Adaptive Local Search Algorithms. *J. Open Source Softw.* **2020**, *5*, 2448. [CrossRef]
31. Cicirello, V.A. JavaPermutationTools: A Java Library of Permutation Distance Metrics. *J. Open Source Softw.* **2018**, *3*, 950. [CrossRef]
32. Cicirello, V.A. The Permutation in a Haystack Problem and the Calculus of Search Landscapes. *IEEE Trans. Evol. Comput.* **2016**, *20*, 434–446. [CrossRef]
33. Knuth, D.E. *The Art of Computer Programming, Volume 1, Fundamental Algorithms*, 3rd ed.; Addison Wesley: Boston, MA, USA, 1997.
34. Junior Mele, U.; Maria Gambardella, L.; Montemanni, R. Machine Learning Approaches for the Traveling Salesman Problem: A Survey. In Proceedings of the 8th International Conference on Industrial Engineering and Applications, Barcelona, Spain, 8–11 January 2021; ACM: New York, NY, USA, 2021; pp. 182–186. [CrossRef]
35. Mele, U.J.; Chou, X.; Gambardella, L.M.; Montemanni, R. Reinforcement Learning and Additional Rewards for the Traveling Salesman Problem. In Proceedings of the 8th International Conference on Industrial Engineering and Applications, Barcelona, Spain, 8–11 January 2021; ACM: New York, NY, USA, 2021; pp. 198–204. [CrossRef]

36. Wang, R.L.; Gao, S. A Co-Evolutionary Hybrid ACO for Solving Traveling Salesman Problem. In Proceedings of the 5th International Conference on Computer Science and Application Engineering, Virtual Conference, Sanya, China, 19–21 October 2021; ACM: New York, NY, USA, 2021; pp. 1–4.
37. Dinh, Q.T.; Do, D.D.; Hà, M.H. Ants Can Solve the Parallel Drone Scheduling Traveling Salesman Problem. In Proceedings of the Genetic and Evolutionary Computation Conference, Lille, France, 10–14 July 2021; ACM: New York, NY, USA, 2021; pp. 14–21. [CrossRef]
38. Varadarajan, S.; Whitley, D. A Parallel Ensemble Genetic Algorithm for the Traveling Salesman Problem. In Proceedings of the Genetic and Evolutionary Computation Conference, Lille, France, 10–14 July 2021; ACM: New York, NY, USA, 2021; pp. 636–643. [CrossRef]
39. Nagata, Y. High-Order Entropy-Based Population Diversity Measures in the Traveling Salesman Problem. *Evol. Comput.* **2020**, *28*, 595–619. [CrossRef]
40. Ibada, A.J.; Tuu-Szabo, B.; Koczy, L.T. A New Efficient Tour Construction Heuristic for the Traveling Salesman Problem. In Proceedings of the 5th International Conference on Intelligent Systems, Metaheuristics & Swarm Intelligence, Victoria, Seychelles, 10–11 April 2021; ACM: New York, NY, USA, 2021; pp. 71–76. [CrossRef]
41. Dell'Amico, M.; Montemanni, R.; Novellani, S. A Random Restart Local Search Matheuristic for the Flying Sidekick Traveling Salesman Problem. In Proceedings of the 8th International Conference on Industrial Engineering and Applications, Barcelona, Spain, 8–11 January 2021; ACM: New York, NY, USA, 2021; pp. 205–209. [CrossRef]
42. Gao, Y.; Shen, Y.; Yang, Z.; Chen, D.; Yuan, M. Immune Optimization Algorithm for Traveling Salesman Problem Based on Clustering Analysis and Self-Circulation. In Proceedings of the 3rd International Conference on Advanced Information Science and System, Sanya, China, 26–28 November 2021; ACM: New York, NY, USA, 2021. [CrossRef]
43. Tong, B.; Wang, J.; Wang, X.; Zhou, F.; Mao, X.; Zheng, W. Optimal Route Planning for Truck–Drone Delivery Using Variable Neighborhood Tabu Search Algorithm. *Appl. Sci.* **2022**, *12*, 529. [CrossRef]
44. Qamar, M.S.; Tu, S.; Ali, F.; Armghan, A.; Munir, M.F.; Alenezi, F.; Muhammad, F.; Ali, A.; Alnaim, N. Improvement of Traveling Salesman Problem Solution Using Hybrid Algorithm Based on Best-Worst Ant System and Particle Swarm Optimization. *Appl. Sci.* **2021**, *11*, 4780. [CrossRef]
45. Rico-Garcia, H.; Sanchez-Romero, J.L.; Jimeno-Morenilla, A.; Migallon-Gomis, H. A Parallel Meta-Heuristic Approach to Reduce Vehicle Travel Time in Smart Cities. *Appl. Sci.* **2021**, *11*, 818. [CrossRef]
46. An, H.C.; Kleinberg, R.; Shmoys, D.B. Approximation Algorithms for the Bottleneck Asymmetric Traveling Salesman Problem. *ACM Trans. Algorithms* **2021**, *17*, 1–12. [CrossRef]
47. Svensson, O.; Tarnawski, J.; Végh, L.A. A Constant-Factor Approximation Algorithm for the Asymmetric Traveling Salesman Problem. *J. ACM* **2020**, *67*, 1–53. [CrossRef]
48. Tsilomitrou, O.; Tzes, A. Mobile Data-Mule Optimal Path Planning for Wireless Sensor Networks. *Appl. Sci.* **2022**, *12*, 247. [CrossRef]
49. He, L.; Liu, Z.Y.; Liu, M.; Yang, X.; Zhang, F.Y. Quadratic Assignment Problem via a Convex and Concave Relaxations Procedure. In Proceedings of the 3rd International Conference on Robotics, Control and Automation, Chengdu, China, 11–13 August 2018; ACM: New York, NY, USA, 2018; pp. 147–153. [CrossRef]
50. Beham, A.; Affenzeller, M.; Wagner, S. Instance-Based Algorithm Selection on Quadratic Assignment Problem Landscapes. In Proceedings of the Genetic and Evolutionary Computation Conference Companion, Berlin, Germany, 15–19 July 2017; ACM: New York, NY, USA, 2017; pp. 1471–1478. [CrossRef]
51. Benavides, X.; Ceberio, J.; Hernando, L. On the Symmetry of the Quadratic Assignment Problem through Elementary Landscape Decomposition. In Proceedings of the Genetic and Evolutionary Computation Conference Companion, Lille, France, 10–14 July 2021; ACM: New York, NY, USA, 2021; pp. 1414–1422. [CrossRef]
52. Baioletti, M.; Milani, A.; Santucci, V.; Tomassini, M. Search Moves in the Local Optima Networks of Permutation Spaces: The QAP Case. In Proceedings of the Genetic and Evolutionary Computation Conference Companion, Prague, Czech Republic, 13–17 July 2019; ACM: New York, NY, USA, 2019; pp. 1535–1542. [CrossRef]
53. Novaes, G.A.S.; Moreira, L.C.; Chau, W.J. Exploring Tabu Search Based Algorithms for Mapping and Placement in NoC-Based Reconfigurable Systems. In Proceedings of the 32nd Symposium on Integrated Circuits and Systems Design, Sao Paulo, Brazil, 26–30 August 2019; ACM: New York, NY, USA, 2019; pp. 1–6. [CrossRef]
54. Thomson, S.L.; Ochoa, G.; Daolio, F.; Veerapen, N. The Effect of Landscape Funnels in QAPLIB Instances. In Proceedings of the Genetic and Evolutionary Computation Conference Companion, Berlin, Germany, 15–19 July 2017; ACM: New York, NY, USA, 2017; pp. 1495–1500. [CrossRef]
55. Irurozki, E.; Ceberio, J.; Santamaria, J.; Santana, R.; Mendiburu, A. Algorithm 989: Perm_mateda: A Matlab Toolbox of Estimation of Distribution Algorithms for Permutation-Based Combinatorial Optimization Problems. *ACM Trans. Math. Softw.* **2018**, *44*, 47. [CrossRef]
56. Cicirello, V.A.; Regli, W.C. An Approach to a Feature-based Comparison of Solid Models of Machined Parts. *Artif. Intell. Eng. Des. Anal. Manuf.* **2002**, *16*, 385–399. [CrossRef]
57. Chen, J.; Zaman, M.; Makris, Y.; Blanton, R.D.S.; Mitra, S.; Schafer, B.C. DECOY: Deflection-Driven HLS-Based Computation Partitioning for Obfuscating Intellectual Property. In Proceedings of the 57th ACM/EDAC/IEEE Design Automation Conference, Virtual Conference, 20–24 July 2020; IEEE Press: Piscataway, NJ, USA, 2020; pp. 1–6.

58. Stoichev, S.; Petrova, D. An Application of an Algorithm for Common Subgraph Detection for Comparison of Protein Molecules. In Proceedings of the International Conference on Computer Systems and Technologies and Workshop for PhD Students in Computing, Ruse, Bulgaria, 18–19 June 2009; ACM: New York, NY, USA, 2009; pp. 1–6. [CrossRef]
59. Zeller, A. Isolating Cause-Effect Chains from Computer Programs. In Proceedings of the 10th ACM SIGSOFT Symposium on Foundations of Software Engineering, Charleston, SC, USA, 18–22 November 2002; ACM: New York, NY, USA, 2002; pp. 1–10. [CrossRef]
60. Wong, J.L.; Kourshanfar, F.; Potkonjak, M. Flexible ASIC: Shared Masking for Multiple Media Processors. In Proceedings of the 42nd Annual Design Automation Conference, Anaheim, CA, USA, 3–17 June 2005; ACM: New York, NY, USA, 2005; pp. 909–914. [CrossRef]
61. Jordan, P.W.; Makatchev, M.; Pappuswamy, U. Understanding Complex Natural Language Explanations in Tutorial Applications. In Proceedings of the Third Workshop on Scalable Natural Language Understanding, Stroudsburg, PA, USA, 8 June 2006; Association for Computational Linguistics: Stroudsburg, PE, USA, 2006; pp. 17–24.
62. Puodzius, C.; Zendra, O.; Heuser, A.; Noureddine, L. Accurate and Robust Malware Analysis through Similarity of External Calls Dependency Graphs. In Proceedings of the 16th International Conference on Availability, Reliability and Security, Vienna, Austria, 17–20 August 2021; ACM: New York, NY, USA, 2021; pp. 1–12. [CrossRef]
63. Wagner, R.A.; Fischer, M.J. The String-to-String Correction Problem. *J. ACM* **1974**, *21*, 168–173. [CrossRef]
64. Levenshtein, V.I. Binary Codes Capable of Correcting Deletions, Insertions and Reversals. *Sov. Phys. Dokl.* **1966**, *10*, 707–710.
65. Sörensen, K. Distance measures based on the edit distance for permutation-type representations. *J. Heuristics* **2007**, *13*, 35–47. [CrossRef]
66. Schiavinotto, T.; Stützle, T. A review of metrics on permutations for search landscape analysis. *Comput. Oper. Res.* **2007**, *34*, 3143–3153. [CrossRef]
67. Ronald, S. More Distance Functions for Order-based Encodings. In Proceedings of the IEEE Congress on Evolutionary Computation, Anchorage, AK, USA, 4–9 May 1998; pp. 558–563.
68. Ronald, S. Distance Functions for Order-based Encodings. In Proceedings of the IEEE Congress on Evolutionary Computation, Indianapolis, IN, USA, 13–16 April 1997; pp. 49–54.
69. Vitter, J.S. Random Sampling with a Reservoir. *ACM Trans. Math. Softw.* **1985**, *11*, 37–57. [CrossRef]
70. Ernvall, J.; Nevalainen, O. An Algorithm for Unbiased Random Sampling. *Comput. J.* **1982**, *25*, 45–47. [CrossRef]
71. Cicirello, V.A. Impact of Random Number Generation on Parallel Genetic Algorithms. In Proceedings of the Thirty-First International Florida Artificial Intelligence Research Society Conference, Melbourne, FL, USA, 21–23 May 2018; AAAI Press: Palo Alto, CA, USA, 2018; pp. 2–7.
72. Caprara, A. Sorting by Reversals is Difficult. In Proceedings of the International Conference on Computational Molecular Biology, Santa Fe, NM, USA, 20–23 January 1997; pp. 75–83.
73. Bafna, V.; Pevzner, P.A. Genome Rearrangements and Sorting by Reversals. *SIAM J. Comput.* **1996**, *25*, 272–289. [CrossRef]
74. Harary, F. *Graph Theory*; Addison-Wesley: Boston, MA, USA, 1967.
75. Cicirello, V.A. Self-Tuning Lam Annealing: Learning Hyperparameters While Problem Solving. *Appl. Sci.* **2021**, *11*, 9828. [CrossRef]
76. Watkins, M.E. A theorem on tait colorings with an application to the generalized Petersen graphs. *J. Comb. Theory* **1969**, *6*, 152–164. [CrossRef]

Article

A Stigmergy-Based Differential Evolution

Valentín Osuna-Enciso [1,†] and Elizabeth Guevara-Martínez [2,*,†]

1 Department of Computer Sciences, Centro Universitario de Ciencias Exactas e Ingenierías-Universidad de Guadalajara, Av. Revolución 1500, Col. Olímpica, Guadalajara 44430, Jalisco, Mexico; valentin.osuna@cucei.udg.mx
2 Department of Engineering, Universidad Anáhuac México, Avenida Universidad Anáhuac 46, Col. Lomas Anáhuac, Huixquilucan 52786, Estado de Mexico, Mexico
* Correspondence: elizabeth.guevara@anahuac.mx
† These authors contributed equally to this work.

Abstract: Metaheuristic algorithms are techniques that have been successfully applied to solve complex optimization problems in engineering and science. Many metaheuristic approaches, such as Differential Evolution (DE), use the best individual found so far from the whole population to guide the search process. Although this approach has advantages in the algorithm's exploitation process, it is not completely in agreement with the swarms found in nature, where communication among individuals is not centralized. This paper proposes the use of stigmergy as an inspiration to modify the original DE operators to simulate a decentralized information exchange, thus avoiding the application of a global best. The Stigmergy-based DE (SDE) approach was tested on a set of benchmark problems to compare its performance with DE. Even though the execution times of DE and SDE are very similar, our proposal has a slight advantage in most of the functions and can converge in fewer iterations in some cases, but its main feature is the capability to maintain a good convergence behavior as the dimensionality grows, so it can be a good alternative to solve complex problems.

Keywords: metaheuristics; differential evolution; stigmergy; CUDA

1. Introduction

Metaheuristics (MH) are computational algorithms inspired by biological phenomena that solve a great diversity of complex scientific and engineering problems, particularly optimization. Some key features of metaheuristics are: (1) robustness, (2) self-organization, (3) adaptation and (4) decentralized control [1]. Two of the most recognized branches in MH, which group several approaches, are Evolutionary Computation (EC) and Swarm Intelligence (SI) algorithms [2]. Even though different metaphors inspire them, all have a set of parameters that the user must tune in a manual, adaptive, or self-adaptive manner to help the algorithm's exploration/exploitation balance [3]. However, in several MH, the use of a global best candidate solution to direct the search process toward the real optimum is common.

For instance, Harmony Search [4–6] utilizes several individuals from the harmony memory (HM) to improvise a new harmony, and this individual is incorporated into the HM only if it is better than the worst individual in the whole population. Differential Evolution [7] uses in most canonical variants [8] a weighted difference of at least two parents to generate a new individual, which interchanges information with the actual individual and becomes the trial solution. The selection step compares the trial individual's fitness against both the actual individual's fitness and the best individual found so far. Another metaheuristic approach is Particle Swarm Optimization [9], which considers the local best and the global best to make a velocity vector added to the actual individual to generate a new individual, whose fitness is compared against the local best and the global best and, if necessary, replaces the global best. The canonical Fireworks algorithm [10,11], whose population is composed of fireworks, modifies the individuals by consecutively applying the explosion, the Gaussian mutation, and the mapping operators. After such alterations, the new candidate solution's fitness is compared against the best individual so far, and the individual with better fitness value is kept in memory. Based on some plants' reproduction strategy, the authors in [12] proposed the Flower Pollination Algorithm (FPA). This metaheuristic uses the abiotic, biotic, and switch operators to either locally or globally modify the actual candidate solution. As in the previous algorithms, FPA keeps tracking the best global individual found at each iteration. Another algorithm is the Gravitational Search algorithm [13], in which the movements of the candidate solutions (called objects) consider their masses and accelerations to actualize their velocities and positions. As in other metaheuristics, the algorithm updates the global worst and the global best at each iteration. Other examples of metaheuristics that actualize the best solution found at each iteration are the Firefly Algorithm [14,15], Cuckoo Search [16], Artificial Bee Colony optimization [17,18] and Chicken Swarm optimization [19].

In order to solve high-dimensional problems and evaluate a large amount of data, it is necessary to improve the efficiency and execution times of evolutionary algorithms. Many authors consider parallel and distributed approaches of metaheuristic algorithms as alternatives to overcome the difficulties derived from the increase in data complexity. A recent review of parallel and distributed approaches for evolutionary algorithms is presented in [20]. In addition, much research reports metaheuristics translated to parallel languages, such as CUDA (Compute Unified Device Architecture) [21], which runs on parallel general-purpose graphic processing units (GPGPU). For example, a survey of GPU parallelization strategies for metaheuristics is conducted in [22]. The authors of [23] propose two approaches to parallelize the Gravitational Search algorithm using CUDA: one that modifies the whole population with a single kernel and another that utilizes multiple kernels to operate the complete population. In the multiple kernel version, the least parallelizable is the update mass kernel. At a given moment, all threads could be trying to write the global best to the same memory location.

With respect to DE, several proposals have been made to parallelize this algorithm, such as the one presented in [24], where opposition-based learning and Differential Evolution are programmed with CUDA to optimize 19 benchmark functions. The authors cannot perform the fittest selection in parallel, and they propose a reduction approach to find the best individual at each iteration. A similar approach is found in [25], where only four functions are tested up to 100 dimensions. In this case, the authors propose the use of an adaptive scaling factor and an adaptive crossover probability; even though the authors utilize the DE/rand/1/bin version, they also keep track of the best individual so far. To represent DE variants, a general pattern DE/$x/y/z$ is used, where x describes the base vector selection type, y represents the number of differentials and z is the type of crossover operator, and if there is an asterisk symbol (*) it means that any valid operator can be used for that process. Thus, for DE/rand/1/bin, "rand" indicates that the individuals are randomly chosen, "1" specifies that only one vector difference is used to form the mutated population and "bin" means that a binomial crossover is applied.

A survey of DE parallelized on GPGPU is found in [26], tracking the research efforts to parallelize DE from computer networks since 2004, passing through the use of GPGPU with first (and therefore, with a lack of features) CUDA SDKs and ending with a brief exploration of DE and GPU-based frameworks. It is worth mentioning that, in most of the articles, the original operators are preserved and only the parallel implementation is performed using one kernel for each process. To have a general idea about the implementation of DE on GPUs, Table 1 shows some proposals highlighting the implemented DE version, whether any operator was modified or not, the number and type of test functions and the maximum dimensionality. It can be observed that the idea of proposing modifications to differential evolution operators has not been widely investigated.

Table 1. GPU differential evolution implementation proposals.

Reference/Year	Version	Modified Operators	Operators Not Modified	Type and Number of Functions	Maximum Dimensionality
[27]/2010	DE/rand/1/bin	None	Mutation, Crossover, Selection	Benchmark functions, 6	100
[28]/2010	DE/rand/1/bin	None	Mutation, Crossover, Selection	Benchmark functions, 12	90
[29]/2011	DE/rand/1/bin	None	Mutation, Crossover, Selection	Task scheduling benchmark, 12	512
[25]/2012	DE/rand/1/bin	None	Mutation, Crossover, Selection	Benchmark functions, 4	100
[30]/2013	DE/rand/1/bin	None	Mutation, Crossover, Selection	Real-world object detection, 2	32, and 9, pertaining to each parametric model
[31]/2012	CDE/rand/1/bin (Coevolution DE)	None	Mutation, Crossover, Selection	Functions with restrictions, 6	10
[32]/2012	EOBDE/rand/1/bin (Elite oppostion-based learning DE)	None	Mutation, Crossover, Selection	Benchmark functions, 10	1000
[24]/2013	GOjDE/rand/1/exp (Generalized opposition j DE)	None	Mutation, Crossover, Selection	Benchmark functions, 19	1000
[33]/2021	DE/*/*/exp	Crossover	Mutation, Selection	Benchmark functions, 3	1000

Some metaheuristics utilize two methods to update the global best (individual and fitness) by: (a) comparing it against the corresponding values of every individual in the population and (b) sorting the corresponding values of the entire population. In the first approach, parallelization implies that $N_p - 1$ individuals will be trying to update the best fitness value in the actual iteration and the best individual to the same memory's positions: this is the worst-case scenario. This could produce race conditions, which are undesirable because they could produce a loss of information. A better method is the sorting-based approach which can only be partially parallelized, in fact, this DE algorithm version is the one used for comparisons in this research.

This paper considers a modification, based on the stigmergy concept, to Differential Evolution (version best/1/bin, in this case the best individual solution of the DE population is selected as the target vector) to eliminate the global best utilization. For this purpose, the mutation and crossover operators are modified. Such adjustment, which we called Stigmergic Differential Evolution (SDE), has a similar performance to the original DE with a lower computational cost in some cases when programmed into GPGPU; mainly, it preserves a good convergence behavior as the dimensionality grows. Therefore, we consider that other metaheuristics that utilize the global best to guide the optimization process could benefit from the proposed methodology.

The remaining paper is as follows: Section 2 explains the concept of stigmergy and explores some of its applications to solve engineering problems. Section 3 gives an account of our proposal and the benchmark of optimization problems for the comparisons. In contrast, Section 4 explains the experimental setup, shows the results and briefly discusses the results. Section 5 concludes this work and gives some future venues around stigmergic metaheuristics.

2. Stigmergy and Some Stigmergic Applications

Some authors consider insect colonies as complex adaptive systems, with properties of: (1) spatial distribution; (2) decentralized control; (3) hierarchical organization of individuals; (4) adaptation to environmental changes, among others [17,34]. For the second property, this means that there is no global control [34], and therefore the interactions among individuals are distributed. To explain the behavior of insect colonies, researchers proposed the term stigmergy. Such a concept comes from entomology, and Pierre-Paul Grassé proposed it to explain the coordination and collaboration by the indirect communication [35,36] of social insects (particularly termites and ants) to complete complicated projects [37]. In the original paper, Grassé considers that every individual in a colony that acts (ergon) on the medium leaves a trace (stigma), which at the same time stimulates another action from the same or another individual (Figure 1). Actions and marks occur on the medium, and as mentioned, the latter only stimulate (but not force) the action of either the same or other individuals.

The concept of stigmergy has served as an inspiration to propose algorithms utilized to solve some problems in engineering [38–40], and it also was used as an inspiration to propose one of the most known metaheuristic algorithms: Ant System (AS) [41], the predecessor of Ant Colony Optimization (ACO) and its variants. In these approaches, it is the pheromone-based construction of trails that represents the stigmergic mechanism [36,42,43].

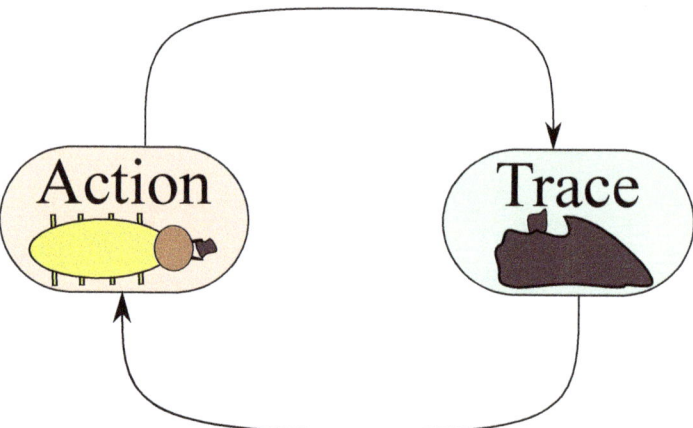

Figure 1. A simple graphic of stigmergy as a closed loop.

ACO considers a population of artificial agents that keep some memory in the form of the pheromone of the best solutions visited so-far; those solutions are paths of a completely connected graph, constructed by selecting every vertex in the path in a probabilistic manner, being that vertices in the graph with the most artificial pheromones are the more likely to be selected. As the algorithm works in the graphs' space to construct the solutions, it is natural that the first applications of this technique are in combinatorial optimization, such as the Traveling Salesman Problem or job-shop scheduling. The application of ACO to solve numerical optimization problems is not straightforward. Often, the proposals include hybridizations with other techniques, as the continuous ACO (CACO) [44], continuous interacting ant colony (CIAC) [45], or continuous ant colony system (CACS) [42]; however, there are some research efforts using direct modifications to the original ACO. For instance, in his doctoral thesis, Korošec proposed two stigmergy-based modifications of the original ACO to solve numerical problems, including the Differential Ant-Stigmergy Algorithm [36]. In the proposal, the authors present a fine-grained discretization of the continuous search space to construct a directed graph, to which they apply the operators Gaussian pheromone aggregation, pheromone evaporation and parameter precision to build a solution for the continuous optimization problem at hand. Korošec and colleagues later improved the original algorithm to tackle high-complexity problems or even dynamic problems [43,46] by using a pheromone operator based on the Cauchy distribution.

A recently proposed stigmergic algorithm to solve numeric problems is the teaching–learning-based optimization [47]. In the paper, students' cooperation (ergon/action) through a board (stigma/mark) simulates the stigmergic process. Algorithmically, the authors create a ranking matrix (of the same size as the original population) for every individual, and they use a roulette wheel to select a guide student. Such guiding students serve as the mean to generate new solutions, which utilize as a standard deviation the weighted average of the remaining individuals. Therefore, the guiding student directs the complete exploration–exploitation process, diluting the stigmergic process as it happens in other metaheuristics.

In the next section, we explain our approach, the Stigmergic Differential Evolution, SDE. The algorithm eliminates the use of global best. Thus, the whole population controls the exploration–exploitation process in a coordinated but decentralized manner; moreover, to explore the efficiency of the proposal and its parallelization feasibility, we programmed it on GPGPU using the CUDA language.

3. Stigmergic Differential Evolution

3.1. Differential Evolution

Differential Evolution (DE) is an algorithm proposed in 1995 to solve continuous optimization problems [7]. Even though there are several variants of DE [8], one of the more popular and powerful variants is the DE/best/1/bin, which since its development has solved several problems in science and engineering [48–50]. The notation in the next explanations considers lower bold letters as vectors, lower italic letters with subindexes as vector components, upper bold letters as matrices and upper or lower letters without subindexes as single real or integer numbers.

The first step in the algorithm is the initialization of a random population:

$$\mathbf{X} = x_{i,j}^k = l_j + rand()(u_j - l_j) \\ j = 1, \cdots, D;\ i = 1, \cdots, N_p;\ k = 0 \tag{1}$$

where N_p is the population size, D represents the problem dimensionality, k is the actual iteration, l and u are the lower and upper limits of the search space and $rand()$ is a random number drawn from a uniform distribution. In this phase the parameters' scaling mutation and crossover factor (F, Cr) are also set. The second stage creates a mutant vector:

$$\mathbf{v}^k = \mathbf{x}_{best}^k + F \cdot \left(\mathbf{x}_{r_1}^k - \mathbf{x}_{r_2}^k\right) \tag{2}$$

considering r_1 and r_2 as random integers uniformly distributed in $[1, N_p]$ with the restriction that $r_1 \neq r_2 \neq i$, and $\mathbf{x}_{best}^k$ is the best individual found so-far. The next step obtains the trial vector:

$$u_j^k = \begin{cases} v_j^k & if\ rand() \leqslant C_r\ or\ j = j_{rand} \\ x_{i,j}^k & otherwise \end{cases} \tag{3}$$

where $j_{rand} \in \{1, 2, \cdots, D\}$ is a random integer taken from a uniform distribution. In the last phase, the algorithm applies the selection operation by evaluating the trial and the original individual with the objective function f:

$$x_i^{k+1} = \begin{cases} \mathbf{u}^k & if\ f\left(\mathbf{u}^k\right) < f\left(\mathbf{x}_i^k\right) \\ x_i^k & otherwise \end{cases} \tag{4}$$

The previously mentioned steps (mutation, crossover and selection) are repeated until a criterion is met, usually a maximum number of iterations.

3.2. The Proposal–Action, Agent, Medium, Trace and Coordination

As in other bioinspired algorithms, in Stigmergic Differential Evolution, we do not try to precisely copy every aspect of the stigmergy concept but only to use it as an inspiration source. Let us consider a stigmergy process [51,52]: the nesting construction/repairing made by termites. This process is parallel, it needs only local information and it has no apparent global leadership. By considering Figure 1, for a single stigmergic loop, it happens that individual$_i$ (previously stimulated by trace$_i$ or other traces) produces an action$_i$ that leaves a trace$_i$ which probabilistically stimulates either individual$_i$ or near individuals. We also consider that, as in the case of real termites, only a few individuals can share a physical space (Figure 2), and only one of them could leave a trace/mark in the medium at a time. Many termites work in the nest, then, they complete the previous steps in parallel, and they repeat the stigmergic cycle until the mound is complete and functional [37].

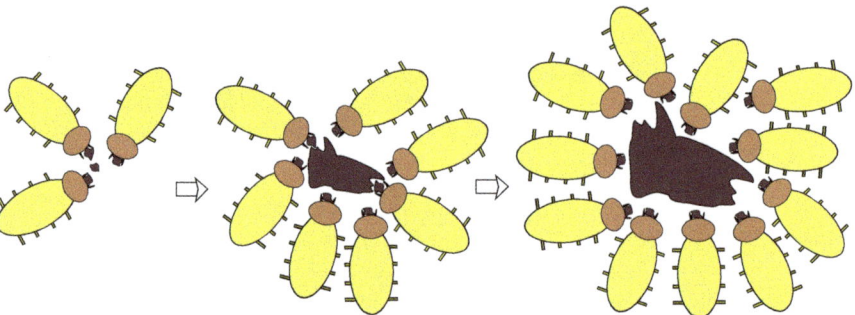

Figure 2. Termites sharing a working area.

As in other metaheuristics, Stigmergic DE considers a population of candidate solutions as the agents, while evaluating the candidate solutions represents the traces. Like its real counterpart, the stigmergic approach considers that few candidate solutions can share consecutive memory spaces with one another. For example, in Table 2, candidate solutions x_{i-1}, x_i and x_{i+1}, and their respective fitness function evaluations, are neighbors, as they share consecutive locations in memory.

Table 2. Population and fitness arrangement in SDE.

Agent	Trace
x_1	$f(x_1)$
...	...
x_{i-1}	$f(x_{i-1})$
x_i	$f(x_i)$
x_{i+1}	$f(x_{i+1})$
...	...
x_{Np}	$f(x_{Np})$

As in other algorithms, Equation (1) serves to initialize the population of candidate solutions in SDE. Prior to the action step, we compare the traces of the actual with the next and the previous individuals in the memory:

$$x_{lead}^k = argmin\left(f\left(x_{i-1}^k\right), f\left(x_i^k\right), f\left(x_{i+1}^k\right)\right) \quad (5)$$

for $i = 2, \ldots, N_{p-1}$; the original index of x_{lead}^k is also stored, e.g., $lead = i - 1$, $lead = i$, or $lead = i + 1$. The remaining two candidate solutions after applying Equation (5) are randomly labeled, with 50% probability, as x_{work1}^k and x_{work2}^k; also, their respective indexes are stored. The action step considers the mutation DE operator with the differential weight F:

$$v_i^k = x_{lead}^k + F \cdot \left(x_{work1}^k - x_{work2}^k \right) \tag{6}$$

together with the original trial DE operator using the crossover rate C_r:

$$u_{i,j}^k = \begin{cases} v_{i,j}^k & if \ rand() < C_r \ or \ j = j_{rand} \\ x_{i,j}^k & otherwise \end{cases} \tag{7}$$

After the trial vector is complete, the next phase of stigmergic DE is the trace update:

$$x_{lead}^{k+1} = \begin{cases} u_i^k & if \ f\left(u_i^k\right) < f\left(x_{lead}^k\right) \\ x_{lead}^k & otherwise \end{cases} \tag{8}$$

$$x_{work1}^{k+1} = \begin{cases} u_i^k & if \ f\left(u_i^k\right) < f\left(x_{work1}^k\right) \\ x_{work1}^k & otherwise \end{cases} \tag{9}$$

$$x_{work2}^{k+1} = \begin{cases} u_i^k & if \ f\left(u_i^k\right) < f\left(x_{work2}^k\right) \\ x_{work2}^k & otherwise \end{cases} \tag{10}$$

It is essential to clarify that Equations (8)–(10) are mutually exclusive in their application; for example, if the condition is true in Equation (8), then Equations (9) and (10) are not applied in this step of the algorithm. As in other metaheuristics, the algorithm repeats the previous steps until it reaches a specific criterion.

One of the main differences between Differential Evolution and Stigmergic Differential Evolution is the complete elimination of the global optimum. Instead, SDE uses a leader's concept, which is a local best concerning the actual individual. Another distinction is the tracing step, which involves the previous, actual and posterior candidate solutions. According to their fitness, SDE could update any of the three individuals in this stage, or none in the worst case. This crucial step enables the algorithm to share information among candidate solutions of the whole population via indirect communication. In that sense, we consider that SDE disseminates the improvements in the search for the optimum in a distributed manner.

3.3. Parallel DE and SDE

In the literature, several proposals for parallelization of DE exist to solve plenty of optimization problems, as reported in [24,25,53–56]. Nevertheless, we utilize a basic CUDA parallelization as mentioned in Section 3.2 of [25] to compare the two algorithms. For the DE version, we programmed three kernels:

1. Generate mutated and trial vectors and evaluate trial vectors.
2. Compare and update the fitness of trial vectors and the original population.
3. Obtain the global best.

This version of DE permits a fine-grained parallelization of the first two kernels. Still, the case is different in the third one, which only can be partially parallelized with reduction steps and using a reduction approach, which is a common way to reduce a search effort in parallel. The complete kernel to obtain the global best is in Algorithm 1; it is essential to notice that this kernel operates only with vector sizes divisible by 2.

In the case of SDE, we programmed only one CUDA kernel containing the steps: comparing previous traces, action and trace updating. In that sense, the nature of Stigmergic DE permits a fine-grained parallelization, meaning all the steps of the stigmergic approach can be applied to every individual in the population in parallel. A pseudocode of the stigmergic kernel is given in Algorithm 2.

Algorithm 1 Kernel of DE to obtain the global best

1: **procedure** _GLOBAL_ VOID GETGLOBALBEST(double* F, double* fbest, double* xbest, double* X, int* indices)
2:    _shared_ double min[Np]; // Declare array in shared memory
3:    _shared_ int indexes[Np]; // Array of indexes in shared memory
4:    int blockId = blockIdx.x + blockIdx.y * gridDim.x;
5:    int arrayIndex = blockId * (blockDim.x * blockDim.y) + (threadIdx.y * blockDim.x) + threadIdx.x, i;
6:    min[threadIdx.x] = F[arrayIndex];
7:    indexes[threadIdx.x] = indices[arrayIndex];
8:    _syncthreads();
9:    int nTotalThreads = blockDim.x; // Total number of active threads
10:   **while** nTotalThreads > 1 **do**
11:     int halfPoint = (nTotalThreads » 1);
12:     // only the first half of the threads will be active.
13:     **if** (threadIdx.x < halfPoint) **then**
14:       // Obtain the shared value stored by another thread
15:       double temp = min[threadIdx.x + halfPoint];
16:       int temp1 = indexes[threadIdx.x + halfPoint];
17:       **if** (temp < min[threadIdx.x]) **then**
18:         min[threadIdx.x] = temp;
19:         indexes[threadIdx.x] = temp1;
20:       **end if**
21:     **end if**
22:     _syncthreads();
23:     nTotalThreads = (nTotalThreads » 1);
24:   **end while**
25:   **if** (threadIdx.x == 0) **then**
26:     fbest[blockIdx.y + blockIdx.x] = min[0];
27:     for (i = 0; i < D; i++) xbest[i] = X[(indexes[0]) * D + i];
28:   **end if**
29:   _syncthreads();
30: **end procedure**

Algorithm 2 Stigmergy-inspired kernel of SDE

1: **procedure** _GLOBAL_ VOID STIGMERGIC(double* X, double* F, double f, double cr, int funcion)
2:   Obtain the thread identity with respect to the block in the grid
3:   Obtain jrand as a uniform integer random number
4:   **if** (arrayIndex > 0 and arrayIndex < Np − 1) **then**
5:     // **Trace comparison**:
6:     Obtain x_{lead}, x_{work1}, x_{work2} according to their fitness
7:     //**Action (Mutant vector, Trial vector)**:
8:     Calculate mutant vector with Equation (6)
9:     Compose trial vector with Equation (7)
10:     //**Trace update**:
11:     Compare and update fitness of trial vector against fitness of x_{lead}, x_{work1}, x_{work2} with Equations (8)–(10)
12:     _syncthreads();
13:   **end if**
14: **end procedure**

A kernel that generates the initial population is the same for both algorithms; also, DE utilizes an extra kernel to evaluate the initial population's fitness only once.

4. Experimental Results

To perform the experiments, the computer has the next features:

- OS: Windows 10 64 bits
- CPU: i7-4770 running at 3.4 GHz
- RAM: 16 GB
- GPU: GTX 960 with 2 GB of RAM
- Compiler: Visual Studio 2019

Both algorithms are programmed in CUDA C by using the SDK version 11.1, reusing some kernel functions such as creating or evaluating the initial population. Moreover, both implementations utilize global memory, except for the reduction approach, to find the global best in DE, which uses shared memory.

The DE version/best/1/bin is the algorithm programmed in this paper, and the SDE operators are a modified version of DE; also, the parameters are set as: $F = 0.8$ and $C_r = 0.3$ in the two approaches. Population size is set as 256 and 512 individuals for every experiment, whereas for every problem the dimension is 100 and 500. The reason for these population sizes is because of the requirements of the kernel to obtain the global best used in DE.

4.1. Benchmark Used in the Experiments

The experimental testbed consists of 27 functions. We used 14 multidimensional objective functions (Table 3) that have different search spaces and different complexity levels. For example, the functions: Step, Ackley, Griewank, Levy, Rastrigin, Schwefel and Schwefel 2.22 have many local minima near the global optimum. The functions: Sphere, Sum Powers, Sum Squares and Trid have only one global optimum and are bowl-shaped. The functions Zakharov, Dixon and Price and Rosenbrock have a valley-shaped form, and the Zakharov function has only one global optimum, whereas the other two functions have several optima. Moreover, 13 two-dimensional objective functions (Table 4) were tested. Some of these functions have an optimal value solution outside the vector 0, ...,0. This is useful for testing a bias towards such a value when new metaheuristic approaches are proposed, as suggested by [57].

Table 3. $D > 2$ objective functions. The optimum value is denoted as f^* and x^* is the candidate solution.

Function	$f(x)$	Limits x_i	$f^*; x^*$
F1: Sphere	$\sum_{i=1}^{d} x_i^2$	$[-5.12, 5.12]$	$0; 0, \ldots, 0$
F2: Step	$\sum_{i=1}^{d} (\lfloor x_i + 0.5 \rfloor)^2$	$[-100, 100]$	$0; [-0.5, 0.5)$
F3: Sum Powers	$\sum_{i=1}^{d} |x_i|^{i+1}$	$[-1, 1]$	$0; 0, \ldots, 0$
F4: Sum Squares	$\sum_{i=1}^{d} i x_i^2$	$[-10, 10]$	$0; 0, \ldots, 0$
F5: Dixon and Price	$(x_i - 1)^2 + \sum_{i=2}^{d} i(2x_i^2 - x_{i-1})^2$	$[-10, 10]$	$0; x_i = 2^{-\frac{2^i-2}{2^i}}$
F6: Rosenbrock	$\sum_{i=1}^{d-1} \left[100(x_{i+1} - x_i^2)^2 + (x_i - 1)^2 \right]$	$[-30, 30]$	$0; 1, \ldots, 1$
F7: Schwefel 2.22	$\sum_{i=1}^{d} |x_i| + \prod_{i=1}^{d} |x_i|$	$[-10, 10]$	$0; 0, \ldots, 0$
F8: Trid	$\sum_{i=1}^{d} (x_i - 1)^2 - \sum_{i=2}^{d} x_i x_{i-1}$	$[-d^2, d^2]$	$-\frac{d(d+4)(d-1)}{6}; x_i = i(d+1-i)$
F9: Zakharov	$\sum_{i=1}^{d} x_i^2 + \left(\sum_{i=1}^{d} 0.5 i x_i \right)^2 + \left(\sum_{i=1}^{d} 0.5 i x_i \right)^4$	$[-5, 10]$	$0; 0, \ldots, 0$

Table 3. Cont.

Function	$f(x)$	Limits x_i	$f^*; x^*$
F10: Ackley	$-a\exp\left(-b\sqrt{\frac{1}{d}\sum_{i=1}^{d} x_i^2}\right) - \exp\left(\frac{1}{d}\sum_{i=1}^{d}\cos(cx_i)\right) + a + \exp(1)$ where $a = 20, b = 0.2$	$[-32.768, 32.768]$	$0; 0, \ldots, 0$
F11: Rastrigin	$10d + \sum_{i=1}^{d}\left[x_i^2 - 10\cos(2\pi x_i)\right]$	$[-5.12, 5.12]$	$0; 0, \ldots, 0$
F12: Schwefel	$418.9829d - \sum_{i=1}^{d} x_i \sin\left(\sqrt{|x_i|}\right)$	$[-500, 500]$	$0; 420.9687, \ldots, 420.9687$
F13: Griewank	$\sum_{i=1}^{d}\frac{x_i^2}{4000} - \prod_{i=1}^{d}\cos\left(\frac{x_i}{\sqrt{i}}\right) + 1$	$[-600, 600]$	$0; 0, \ldots, 0$
F14: Levy	$\sin^2(\pi w_i) + \sum_{i=1}^{d-1}(w_i - 1)^2\left[1 + 10\sin^2(\pi w_i + 1)\right] + (w_d - 1)^2 \cdot \left[1 + \sin^2(2\pi w_d)\right]$ where $w_i = 1 + \frac{x_i - 1}{4}$	$[-10, 10]$	$0; 1, \ldots, 1$

Table 4. $D = 2$ objective functions. The optimum value is denoted as f^* and x^* is the candidate solution.

Function	$f(x)$	Limits x_i	$f^*; x^*$
F15: Treccani	$x_1^4 + 4x_1^3 + 4x_1^2 + x_2^2$	$[-5, 5]$	$0; \{(-2,0), (0,0)\}$
F16: Beale	$(1.5 - x_1 + x_1 x_2)^2 + (2.25 - x_1 + x_1 x_2^2)^2 + (2.625 - x_1 + x_1 x_2^3)^2$	$[-4.5, 4.5]$	$0; (3, 0.5)$
F17: Booth	$(x_1 + 2x_2 - 7)^2 + (2x_1 + x_2 - 5)^2$	$[-10, 10]$	$0; (1, 3)$
F18: Brent	$(x_1 + 10)^2 + (x_2 + 10)^2 + \exp(-x_1^2 - x_2^2)$	$[-10, 10]$	$0; (-10, 10)$
F19: Cube	$100(x_2 - x_1^3)^2 + (1 - x_1)^2$	$[-10, 10]$	$0; (1, 1)$
F20: Davis	$(x_1^2 + x_2^2)^{0.25}\left[\sin^2(50(3x_1^2 + x_2^2)^{0.1}) + 1\right]$	$[-100, 100]$	$0; (0, 0)$
F21: Matyas	$0.26(x_1^2 + x_2^2) - 0.48 x_1 x_2$	$[-10, 10]$	$0; (0, 0)$
F22: Schaffer	$0.5 + \frac{\sin^2\left(\sqrt{x_1^2 + x_2^2}\right) - 0.5}{\left(1 + 0.001(x_1^2 + x_2^2)\right)^2}$	$[-100, 100]$	$0; (0, 0)$
F23: Bohachevsky1	$x_1^2 + 2x_2^2 - 0.3\cos(3\pi x_1) - 0.4\cos(4\pi x_2) + 0.7$	$[-100, 100]$	$0; (0, 0)$
F24: Egg Crate	$x_1^2 + x_2^2 + 25\left[\sin^2(x_1) + \sin^2(x_2)\right]$	$[-5, 5]$	$0; (0, 0)$
F25: Bohachevsky2	$x_1^2 + 2x_2^2 - 0.3\cos(3\pi x_1)\cos(4\pi x_2) + 0.3$	$[-100, 100]$	$0; (0, 0)$
F26: Bohachevsky3	$x_1^2 + 2x_2^2 - 0.3\cos(3\pi x_1 + 4\pi x_2) + 0.3$	$[-100, 100]$	$0; (0, 0)$
F27: Three-Hump Camel	$2x_1^2 - 1.05 x_1^4 + \frac{x_1^6}{6} + x_1 x_2 + x_2^2$	$[-5, 5]$	$0; (0, 0)$

4.2. Execution Time

To compare the computational time of DE and SDE, we stored the execution time after 5000 iterations of every algorithm and calculated the average and standard deviation after 30 runs. In this case, we measured the execution time since the initial population's creation until the last iteration; the time counter is reset to zero only after one run is complete. The problem dimensionality has either 100 or 500 dimensions, with 256 and 512 individuals; therefore, this experiment has four instances. The calculus of speedup considers the average time of each function $\left(speedup = \frac{T_{DE}}{T_{SDE}}\right)$. The results are in Tables 5 and 6, with the best ones in bold letters.

SDE has a slightly better averaged execution time in most of the cases, except for the Griewank function and in some experiments for the Brent function. However, a closer review of the speedups reveals that both algorithms have a similar performance regarding the execution time. To further clarify such an issue, we used the Wilcoxon test [58] to probe the null hypothesis that the execution times of both algorithms come from distributions with equal medians. The p-values for DE vs SDE are: (1) $p = 0.1215$ for $N_p = 256$ and $D = 100$; (2) $p = 0.09$ for $N_p = 512$ and $D = 100$; (3) $p = 0.1323$ for $N_p = 256$ and $D = 500$; (4) $p = 0.1115$ for $N_p = 512$ and $D = 500$. According to these results, in every case, the null hypothesis cannot be rejected, suggesting (with 5% certainty) that the execution time of both algorithms is similar for each experimental instance.

Table 5. Mean and averaged execution times in seconds of DE and SDE ($D = 100$).

Function	Execution Time $\mu(\sigma)$ for 256 Individuals			Execution Time $\mu(\sigma)$ for 512 Individuals		
	DE	SDE	Speedup	DE	SDE	Speedup
F1: Sphere	9.2106 (0.0782)	**8.7910 (0.0708)**	1.05	10.3554 (0.0606)	**9.2490 (0.0638)**	1.12
F2: Step	11.4135 (0.0608)	**10.6315 (0.0899)**	1.07	12.6158 (0.0.0241)	**11.3259 (0.0703)**	1.11
F3: Sum Powers	9.2367 (0.0949)	**8.8988 (0.0745)**	1.04	10.0024 (0.0833)	**9.4303 (0.0487)**	1.06
F4: Sum Squares	9.2517 (0.1358)	**8.7713 (0.0808)**	1.05	10.0787 (0.0611)	**9.3388 (0.0978)**	1.08
F5: Dixon and Price	14.4088 (0.1586)	**13.9450 (0.0560)**	1.03	15.7625 (0.1041)	**14.8474 (0.0390)**	1.06
F6: Rosenbrock	19.2443 (0.0919)	**18.8281 (0.0662)**	1.02	21.1980 (0.1622)	**19.9684 (0.0738)**	1.06
F7: Schwefel 2.22	6.1812 (0.0072)	**5.7724 (0.0417)**	1.07	6.3052 (0.0066)	**6.0008 (0.0449)**	1.05
F8: Trid	9.6917 (0.1798)	**8.8270 (0.0528)**	1.10	10.3362 (0.0354)	**9.4783 (0.0824)**	1.09
F9: Zakharov	9.6774 (0.1288)	**9.1474 (0.0544)**	1.06	10.4172 (0.0930)	**9.7241 (0.0509)**	1.07
F10: Ackley	12.0315 (0.0585)	**11.3902 (0.0685)**	1.06	13.4087 (0.1139)	**12.1660 (0.0722)**	1.10
F11: Rastrigin	11.8263 (0.0988)	**11.3737 (0.1139)**	1.04	12.7528 (0.0724)	**12.1194 (0.1042)**	1.05
F12: Schwefel	8.1851 (0.0569)	**7.9648 (0.0890)**	1.03	8.7612 (0.0524)	**8.2109 (0.0296)**	1.06
F13: Griewank	**13.7655 (0.0846)**	14.2824 (0.1054)	0.96	**15.3910 (0.0995)**	15.6216 (0.0842)	0.98
F14: Levy	17.9642 (0.0813)	**16.7896 (0.1229)**	1.07	20.6549 (0.1170)	**18.4108 (0.0842)**	1.12
F15: Treccani	4.6935 (0.1371)	**4.2295 (0.0598)**	1.11	4.9562 (0.1360)	**4.3714 (0.0289)**	1.13
F16: Beale	4.8087 (0.0155)	**4.6405 (0.0527)**	1.04	5.0649 (0.0409)	**4.8098 (0.0421)**	1.05
F17: Booth	4.4926 (0.0541)	4.4926 (0.0541)	1.00	4.7859 (0.0215)	**4.6785 (0.0529)**	1.02
F18: Brent	**4.5868 (0.0186)**	4.7189 (0.0327)	0.97	4.9435 (0.0130)	**4.9202 (0.0297)**	1.01
F19: Cube	4.6144 (0.0115)	**4.1862 (0.0413)**	1.10	4.9616 (0.0120)	**4.3657 (0.0313)**	1.14
F20: Davis	5.0567 (0.0565)	**4.4333 (0.0415)**	1.14	5.4204 (0.0099)	**4.6649 (0.0628)**	1.16
F21: Matyas	5.0291 (0.0124)	**4.0977 (0.0641)**	1.23	5.2439 (0.0023)	**4.2874 (0.0416)**	1.22
F22: Schaffer	4.8074 (0.0324)	**4.4074 (0.0373)**	1.09	5.0514 (0.0506)	**4.6265 (0.0282)**	1.09
F23: Bohachevsky1	4.5953 (0.0241)	**4.2020 (0.0644)**	1.09	4.8132 (0.0287)	**4.3676 (0.0325)**	1.10
F24: Egg Crate	4.8835 (0.0081)	**4.2803 (0.0422)**	1.14	5.2487 (0.0579)	**4.4989 (0.0353)**	1.17
F25: Bohachevsky2	4.5948 (0.0202)	**4.1841 (0.0447)**	1.10	4.8038 (0.0175)	**4.3554 (0.0375)**	1.10
F26: Bohachevsky3	4.6091 (0.0156)	**4.1864 (0.0448)**	1.10	4.8021 (0.0221)	**4.3475 (0.0505)**	1.10
F27: Three-Hump Camel	4.8569 (0.0200)	**4.1695 (0.0477)**	1.16	5.2030 (0.0.0237)	**4.3531 (0.0377)**	1.20

Table 6. Mean and averaged execution times in seconds of DE and SDE ($D = 500$).

Function	Execution Time $\mu(\sigma)$ for 256 Individuals			Execution Time $\mu(\sigma)$ for 512 Individuals		
	DE	SDE	Speedup	DE	SDE	Speedup
F1: Sphere	45.1006 (0.1674)	**43.5942 (0.1625)**	1.03	47.8527 (0.0200)	**46.3508 (0.0062)**	1.03
F2 Step	55.7603 (0.0492)	**55.1497 (0.1864)**	1.01	59.7023 (0.0347)	**57.5648 (0.2401)**	1.04
F3: Sum Powers	47.3859 (0.1270)	**44.0992 (0.2072)**	1.07	51.0338 (0.0389)	**47.4896 (0.0307)**	1.07
F4: Sum Squares	45.0150 (0.1787)	**43.3325 (0.1482)**	1.04	48.2126 (0.0252)	**46.6274 (0.0068)**	1.03
F5: Dixon and Price	70.2937 (0.2209)	70.2279 (0.1731)	1.00	74.9822 (0.0298)	**71.0612 (0.0428)**	1.06
F6: Rosenbrock	95.6219 (0.2426)	**94.9924 (0.1544)**	1.01	101.3595 (0.036)	**95.4034 (0.0300)**	1.06
F7: Schwefel 2.22	31.7049 (0.0719)	**30.0035 (0.1929)**	1.06	33.6836 (0.2565)	**31.4718 (0.1376)**	1.07
F8: Trid	47.2107 (0.1703)	**43.6624 (0.1782)**	1.08	50.2172 (0.0326)	**48.7325 (0.0232)**	1.03
F9: Zakharov	46.9736 (0.1404)	**45.1113 (0.1838)**	1.04	50.1629 (0.0517)	**47.0111 (0.0762)**	1.07
F10: Ackley	59.2495 (0.1712)	**56.5934 (0.1743)**	1.05	62.5976 (0.1638)	**60.1282 (0.0707)**	1.04
F11: Rastrigin	58.9961 (0.0955)	**56.4700 (0.1942)**	1.04	62.2493 (0.0957)	**59.5437 (0.3830)**	1.05
F12: Schwefel	39.6069 (0.1385)	**39.0363 (0.1085)**	1.01	41.7337 (0.0551)	**40.8717 (0.0721)**	1.02
F13: Griewank	**67.4986 (0.1488)**	71.4827 (0.1742)	0.94	**71.7673 (0.0992)**	75.6721 (0.0595)	0.95
F:14 Levy	88.1310 (0.1206)	**83.0872 (0.1613)**	1.06	93.9297 (0.1430)	**89.1134 (0.0781)**	1.05
F15: Treccani	23.8224 (0.6603)	**21.2687 (0.2196)**	1.12	25.3578 (0.6672)	**22.4922 (0.2129)**	1.13
F16: Beale	23.5141 (0.0369)	**23.1105 (0.3537)**	1.02	25.1768 (0.0694)	**24.5135 (0.2085)**	1.03
F17: Booth	23.2668 (0.0490)	**23.0368 (0.2329)**	1.01	24.8573 (0.0480)	**24.4406 (0.1868)**	1.02
F18: Brent	**23.1873 (0.0809)**	23.8490 (0.1661)	0.97	**24.7574 (0.0653)**	25.2045 (0.1205)	0.98
F19: Cube	23.6192 (0.0523)	**21.1600 (0.2700)**	1.12	25.1667 (0.0651)	**22.5187 (0.2063)**	1.12
F20: Davis	24.6654 (0.0647)	**21.3285 (0.2838)**	1.16	26.2479 (0.0403)	**22.7148 (0.1962)**	1.16
F21: Matyas	25.3824 (0.0144)	**21.1435 (0.1936)**	1.20	27.0852 (0.0140)	**22.3725 (0.1781)**	1.21
F22: Schaffer	23.3038 (0.0899)	**21.3305 (0.1633)**	1.09	24.9515 (0.0555)	**22.7985 (0.2195)**	1.09

Table 6. Cont.

Function	Execution Time $\mu(\sigma)$ for 256 Individuals			Execution Time $\mu(\sigma)$ for 512 Individuals		
	DE	SDE	Speedup	DE	SDE	Speedup
F23: Bohachevsky1	23.0564 (0.0598)	**21.3298 (0.2496)**	1.08	24.6339 (0.0576)	**22.3473 (0.1723)**	1.10
F24: Egg Crate	24.4434 (0.0727)	**21.1914 (0.1658)**	1.15	26.0469 (0.0843)	**22.5681 (0.2205)**	1.15
F25: Bohachevsky2	23.0607 (0.1270)	**21.1913 (0.2459)**	1.09	24.7679 (0.0580)	**22.3699 (0.1505)**	1.11
F26: Bohachevsky3	23.1568 (0.0570)	**21.1915 (0.2585)**	1.09	24.8482 (0.0853)	**22.3834 (0.2782)**	1.11
F27: Three-Hump Camel	24.6552 (0.1462)	**21.1373 (0.2239)**	1.17	26.3881 (0.2053)	**22.2933 (0.2343)**	1.18

4.3. Convergence Speed

For the convergence comparison between DE and SDE, we used the same four instances as in the previous experiment: dimensions of 100 and 500 and populations of 256 and 512 individuals. In this case, the iterations per run are increased to 20,000, with the same number of runs per experimental instance; nevertheless, both algorithms iterate either until the algorithm is near to the optimum (e.g., $f(x^*) = -0.99 \frac{d(d+4)(d-1)}{6}$ for Trid), or until reaching 20,000 iterations. After either of the two conditions is true, then the best global fitness and the number of iterations are stored, in the case of DE. However, as SDE does not utilizes a global best, we use the fitness of the first individual in the population as the 'global best' to demonstrate the capability of the algorithm to distribute the improvements through all the candidate solutions but with decentralized control. Therefore, when either the fitness of the first individual is close to the real global optimum, or 20,000 iterations are achieved, a run is completed in SDE. The results are in Tables 7–10 with the best results in bold letters.

Table 7. Mean and averaged final fitness and iterations of DE and SDE ($N_p = 256$; $D = 100$).

Function	Final Fitness $\mu(\sigma)$		Final Iterations $\mu(\sigma)$	
	SDE	DE	SDE	DE
F1: Sphere	0.0096 (3.5238×10^{-4})	0.0096 (3.2083×10^{-4})	857.26 (21.30)	3832.63 (106.76)
F2: Step	0.00 (0)	0.00 (0)	829.8 (28.91)	4612.2 (143.3675)
F3: Sum Powers	0.0074 (0.0022)	**0.0073 (0.0021)**	53.16 (9.05)	**29.86 (5.96)**
F4: Sum Squares	0.0097 (3.0754×10^{-4})	**0.0095 (4.5164×10^{-4})**	1266.10 (18.23)	5562.43 (122.29)
F5: Dixon and Price	39.2690 (17.583)	**0.6667 (8.0515×10^{-7})**	20,000.00 (0)	20,000.00 (0)
F6: Rosenbrock	366.7301 (130.1247)	**90.1404 (0.1371)**	20,000.00 (0)	20,000.00 (0)
F7: Schwefel 2.22	0.0098 (1.2780×10^{-4})	0.0098 (2.2376×10^{-4})	1227.1 (19.36)	6314.47 (136.45)
F8: Trid	7.889×10^6 (9.809×10^6)	**4.821×10^5 (1.97×10^5)**	20,000.00 (0)	20,000.00 (0)
F9: Zakharov	**0.2932 (0.13712)**	262.7573 (19.5592)	20,000.00 (0)	20,000.00 (0)
F10: Ackley	0.0098 (9.9644×10^{-5})	**0.0097 (1.7791×10^{-4})**	1367.93 (24.07)	6499.70 (131.53)
F11: Rastrigin	**4.3725 (2.6246)**	592.3126 (17.7795)	18841.96 (4407.1)	20,000.00 (0)
F12: Schwefel	**55.5158 (74.2989)**	3723.5101 (5356.9)	10405.80 (9128.0)	20,000.00 (0)
F13: Griewank	0.0097 (3.3617×10^{-4})	**0.0095 (3.5729×10^{-4})**	1307.30 (19.59)	5935.20 (107.78)
F14: Levy	0.0097 (2.7136 × 10^{-4})	**0.0095 (4.6494×10^{-4})**	988.06 (18.48)	6339.90 (175.04)
F15: Treccani	0.0062 (0.0032)	**0.0048 (0.0032)**	32.53 (11.66)	**6.53 (3.53)**
F16: Beale	0.0058 (0.0034)	0.0058 (0.0025)	92.83 (49.57)	**7.73 (4.69)**
F17: Booth	0.0071 (0.0032)	**0.0044 (0.0032)**	89.83 (40.78)	**21.4667 (7.64)**
F18: Brent	0.0076 (0.0021)	**0.0048 (0.0029)**	69.4 (22.60)	**15.7 (4.66)**
F19: Cube	0.0062 (0.0031)	**0.0054 (0.0024)**	243.23 (179.04)	**37 (12.16)**
F20: Davis	0.0709 (0.0703)	**0.0064 (0.0018)**	453.33 (59.00)	**153.03 (10.76)**
F21: Matyas	0.0065 (0.0032)	**0.0050 (0.0034)**	47.46 (24.72)	**3.16 (2.61)**
F22: Schaffer	**0.0046 (0.0031)**	0.0048 (0.0029)	193.36 (97.32)	**19.07 (10.65)**
F23: Bohachevsky1	0.0067 (0.0029)	**0.0049 (0.0025)**	84.83 (40.78)	**36.23 (5.48)**
F24: Egg Crate	0.0061 (0.0029)	**0.0039 (0.0029)**	43.86 (11.56)	**15.8 (5.67)**
F25: Bohachevsky2	0.0060 (0.0033)	**0.0058 (0.0026)**	126.66 (45.06)	**37.1 (5.87)**
F26: Bohachevsky3	**0.0050 (0.0035)**	0.0041 (0.0029)	122.2 (42.91)	**33.93 (5.64)**
F27: Three-Hump Camel	0.0058 (0.0028)	**0.0056 (0.0027)**	39.1 (17.82)	**7.5 (3.42)**

It can be seen from the results of these experiments (Tables 7–10) that the proposed algorithm does not outperform DE on two-dimensional problems. Although SDE reaches the optimum, it performs a higher number of iterations than DE, however, the performance of SDE in high dimensions is remarkable compared with DE.

Table 7 shows the experimental instance when $D = 100$ and $N_p = 256$. In this case, both algorithms have a good performance with the Sphere, Step, Sum Powers, Sum Squares, Schwefel 2.22, Ackley, Griewank and Levy functions, as they both achieve a good average fitness; however, except for the Sum Powers function where DE converges in fewer iterations (almost in half of the iterations), SDE is faster than DE in all other cases, with a reduction of more than 84% in the number of iterations for the best result (Levy function), as shown in Figure 3. For the functions Trid, Zakharov, Rastrigin and Schwefel, both algorithms have problems finding the global best, but sometimes SDE obtains better fitness values than DE, with fewer iterations on average.

Table 8 shows that the SDE behavior for more individuals ($N_p = 512$) is as reported in previous lines.

Table 8. Mean and averaged final fitness and iterations of DE and SDE ($N_p = 512$; $D = 100$).

Function	Final Fitness $\mu(\sigma)$		Final Iterations $\mu(\sigma)$	
	SDE	DE	SDE	DE
F1: Sphere	0.0096 (3.1580 × 10^{-4})	**0.0095 (4.3491 × 10^{-4})**	**861.00 (20.67)**	4344.16 (114.73)
F2: Step	0.0 (0)	0.0 (0)	**828.56 (24.44)**	5486.53 (193.96)
F3: Sum Powers	0.0077 (0.0020)	**0.0070 (0.0023)**	53.00 (8.47)	**26.86 (6.14)**
F4: Sum Squares	0.0097 (2.0231 × 10^{-4})	**0.0095 (4.1157 × 10^{-4})**	**1263.26 (17.43)**	6216.26 (113.12)
F5: Dixon and Price	36.3413 (16.9494)	**0.6667 (1.8257 × 10^{-7})**	20,000.00 (0)	20,000.00 (0)
F6: Rosenbrock	348.7744 (127.6737)	**90.7098 (0.1733)**	20,000.00 (0)	20,000.00 (0)
F7: Schwefel 2.22	0.0098 (1.3374 × 10^{-4})	0.0098 (2.0745 × 10^{-4})	**1227.43 (17.91)**	6962.76 (111.36)
F8: Trid	7.746 × 10^6 (6.448 × 10^6)	3.087 × 10^6 (8.365 × 10^5)	20,000.00 (0)	20,000.00 (0)
F9: Zakharov	**0.2058 (0.0850)**	257.4497 (14.3462)	20,000.00 (0)	20,000.00 (0)
F10: Ackley	0.0098 (2.0868 × 10^{-4})	**0.0097 (2.2258 × 10^{-4})**	**1373.53 (21.39)**	7401.03 (159.97)
F11: Rastrigin	**0.4355 (0.6197)**	600.0198 (16.0965)	**9035.86 (8486.8)**	20,000.00 (0)
F12: Schwefel	**52.4709 (65.0721)**	5852.3797 (5544.0)	**1981.50 (32.62)**	20,000.00 (0)
F13: Griewank	0.0097 (2.8136 × 10^{-4})	**0.0095 (3.6761 × 10^{-4})**	**1303.60 (22.27)**	6690.20 (152.02)
F14: Levy	0.0097 (2.1518 × 10^{-4})	**0.0096 (2.4228 × 10^{-4})**	**989.70 (25.23)**	7744.56 (209.23)
F15: Treccani	0.0059 (0.0031)	**0.0049 (0.0028)**	33.5 (14.83)	**3.8 (3.14)**
F16: Beale	0.0066 (0.0032)	**0.0048 (0.0031)**	76.16 (36.74)	**4.73 (2.99)**
F17: Booth	0.0068 (0.0029)	**0.0051 (0.0031)**	82.66 (34.35)	**15.86 (6.83)**
F18: Brent	0.0068 (0.0025)	**0.0044 (0.0026)**	61.4 (21.82)	**13.63 (4.77)**
F19: Cube	0.0051 (0.0040)	**0.0048 (0.0028)**	329.83 (205.35)	**27.47 (11.98)**
F20: Davis	**0.0038 (0.0346)**	0.0065 (0.0017)	587.36 (99.63)	**149.76 (9.49)**
F21: Matyas	0.0059 (0.0035)	**0.0038 (0.0026)**	48.13 (23.34)	**2.1 (1.56)**
F22: Schaffer	**0.0048 (0.0033)**	0.0048 (0.0027)	258.2 (140.22)	**16.56 (8.22)**
F23: Bohachevsky1	0.0070 (0.0028)	**0.0055 (0.0031)**	88.13 (24.36)	**33.8 (5.88)**
F24: Egg Crate	0.0067 (0.0029)	**0.0053 (0.0028)**	45.67 (15.39)	**13.23 (4.78)**
F25: Bohachevsky2	0.0053 (0.0033)	**0.0049 (0.0029)**	108.23 (61.14)	**36.1 (4.79)**
F26: Bohachevsky3	0.0059 (0.0032)	**0.0039 (0.0026)**	102.3 (34.34)	**30.86 (3.88)**
F27: Three-Hump Camel	**0.0053 (0.0033)**	0.0070 (0.0023)	53.00 (8.47)	**26.86 (6.14)**

From Tables 9 and 10 ($D = 500$), it can be seen that DE struggles with high dimensionality, except for the Sum Powers function. In fact, DE no longer reaches the optimum in 20,000 iterations for the Sum Squares, Schwefel 2.22, Ackley, Griewank and Levy functions, whereas SDE has no problems achieving the optimum for these functions. Another important detail is that with a larger number of individuals ($Np = 512$), SDE finds the optimum in slightly fewer iterations than when the population is smaller (see Tables 9 and 10). This highlights the advantage of using the stigmergy concept for the indirect communication among the population.

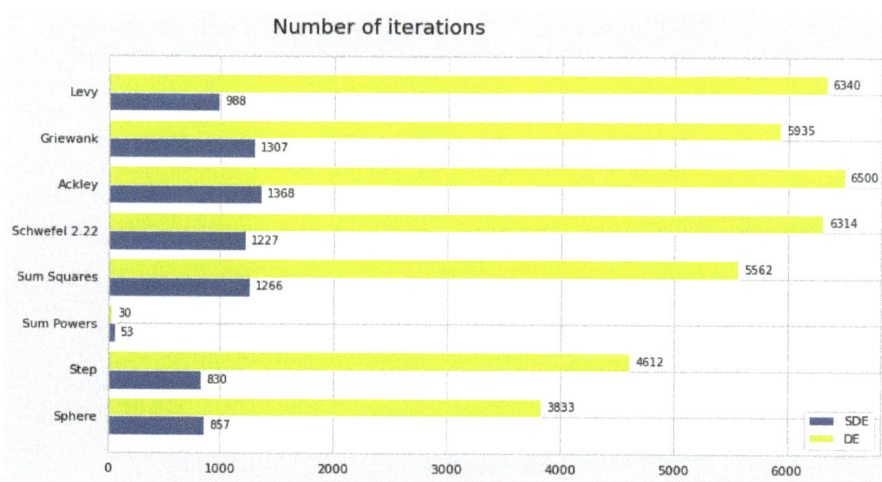

Figure 3. Comparison between the number of SDE and DE iterations.

Table 9. Mean and averaged final fitness and iterations of DE and SDE ($N_p = 256$; $D = 500$).

Function	Final Fitness $\mu(\sigma)$		Final Iterations $\mu(\sigma)$	
	SDE	DE	SDE	DE
F1: Sphere	**0.0099 (7.1685×10^{-5})**	1.5596 (0.1287)	**3915.83 (37.75)**	20,000.00 (0)
F2: Step	**0.00(0)**	648.46 (49.02)	**3796.33 (208.28)**	20,000.00 (0)
F3: Sum Powers	0.0073 (0.0018)	**0.0070 (0.0022)**	68.66 (14.84)	**65.60 (15.11)**
F4: Sum Squares	**0.0099 (6.1043×10^{-5}))**	1114.1061 (97.5739)	**6257.00 (48.87)**	20,000.00 (0)
F5: Dixon and Price	1116.7 (411.7089)	1.1704×10^4 (2084.0)	20,000.00 (0)	20,000.00 (0)
F6: Rosenbrock	2015.3 (675.7089)	3656.1 (321.2327)	20,000.00 (0)	20,000.00 (0)
F7 Schwefel 2.22	**0.0099 (5.7235×10^{-5})**	39.2896 (2.1531)	**5416.53 (37.09)**	20,000.00 (0)
F8: Trid	1.4111×10^{10} (1.662×10^{10})	3.6207×10^{10} (4.049×10^9)	20,000.00 (0)	20,000.00 (0)
F9: Zakharov	3137.0 (397.5563)	4175.8 (112.141)	20,000.00 (0)	20,000.00 (0)
F10: Ackley	**0.0099 (4.3961×10^{-5})**	3.1094 (0.0798)	**5680.50 (117.0)**	20,000.00 (0)
F11: Rastrigin	1568.87 (252.57)	4620.0471 (101.5107)	20,000.00 (0)	20,000.00 (0)
F12: Schwefel	5.0077×10^4 (1.9810×10^4)	1.7129×10^5 (776.5318)	20,000.00 (0)	20,000.00 (0)
F13: Griewank	**0.0099 (7.9325×10^{-5})**	6.3713 (0.5136)	**5270.16 (42.23)**	20,000.00 (0)
F14: Levy	0.0247 (0.0811)	17.0589 (1.7703)	**5233.73 (2780.0)**	20,000.00 (0)
F15: Treccani	0.0049 (0.0034)	**0.0039 (0.0026)**	33.13 (7.89)	**6.86 (3.77)**
F16: Beale	0.0057 (0.0032)	**0.0052 (0.0029)**	67.76 (35.51)	**8.26 (4.40)**
F17: Booth	0.0064 (0.0036)	**0.0051 (0.0030)**	80.5 (41.65)	**21.63 (10.99)**
F18: Brent	0.0074 (0.0021)	**0.0049 (0.0029)**	63.8 (17.98)	**14.43 (5.87)**
F19: Cube	0.0049 (0.0038)	**0.0048 (0.0031)**	271.63 (117.09)	**38.03 (15.18)**
F20: Davis	0.0069 (0.0033)	**0.0065 (0.0018)**	584.75 (133.71)	**154.3 (9.18)**
F21: Matyas	0.0064 (0.0027)	**0.0047 (0.0028)**	41.63 (19.11)	**4 (3.36)**
F22: Schaffer	0.0047 (0.0029)	**0.0047 (0.0031)**	214.08 (107.21)	**21.7 (11.65)**
F23: Bohachevsky1	0.0074 (0.0025)	**0.0047 (0.0028)**	94.2 (22.08)	**35.93 (5.19)**
F24: Egg Crate	**0.0057 (0.0026)**	0.0060 (0.0027)	48.23 (14.61)	**16.36 (3.92)**
F25: Bohachevsky2	0.0047 (0.0034)	**0.0045 (0.0027)**	145.2 (68.44)	**37.3 (5.75)**
F26: Bohachevsky3	0.0061 (0.0031)	**0.0048 (0.0026)**	123.07 (42.16)	**34.3 (4.06)**
F27: Three-Hump Camel	0.0070 (0.0026)	**0.0042 (0.0026)**	41.03 (21.40)	**7.4 (4.61)**

Table 10. Mean and averaged final fitness and iterations of DE and SDE ($N_p = 512$; $D = 500$).

Function	Final Fitness $\mu(\sigma)$		Final Iterations $\mu(\sigma)$	
	SDE	DE	SDE	DE
F1: Sphere	**0.0099 (6.0612×10^{-5})**	27.3649 (1.4603)	**3685.16 (30.18)**	20,000.00 (0)
F2: Step	**0.0 (0)**	1.0338×10^4 (696.03)	**3311.96 (107.64)**	20,000.00 (0)
F3: Sum Powers	0.0077 (0.0017)	**0.0075 (0.0019)**	68.86 (13.06)	**62.63 (13.32)**
F4: Sum Squares	**0.0099 (1.0435×10^{-4})**	1.9638×10^4 (1214.0749)	**5884.63 (30.66)**	20,000.00 (0)
F5: Dixon and Price	**1186.7 (455.0704)**	1.1704×10^4 (2084.0)	20,000.00 (0)	20,000.00 (0)
F6: Rosenbrock	**2266.0 (683.9547)**	3656.1 (321.2327)	20,000.00 (0)	20,000.00 (0)
F7: Schwefel 2.22	**0.0099 (4.9013×10^{-5})**	202.2585 (8.5413)	**5181.37 (27.57)**	20,000.00 (0)
F8: Trid	1.4072×10^{10} (1.775×10^{10})	1.8198×10^{11} (1.070×10^{10})	20,000.00 (0)	20,000.00 (0)
F9: Zakharov	**3109.1 (348.8842)**	4175.8 (112.1416)	20,000.00 (0)	20,000.00 (0)
F10: Ackley	**0.0099 (3.4051×10^{-5})**	7.0142 (0.1369)	**5229.16 (48.38)**	20,000.00 (0)
F11: Rastrigin	**1567.5246 (388.10)**	4990.5538 (64.7224)	20,000.00 (0)	20,000.00 (0)
F12: Schwefel	4.0269×10^4 (2.1107×10^4)	1.7120×10^5 (760.9840)	20,000.00 (0)	20,000.00 (0)
F13: Griewank	**0.0098 (1.2733×10^{-4})**	94.8377 (5.1547)	**4964.13 (42.25)**	20,000.00 (0)
F14: Levy	**0.0099 (8.3354×10^{-5})**	96.7644 (6.7583)	**4355.60 (48.12)**	20,000.00 (0)
F15: Treccani	0.0061 (0.0024)	**0.0049 (0.0032)**	33.16 (13.69)	**4.1 (2.23)**
F16: Beale	0.0063 (0.0032)	**0.0044 (0.0025)**	69.66 (31.39)	**7.4 (3.59)**
F17: Booth	0.0075 (0.0026)	**0.0056 (0.0029)**	87.63 (36.26)	**16.56 (9.58)**
F18: Brent	0.0075 (0.0024)	**0.0048 (0.0029)**	65.4 (19.48)	**14.23 (4.27)**
F19: Cube	**0.0058 (0.0037)**	0.0059 (0.0030)	252.03 (168.68)	**26.67 (11.38)**
F20: Davis	**0.0037 (0.0034)**	0.0067 (0.0015)	571.11 (133.95)	**149.2 (12.54)**
F21: Matyas	0.0049 (0.0032)	**0.0035 (0.0028)**	47 (16.06)	**2.13 (1.63)**
F22: Schaffer	**0.0051 (0.0031)**	0.0055 (0.0031)	218.36 (168.65)	**14.96 (7.87)**
F23: Bohachevsky1	0.0072 (0.0027)	**0.0043 (0.0032)**	92.13 (27.13)	**32.63 (6.32)**
F24: Egg Crate	0.0063 (0.0027)	**0.0050 (0.0031)**	53.67 (17.51)	**13.4 (3.63)**
F25: Bohachevsky2	0.0058 (0.0034)	**0.0043 (0.0031)**	120.43 (52.24)	**31.83 (6.90)**
F26: Bohachevsky3	0.0073 (0.0027)	**0.0050 (0.0032)**	140.73 (64.64)	**30.9 (5.54)**
F27: Three-Hump Camel	0.0062 (0.0028)	**0.0051 (0.0030)**	49.9 (19.74)	**4.93 (3.31)**

5. Discussion

Stigmergic DE implements slight modifications to the original DE operators, so the computational time required to complete several iterations is practically the same, as suggested by the results in Tables 5 and 6 and after applying the Wilcoxon test to such data. Figure 4 shows a comparison between the execution time of DE and SDE for each test function, considering $D = 500$ and $N_p = 512$. The comparison considers the difference in percentage between the execution times of each algorithm. In most cases (with the exception of the Brent and Griewank functions in these results), the proposal has a better time performance than its DE counterpart of about 17% in the best case. This behavior is similar in all these experiments; however, as already mentioned, SDE has a marginal advantage.

Moreover, both algorithms present similar problems in finding the global optimum in some functions. However, as can be seen from Tables 7 and 8, in the case of the Rastrigin, Schwefel and Zacharov functions, the difference between the final fitness values obtained for SDE and DE is very large. SDE obtains fitness values lower than DE by an order of magnitude 2, and for the case of the Trid, Dixon and Price and Rosenbrock functions, DE performs better than SDE, but the difference is less significant. If the dimension increases (Tables 9 and 10), only in the case of the Sum Powers function does DE have a better performance than SDE, but it is minimal, on the order of 0.0001. These results show the advantage of the proposal in high dimensions; empirically, this is due to the intrinsic parallelism of the stigmergy concept that was used as an inspiration to modify the original operators of a metaheuristic (DE, version best/1/bin in this case). The proposal avoids using the global best in the construction of new candidate solutions, thus simulating indirect communication among all individuals in the population.

In summary, under the same conditions for both metaheuristics, our proposal can find a similar result as DE (and better outcomes for some objective functions) but using fewer iterations on average (Tables 7–10), especially when the problems are high-dimensional.

Even if the iterations are insufficient to find the global optimum, SDE performs better than DE, so it can be a good alternative for solving multiple engineering optimization problems. As mentioned in [59], applications can be in industrial engineering for the job-shop scheduling problem [60,61], in image processing to perform a multilevel thresholding process on a 2D histogram to segment images [62,63], in path planning to solve robotics challenges [64], or to solve area coverage problems for wireless sensor networks (WSNs) [65], to mention a few.

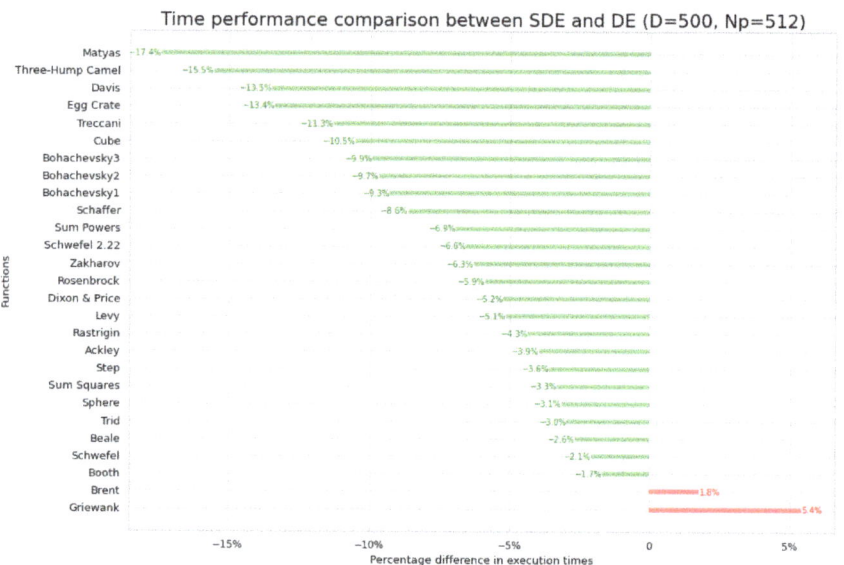

Figure 4. Percentage difference in execution times of DE and SDE for test functions with $D = 500$ and $N_p = 512$.

6. Conclusions

In this paper, we presented a modification to Differential Evolution that uses the global best to construct new candidate solutions. The adjustment considers stigmergy, a term taken from entomology to explain the indirect communication used by social insects to achieve complex projects. In the proposal, we eliminate the use of global best and instead we employ a local best. Like the real ants or termites, we utilize two neighbors of that local leader in the approach. The new candidate solution uses the leader's information and two neighbors in conjunction with a crossover operation as in the original DE. The final step in the approach is trace updating, which could modify any of the three individuals and their fitness values, simulating distributed and indirect communication with all the individuals in the artificial colony. We called this changed metaheuristic Stigmergic Differential Evolution (SDE). Due to the elimination of the global best, SDE has features that make it more parallelizable than DE, therefore, both algorithms were programmed in the CUDA-C language.

We compared DE and SDE with 27 objective functions under the same circumstances. The results suggest that SDE achieves faster convergence than DE when the dimensionality increases and there are a larger number of individuals in the population, and it performs similarly to DE for 2D test functions, but DE converges in fewer iterations. It was demonstrated in [66] that canonical DE is adversely altered by increments in dimensionality, which affects in a super-linear fashion the convergence to the global optimum of the metaheuristic. In that sense, even though SDE has average convergence properties (e.g., Tables 5 and 6), we claim that its main feature is the capability to maintain a good convergence behavior as dimensionality grows.

The proposed modification could apply to other metaheuristics that use a global best to construct a candidate solution. A possible future work would consider the stigmergy-based adaptation of the original operators in other metaheuristic algorithms and their comparison.

Author Contributions: Conceptualization, V.O.-E.; methodology, V.O.-E.; software, V.O.-E. and E.G.-M.; validation, V.O.-E. and E.G.-M.; formal analysis, V.O.-E.; investigation, V.O.-E. and E.G.-M.; resources, V.O.-E. and E.G.-M.; writing—original draft preparation, V. O.-E.; writing—review and editing, E.G.-M.; visualization, V.O.-E. and E.G.-M.; supervision, V.O.-E. and E.G.-M.; project administration, V.O.-E. and E.G.-M.; funding acquisition, V.O.-E. and E.G.-M. All authors have read and agreed to the published version of the manuscript.

Funding: This research received no external funding.

Institutional Review Board Statement: Not applicable.

Informed Consent Statement: Not applicable.

Data Availability Statement: Not applicable.

Acknowledgments: The authors wish to thank the Information and Communications Technology Laboratory of the Universidad Anáhuac México for allowing the use of computer equipment.

Conflicts of Interest: The authors declare no conflict of interest.

References

1. Cui, Z.; Alex, R.; Akerkar, R.; Yang, X.S. Recent Advances on Bioinspired Computation. *Sci. World J.* **2014**, *2014*, 934890. [CrossRef] [PubMed]
2. Ser, J.D.; Osaba, E.; Molina, D.; Yang, X.S.; Salcedo-Sanz, S.; Camacho, D.; Das, S.; Suganthan, P.N.; Coello, C.A.C.; Herrera, F. Bio-inspired computation: Where we stand and what's next. *Swarm Evol. Comput.* **2019**, *48*, 220–250. [CrossRef]
3. Črepinšek, M.; Liu, S.H.; Mernik, M. Exploration and exploitation in evolutionary algorithms. *ACM Comput. Surv.* **2013**, *45*, 1–33. [CrossRef]
4. Geem, Z.W.; Kim, J.H.; Loganathan, G. A New Heuristic Optimization Algorithm: Harmony Search. *Simulation* **2001**, *76*, 60–68. [CrossRef]
5. Lee, K.S.; Geem, Z.W. A new meta-heuristic algorithm for continuous engineering optimization: Harmony search theory and practice. *Comput. Methods Appl. Mech. Eng.* **2005**, *194*, 3902–3933. [CrossRef]
6. Geem, Z.W. *Music-Inspired Harmony Search Algorithm: Theory and Applications*; Springer: Berlin/Heidelberg, Germany, 2009; Volume 191.
7. Storn, R.; Price, K. Differential Evolution—A Simple and Efficient Heuristic for Global Optimization over Continuous Spaces. *J. Glob. Optim.* **1997**, *11*, 341–359.:1008202821328. [CrossRef]
8. Mezura-Montes, E.; Velázquez-Reyes, J.; Coello, C.A.C. A comparative study of differential evolution variants for global optimization. In Proceedings of the GECCO'06—Genetic and Evolutionary Computation Conference, Seattle, WA, USA, 8–12 July 2006; ACM Press: New York, NY, USA, 2006. [CrossRef]
9. Kennedy, J.; Eberhart, R. Particle swarm optimization. In Proceedings of the ICNN'95—International Conference on Neural Networks, Perth, WA, Australia, 27 November–1 December 1995; IEEE: Piscataway, NJ, USA, 1995. [CrossRef]
10. Tan, Y.; Zhu, Y. Fireworks Algorithm for Optimization. In *LNCS*; Springer: Berlin/Heidelberg, Germany, 2010; pp. 355–364. [CrossRef]
11. Tan, Y. *Fireworks Algorithm*; Springer: Berlin/Heidelberg, Germany, 2015.
12. Yang, X.S. Flower Pollination Algorithm for Global Optimization. In *Unconventional Computation and Natural Computation*; Springer: Berlin/Heidelberg, Germany, 2012; pp. 240–249. [CrossRef]
13. Rashedi, E.; Nezamabadi-pour, H.; Saryazdi, S. GSA: A Gravitational Search Algorithm. *Inf. Sci.* **2009**, *179*, 2232–2248. [CrossRef]
14. Łukasik, S.; Żak, S. Firefly Algorithm for Continuous Constrained Optimization Tasks. In *Computational Collective Intelligence. Semantic Web, Social Networks and Multiagent Systems*; Springer: Berlin/Heidelberg, Germany, 2009; pp. 97–106. [CrossRef]
15. Yang, X.S. Firefly Algorithms for Multimodal Optimization. In *Stochastic Algorithms: Foundations and Applications*; Springer: Berlin/Heidelberg, Germany, 2009; pp. 169–178. [CrossRef]
16. Yang, X.S.; Deb, S. Cuckoo Search via Lévy flights. In Proceedings of the 2009 World Congress on Nature Biologically Inspired Computing (NaBIC), Coimbatore, India, 9–11 December 2009; pp. 210–214. [CrossRef]
17. Karaboga, D. *An Idea Based on Honey Bee Swarm for Numerical Optimization*; Technical Report—TR06; Erciyes University: Kayseri, Turkey, 2005.
18. Karaboga, D.; Basturk, B. A powerful and efficient algorithm for numerical function optimization: Artificial bee colony (ABC) algorithm. *J. Glob. Optim.* **2007**, *39*, 459–471. [CrossRef]

19. Meng, X.; Liu, Y.; Gao, X.; Zhang, H. A New Bio-inspired Algorithm: Chicken Swarm Optimization. In *LNCS*; Springer: Berlin/Heidelberg, Germany, 2014; pp. 86–94. [CrossRef]
20. Raghul, S.; Jeyakumar, G. Parallel and Distributed Computing Approaches for Evolutionary Algorithms—A Review. In *Advances in Intelligent Systems and Computing*; Springer: Singapore, 2021; pp. 433–445. [CrossRef]
21. Buck, I. GPU computing with NVIDIA CUDA. In *ACM SIGGRAPH 2007 Courses on—SIGGRAPH'07*; ACM Press: New York, NY, USA, 2007. [CrossRef]
22. Essaid, M.; Idoumghar, L.; Lepagnot, J.; Brévilliers, M. GPU parallelization strategies for metaheuristics: A survey. *Int. J. Parallel Emergent Distrib. Syst.* **2018**, *34*, 497–522. [CrossRef]
23. Zarrabi, A.; Samsudin, K.; Karuppiah, E.K. Gravitational search algorithm using CUDA: A case study in high-performance metaheuristics. *J. Supercomput.* **2014**, *71*, 1277–1296. [CrossRef]
24. Wang, H.; Rahnamayan, S.; Wu, Z. Parallel differential evolution with self-adapting control parameters and generalized opposition-based learning for solving high-dimensional optimization problems. *J. Parallel Distrib. Comput.* **2013**, *73*, 62–73. [CrossRef]
25. Qin, A.K.; Raimondo, F.; Forbes, F.; Ong, Y.S. An improved CUDA-based implementation of differential evolution on GPU. In Proceedings of the GECCO'12: 14th International Conference on Genetic and Evolutionary Computation, Philadelphia, PE, USA, 7–11 July 2012; ACM Press: New York, NY, USA, 2012. [CrossRef]
26. Krömer, P.; Platoš, J.; Snášel, V. A brief survey of differential evolution on Graphic Processing Units. In Proceedings of the 2013 IEEE Symposium on Differential Evolution (SDE), Singapore, 16–19 April 2013; pp. 157–164. [CrossRef]
27. de P. Veronese, L.; Krohling, R.A. Differential evolution algorithm on the GPU with C-CUDA. In Proceedings of the IEEE Congress on Evolutionary Computation, Barcelona, Spain, 18–23 July 2010; IEEE: Piscataway, NJ, USA, 2010. [CrossRef]
28. Zhu, W. Massively parallel differential evolution—Pattern search optimization with graphics hardware acceleration: An investigation on bound constrained optimization problems. *J. Glob. Optim.* **2010**, *50*, 417–437. [CrossRef]
29. Krömer, P.; Snášel, V.; Platoš, J.; Abraham, A. Many-threaded implementation of differential evolution for the CUDA platform. In Proceedings of the 13th annual conference on Genetic and evolutionary computation—GECCO'11, Dublin, Ireland, 12–16 July 2011; ACM Press: New York, NY, USA, 2011. [CrossRef]
30. Ugolotti, R.; Nashed, Y.S.; Mesejo, P.; Ivekovič, Š.; Mussi, L.; Cagnoni, S. Particle Swarm Optimization and Differential Evolution for model-based object detection. *Appl. Soft Comput.* **2013**, *13*, 3092–3105. [CrossRef]
31. Fabris, F.; Krohling, R.A. A co-evolutionary differential evolution algorithm for solving min–max optimization problems implemented on GPU using C-CUDA. *Expert Syst. Appl.* **2012**, *39*, 10324–10333. [CrossRef]
32. Zhou, X.; Wu, Z.; Wang, H. Elite Opposition-Based Differential Evolution for Solving Large-Scale Optimization Problems and Its Implementation on GPU. 2012 13th International Conference on Parallel and Distributed Computing, Applications and Technologies, Beijing, China, 14–16 December 2012; IEEE: Piscataway, NJ, USA, 2012. [CrossRef]
33. Zibin, P. Performance Analysis and Improvement of Parallel Differential Evolution. *arXiv* **2021**, arXiv:2101.06599.
34. Bonabeau, E. Social Insect Colonies as Complex Adaptive Systems. *Ecosystems* **1998**, *1*, 437–443. [CrossRef]
35. Feltell, D.; Bai, L.; Jensen, H.J. An individual approach to modelling emergent structure in termite swarm systems. *Int. J. Model. Identif. Control* **2008**, *3*, 29. [CrossRef]
36. Korošec, P.; Tashkova, K.; Šilc, J. The differential Ant-Stigmergy Algorithm for large-scale global optimization. In Proceedings of the IEEE Congress on Evolutionary Computation, Barcelona, Spain, 18–23 July 2010; pp. 1–8. [CrossRef]
37. Oberst, S.; Lai, J.C.; Martin, R.; Halkon, B.J.; Saadatfar, M.; Evans, T.A. Revisiting stigmergy in light of multi-functional, biogenic, termite structures as communication channel. *Comput. Struct. Biotechnol. J.* **2020**, *18*, 2522–2534. [CrossRef]
38. Cimino, M.G.; Minici, D.; Monaco, M.; Petrocchi, S.; Vaglini, G. A hyper-heuristic methodology for coordinating swarms of robots in target search. *Comput. Electr. Eng.* **2021**, *95*, 107420. [CrossRef]
39. Amorim, K.S.; Pavani, G.S. Ant Colony Optimization-based distributed multilayer routing and restoration in IP/MPLS over optical networks. *Comput. Netw.* **2021**, *185*, 107747. [CrossRef]
40. Upeksha, R.G.C.; Pemarathne, W.P.J. Ant Colony Optimization Algorithms for Routing in Wireless Sensor Networks: A Review. In *Lecture Notes in Electrical Engineering*; Springer: Singapore, 2022; pp. 47–57. [CrossRef]
41. Dorigo, M.; Maniezzo, V.; Colorni, A. Ant system: Optimization by a colony of cooperating agents. *IEEE Trans. Syst. Man Cybern. Part B Cybern.* **1996**, *26*, 29–41. [CrossRef] [PubMed]
42. Karimi, A.; Nobahari, H.; Siarry, P. Continuous ant colony system and tabu search algorithms hybridized for global minimization of continuous multi-minima functions. *Comput. Optim. Appl.* **2008**, *45*, 639–661. [CrossRef]
43. Brest, J.; Korošec, P.; Šilc, J.; Zamuda, A.; Bošković, B.; Maučec, M.S. Differential evolution and differential ant-stigmergy on dynamic optimisation problems. *Int. J. Syst. Sci.* **2013**, *44*, 663–679. [CrossRef]
44. Bilchev, G.; Parmee, I.C. The ant colony metaphor for searching continuous design spaces. In *Evolutionary Computing*; Springer: Berlin/Heidelberg, Germany, 1995; pp. 25–39. [CrossRef]
45. Dréo, J.; Siarry, P. A New Ant Colony Algorithm Using the Heterarchical Concept Aimed at Optimization of Multiminima Continuous Functions. In *Ant Algorithms*; Springer: Berlin/Heidelberg, Germany, 2002; pp. 216–221. [CrossRef]
46. Korošec, P.; Šilc, J. A Stigmergy-Based Algorithm for Continuous Optimization Tested on Real-Life-Like Environment. In *Lecture Notes in Computer Science*; Springer: Berlin/Heidelberg, Germany, 2009; pp. 675–684. [CrossRef]

47. Meghdadi, A.; Akbarzadeh-T, M. A stigmergic approach to teaching-learning-based optimization for continuous domains. *Swarm Evol. Comput.* **2021**, *62*, 100826. [CrossRef]
48. Jebaraj, L.; Venkatesan, C.; Soubache, I.; Rajan, C.C.A. Application of differential evolution algorithm in static and dynamic economic or emission dispatch problem: A review. *Renew. Sustain. Energy Rev.* **2017**, *77*, 1206–1220. [CrossRef]
49. Mashwani, W.K. Enhanced versions of differential evolution: State-of-the-art survey. *Int. J. Comput. Sci. Math.* **2014**, *5*, 107. [CrossRef]
50. Wang, Y.; Cai, Z. Combining Multiobjective Optimization With Differential Evolution to Solve Constrained Optimization Problems. *IEEE Trans. Evol. Comput.* **2012**, *16*, 117–134. [CrossRef]
51. Heylighen, F. Stigmergy as a universal coordination mechanism I: Definition and components. *Cogn. Syst. Res.* **2016**, *38*, 4–13. [CrossRef]
52. Heylighen, F. Stigmergy as a universal coordination mechanism II: Varieties and evolution. *Cogn. Syst. Res.* **2016**, *38*, 50–59. [CrossRef]
53. Nasim, A.; Burattini, L.; Fateh, M.F.; Zameer, A. Solution of Linear and Non-Linear Boundary Value Problems Using Population-Distributed Parallel Differential Evolution. *J. Artif. Intell. Soft Comput. Res.* **2019**, *9*, 205–218. [CrossRef]
54. Solis-Muñoz, F.J.; Osornio-Rios, R.A.; Romero-Troncoso, R.J.; Jaen-Cuellar, A.Y. Differential Evolution Implementation for Power Quality Disturbances Monitoring using OpenCL. *Adv. Electr. Comput. Eng.* **2019**, *19*, 13–22. [CrossRef]
55. Zuo, L.; Liu, B.; Wen, Z.; Sun, H.; Di, R.; Wu, P. Research of Dynamic Economic Emission Dispatch Based on Parallel Molecular Differential Evolution Algorithm. *IOP Conf. Ser. Earth Environ. Sci.* **2018**, *170*, 032003. [CrossRef]
56. Laguna-Sánchez, G.A.; Olguín-Carbajal, M.; Cruz-Cortés, N.; Barrón-Fernández, R.; Martínez, R.C. A differential evolution algorithm parallel implementation in a GPU. *J. Theor. Appl. Inf. Technol.* **2016**, *86*, 184–195.
57. Davarynejad, M.; van den Berg, J.; Rezaei, J. Evaluating center-seeking and initialization bias: The case of particle swarm and gravitational search algorithms. *Inf. Sci.* **2014**, *278*, 802–821. [CrossRef]
58. Hollander, M.; Wolfe, D.A.; Chicken, E. *Nonparametric Statistical Methods*; John Wiley & Sons: Hoboken, NJ, USA, 2013.
59. Ahmad, M.F.; Isa, N.A.M.; Lim, W.H.; Ang, K.M. Differential evolution: A recent review based on state-of-the-art works. *Alex. Eng. J.* **2022**, *61*, 3831–3872. [CrossRef]
60. Wisittipanich, W.; Kachitvichyanukul, V. Differential Evolution Algorithm for Job Shop Scheduling Problem. *Ind. Eng. Manag. Syst.* **2011**, *10*, 203–208. [CrossRef]
61. Santucci, V.; Baioletti, M.; Milani, A. Algebraic Differential Evolution Algorithm for the Permutation Flowshop Scheduling Problem With Total Flowtime Criterion. *IEEE Trans. Evol. Comput.* **2016**, *20*, 682–694. [CrossRef]
62. Sarkar, S.; Das, S. Multilevel Image Thresholding Based on 2D Histogram and Maximum Tsallis Entropy—A Differential Evolution Approach. *IEEE Trans. Image Process.* **2013**, *22*, 4788–4797. [CrossRef]
63. Osuna-Enciso, V.; Cuevas, E.; Sossa, H. A comparison of nature inspired algorithms for multi-threshold image segmentation. *Expert Syst. Appl.* **2013**, *40*, 1213–1219. [CrossRef]
64. Jain, S.; Sharma, V.K.; Kumar, S. Robot Path Planning Using Differential Evolution. In *Advances in Computing and Intelligent Systems*; Springer: Singapore, 2020; pp. 531–537. [CrossRef]
65. Qin, N.; Chen, J. An area coverage algorithm for wireless sensor networks based on differential evolution. *Int. J. Distrib. Sens. Netw.* **2018**, *14*, 1–11. [CrossRef]
66. Chen, S.; Montgomery, J.; Bolufé-Röhler, A. Measuring the curse of dimensionality and its effects on particle swarm optimization and differential evolution. *Appl. Intell.* **2014**, *42*, 514–526. [CrossRef]

Article

Optimizing the Layout of Run-of-River Powerplants Using Cubic Hermite Splines and Genetic Algorithms

Alejandro Tapia Córdoba [1], Pablo Millán Gata [1] and Daniel Gutiérrez Reina [2,*]

[1] Departamento de Ingeniería, Universidad Loyola, 41704 Seville, Spain
[2] Departamento de Ingeniería Electrónica, Universidad de Sevilla, 41092 Sevilla, Spain
* Correspondence: dgutierrezreina@us.es

Abstract: Despite the clear advantages of mini hydropower technology to provide energy access in remote areas of developing countries, the lack of resources and technical training in these contexts usually lead to suboptimal installations that do not exploit the full potential of the environment. To address this drawback, the present work proposes a novel method to optimize the design of mini-hydropower plants with a robust and efficient formulation. The approach does not involve typical 2D simplifications of the terrain penstock layout. On the contrary, the problem is formulated considering arbitrary three-dimensional terrain profiles and realistic penstock layouts taking into account the bending effect. To this end, the plant layout is modeled on a continuous basis through the cubic Hermite interpolation of a set of key points, and the optimization problem is addressed using a genetic algorithm with tailored generation, mutation and crossover operators, especially designed to improve both the exploration and intensification. The approach is successfully applied to a real-case scenario with real topographic data, demonstrating its capability of providing optimal solutions while dealing with arbitrary terrain topography. Finally, a comparison with a previous discrete approach demonstrated that this algorithm can lead to a noticeable cost reduction for the problem studied.

Keywords: micro-hyropower plant; layout optimization; Hermite splines; genetic algorithm

1. Introduction

1.1. Electrification in Developing Countries and Mini-Hydropower Plants

The use of Mini Hydro-Power Plants (MHPP) constitutes one of the most efficient solutions for the problem of energy access in remote rural areas, especially in the context of developing countries [1]. These small installations are capable of exploiting the potential energy of a natural water flow to generate electricity with minimal environmental impact and simple equipment. Nevertheless, the context of poverty of these emplacements conditions the design strategies, which are usually based on the personal experience of local technicians and the use of hand rules. For this reason, the development of efficient and robust design strategies can play an important role in the application of these technologies to combat energy poverty.

An MHPP basically consists of extracting a fraction of the water flow from a natural course and conducting it downhill through a long pipe, called penstock, at the end of which the water interacts with a generation unit and transforms its kinetic energy into electrical energy (see Figure 1). It is relevant to note that the water is returned to its natural course, and thus the environmental impact is almost zero. The potential and cost efficiency of an MHPP is determined by the correct use of the terrain or, in other words, the optimal emplacement of its elements and of the penstock layout to achieve the maximum height difference with the shortest pipe length (as pipe friction lowers the efficiency). When the price of the equipment is low, the costs of the deployment can become relevant with respect to the overall cost, and thus the civil works involved become a variable of interest. This, together with the penstock bending and the arbitrary profiles of the terrain and the river, increases the complexity of the problem and motivates the use of numerical and heuristic optimization approaches.

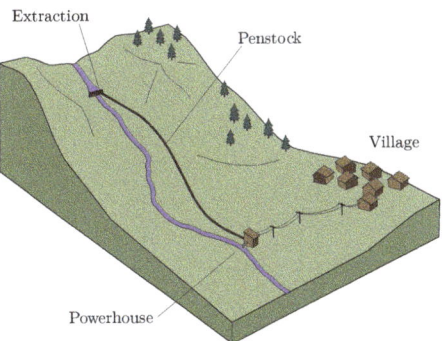

Figure 1. Basic scheme of an MHPP.

1.2. Optimization of MHPP

The problem of optimizing MHPP is addressed in the literature from a wide variety of perspectives. For example, some works propose solutions to reduce the environmental impact of the plants [2,3], to quantify the potential of these installations [4], to improve the MHPP viability [5], or to operate the plants under water scarcity conditions [6,7].

Although the seasonal character of water availability induces variations in the energy production of MHPP, these plants remain one of the best solutions for isolated, rural locations with no access to the power grid, and therefore a large effort has been made in the literature to improve their design and viability. In line with this, some studies have focused on the overview and characterization of the main elements and variables of these installations [8], the design of the turbines (see [9] for a nice survey about this topic), and the analysis of the most important economic indexes [10].

A very relevant line of work is that related to the optimal components selection and/or operation of MHPP in situations in which the plant layout, that is, the location of its main elements (namely water intake, penstock profile, and powerhouse), has been fixed. For example, Ref. [11] presents a numerical tool to maximize power production or economic profit by selecting the adequate penstock diameter and type of turbines. The work takes into account the dependence of the turbine efficiency on variables such as water flow, suction head, and rotational speed. Moreover, ref. [12] presents the most important features of the turbine model from a control point of view, and, in [13], the use of model predictive control to deal with prediction errors in the available water in the reservoir and is proposed.

In general, the number of variables and constraints and its nonlinear character involve a degree of complexity that makes the use of traditional gradient-based optimization strategies difficult. For this reason, the utilization of computational intelligence methods has gained great attention to deal with the previous optimization problems. Some examples of this include [14], where the authors present a stochastic evolutionary algorithm to find the blade of a Turgo water turbine that maximizes the hydraulic efficiency. Similarly, in [15], the bucket shape of Pelton turbines in various operation conditions is optimized using a Lagrangian formulation and solving the problem with evolutionary algorithms. In addition, ref. [16] proposes a complex simulation algorithm to analyze plant operation yearly and compute energy production and economic indices. The authors' findings show that the optimal sizing in terms of some economic indicators is not the same as the one maximizing the exploitation of the hydraulic potential. Moreover, ref. [17] investigates the potential of using a proposed metaheuristic method to provide optimal operations for multireservoir systems, with the aim of optimally improving hydro-power generation.

This work addresses the problem of optimizing the layout of a MHPP. The problem can be roughly stated as follows: given the profile of a river and its surrounding terrain, find the location of the water intake, the turbine, and the penstock layout in such a way that the mechanical stress in the pipe is below a certain level, σ_{max}; the obtained power is above a desired value, P_{min}; and the cost of the installation is minimized. Typically, this problem has been tackled, making some simplifications to use traditional analysis or optimization tools. For example, in [18], the authors propose a theoretical analysis to find the optimal penstock layout and diameter for low head plants. The optimization problem considers a 2D formulation of the MHPP layout, which, in addition, is considered composed of a unique straight segment. Despite the elegance of the obtained analytical results, these simplifications limit the practical implementation, for example, in the (frequent) situation in which the river profile is irregular. Removing this kind of simplification improves the applicability of the developed methodologies at the cost of a dramatic increase in the problem complexity. This results in optimization problems that cannot be solved resorting to linear programming or convex optimization tools, and this is when machine learning techniques and metauristic optimization come into play. For example, ref. [19] presents a method to select an adequate turbine and to compute the optimal and penstock diameter based on Honey Bee Mating algorithm, ref. [20] introduces the application of a genetic algorithm to optimize the flow rate and number of generators in a multi-objective problem where generated energy and investment cost are the objective functions, ref. [16] develops a stochastic evolutionary algorithm to select the optimal turbine capacity and lengths/diameters of the penstock segments, and [11] presents an evolutionary algorithms to maximize power production or net economic profit by optimizing, the penstock diameter, and the type and configuration of the turbines. In all previous works, the intake and turbine locations are assumed to be given. On the contrary, some works include the penstock layout and the intake and turbine locations in the optimization problems. Some examples of these are presented in [21,22], where arbitrary 2D river profiles are considered and integer programming and evolutionary methods are employed, respectively. The extension of these approaches to (discretized) 3D profiles is presented in [23]. Another interesting work presented in [24] makes use of three optimization modules and genetic algorithms to simultaneously determine the optimal intake location, penstock length and diameter, and turbine number, capacity, and discharge schedule.

As a significant improvement with respect to previous approaches, this work proposes the optimization of the layout considering a continuous, 3D formulation of the problem, capable of dealing with an arbitrary terrain and river profile, using a Genetic Algorithm (GA).

1.3. Contributions

This paper presents a new and improved method for computing optimal MHPP layouts considering cost, power, and flow constraints. As in [23], the 3D terrain is discretized based on the available topographic data, and civil work costs to deploy and install the penstock are taken into account. However, in the new developed method, the penstock is not assumed to be composed of straight lengths, which is an unrealistic constraint in real installations where the pipe length bends considerably in the horizontal plane to accommodate the penstock layout (see Figure 2).

The new method is formulated based on three main components. First of all, the main equations characterizing the MHPP power, flows, costs, and constraints are derived. Secondly, the penstock profile is approximated and used suitable approximation cubic Hermite splines for the vertical dimension. This choice is made to preserve, on the one hand, the monotonicity of the curve that describes the penstock (required to avoid air entrapment) and, on the other hand, to provide an easy formulation for the bending radius of the penstock length, as it can be determined in terms of the derivatives of the cubic polynomials. This consideration makes it possible to account for an extra degree of freedom that can be limited according to the material and diameter of the penstock. Finally, a specifically designed genetic algorithm is applied to characterize the optimal MHPP layout, studying both single-objective and multiple-objective problems.

Figure 2. Bending in a real MHPP penstock length.

The constraint relaxation based on spline interpolation not only involves a better approximation of the penstock layout, but also makes it possible, as it will be shown in the paper, to obtain designs that are better in terms of cost and attainable power.

2. Description of the Problem
2.1. Power Generation

The obtainable power P of a hydropower plant can be estimated in terms of the net head, H_t, and the water flow rate, Q, as

$$P = \rho g Q h_t \eta, \tag{1}$$

where ρ is the density of the water, g is the acceleration of gravity, and η is the overall efficiency of the generation equipment. The net head, H_t, represents the net height of the water at the entrance of the turbine, which can be written as the gross height, H_g minus the friction loss h_{fric}:

$$h_t = H_g - h_{fric}, \tag{2}$$

Considering an action turbine, in which the energy to be transformed in the turbine is entirely kinetic (through the formation of a jet), it can be written that

$$h_t = \frac{1}{2g} v_{jet}^2 \tag{3}$$

The velocity of the jet, v_{jet}, can be written in terms of the water flow rate Q and the nozzle injector section area, S_{noz}, and after substituting the expression for h_t can be rewritten as

$$h_t = \frac{Q^2}{2g C_D S_{noz}^2}, \tag{4}$$

where a discharge coefficient C_D has been introduced to model the formation of a *vena contracta* in the jet. Finally, a simple approximation for the height loss term from [25] as:

$$h_{fric} \approx k_p \frac{L_p}{D_p^5} Q^2 \tag{5}$$

with k_p being a constant of the pipe material and D_p its internal diameter. Now, substituting this last expression, together with Equation (4) in Equation (2), an expression for the water flow rate that will feed the plant can be obtained:

$$Q = \left[\frac{H_g}{\frac{1}{2gC_D^2 S_{noz}} + k_p \frac{L_p}{D_p^5}} \right]^{\frac{1}{2}} \quad (6)$$

Finally, introducing Equation (6) in Equation (4), and the resultant expression in Equation (1), it yields:

$$P = \frac{\eta \rho}{2C_D^2 S_{noz}^2} \left[\frac{H_g}{\frac{1}{2gC_D^2 S_{noz}^2} + k_p \frac{L_p}{D_p^5}} \right]^{\frac{3}{2}} \quad (7)$$

It is relevant to note that this expression allows for estimating the power generated by any hydropower plant on the basis of the gross head H_g, and penstock length and diameter, L_p and D_p. These three variables, indeed, are determined by the spatial layout of the plant over the terrain.

2.2. Terrain

To model the layout of an MHPP, the height map of the terrain is required to be properly characterized. To this end, a set of experimental topographic data points, $T_i(x_i, y_i, z_i)$, are considered to be obtained through a topographic survey. On the basis of these points, a continuous height function $z = f(x, y)$ is built through a linear interpolation. The river layout is determined on the $x - y$ plane using the aerial imagery from the topographic survey. This provides a second set of data points $R_i(x_i, y_i, f(x_i, y_i))$ for the river, which is transformed into a continuous function using a cubic spline interpolation.

2.3. Spline-Based Penstock Layout

As shown in Figure 3, possible MHPP layouts are modeled as a parametrized continuous curves, Γ, connecting two points of the river: the water extraction and the turbine emplacement points. For an enhanced formulation, these curves Γ (candidate solutions) are spline-based interpolations of a subset of spatial points belonging to $T_i(x_i, y_i, z_i)$, named nodes (see Figure 4). This way, any given solution can be built through the interpolation of n nodes in the form of (x_i, y_i, z_i), as long as the first and last belong to the river. Thus, the solution Γ is written in terms of the cited coefficients as

$$\Gamma(t) = [S_x(t), S_y(t), S_z(t)], \quad (8)$$

where $S_x(t)$, $S_y(t)$ and $S_z(t)$ are the interpolation functions for x, y, and z coordinates of the nodes, respectively. The parameter t is trivially defined to match the nodes index, and thus $\Gamma(i) = (x_i, y_i, z_i)$.

It is relevant to note that, for any layout to be feasible, it is required for the slope of the penstock not to change its sign [26]. For this reason, a Hermite cubic interpolation has been employed for $S_z(t)$, while natural cubic splines have been employed for $S_x(t)$ and $S_y(t)$ functions. This strategy guarantees that, as long as the nodes are ordered in height, the sign of the slope of the penstock will not change the layout. An illustrative example of this is shown in Figure 5 for the sake of understanding.

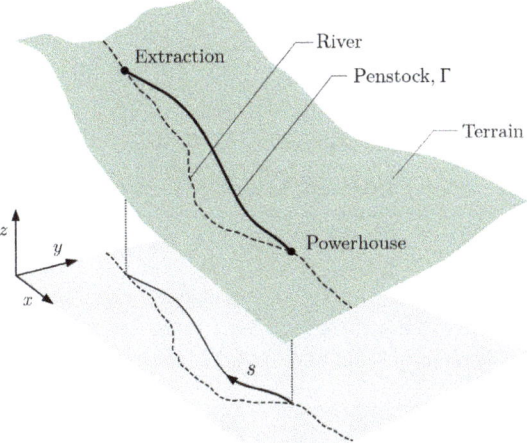

Figure 3. MHPP simplified model through 3D spatial curves.

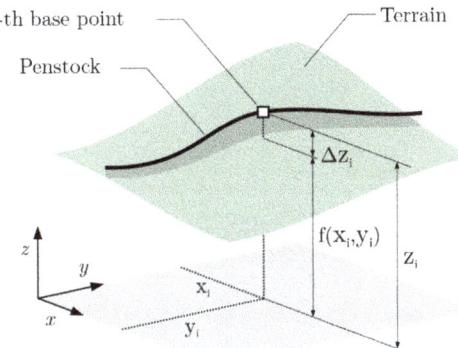

Figure 4. Example of a node, acting as a base point for the spline interpolation. The height of the candidate solutions at the nodes are denoted as z_i, in its difference with the height of the terrain at this point is $\Delta z_i = z_i - z(x_i, y_i) = z_i - f(x_i, y_i)$.

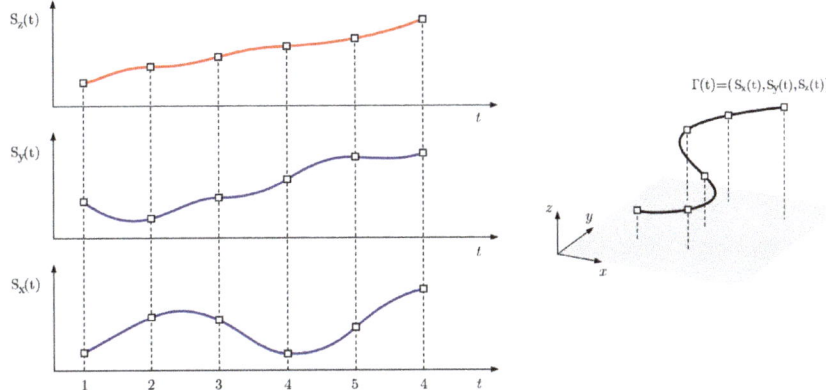

Figure 5. Example of interpolation functions for a solution with five nodes, whose coordinates are represented using squares. Note that $S_x(t)$ and $S_y(t)$ (in blue) are natural splines, while $S_z(t)$ (in red) is a monotone Hermite spline.

Finally, to ease further calculations related to the length of the penstock, an arc-length reparametrization $s(t)$ such that

$$s(t) = \int_0^t ||\Gamma(t)'|| dt \xrightarrow{t(s)=s(t)^{-1}} \Gamma = \Gamma(t(s)) \qquad (9)$$

Note that, with this reparametrization, the curve $\Gamma(s)$ is defined for $s \in [0, L_p]$. Once the curve $\Gamma(s)$ has been built, the performance of the resulting MHPP (power and water flow rate) can be determined using Equations (6) and (7). In addition, this approach permits a simple estimation of the cost, as will be discussed in the next section.

2.4. Pipe Curvature

As it was mentioned in Section 1.3, in this work, the penstock is not considered to be composed of straight segments, as in previous approaches, but able to slightly bend along its layout, as a result of the real deformation that can be observed in long pipes (as shown in Figure 2). Nevertheless, these curvatures are required to be compatible with the material properties (in particular, its stiffness and resistance), and thus a limitation is required to be imposed for the curvature radius r.

Considering the pipe as a simple Euler beam under pure bending [27], a direct relation can be written between the radius or curvature r and the maximum mechanical stress σ, involving the Young modulus of the pipe material (typically mild steel [28]), E, and the diameter, D_p:

$$\sigma_{max} = E \frac{D_p}{2r} \qquad (10)$$

The Von-Mises yielding criterion [27] can be used to impose the safety of the pipe as follows:

$$\frac{\sigma}{S_y} \leq 1, \qquad (11)$$

where S_y is the tensile strength of the material. Substituting and reordering, it can be written as:

$$r(s) \geq \frac{E}{2S_y} D_p \qquad (12)$$

It is relevant to note that the minimum curvature radius allowable depends on the diameter of the pipe.

2.5. Cost of the MHPP

The minimization of the cost of the plant is considered as the objective of the optimization problem addressed in this work, and thus an appropriate model is required to be formulated. First, as the generation equipment is sized for a range around the objective power generation level, the penstock layout is the main variable in the optimization problem, which includes not only the cost of the pipe itself, C_p, but also the costs of the labor involved in its deployment, C_{cw}, associated with the required excavation and supports.

For the cost of the pipe, the following polynomial expression is proposed:

$$C_p = L_p \sum_{i=0}^{m} a_i D_p^i, \qquad (13)$$

where a_i and b_i are experimentally adjusted coefficients to be fitted to the costs of the local manufacturers. The costs of the civil works can now be evaluated as the sum of those related to the supports, C_{sup}, and those related to the excavations, C_{exc}:

$$C_{cw} = C_{sup} + C_{exc} \qquad (14)$$

To calculate these two costs, the height difference between the terrain and the penstock is required to be evaluated. This variable will be denoted with $\epsilon(s)$, which is written in terms of the arc-length variable s, shown in Figure 3:

$$\epsilon(s) = S_z(s) - f(S_x(s), S_y(s)) \qquad (15)$$

This gap (shaded in Figure 4) can reach positive and negative values along the path of the penstock, having different implications: positive gaps (the penstock lies over the terrain surface) results in the need for supports, while negative ones (the penstock lies under the terrain surface) result in the need for excavations. Now, both C_{sup} and C_{exc} can be calculated. The first is calculated through the integration along the path of the cost of a single support (that depends on its height) times the linear density of supports μ_{sup}. The cost of a single support has been defined by a constant k_{sup} times its squared height. This expression has been proposed to fit the average costs following the local technicians indications. On the other hand, the excavation cost is calculated as a unitary volumetric cost, k_{exc}, multiplied by the total volume to be excavated, which is determined by integrating along the path a certain digging cross-section, defined by a cut angle β_{exc}. These approximations result in the following expressions:

$$C_{sup}(s) = \begin{cases} \mu_{sup} \int_{\Gamma(s)} k_{sup} \epsilon_{sup}^2 ds & where\ \epsilon \geq 0 \\ 0 & otherwise \end{cases} \qquad (16)$$

$$C_{exc}(s) = \begin{cases} k_{exc} \int_{\Gamma(s)} (tan(\beta_{exc}) \epsilon_{exc}^2 - D_p \epsilon_{exc}) ds & where\ \epsilon < 0 \\ 0 & otherwise \end{cases} \qquad (17)$$

The scheme of these two models is shown in Figure 6 for a clearer understanding. Note that the extra distance ϵ_0 to be nailed to the ground has been considered proportional to ϵ, and thus can be grouped in constant k_{exc}.

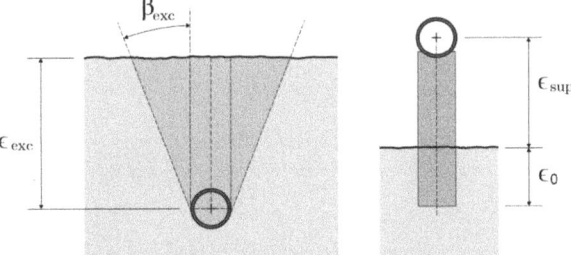

Figure 6. Scheme of the model proposed for the excavations (**left**) and supports (**right**).

2.6. Problem Formulation

Finally, the problem is formulated as the minimization of the cost, C, such that the solution represents a feasible MHPP layout, with a generated power P above a certain required level, P_{min}. This is:

$$\begin{aligned} & min & & (13) \\ & s.t. & & (7) > P_{min} \\ & & & (10) < S_y \end{aligned}$$

2.7. Complexity of the Problem

The target optimization problem is a non-convex one, where classical gradient-based solvers would fail to find optimal solutions. Genetic algorithms like the one used in this work are suitable for non-convex scenarios with restrictions. The complexity of the problem is high, since the number of design variables (nodes in the penstock layout) may be variable, and the variables are continuous. Consequently, the search space is both infinite and nonconvex, making it impossible to use brute force or exhaustive algorithms due to the computational time required. Therefore, a metaheuristic algorithm such as a Genetic Algorithm is a suitable approach to obtain optimal solutions in a reasonable computational time. Furthermore, Genetic Algorithms are appropriate for dealing with the restrictions of the target problem and for solutions of variable lengths like the ones used in this work.

3. Genetic Algorithm

Genetic algorithms (GA) are metaheuristic optimization algorithms that are widely employed to solve complex engineering problems [29,30]. GAs are population-based approaches. This means that they seek the optimum values (maximum and/or minimum) of a given problem from a population of random solutions. These random solutions evolve over generations improving at each step by genetic operators: selection, crossover, and mutation. Genetic operators are inspired by the Darwinian theory [31] in which those individuals who better adapt to their ecosystem are the ones that will have more probability to survive over time. In an optimization problem, the individuals are given by potential solutions, and the adaptation to the ecosystem is obtained by the quality of the potential solution in the fitness function. As a rule, the higher the quality of the solution, the better the adaptation. Thus, the population of the potential solutions evolves towards the optimum value until the stop criterion is reached, which is normally fixed as a number of generations.

3.1. Individual Encoding

The individual chromosome represents a potential solution of the target problem. The chromosome is composed of genes, each one representing a design variable of the problem. In this case, the variables consist of (see Figure 7):

- The coordinates s_1 and s_n corresponding to the end nodes;
- The coordinates (x_i, y_i) of the interior nodes of the penstock, being i the number of each node;
- The height of the nodes relative to the surface of the terrain ΔZ_i. Notice that it is a relative value with respect to $f(x_i, y_i)$, which determines the actual height of the terrain;
- The diameter of the penstock D_p.

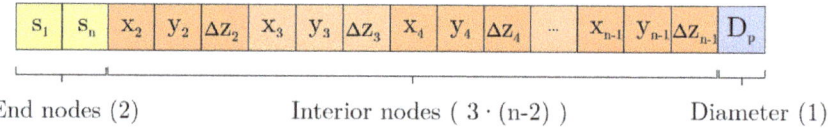

Figure 7. Chromosome encoding.

3.2. Individual Generation

Given the complexity of the problem and the constraints, the generation of individuals on the basis of a purely random generation of nodes through the space might cause a high rate of either unfeasible or low-quality individuals. For this reason, a customized generation routine has been developed to guarantee the feasibility of the individuals and improve their initial fitness. The generation scheme developed is based on the generation of selection an arbitrary number of points (x, y) chosen from the river x, y layout. The highest and lower of these nodes are considered the location of the dam and the powerhouse, respectively, while the rest are considered the interior nodes. The positions of the interior nodes are displaced from the original emplacement through a Gaussian probability function in both x and y directions. The Gaussian distribution is the same for both axes, centered at zero and with a scale hyperparameter σ to be tuned. Finally, the inner nodes are assigned a z_i coordinate as the terrain height at that point plus a distance Δz randomly generated from this same Gaussian function. In Figure 8, an example of a generated individual with five nodes is shown.

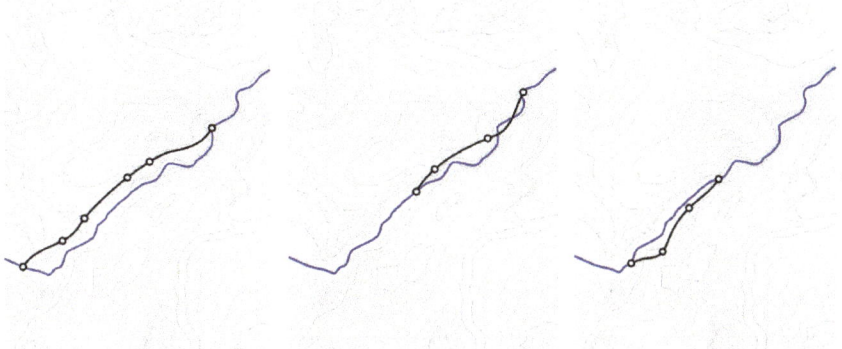

Figure 8. Example of three arbitrary individuals (black line) created for the given terrain (river is represented in blue). Note how each individual can be created with a different number of nodes.

3.3. Mutation Operator

The mutation operator used in this work is inspired by that from [23], as it demonstrated its efficiency in a related (discrete) problem. This operator is based on performing three different actions that are susceptible to being applied to individuals. These are:

- With a certain probability, $p_{mut,0}$, one of the internal nodes can be removed from the individual (see node 4 in Figure 9);
- With a certain probability, $p_{mut,1}$, a new node is attached to the individual. This new node is generated through a procedure similar to that in the generation scheme, beginning with the selection of an arbitrary point of the river between the dam and the powerhouse and its later Gaussian displacement (see node 6 in Figure 9).
- With a certain probability, $p_{mut,d}$, a gene changes its emplacement by means of a Gaussian displacement. This displacement is performed in the three dimensions of space if the node is interior (as nodes 2 and 3 in Figure 9), or constrained to the river profile if it is not (as nodes 1 and 5 in Figure 9).

In addition, the value of the diameter (stored in the last gene) is modified following the same Gaussian displacement cited before.

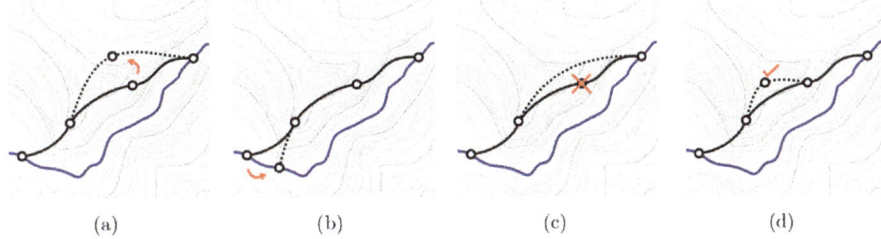

(a)   (b)   (c)   (d)

Figure 9. Example of the different actions of the mutation operator on an arbitrary individual: (**a**) displacement of an interior node, (**b**) displacement of an end node, (**c**) elimination of an existing node, and (**d**) creation of a new node. The black continuous and dashed line represents, respectively, the individual before and after the mutation. The blue line represents the river, and the black circles represent the nodes.

3.4. Crossover Operator

For the crossover operator, a tailored operator, based on an exchange of nodes between the two parents, is proposed as follows:

1. The offspring is initialized by defining their two end nodes as the lowest and highest nodes contained by the parents;

2. The interior nodes of both parents are grouped together, and then each of these are being randomly assigned to one offspring;
3. The diameters of the offspring are determined using a simulated binary crossover between the diameters of the parents. This is being D_1 and D_2, respectively, the diameter of the two parents, the diameter of the offspring, D_1^* and D_2^* are calculated as:

$$D_1^* = (1-\phi)D_1 + \phi D_2 \qquad (18)$$
$$D_2^* = \phi D_1 + (1-\phi)D_2 \qquad (19)$$

where ϕ is generated through a random variable $\alpha \sim \mathcal{U}(0,1)$ as:

$$\phi = (1+2\alpha)x - \alpha \qquad (20)$$

from where it can be seen that ϕ is generated between $-\alpha$ and $1+\alpha$. With this approach, the blend crossover of both diameters is not limited to result in values within the range of the diameters of the parents.

In Figure 10, an example of the node exchange between two individuals during a crossover operation is represented for a better understanding.

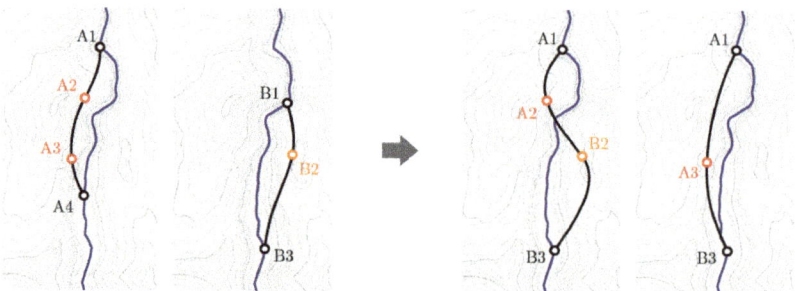

Figure 10. Example of the exchange of nodes between two individuals during a crossover operation. Note how the highest and lowest nodes (A1 and B3) are directly inherited by the offspring, while the internal nodes (A2, A3, and B2) are randomly distributed among them. End nodes A4 and B1 are thus not used.

3.5. Fitness Function

Death penalty is employed to deal with unfeasible individuals. Thus, as the objective is the minimization of the cost C of the plant, the fitness function, F, is then defined as:

$$\begin{cases} if \quad \textbf{solution valid} & F = (13), \\ else & F = \infty. \end{cases} \qquad (21)$$

Note that the only two reasons for which an individual may not be feasible are (i) an excessive curvature of the pipe at any point of its layout or (ii) an insufficient power generation level.

4. Simulation Examples and Results

This section summarizes the results of the application of the method proposed to a real case.

4.1. Scenario Parameters

To evaluate the performance of the proposed approach, a real application case is proposed. The case study is based on the rural community of San Miguelito, in the town of Quimistán (Honduras), that lacks access to the electrical grid because of the geographical limitations. This community was chosen by the Honduran Foundation for Agronomic

Research (FHIA) as a candidate for the installation of a micro-hydropower plant. In a first evaluation by the technicians, a minimum generation of 8 kW was established as the power required to supply the approximately 40 families that live in the community. An aerial drone topographic survey was performed to provide the dataset, composed of a grid of 2900 terrain data points and a set of aerial images, based on which an additional set of 59 points of the river profile were determined. Regarding the problem constants, those from [23] have been considered in this work, for the sake of comparison. These parameters are summarized in Table 1.

Table 1. Summary of the scenario parameters considered for the example.

Parameter	Value
P_{min} (KW)	7
D_{noz} (m)	0.022
E (GPa)	200
S_y (MPa)	250
K_{sup} (c.u.)	9
K_{exc} (c.u./m^3)	8
X_{sup} (1/m)	0.2
β_{exc} (°)	10
a_0 (c.u./m)	13.14
a_1 (c.u./m^2)	99.76
a_2 (c.u./m^3)	616.10

For a further analysis of the performance of the method, two additional study cases have been proposed by modifying the example presented in this section:

1. Modified problem 1 is based on reducing the power output constraint. This models the design of a plant for a much smaller village, with a consequently lower power supply requirement. In particular, the required power output, P_{min}, has been set to 4-kW.
2. Modified problem 2 is based on (i) increasing the required power output of the plant and (ii) modifying the costs associated with the pipe and its deployment. The first of these modifications represents the application of the method to supply a more populated village. In particular, the required power output, P_{min}, has been set to 14-kW. The second of these modifications is based on considering a low quality terrain that strongly makes the transport of the pipe difficult, translating into an increase of its cost per unit length, but eases the excavations labor involved. This has been modeled by using a 1.5 multiplier for C_p, on one hand, and reducing the required cut angle of the excavations, β_{exc}, to 10° and the unitary volumetric cost, K_{exc}, to 2 c.u./m^3, on the other.

4.2. Algorithm Parameters

A $\mu + \lambda$ Genetic Algorithm has been employed to address the optimization problem developed (The complete code can be found in https://github.com/atapiaco/run_of_river_plant_3D_optimization), using the coordinates of the nodes of the layout Γ as genes. To find an adequate value of the hyperparameter on the generation scale σ, a search has been performed. A total of 10 thousand individuals have been generated and evaluated for values of σ ranging from 0 to 10, with the results shown in Figure 11. The best fitness was obtained with $\sigma = 4.2$, and thus this value has been employed for the simulation.

Given the similarity of the mutation scheme to that proposed in the discrete version of the problem [23], the optimal probabilities that were determined have been considered in this work. These are

$$p_{mut,mov} = 0.01, \quad p_{mut,01} = 0.05, \quad p_{mut,10} = 0.20, \quad \beta_{mut} = 0.5$$

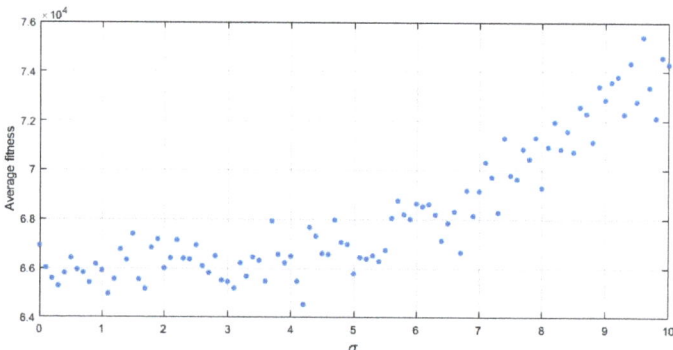

Figure 11. Influence of the scale parameter σ on the average fitness of the generated populations. Each dot represents the average fitness of a population of 10,000 individuals.

Finally, the algorithm has been executed for different values of the mutation and crossover probabilities, with 10 trials for combination. A summary of the main parameters of the GA is shown in Table 2.

Table 2. Parameters of the Genetic Algorithm.

Parameter	Value
λ	2000
μ	2000
Generations	100
Selection	Tournament size = 3
Generation	Custom generation scheme
	$\sigma = 4.20$
Crossover	Custom crossover scheme
	$\phi = 0.50$
Mutation	Custom mutation scheme
	$p_{mut,mov} = 0.01$
	$p_{mut,01} = 0.05$
	$p_{mut,10} = 0.20$
	$p_{cx} = [0.2, 0.3, 0.4, 0.5, 0.6, 0.7]$
	$\mu_{cx} = [0.2 0.4 0.5 0.6 0.8]$
Number of trials	10
Diameter range (m)	0.01–0.33

4.3. Results

The results obtained for the different combinations of mutation and crossover probabilities obtained for the proposed reference case are summarized in Table 3. It can be observed that the best individual has been obtained using 0.7 and 0.3 for the crossover (p_{cx}) and mutation (p_{mut}) probabilities, respectively. The execution of the algorithm exhibited good convergence after approx. 500 generations for almost all trials, with slight improvements during the next hundred generations. In Figure 12, the best fitness of each generation has been represented by one of the trials of the cited probability values.

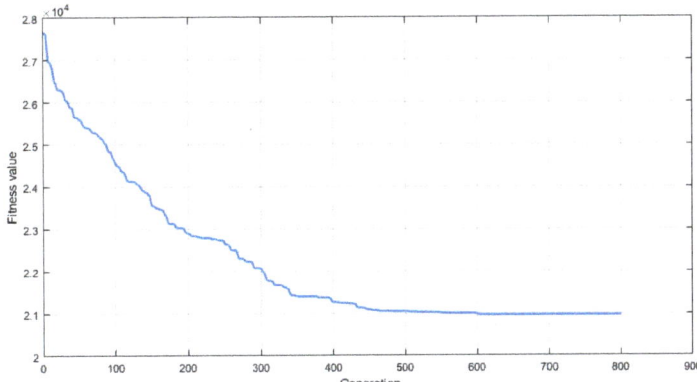

Figure 12. Evolution of the best fitness obtained through generations for the best case evaluated ($p_{mut} = 0.30$, $p_{cx} = 0.70$).

Table 3. Summary of the last population and best individuals obtained. Lower cost has been highlighted in bold.

	Hyper-parameters				
p_{mut}	0.70	0.60	0.50	0.40	0.30
p_{cx}	0.30	0.40	0.50	0.60	0.70
	Final population fitness				
Mean	22,066.77	23,031.35	23,397.92	22,551.43	22,133.37
Std. dev.	991.25	876.34	345.59	583.70	828.1163
Min	21,193.04	21,836.64	23,216.90	21,787.5	20,966.11
Max	41,083.82	48,206.39	41,396.83	33,011.61	25,811.00
	Best individual				
Gross h. (m)	80.74	84.04	86.98	82.40	79.98
Power (W)	7017.69	7000.24	7009.17	7000.49	7003.06
Pens. length (m)	534.56	562.24	597.30	552.59	532.42
Min, bending radius (m)	80.00	67.36	60.00	58.07	58.99
Bending radius allowed (m)	54.23	53.08	52.50	53.67	54.55
Pipe diam. (m)	0.14	0.13	0.13	0.13	0.14
Cost (c.u.)	21,193.04	1836.64	23,216.90	21,787.50	**20,966.11**

This individual represents a plant layout with a total cost of 20,966.11 (c.u.), significantly better than the one obtained by other configurations. The solution provides enough power for the electrical demand required (power generation is slightly superior to the 7 kW required) and satisfies the curvature constraint. It can also be seen how the minimal bending radius depends on the diameter of the pipe, according to Equation (12). Observing the layout that corresponds to this solution (see Figure 13), it can be seen how the penstock, with a 14 cm diameter, cuts through rough terrain and avoids local curvatures of the river.

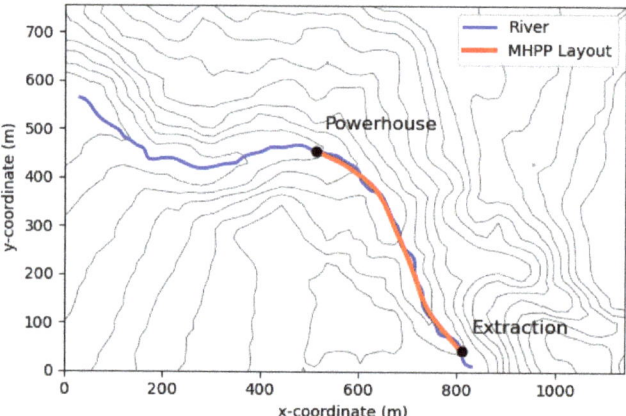

Figure 13. Layout of the optimal solution obtained for the reference problem.

Once the optimal hyper-parameters have been determined, these have been used to solve the modified versions of the reference case. The best solutions obtained are represented in Figure 14, and the main variables associated with these are summarized in Table 4, together with the reference ones.

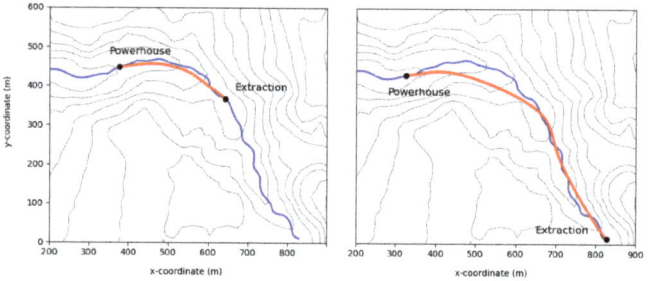

Figure 14. Layout of the optimal solutions obtained for the modified problems 1 (**left**) and 2 (**right**).

Table 4. Summary of the optimal solutions obtained for the reference case and the two modifications.

	Ref. Problem	Modification 1	Modification 2
Gross height (m)	79.98	48.98	109.49
Power (W)	7003.46	4000.19	14,000.67
Pens. length (m)	532.42	298.75	735.37
Min. bending radius (m)	58.99	127.72	100.00
Bending radius allowed (m)	54.55	54.55	68.79
Pipe diam. (m)	0.14	0.14	0.17
Cost (c.u.)	20,966.11	11,769.02	42,191.30

A few interesting comments can be made after observing these results. With respect to the first modification of the problem, it can be seen that, as the required penstock length is smaller, the layout can be drawn with a much smaller curvature (the minimum bending radius is more than twice the minimum allowed). It can also be noted that, in comparison with the reference case, the cost reduction is significantly higher than the power reduction. This is understandable, given that the shorter the length of the penstock, the easier to find a place to take advantage of the benefits of a region of the terrain (high river slope, low terrain bumps, etc.). With respect to the second modified problem, the most evident fact is the need of a larger penstock, which also has increased its diameter from 14 to

17 cm. As was expected, increasing the cost of the penstock and reducing the cost of the excavations causes the algorithm to take advantage of cutting through rough terrain at the curved part of the river, as the required excavations are now preferable with respect to a longer penstock laying over the ground. These two modified cases demonstrate the goodness of the approach and its capability to provide different solutions in accordance with the particularities of both the environment and the economic factors.

4.4. Comparison with Previous Approaches

As indicated in Section 1.2, the proposed method constitutes a substantial novelty relative to previous approaches in the literature, fundamentally due to (i) the 3D nature of the approach, which considers the real topography of the terrain avoiding simplification errors that are present in 2D-based traditional methods (such as those in [22,32], and (ii) its continuous formulation, which improves the search space and avoids the strong conditioning between the resolution of the terrain mesh and the efficiency of the method that arises from discrete methods (such as [23]. Regarding approaches based on 2D simplification, it is clear that their practical implementation is limited to those cases in which the curvature of the river is negligible. In those cases where this is not met, not only might the performance of the plant differ, but the difference in length of the optimal layout projected (2D) and the real one (3D) can cause incompatibility problems during the installation. These issues have been numerically demonstrated by the authors in [32], where the authors propose a 3D approach, in which the plant layout is defined as a connection of straight pipe lengths, connected to each other using elbows. For the sake of a fair comparison, the approach proposed in this work cited has been used to solve the reference example in Section 4.1. The method considered is defined on a discrete basis, the raw topographic data points considered as candidate positions for these elbows to be deployed on the terrain. It is relevant to note that, in addition to the different framework of the discrete approach, the cost function is slightly different, as it includes an additional term to account for the cost of the elbows and their installation. For this approach, Equation (13) is transformed into:

$$C_p = L_p \sum_{i=0}^{m} a_i D_p^i + n_c \sum_{i=0}^{n} b_i D_p^i \tag{22}$$

where n_c represents the total number of penstock pipe elbows, and the constants b_i are adjusted to match the real costs from local manufacturers. For this problem, the values from [32] have been used; these are:

$$b_0 = 50, \quad b_3 = 1200, \quad b_i = 0 \; \forall i \neq \{0,3\}$$

The optimal solution obtained with this method is represented in Figure 15, and numerically summarized in Table 5. When these results are observed, it can be seen that the continuous approach proposed in this work provides a better solution. Nevertheless, some interesting additional comments can be made. First, it is relevant to note that the cost reduction is not very high (about 3% reduction), as both layouts are qualitatively similar. It is also interesting to note that the penstock obtained with the discrete approach is 11 m shorter with the same diameter, but it is the cost of the elbows that increases the cost. Finally, the fact that the continuous approach provided a power generation closer to the minimum constraint (7 kW) demonstrates how the discrete is strongly conditioned by the refinement of the topographic survey, that is, the number of terrain points. The fewer the data points, the lower the probability of finding a solution that satisfies the constraints more tightly.

To conclude the comparison between these two approaches, it is mandatory to note that these actually refer to two different alternatives for the physical installation of the penstock, and thus the superiority of the continuous approach is not to be taken for granted, as some particular cases (such very steep, irregular terrains) might be better suited to layouts with a penstock composed of straight segments connected through elbows, avoiding the curvature limitations of the continuous approach. For these reasons, both methods are recommended to be considered for the optimization of these plants.

Table 5. Comparison between the current (continuous) and the previous (discrete) approach.

	Continuous Approach	Discrete Approach [32]
Gross height (m)	79.98	78.56
Power (W)	7003.06	7173.34
Pens. length (m)	532.42	523.78
Pipe diam. (m)	0.14	0.14
Cost (c.u.)	209,66.11	215,90.50

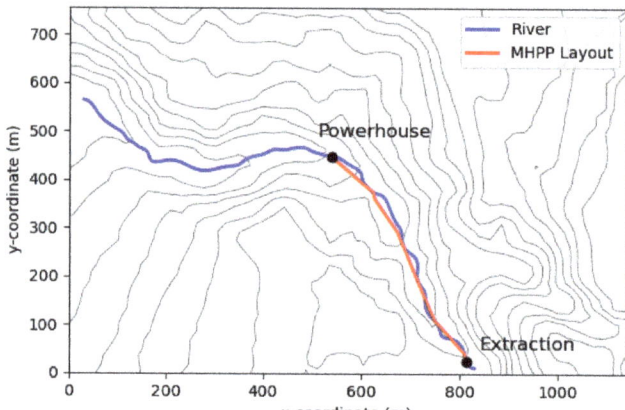

Figure 15. Optimal solution obtained using the discrete approach from [32].

5. Conclusions and Further Work

This work proposes the optimization of an MHPP using a continuous formulation of the problem from a three-dimensional approach. The problem is formulated as the minimization of the cost of the plant with a minimum power generation constraint. The problem considers not only the cost of the equipment, but also the cost of the civil works involved in its deployment, in such a way that the strong dependence of the path of the penstock through the terrain and the labor involved in terrain excavation and installation of supports is considered. A GA has been developed to solve the optimization problem, for which initial population generation, mutation, and crossover tailored operators have been designed, given the complexity of the constraints involved. The algorithm has been applied to an illustrative case study based on a real-case scenario in a small remote community in Honduras, and two additional modifications of this problem, to further evaluate its performance. The real topography of the terrain and the river profile have been determined through an aerial topographic survey, and an optimal layout for the MHPP has been precisely determined. The solution obtained permits the generation of 7 kW, with a total cost of 20,966 c.u. The analysis of the solution obtained demonstrates how the algorithm builds a layout that cuts through rough terrain, thus demonstrating the benefits of using this approach. Finally, this approach has been numerically compared with a previous approach published in the literature, which showed that the continuous approach proposed can lead to a 2.8% cost reduction of the installation.

Author Contributions: Conceptualization, A.T.C.; Methodology, A.T.C., D.G.R.; Software and Validation, A.T.C.; Writing—original draft preparation, A.T.C., D.G.R., P.M.G.; writing—review and editing, A.T.C., D.G.R., P.M.G. All authors have read and agreed to the published version of the manuscript.

Funding: This research received no external funding.

Institutional Review Board Statement: Not applicable.

Informed Consent Statement: Not applicable.

Conflicts of Interest: The authors declare no conflict of interest.

References

1. Couto, T.B.; Olden, J.D. Global proliferation of small hydropower plants-science and policy. *Front. Ecol. Environ.* **2018**, *16*, 91–100. [CrossRef]
2. Apichonnabutr, W.; Tiwary, A. Trade-offs between economic and environmental performance of an autonomous hybrid energy system using micro hydro. *Appl. Energy* **2018**, *226*, 891–904. [CrossRef]
3. Kelly-Richards, S.; Silber-Coats, N.; Crootof, A.; Tecklin, D.; Bauer, C. Governing the transition to renewable energy: A review of impacts and policy issues in the small hydropower boom. *Energy Policy* **2017**, *101*, 251–264. [CrossRef]
4. Samora, I.; Manso, P.; Franca, M.J.; Schleiss, A.J.; Ramos, H.M. Energy recovery using micro-hydropower technology in water supply systems: The case study of the city of Fribourg. *Water* **2016**, *8*, 344. [CrossRef]
5. Berrada, A.; Bouhssine, Z.; Arechkik, A. Optimisation and economic modeling of micro hydropower plant integrated in water distribution system. *J. Clean. Prod.* **2019**, *232*, 877–887. [CrossRef]
6. Kuriqi, A.; Pinheiro, A.N.; Sordo-Ward, A.; Garrote, L. Flow regime aspects in determining environmental flows and maximising energy production at run-of-river hydropower plants. *Appl. Energy* **2019**, *256*, 113980. [CrossRef]
7. Hoseinzadeh, S.; Ghasemi, M.H.; Heyns, S. Application of hybrid systems in solution of low power generation at hot seasons for micro hydro systems. *Renew. Energy* **2020**, *160*, 323–332. [CrossRef]
8. Anaza, S.; Abdulazeez, M.; Yisah, Y.; Yusuf, Y.; Salawu, B.; Momoh, S. Micro hydro-electric energy generation—An overview. *Am. J. Eng. Res. (AJER)* **2017**, *6*, 5–12.
9. Jawahar, C.; Michael, P.A. A review on turbines for micro hydro power plant. *Renew. Sustain. Energy Rev.* **2017**, *72*, 882–887. [CrossRef]
10. Hosseini, S.; Forouzbakhsh, F.; Rahimpoor, M. Determination of the optimal installation capacity of small hydro-power plants through the use of technical, economic and reliability indices. *Energy Policy* **2005**, *33*, 1948–1956. [CrossRef]
11. Yildiz, V.; Vrugt, J.A. A toolbox for the optimal design of run-of-river hydropower plants. *Environ. Model. Softw.* **2019**, *111*, 134–152. [CrossRef]
12. Borkowski, D.; Majdak, M. Small Hydropower Plants with Variable Speed Operation—An Optimal Operation Curve Determination. *Energies* **2020**, *13*, 6230. [CrossRef]
13. Payet-burin, R.; Kromann, M.T.; Pereira-Cardenal, S.; Strzepek, K.M.; Bauer-Gottwein, P. Using model predictive control in a water infrastructure planning model for the Zambezi river basin. In Proceedings of the 38th IAHR World Congress: Water–Connecting the World, Panama City, Panama, 1–6 September 2019; CRC Press: Boca Raton, FL, USA, 2019; pp. 4057–4062.
14. Anagnostopoulos, J.S.; Papantonis, D.E. Flow Modeling and Runner Design Optimization in Turgo Water Turbines. *Int. J. Mech. Aerosp. Ind. Mechatron. Eng.* **2007**, *1*, 204–209.
15. Anagnostopoulos, J.S.; Papantonis, D.E. A fast Lagrangian simulation method for flow analysis and runner design in Pelton turbines. *J. Hydrodyn.* **2012**, *24*, 930–941. [CrossRef]
16. Anagnostopoulos, J.S.; Papantonis, D.E. Optimal sizing of a run-of-river small hydropower plant. *Energy Convers. Manag.* **2007**, *48*, 2663–2670. [CrossRef]
17. Ehteram, M.; Binti Koting, S.; Afan, H.A.; Mohd, N.S.; Malek, M.; Ahmed, A.N.; El-shafie, A.H.; Onn, C.C.; Lai, S.H.; El-Shafie, A. New evolutionary algorithm for optimizing hydropower generation considering multireservoir systems. *Appl. Sci.* **2019**, *9*, 2280. [CrossRef]
18. Alexander, K.V.; Giddens, E.P. Optimum penstocks for low head microhydro schemes. *Renew. Energy* **2008**, *33*, 507–519. [CrossRef]
19. Bozorg Haddad, O.; Moradi-Jalal, M.; Marino, M.A. Design–operation optimisation of run-of-river power plants. In *Proceedings of the Institution of Civil Engineers-Water Management*; Thomas Telford Ltd.: London, UK, 2011; Volume 164, pp. 463–475.
20. Hounnou, A.H.; Dubas, F.; Fifatin, F.X.; Chamagne, D.; Vianou, A. Multi-objective optimization of run-of-river small-hydropower plants considering both investment cost and annual energy generation. *Int. J. Energy Power Eng.* **2019**, *13*, 17–21.
21. Tapia, A.; Millán, P.; Gómez-Estern, F. Integer programming to optimize Micro-Hydro Power Plants for generic river profiles. *Renew. Energy* **2018**, *126*, 905–914. [CrossRef]
22. Tapia, A.; Reina, D.G.; Millán, P. An Evolutionary Computational Approach for Designing Micro Hydro Power Plants. *Energies* **2019**, *12*, 878. [CrossRef]
23. Tapia, A.; del Nozal, A.; Reina, D.; Millán, P. Three-dimensional optimization of penstock layouts for micro-hydropower plants using genetic algorithms. *Appl. Energy* **2021**, *301*, 117499. [CrossRef]
24. Abdelhady, H.U.; Imam, Y.E.; Shawwash, Z.; Ghanem, A. Parallelized Bi-level optimization model with continuous search domain for selection of run-of-river hydropower projects. *Renew. Energy* **2021**, *167*, 116–131. [CrossRef]
25. Leon, A.S.; Zhu, L. A dimensional analysis for determining optimal discharge and penstock diameter in impulse and reaction water turbines. *Renew. Energy* **2014**, *71*, 609–615. [CrossRef]

26. Thake, J. *Micro-Hydro Pelton Turbine Manual: Design, Manufacture and Installation for Small-Scale Hydropower*; ITDG Publishing: Rugby, UK, 2000.
27. Bauchau, O.A.; Craig, J.I. Euler-Bernoulli beam theory. In *Structural Analysis*; Springer: Berlin/Heidelberg, Germany, 2009; pp. 173–221.
28. Kumar, R.; Singal, S.K. Penstock material selection in small hydropower plants using MADM methods. *Renew. Sustain. Energy Rev.* **2015**, *52*, 240–255. [CrossRef]
29. Reina, D.; Tawfik, H.; Toral, S. Multi-subpopulation evolutionary algorithms for coverage deployment of UAV-networks. *Ad. Hoc. Netw.* **2018**, *68*, 16–32. [CrossRef]
30. Alvarado-Barrios, L.; Rodríguez del Nozal, A.; Tapia, A.; Martínez-Ramos, J.L.; Reina, D. An evolutionary computational approach for the problem of unit commitment and economic dispatch in microgrids under several operation modes. *Energies* **2019**, *12*, 2143. [CrossRef]
31. Holland, J.H. *Genetic Algorithms and Adaptation*; Springer US: New York, NY, USA, 1984; pp. 317–333. [CrossRef]
32. Tapia, A.; Reina, D.; Millán, P. Optimized Micro-Hydro Power Plants Layout Design Using Messy Genetic Algorithms. *Expert Syst. Appl.* **2020**, *159*, 113539. [CrossRef]

Article

Predicting the Risk of Overweight and Obesity in Madrid—A Binary Classification Approach with Evolutionary Feature Selection

Daniel Parra [1], Alberto Gutiérrez-Gallego [1], Oscar Garnica [1], Jose Manuel Velasco [1], Khaoula Zekri-Nechar [2], José J. Zamorano-León [3], Natalia de las Heras [4] and J. Ignacio Hidalgo [1,*]

1. Computer Architecture and Automation Department, Faculty of Computer Science, Universidad Complutense de Madrid, 28040 Madrid, Spain
2. Department of Medicine, Faculty of Medicine, Universidad Complutense de Madrid, 28040 Madrid, Spain
3. Public Health and Maternal and Child Health Department, Faculty of Medicine, Universidad Complutense de Madrid, 28040 Madrid, Spain
4. Department of Physiology, Faculty of Medicine, Universidad Complutense de Madrid, 28040 Madrid, Spain
* Correspondence: hidalgo@ucm.es

Abstract: In this paper, we experimented with a set of machine-learning classifiers for predicting the risk of a person being overweight or obese, taking into account his/her dietary habits and socioeconomic information. We investigate with ten different machine-learning algorithms combined with four feature-selection strategies (two evolutionary feature-selection methods, one feature selection from the literature, and no feature selection). We tackle the problem under a binary classification approach with evolutionary feature selection. In particular, we use a genetic algorithm to select the set of variables (features) that optimize the accuracy of the classifiers. As an additional contribution, we designed a variant of the Stud GA, a particular structure of the selection operator of individuals where a reduced set of elitist solutions dominate the process. The genetic algorithm uses a direct binary encoding, allowing a more efficient evaluation of the individuals. We use a dataset with information from more than 1170 people in the Spanish Region of Madrid. Both evolutionary and classical feature-selection methods were successfully applied to Gradient Boosting and Decision Tree algorithms, reaching values up to 79% and increasing the average accuracy by two points, respectively.

Keywords: feature selection; classification; genetic algorithm; evolutionary computing; overweight; obesity

1. Introduction

A person is considered overweight when her/his Body Mass Index (BMI) is higher than 25 and obese if it is over 30. BMI is computed by dividing the weight of the person (in kilograms) by the square of her/his height (in meters) [1]. According to the Spanish National Institute of Statistics [2], the rate of individuals with obesity has increased from 7.4% in 1987 to 17.4% in 2017. In just 30 years, it has multiplied by 2.4 times, taking the data for Spain as a reference. This health problem affects all sectors of the population, although it is not equally distributed. In particular, men are more prone to develop overweight/obesity than women. The number of cases of childhood obesity has also increased, reaching 10.3% of children between 2 and 17 years old. The World Health Organization [3] also shows that overweight people are more prone to developing cerebrovascular and respiratory problems, gallbladder disease and may increase the risk of different types of cancer. Hence, it is necessary to prevent future cases of overweight or obesity.

Some people may have a certain predisposition to suffer from overweight/obesity due to genetics. Overweight and obesity are usually a consequence of social behaviors, such as high-fat meals and sedentary lifestyles. Prevention should begin at early ages to consolidate healthy lifestyle habits, and those who suffer from any of these disorders should seek the help and advice of medical staff. Regarding prevention, it would be essential to know the relationships among lifestyle indicators, overweight, and obesity. In this context, a system predicting the risk of developing overweight and obesity could be very useful. In this paper, we show that it is possible to analyze the different factors and habits of a person and obtain the relationship among them by machine-learning techniques. In particular, we investigate the performance of a set of machine-learning classification algorithms when classifying people as overweight/obese versus non-overweight/non-obese using information about lifestyle habits. This information was collected by a pool of 14 different institutions participating in the consortium of the GenObIA project (work financed by the Community of Madrid and the European Social Fund through the GENOBIA-CM project with reference S2017/BMD-3773).

In machine learning (ML), classification is related to the assignment of class labels to data in the problem domain. A critical element in this type of problem is the selection of the variables used as predictors (feature selection). Using all variables is not always the best option; sometimes it is necessary to discard part of them in order to avoid noise in the data and also to reduce computational load. Feature-selection (FS) methods aim to reduce dimensionality, reduce execution times, and improve model results. In this work, four different approaches for FS have been tested, firstly, without applying FS, secondly by classical FS, and finally two FS methods using genetic algorithms (GAs). In particular, we investigate FS by a GA, using two selection operators: a classical tournament selection, and *Stud* selection, a new method proposed in this paper. Experimental results show that machine-learning algorithms are good classifiers when combined with evolutionary feature-selection methods for this particular problem.

The main contributions of this work are:

- We tried different configurations for a set of ML classifiers for people being overweight or obese based on information concerning lifestyle and dietary habits.
- We use a dataset with data of more than 1170 people, which is the largest study in Spain to the best of our knowledge.
- We explore four different feature-selection methods.
- We propose the *Stud* selection operator, a variant of the *stud GA* algorithm presented in [4], that could be adapted to another evolutionary algorithm.

The rest of the paper is organized as follows. Section 2 provides a brief review of related work. Section 3 describes the evolutionary algorithms for feature selection and implementation details. In Section 4, we explain the experimental setup and Section 5 collects the experimental results. In Section 6, the statistical analysis can be found and finally Section 7 contains the conclusions and future work.

2. Literature Review

Machine learning [5] is a constantly evolving branch of artificial intelligence related to those algorithms that try to simulate human intelligence using information from their environment. Techniques based on machine learning have been used in different fields, such as finance [6], pattern recognition [7,8], and medical applications [9,10].

In machine learning, it is important to carefully choose which features should be used and which should be discarded to construct the models. Medical datasets commonly present a small number of cases with a large number of variables, thus introducing different problems such as dimensionality and high computational requirements [11]. To deal with these obstacles, the use of feature-selection techniques has been proposed to select the variables that provide the greatest value. According to [12], three common FS categories are: filters, wrappers, and embedded methods.

Filter methods stand out mainly for their speed and scalability, being of great help in extremely large datasets. Through a series of statistical processes, scores are assigned to the different variables; the ones with the highest scores will be used to create the model.

The great limitation of these methods is that they do not consider the relationship between variables. Two examples of this category are Pearson's correlation coefficient, which allows us to quantify the linear dependence between two variables, and mutual information, which seeks to reduce the uncertainty of a random variable through knowledge of another random variable [12,13].

Wrapper methods search for the most appropriate subset of variables using the selected predictor as a black box to score the different subsets of variables generated throughout the different iterations. Wrapper methods [13] have a high computational cost since they require training and testing for each possible subset. Sequential Feature Selection (SFS) [14] starts with an empty set and adds the different variables individually, looking for the one that contributes the most value to the set. Once identified, this variable will be permanently incorporated into the subset, and then the next iteration will follow, repeating the same process until the desired number of variables is obtained. Based on this implementation, it is possible to find different variants, such as Sequential Backward Selection (SBS), which starts with the complete set of variables and reduces it through iterations, or Sequential Floating Forward Selection (SFFS) [15], which is based on the SFS method and incorporates a backward component with the SBS.

An example of the use of this type of technique is [16], where the authors apply a wrapper method, called Recursive Feature Elimination with Cross-Validation (RFECV), to select the best variables for a classification problem in the medical domain and obtain an accuracy improvement. In our work, RFECV is one of the techniques against which we compare the performance of our proposal, since it is a good FS alternative and has the benefits of wrapper methods. Heuristic search methods applied to feature selection can also be considered part of the wrapper methods. Examples of the use of genetic algorithms focused on feature selection can be found in the medical field, these methods being interesting for large datasets. In [17], an optimization algorithm based on a genetic algorithm is proposed, which allows to optimize the values of the SVM parameters, obtaining an optimal subset of features, and improving the classification accuracy. Embedded methods aim to reduce the computational time required to reclassify subsets of distinct variables. In order to do that, it tries to combine the advantages of the filter and wrapper methods. One of the main characteristics of embedded methods is the introduction of feature selection as part of the training process rather than as a separate phase, i.e., the feature-selection process becomes an integral part of the model.

Our proposal, *Stud* selection, is a variant of the *stud GA* algorithm presented in [4]. In *stud GA*, the fittest individual is considered a Stud and the rest of the population is crossed with it to obtain the offspring. On the other hand, in *stud GA*, diversity is maintained using the hamming distance between the Stud and the individual that will serve as the second parent. If the diversity is above a set threshold, the crossing is made to produce an offspring; otherwise the current second parent is mutated to produce the offspring.

In this work we have adapted this strategy to the particularities of our data. We found that crossover based on hamming distance produced a large computational load that did not mean significantly improved solutions. Therefore, we decided to use a simple one-point crossover and compensate for the possible loss of diversity with four studs.

In relation to the study of factors related to obesity, Hudson Reddon [18] deals with physical activity and genetic predisposition to obesity. With this purpose, the impact of exercise on 14 variants predisposing to obesity is analyzed. Physical activity is able to reduce the impact of the fat mass and obesity-associated (*FTO*) gene variation and obesity genetic risk scores. The results include the identification of an interaction between physical activity and *FTO SNP rs1421085*, a single nucleotide polymorphism of the gene associated with fat mass, and obesity, in a prospective cohort of six ethnic groups. According to this study, prevention programs with a heavy physical activity load can be a very important resource to combat obesity.

In [19], the authors try to identify risk factors for overweight and obesity using machine-learning techniques (regression and classification). Their main contributions are the identification of factors related to obesity/overweight, the analysis of these factors, and their respective variable analysis. The issue we find with this work is the use of the variable "weight" since it is a variable that is unknown in the future and is part of the BMI formula (the value to be predicted). In our opinion, weight cannot be used in a classification method.

There are examples of the use of evolutionary computation in similar environments. Ref. [20] deals with the prediction of obesity in children using a hybrid approach, combining supervised learning algorithms based on applying Bayes' theorem with the "naive" assumption of conditional independence between every pair of features, a.k.a. Naïve Bayes, and a GA. In this case, the use of Naïve Bayes in prediction presented problems when dealing with zero-value parameters, and as a solution, the author proposes to use GA for parameter optimization. The initial experiment to identify the usability of their approach indicated a 75% improvement in accuracy. Similarly, this work proposes creating a genetic algorithm to support the classification model in use by selecting the most useful features.

Among the studies dealing with overweight or obesity, datasets have few cases and limited information. There are also occasions where the decisions made may be questionable, such as the use of weight in the dataset.

The work presented here seeks to predict the risk of overweight and obesity in Madrid. With this aim, a binary classification approach with evolutionary feature selection is proposed. Hence, we provide the most relevant variables for the classification algorithm through an evolutionary process. A particular structure of the feature-selection process has been developed. Additionally, a high-quality dataset has been used, composed of detailed information about the habits of the individuals and their health.

3. Methodology

This section explains the methodology applied in this work and how the feature selection for the classification problem was performed using genetic algorithms. Figure 1 shows a diagram of the feature-selection process and the generation of the classification model.

In order to apply the methodology, we need to select three main items:

- The machine-learning technique.
- The dataset, which defines the features.
- The FS method, and if it applies, the parameters of the genetic algorithm.

From the original dataset, a curation process is performed, which also defines the initial dataset to be used. After that, the FS process is applied, and the selected features will be used to train the ML algorithm. The best classification ML model will be chosen after analyzing the results, their accuracy, and the number of false negatives.

The objective of a GA is to find the best solution of a problem through the iterative transformation (using crossover and mutation operators [21]) of an initial set of potential solutions (population). For each solution (individual), its performance (fitness) is evaluated, and, based on this value, the fittest ones will have a higher probability of passing to the next iteration. After a certain number of iterations, one of the candidates will be selected as the solution to the problem.

Four key concepts need to be considered when designing GAs:

- The encoding of the problem.
- The size of the population and the initialization method.
- The selection method including the fitness function.
- The processes by which the changes are introduced in the next iteration, including the probabilities and parameters [22].

After the execution of the GA, the fittest individual of the last generation represents the set of features (variables) selected to train the ML models.

Figure 1. Workflow diagram.

Evaluation Metrics

Table 1 shows a description of a confusion matrix. It is a numeric matrix where we can see the number of successes of the model for the different classes. The confusion matrix of Table 1 is for algorithms classifying data into two classes (binary classification): Positive and Negative. For constructing it, it is necessary to compute the following values:

- Number of True Positives (*TPs*): the class assigned to the sample by the model is Positive and it is also the real class.
- Number of False Negatives (*FNs*): the class assigned to the sample by the model is Negative, but the real class is Positive.
- Number of False Positives *(FPs)*: the class assigned to the sample by the model is Positive, but the real class is Negative.
- Number of True Negatives (*TNs*): the class assigned to the sample by the model is Negative and it is also the real class.
- Number of Total Samples (*Total*).

Table 1. Example of confusion matrix structure for binary classification.

	Positive Prediction	*Negative Prediction*
Positive Class	(TP)	(FN)
Negative Class	(FP)	(TN)

From those metrics, we can also obtain different metrics to measure the model's performance.

- *Accuracy*: Percentage of data correctly classified.

$$Accuracy = \frac{TP + TN}{Total} \quad (1)$$

- *Misclassification Rate*: Percentage of misclassified data.

$$Misclassification\ Rate = \frac{FP + FN}{Total} \quad (2)$$

- *Precision*: The percentage of correct predictions obtained.

$$Precision = \frac{TP}{TP + FP} \quad (3)$$

- *Recall*: True positive rate, the percentage of data that manages to be classified from the positive class.

$$Recall = \frac{TP}{TP + FN} \quad (4)$$

Precision and recall can be associated with the positive and negative classes. In our case, we will focus on reducing the number of false negatives since these would be cases of overweight or obesity that our model does not detect. In other words, we seek to obtain a high recall of the positive class.

The GA selects the best set of features for prediction. A solution is represented by a binary chain (chromosome), with as many positions as features available in the curated dataset. Each of the positions (genes) in the chromosome applies to a feature. The value of the gene will indicate if the feature is selected (1) or not (0) as the prediction variable. The initial population will be generated randomly.

Due to the fact that we have a balanced dataset, to evaluate an individual (a model), we used a classical cross-validation scheme, with stratification [23]. Each model is trained using only the features expressed as 1s in the individual genotype. The average accuracy rate of the 10 folds, F, is used as fitness function:

$$F = \frac{1}{10} \sum_{i=1}^{10} \text{Accuracy}_i$$

where Accuracy_i is the accuracy obtained for each one of the cross-validation folds [24].

In this paper, we propose a variation of the *Stud GA* method and we compare its performance with a traditional tournament implementation.

The *Stud* selection method works as follows. First, the four best individuals of the generation are selected, and form the *Stud candidates group*. Second, the two best individuals pass to the following iteration without crossover. Finally, the rest of the population is completed by applying the crossover operator to a pair formed by: (i) a member of the *Stud candidates group* and (ii) another individual of the population in the event that the probability of crossover is met.

For the tournament selection, we use a simple implementation with selection pressure of five [25]. As usual in the literature [26], we denote selective pressure to the size of the tournament pool. Adjusting this parameter allows us to find a trade-off between exploration and exploitation of the fitness landscape of the algorithm. In this study, we use a selection pressure of five, which is a value that prioritizes exploitation over exploration and seemed to work well in the preliminary experiments with our datasets.

After the selection of the individuals, we apply a single-point crossover, choosing a point in the chromosome of the two selected individuals and generating one offspring. With this purpose, we combine the information from one of them up to the crossover point and complete it with the remaining information from the other individual.

A random mutation is introduced in the individual with a very low probability. This mutation affects a gene of the individual, flipping its value.

The parameters used for the GA were a crossover probability of 0.82, a mutation probability of 0.09, a population of 50 individuals, and 100 generations.

4. Experimental Setup

4.1. Dataset

The original dataset is the result of surveys carried out in different center parts of the GenObIA consortium, including universities and hospitals. The information collected in these surveys includes lifestyle, nutrition habits, and information about pathologies suffered by the person in the past.

4.1.1. Data Curation

The original dataset, Appendix A.2, Table A2, is composed of a total of 93 variables and 1179 subjects, among which we find:

- One subject identifier;
- Thirteen variables of general information about the subject such as weight, age, education, stress, etc.;
- Seven variables related to alcoholic drinks, distinguishing between distilled and fermented drinks;
- Seven variables on smoking habits, such as number of cigarettes, pipes, cigars. For ex-smokers: time since a smoker quit smoking;
- Fifteen variables related to pathologies, such as types of cancer, sleep apnea, and type 2 diabetes mellitus, among others;

- Thirty-four variables on nutritional habits; we found information on the portions of different types of food and the points of adherence to the Mediterranean diet according to these portions;
- Sixteen variables related to physical exercise and its intensity.

The dataset is balanced in terms of the predicted variable, overweight/obesity (BMI $\geq$ 25), with 48% being obese/overweight individuals and 52% being non-obese/non-overweight individuals. Therefore, we considered unnecessary the use of classification techniques focused on imbalanced datasets.

In order to avoid repeated information that may introduce noise in the system, a reduction in the number of variables was performed. An example of a reduction is the case of the variables referring to the food intake, which were replaced by a unique variable, namely adherence to the Mediterranean Diet (ADH). The original dataset, contains a set of variables related to food, 16 associated with the servings, and derived from these, another 16 measuring their adherence to the Mediterranean diet using points. If there are more than eight points in total, the subject is considered to have a high ADH.

In addition, some redundant variables were eliminated. For instance, the dataset initially contains two variables referring to exercise that were computed with a set of features of the pool: Cal_IPAQ, which reports the calories burned as a function of physical exercise; and IPAQ, which reports the information on the exercise performed and its intensity. IPAQ takes into account the duration and intensity of exercise, pondering the value of sedentary, moderate, and vigorous exercise. Cal_IPAQ includes weight as a variable for its calculation and therefore cannot be used since weight is also present in the close form for computing BMI. Hence, we use only IPAQ as a training variable.

There are also some features, such as the place or institution, where the sample was obtained (*center*), that were removed from the dataset since a high correlation with the BMI was observed due to the differences in the nature of the population of the places (policeman, sport teams, retired people, etc.).

After processing the data, which in our case was supervised by the medical staff that participated in the project, the total number of variables was reduced, from 93 to 41, as shown in Appendix A.1, Table A1. This table shows the different features of the study, providing its identifier, name, short description, and type.

4.1.2. Dataset with Pathologies

Two datasets were generated using the variables of Appendix A.1. One of them, called dataset with pathologies, includes the variables related to pathologies. This dataset includes most of the features. In particular, all the variables, excluding variables 37 to 40. The objective of this dataset is to evaluate the classifiers with standard data of the pools, which includes information on the health record of the person.

4.1.3. Dataset without Pathologies

There is a set of variables related to pathologies. When dealing with this type of variable, it is necessary to consider whether a pathology is a cause or a consequence of overweight/obesity. An example is the variable number 33, *Apnea*, which indicates whether the subject suffers from sleep apnea. Usually, overweight or obese people suffer from this problem. However, it is not necessarily true that because they suffer from sleep apnea, they are suffering from overweight/obesity. The same applies to other pathologies. In order to evaluate this kind of artifact, we create a new dataset, selecting all the variables in Appendix A.1, but excluding those of pathological type and variable 11.

5. Experimental Results

We performed experiments combining ten different algorithms as classifiers and four different feature-selection strategies (two evolutionary feature-selection methods, one feature selection from the literature, and no feature selection) on the two datasets explained above (With and Without Pathologies).

Tables 2 and 3 present the experimental results. These tables contain one row for each configuration, identified by an acronym (*ID*), and 11 additional columns including: the name of the algorithm (*ALGORITHM*), the feature-selection strategy (*FS*), the number of variables of the dataset (*VARIABLES*), the accuracy of the best solution (*BEST*), the accuracy of the worst solution (*WORST*), the mean (*MEAN*), and the standard deviation (*STD*) of 30 runs for the configuration in the row. In addition, the four last columns show precision (*PRECISION_0* and *PRECISION_1*)) and recall values (*RECALL_0* and *RECALL_1*) for class 0 (non-overweight/non-obese) and 1 (overweight or obese) for the best solution with this algorithm.

Table 2. Results of the different algorithms for the set of variables with pathologies. The table shows the algorithm, selection method, number of variables used, best case, worst case, mean, and standard deviation. The algorithm with the highest mean is marked with bold font.

ID	ALGORITHM	FS	VARIABLES	BEST	WORST	MEAN	STD +/−	PRECISION_0	PRECISION_1	RECALL_0	RECALL_1
DT-S	Decision Tree	Stud-GA	14	0.7661	0.6407	0.7150	0.0309	0.7682	0.7639	0.7733	0.7586
DT-T	Decision Tree	Tournament-GA	16	0.7492	0.6542	0.7103	0.0238	0.7255	0.7746	0.7762	0.7237
DT-RFECV	Decision Tree	RFECV	13	0.7559	0.6373	0.6962	0.0257	0.7597	0.7518	0.7697	0.7413
DT	Decision Tree	No FS	37	0.7254	0.6237	0.6914	0.0232	0.7035	0.7561	0.8013	0.6458
XGB-S	XGBOOST	Stud-GA	20	0.7831	0.6780	0.7216	0.0245	0.7451	0.8239	0.8201	0.7500
XGB-T	XGBOOST	Tournament-GA	19	0.7559	0.6746	0.7029	0.0187	0.7614	0.7479	0.8171	0.6794
XGB-RFECV	XGBOOST	RFECV	20	0.7424	0.6746	0.7085	0.0201	0.7677	0.7143	0.7484	0.7353
XGB	XGBOOST	No FS	37	0.7593	0.6441	0.6981	0.0225	0.7669	0.7500	0.7911	0.7226
GB-S	**Gradient Boosting**	**Stud-GA**	**19**	**0.7966**	**0.6881**	**0.7382**	**0.0231**	**0.8204**	**0.7656**	**0.8204**	**0.7656**
GB-T	Gradient Boosting	Tournament-GA	23	0.7864	0.6644	0.7332	0.0251	0.7738	0.8031	0.8387	0.7286
GB-RFECV	Gradient Boosting	RFECV	17	0.7797	0.6915	0.7324	0.0199	0.7727	0.7899	0.8447	0.7015
GB	Gradient Boosting	No FS	37	0.7797	0.6814	0.7305	0.0211	0.7683	0.7939	0.8235	0.7324
ADB	ADABOOST	No FS	37	0.6746	0.6000	0.6424	0.0189	0.6690	0.6800	0.6690	0.6800
BG	BAGGING	No FS	37	0.7458	0.6678	0.7114	0.0210	0.7486	0.7411	0.8253	0.6434
BNB	BERNOULLI NB	No FS	37	0.7220	0.6271	0.6675	0.0198	0.7429	0.6917	0.7784	0.6484
ET	EXTRA TREES	No FS	37	0.7627	0.6508	0.7119	0.0276	0.7419	0.7857	0.7931	0.7333
GNB	GAUSSIAN NB	No FS	37	0.7051	0.5932	0.6603	0.0230	0.6792	0.7711	0.8834	0.4848
LR	LOGISTIC REGRESSION	No FS	37	0.7627	0.6780	0.7098	0.0190	0.7603	0.7651	0.7603	0.7651
RFC	RANDOM FOREST	No FS	37	0.7763	0.6644	0.7292	0.0240	0.7582	0.7958	0.8000	0.7533

Table 3. Results of the different algorithms for the set of variables without pathologies. The table shows the algorithm, selection method, number of variables used, best case, worst case, mean, and standard deviation. The algorithm with the highest mean is marked with bold font.

ID	ALGORITHM	FS	VARIABLES	BEST	WORST	MEAN	STD +/−	PRECISION_0	PRECISION_1	RECALL_0	RECALL_1
DT-S	Decision Tree	Stud-GA	14	0.7322	0.6407	0.6934	0.0212	0.7419	0.7214	0.7468	0.7163
DT-T	Decision Tree	Tournament-GA	15	0.7525	0.6169	0.6862	0.0307	0.7548	0.75	0.7697	0.7343
DT-RFECV	Decision Tree	RFECV	1	0.7186	0.6373	0.6821	0.0195	0.7200	0.7172	0.7248	0.7123
DT	Decision Tree	No FS	26	0.7153	0.6203	0.6799	0.0252	0.7533	0.6759	0.7062	0.7259
XGB-S	XGBOOST	Stud-GA	12	0.7424	0.6237	0.6989	0.026	0.7702	0.709	0.7607	0.7197
XGB-T	XGBOOST	Tournament-GA	11	0.7288	0.6542	0.6948	0.0223	0.7235	0.736	0.7885	0.6619
XGB-RFECV	XGBOOST	RFECV	2	0.7186	0.6475	0.6850	0.0164	0.6746	0.7778	0.8028	0.6405
XGB	XGBOOST	No FS	26	0.7254	0.6068	0.6803	0.0295	0.7024	0.7559	0.7919	0.6575
GB-S	Gradient Boosting	Stud-GA	12	0.7797	0.6814	0.7295	0.0230	0.7578	0.8060	0.8243	0.7347
GB-T	Gradient Boosting	Tournament-GA	12	0.7763	0.6610	0.7307	0.0250	0.7636	0.7923	0.8235	0.7254
GB-RFECV	Gradient Boosting	RFECV	9	0.7695	0.661	0.7171	0.0236	0.7471	0.8	0.8355	0.6993
GB	Gradient Boosting	No FS	26	0.7695	0.6678	0.7169	0.0280	0.7857	0.7518	0.7756	0.7626
ADB	ADABOOST	No FS	26	0.6678	0.5661	0.6236	0.0262	0.6883	0.6454	0.6795	0.6547
BG	BAGGING	No FS	26	0.7695	0.6644	0.7086	0.0254	0.7709	0.7672	0.8364	0.6846
BNB	BERNOULLI NB	No FS	26	0.6881	0.5695	0.6154	0.0295	0.7048	0.6667	0.7312	0.6370
ET	EXTRA TREES	No FS	26	0.7661	0.6542	0.7125	0.0251	0.7197	0.8188	0.8188	0.7197
GNB	GAUSSIAN NB	No FS	26	0.6881	0.5763	0.6488	0.0279	0.6550	0.7339	0.7724	0.6067
LR	LOGISTIC REGRESSION	No FS	26	0.7559	0.6644	0.7060	0.0232	0.7697	0.7385	0.7888	0.7164
RFC	**RANDOM FOREST**	**No FS**	**26**	**0.7695**	**0.6983**	**0.7331**	**0.0188**	**0.7987**	**0.7353**	**0.7791**	**0.7576**

The interpretation of the FS column is:

- **Stud-GA**: Evolutionary feature selection with *Stud* selection operator.
- **Tournament-GA**: Evolutionary feature selection with tournament selection operator.
- **RFECV**: Feature selection with recursive feature elimination (RFE) with cross-validation (CV).
- **No-FS**: No feature selection applied in the configuration

As mentioned, ten classification algorithms were used, using the implementation available at the Scikit-learn python library [27]:

- **Decision Tree (DT)**: Its objective is to create a model to predict the value of a target variable by learning simple decision rules from data characteristics.
- **Gradient Boosting (GB)**: An additive model is created in a stepwise way, allowing the optimization of arbitrary differentiable loss functions. At each step, a regression tree is fitted to the negative gradient of the given loss function
- **Adaboost (ADB)**: A meta-estimator that starts by fitting a classifier on the original dataset and next fits additional copies of the classifier for the same dataset where the weights of the misclassified instances are adjusted in order to make the subsequent classifiers concentrate on the difficult cases.
- **Bagging (BG)**: Fits base classifiers on random subsets of the original dataset and aggregates its individual predictions into a final prediction.
- **Bernoulli Naive Bayes (BNB)**: This classifier is useful for discrete data and is designed to handle binary/boolean features.
- **Extra Trees (ET)**: A meta estimator that fits random Decision Trees on multiple sub-samples of the dataset and uses the average to improve predictive accuracy and control overfitting.
- **Gaussian Naive Bayes (GNB)**: Another Naïve Bayes model. This classifier is used when the values of the predictor are continuous
- **Logistic Regression (LR)**: This algorithm attempts to predict the probability that a given data entry will belong to a category. Just as linear regression assumes that the data follow a linear function, logistic regression models the data using the sigmoid function.
- **Random Forest (RFC)**: This technique fits several Decision Tree classifiers on multiple subsamples of the dataset and uses averaging to improve predictive accuracy and control overfitting.
- **XGBoost (XGB)**: A tree boosting system that stands out for its scalability and is widely used by data scientists.

Appendix A.3, Table A3 provides a table with information on the parameters used for each model.

Regarding evolutionary feature-selection methods, we focus their application on three models. The first one is Gradient Boosting [28] as it obtained consistently good results among the set of classifiers without feature selection. Next is XGBoost, [29], which is a state-of-the-art machine-learning algorithm that allows us to measure the goodness of results from the other algorithms. Finally, there is Decision Tree [30], because the use of models based on trees provides a solution with a straightforward interpretation by the medical staff. The development of understandable solutions for medical doctors is one of the main objectives of our research. Understanding why a model makes a particular prediction can be as crucial as prediction accuracy in the medical field. In some cases, the best results are obtained with complex models that are difficult to interpret. Thanks to the SHAP library [31], it is possible to obtain from each feature a value of importance for a particular predictor. The SHAP algorithm aims to explain the outcome of machine-learning models, representing the results by means of graphs. It is based on the Shapley values of game theory. In particular, in this paper we will focus on the ones that allow us to show the impact of the different variables in the model. To understand these graphs, two factors must be taken into account: the position of the points on the horizontal axis and the color. Let us take Figure 2 as an example.

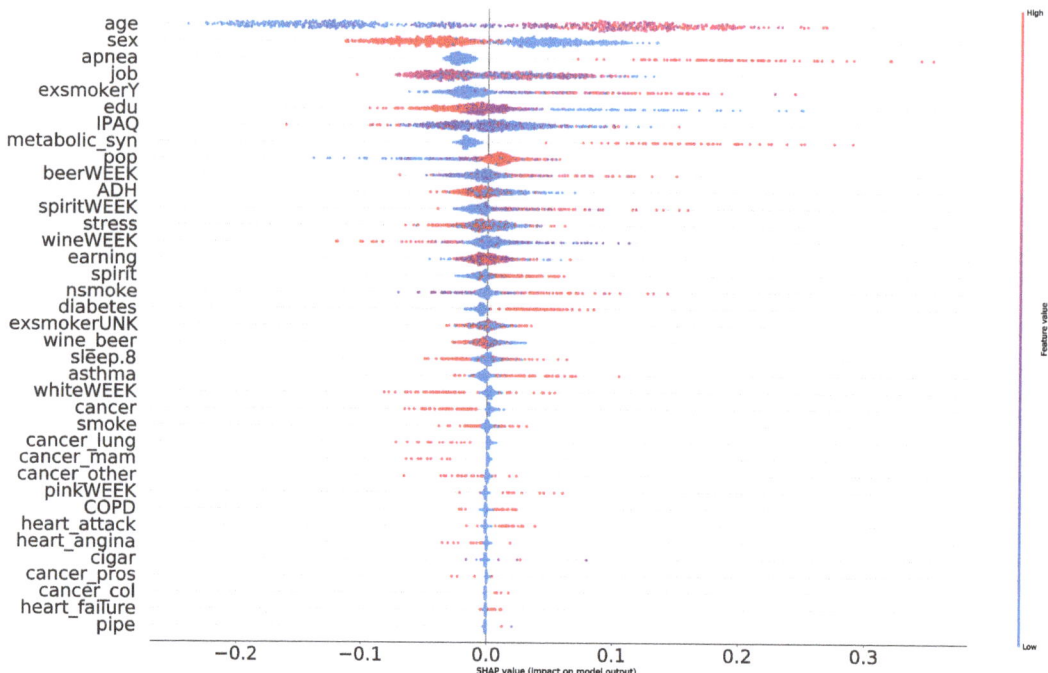

Figure 2. Impact of variables with SHAP, Random Forest with pathologies, and no FS.

- The color of the dots denotes the numerical value of the variable. In the case of age, the redder the dot the higher the age and the bluer the dot the lower the age. In the case of sex. The red color represents the female sex and the blue color represents the male sex. In the case of education, the red color represents higher levels of education while the blue color represents people with low levels of education or no education.
- The position of the points on the horizontal axis in our study indicates the probability of overweight/obesity. The further to the right the point is (positive values), the greater the probability that the person suffers from this problem. The further to the left (negative values), the lower the probability.

Thus, we can see as an example that Figure 2 provides us with the following information:

- Age is an important factor for the probability of being overweight/obese. We can clearly see that higher ages (red dots) have a higher probability than lower ages (blue dots).
- Gender is an unequivocal factor, with male gender (blue dots) being related to a higher probability of being overweight/obese while female gender (red dots) is similarly distributed but in the negative range.
- Education is another important factor to take into account. Low levels of education (blue dots) have a higher probability while high levels (red dots) are related to a lower probability.
- In the case of other variables, such as job, for example, we can see that the color of the points is intermixed, indicating no correlation with the probability of being overweight/obese.

5.1. Results Using the Dataset with Pathologies

5.1.1. Results without Feature Selection

In the scenario with pathologies, Table 2 summarizes the results of the algorithms with and without feature selection (No-FS). The average accuracy rate is 0.6953, and the standard deviation is 0.0297, with DT, ADB, BNB, and GNB being below this average. The algorithms with the best results are GB and RFC, with a mean that is close to 0.74.

Figure 2 shows the impact of the different variables of the RFC model. The age variable is in first place, in the case of younger people, represented by the lowest values (blue), which are mainly found in the left part of the graph, indicating a lower risk of being overweight/obese. In contrast, the higher values of this variable, the older people (red), have a higher risk of suffering from these health problems. In the case of sex, being male has a higher probability of being classified in the overweight/obesity class. Third, the first variable associated with the type of pathology, *apnea*, appears. The variable apnea has all the blue points very close to the zero point, while the red points extend along the positive part of the axis. This indicates that in cases of suffering from this pathology it is likely to suffer from overweight/obesity, but otherwise, it does not have a significant impact on the prediction.

5.1.2. Results for Gradient Boosting with Classical Feature Selection (GB-RFECV)

Moreover, in Table 2, using RFECV, an average accuracy rate of 0.7324 was achieved, reaching a maximum of 0.7797. A total of 17 variables were selected, including stress; some of the variables related to alcoholic drinks, education, time since quitting smoking, and physical exercise. Some of the most selected variables, such as sex, age, or apnea, are among the variables obtained. Figure 3 shows the evolution of the accuracy rate for different algorithms and datasets, taking the average of the cross-validation runs with RFECV in relation to the number of variables chosen. The average accuracy rate increases until it reaches 17 variables and then starts to decrease, probably due to the selection of variables that introduce noise.

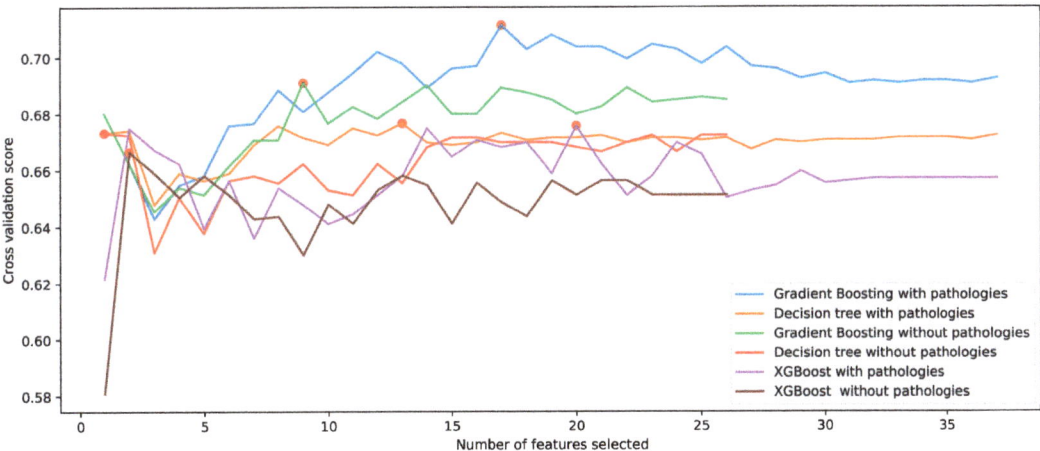

Figure 3. Mean accuracy evolution graph with RFECV.

5.1.3. Results for Gradient Boosting with Evolutionary Feature Selection

The results presented below are obtained with GB using evolutionary FS.

GA with *Stud* selection for Gradient Boosting (GB-S): In this case, the selection method reduced the number of variables used from 37 to 19, keeping its classification rate, and even obtaining a slight improvement, reaching an average accuracy of 0.7382, as shown in Table 2. Three of the variables selected were age, sex, and apnea, these being the most frequently chosen. Other variables to be highlighted are those related to smoking, education, earning, and adherence to the Mediterranean diet.

Figure 4 shows the graph corresponding to the Gradient Boosting model with *Stud* selection for the dataset with pathologies. The first variables that appear are age, sex, and apnea (as in the previous cases). In the case of education, it is shown that a lower level of education increases the probability of being overweight or obese. On the other hand, those individuals who suffer from metabolic syndrome are also more likely to be overweight or obese, but as in the case of apnea, if the individual does not suffer from this pathology, the variable does not have such a strong impact. Again, this can be seen in that the blue dots are clustered next to the zero point, while the red dots are spread over the positive values.

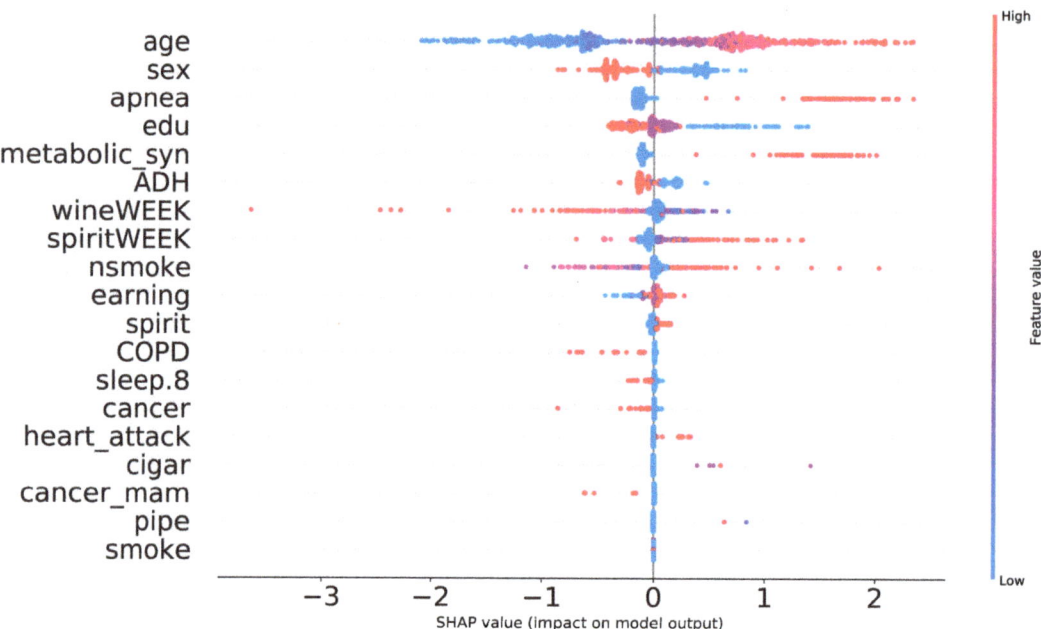

Figure 4. Impact of variables with SHAP, Gradient Boosting with *Stud* selection, with pathologies.

GA with tournament selection for Gradient Boosting (GB-T): Using the tournament selection method with selection pressure of five, 23 variables were selected by the GA, achieving an average accuracy of 0.7332, as can be found in Table 2. Similar to the previous case, among the features selected, some of the most common ones are sex, age, adherence to the Mediterranean diet, and some new additions, which were variables related to heart disease and different types of cancer. Other variables to note are the appearance of distilled/fermented beverages, education, earnings, and stress.

5.1.4. Results for Decision Tree with Classical Feature Selection (DT-RFECV)

Using RFECV, a total of 13 variables were selected, reaching an accuracy rate of 0.6962 and, in the best case, up to 0.7559, as shown in Table 2. Some of the variables are education, alcoholic drinks, physical exercise, and diabetes, among others. Again the most common variables are included: sex, age, and apnea. Figure 3 shows the evolution of the accuracy rate, in this case, the maximum working with 13 variables, reducing the accuracy when the number of variables increases.

5.1.5. Results for Decision Tree with Evolutionary Feature Selection

The results obtained with DT using evolutionary FS are presented below.

GA with *Stud* selection for Decision Tree (DT-S): Fourteen variables were selected, obtaining an average accuracy of 0.7150 and, in the best case, up to 0.7661, as can be verified in Table 2. Among the variables used were sex, age, and seven variables related to pathologies.

Figure 5 shows the impact of the different variables in the model Stud with Decision Tree with pathologies. Once again, age, sex and apnea are at the top of the list. In this case, diabetes appears with a similar distribution to apnea, although with less impact. The appearance of fermented beverages is also interesting.

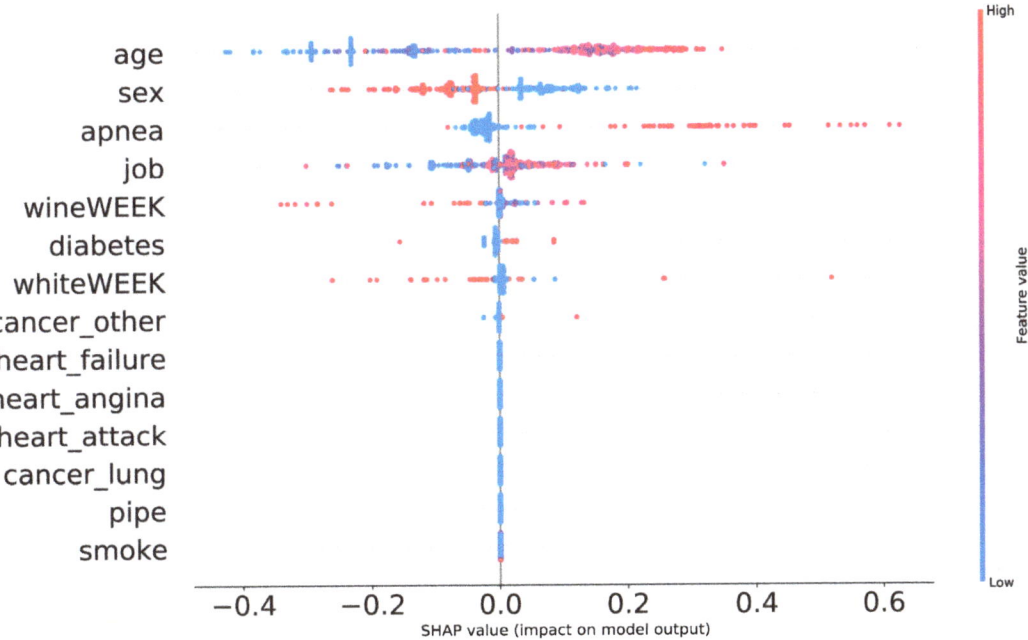

Figure 5. Impact of variables with SHAP, Decision Tree with *Stud* selection, with pathologies.

GA with tournament selection for Decision Tree (DT-T): Thanks to the feature selection, with only 16 variables, an accuracy reaching 0.7492 was achieved in the best cases and an average of 0.7103, as shown in Table 2.

Some of the variables used for this case are metabolic syndrome, those related to cancer, fermented drinks, specifically those related to wine, and the variable heart_angina. Moreover, as in the previous cases, sex and age appear.

Figure 6 shows a heat map representing the frequency with which the different FS models (columns) selected each of the variables (rows). Depending on the color of each cell, it is possible to get an idea of the number of times a variable was selected, with the cool colors being those cases with the lowest number of occurrences and the warm colors being those that appeared most frequently. In the case of the Decision Tree, there is greater diversity in the variables chosen by the models, since the colors are not so intense, unlike Gradient Boosting. Despite the differences between the variables chosen for the models, the extremes (top and bottom) of the heat map show a similar range of colors in most cases. The most frequent variables were sex, age, education, apnea, alcoholic beverages, both fermented and distilled, and metabolic syndrome.

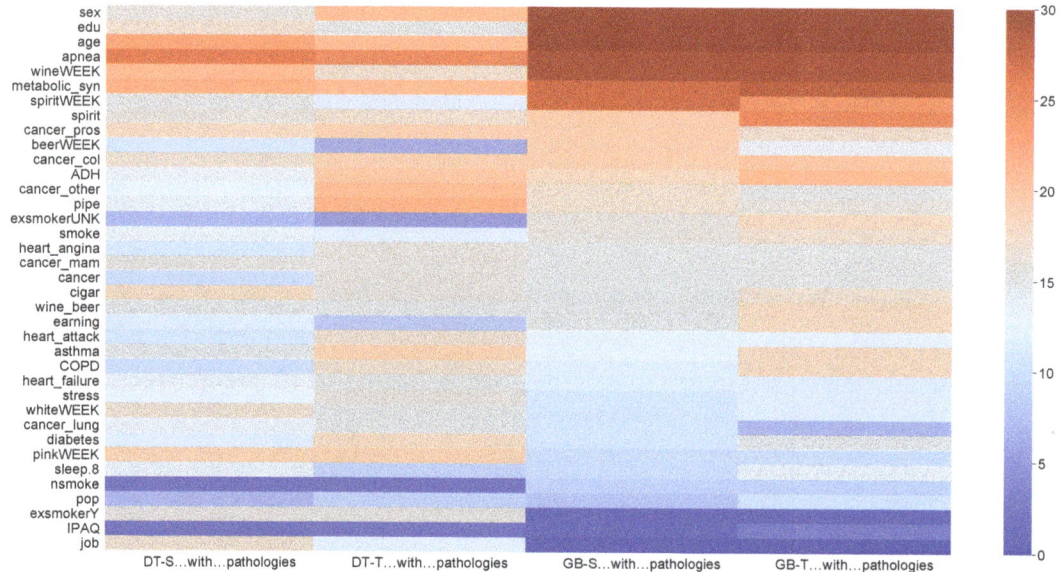

Figure 6. Heat map for frequency of the different variables using FS with GA, dataset with pathologies.

5.1.6. Results for XGBoost with Classical Feature Selection (XGB-RFECV)

This time the average accuracy rate achieved was 0.7085, reaching a maximum of 0.7424, as shown in Table 2. A total of 20 variables were used, including sex, age, education, metabolic syndrome, apnea, IPAQ, and different variables associated with alcoholic beverages, among others. Figure 3 shows the evolution of the accuracy rate with the number of variables chosen. In this case, the maximum value is obtained with 20 variables.

5.1.7. Results for XGBoost with Evolutionary Feature Selection

The results obtained by applying the evolutionary FS to XGBoost are presented below.

GA with *Stud* selection for XGBoost (XGB-S): As shown in Table 2, an average accuracy rate of 0.7216 and a maximum of 0.7831 was achieved using 20 variables. Similar to other cases, variables such as sex, age, edu, and those related to alcoholic beverages appear. Nine of the selected variables belong to the group of pathologies. Among them, we find several types of cancer, apnea, diabetes, and metabolic syndrome.

GA with tournament selection for XGBoost (XGB-T): In this case, using a total of 19 variables, a mean accuracy ratio of 0.7029 was obtained, and a maximum of 0.7559 was reached, as seen in Table 2. Among the most common variables, we found sex, edu, and apnea, but age did not appear. In addition, five variables related to smoking and up to nine related to pathologies were selected.

5.2. Results without Pathologies

A new batch of experiments was performed with the dataset without pathologies, using the same algorithms as in the previous section. The results with this new dataset are worse than those obtained using the dataset with pathologies. It may indicate that the models obtained using the dataset with pathologies gave more importance to these variables, which may be considered a consequence of overweight/obesity rather than a cause.

5.2.1. Results without Feature Selection

After testing the dataset without pathologies with the algorithms without feature selection, a reduction in the accuracy rate of the models was observed with respect to the results with pathologies, obtaining an average accuracy rate of 0.6825 and a standard deviation of 0.0408. RFC obtained the highest result, reaching 0.7331 accuracy, as can be seen in Table 3.

Figure 7 shows the impact of the variables for the RFC model, using the dataset without pathologies. Age and sex maintain the highest positions with the same characteristics seen in previous cases. Among the variables referring to eating habits, vegetables and soda stand out. The values of vegetables seem to have an inversely proportional relationship with the possibility of being overweight or obese while in the case of soda this relationship is direct. In Figure 8, we can see the Pearson Correlation Matrix of the variables in Figure 7. As we can see, no significant correlation can be appreciated between variables with the exception of *smoke* (smoker or non-smoker) and *nsmoke* (number of cigarettes per day).

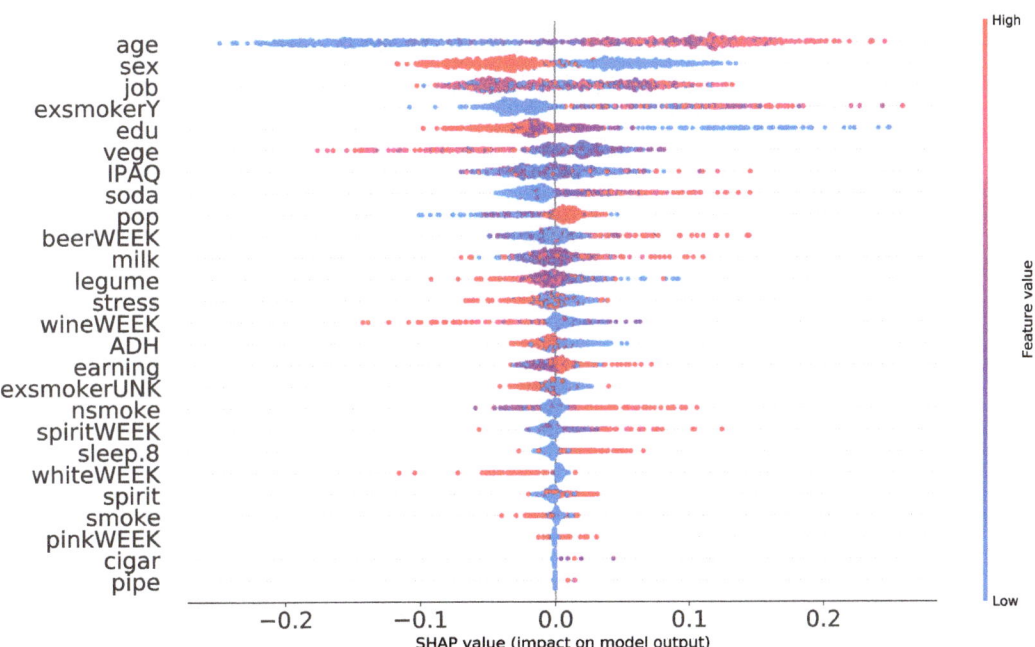

Figure 7. Impact of variables with SHAP, Random Forest without pathologies, and no FS.

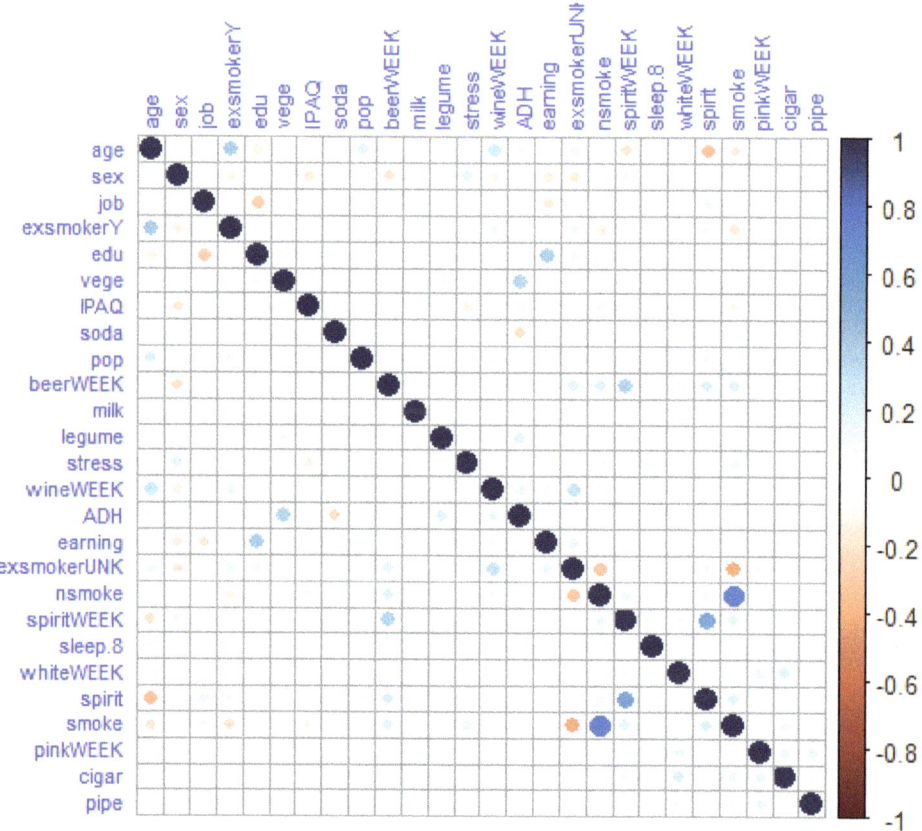

Figure 8. Pearson Correlation Matrix of variables in Figure 7.

5.2.2. Results for Gradient Boosting with Classical Feature Selection (GB-RFECV)

In this configuration, the number of variables selected was 9, reaching an average accuracy rate of 0.7171 and reaching 0.7695 in the best case, as shown in Table 3. Legume intake, vegetable intake, physical exercise, and time as an ex-smoker are some of the selected variables. On the other hand, there are also those more common ones such as sex, age, and education. In Figure 3, the accuracy reaches its maximum value at nine variables.

5.2.3. Results for Gradient Boosting with Evolutionary Feature Selection

The results obtained using GB with evolutionary FS are presented below.

GA with *Stud* selection for Gradient Boosting (GB-S): Using only 12 variables, it has been possible to achieve an average accuracy of 0.7295 and 0.7797 for the best case, achieving a slight improvement compared to the version without feature selection, as shown in Table 3. In this new set of variables, sex and age continue to dominate similarly to the set with pathologies. Other variables used are those related to hours of sleep, fermented/distilled drinks, soft drinks, legume portions, and education.

The graph of the impact of the variables corresponding to the Gradient Boosting model with *Stud* selection for the case without pathologies is shown in Figure 9. In the highest positions we find sex, age, education, and soda with the same performance as above. A new variable to highlight is legumes, whose highest values appear in the left zone of the graph. As for alcoholic beverages, we find apparent differences between distilled and fermented beverages; for the higher values of the variable wineWEEK, the probability of being overweight/obese is lower. In the case of spirits, it seems that their intake may be associated with overweight/obesity.

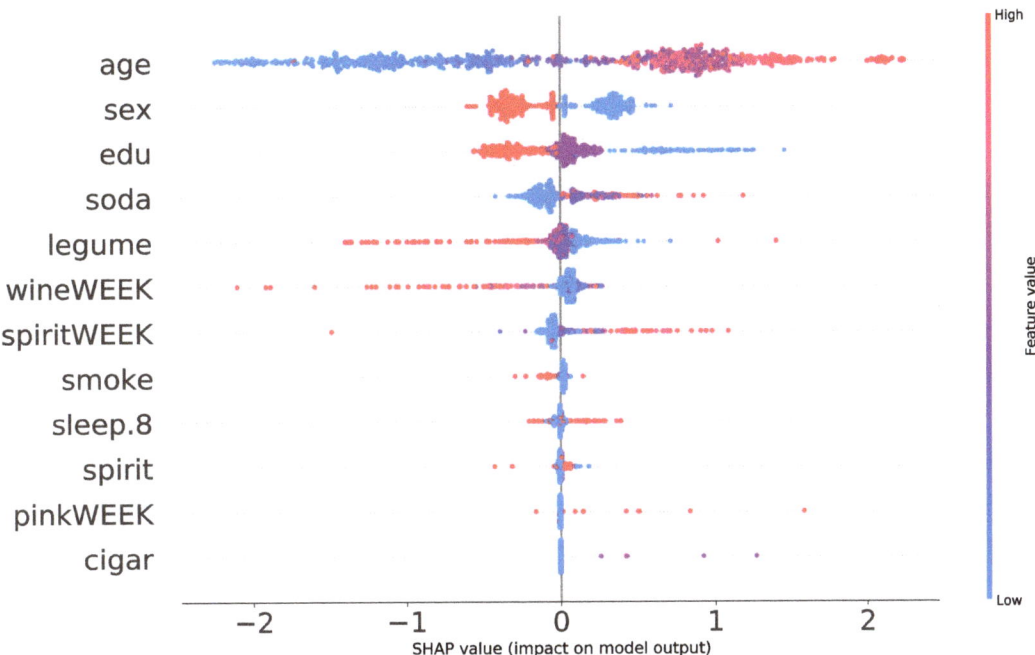

Figure 9. Impact of variables with SHAP, Gradient Boosting with *Stud* selection, without pathologies.

GA with tournament selection for Gradient Boosting (GB-T): Using the tournament selection method, the number of variables used was also 12 with an average accuracy of 0.7307 and for the best-case scenario up to 0.7763, as shown in Table 3. New variables were included: adherence to the Mediterranean diet and the population. Other variables seen previously such as education, soft drinks, distilled/fermented beverages, sex, age, and hours of sleep, among others, also appear.

5.2.4. Results for Decision Tree with Classical Feature Selection (DT-RFECV)

For this case, RFECV has selected a single variable, age, achieving an average accuracy rate of 0.6821 and a maximum of 0.7186, as shown in Table 3. Classical Feature Selection algorithms have solutions with very few variables; in our case, we are looking for models that explain more extensively the causes of overweight and obesity, so we explore other solutions such as those based on evolutionary algorithms that allow us to obtain models with a greater number of variables and similar accuracy. Looking at Figure 3, the accuracy reaches its maximum value with only one variable, but here we have to consider two facts:

1. This value is not very far from the case of using all the variables;
2. From a medical point of view, the use of a single variable model does not provide any help to clinical practice.

Figure 3 shows the variation of the cross-validation score (inside RFECV strategy) for the different algorithms for both the pathology and non-pathology cases. This value is of little interest if we look at the small difference in the values on the horizontal axis and at the fact that we are comparing two different datasets. What really interests us is to see the differences in the number of selected features (peaks marked with a red dot) and how these values vary depending on the algorithms and the datasets.

5.2.5. Results for Decision Tree with Evolutionary Feature Selection

The following results were obtained with the DT using the evolutionary GA.

GA with *Stud* selection for Decision Tree (DT-S): Using the *Stud* selection method, a total of 14 variables were chosen, with an average accuracy of 0.6934 and 0.7322 for the best case, as can be seen in Table 3. In this case, the age variable was taken but not the sex variable. The variables chosen include education, population, hours of sleep, alcoholic drinks, soft drinks, legumes rations, vegetable rations, and the stress variable as a new incorporation.

Figure 10 shows the impact of the different variables for the Decision Tree model with Stud for the set without pathologies. First, age appears again, the next variable is education and it is observed that, for higher values and higher level of studies, the probability of being overweight/obese is lower. In the case of soft drinks, higher consumption implies a higher probability of being overweight or obese. Another variable to note is vegetable, where it is observed that most of the red dots are on the left side of the graph, so it can be considered that it is less likely to be overweight/obese. It would be interesting to study in the future the causes of the behaviors of the variables stress and ex-smoker.

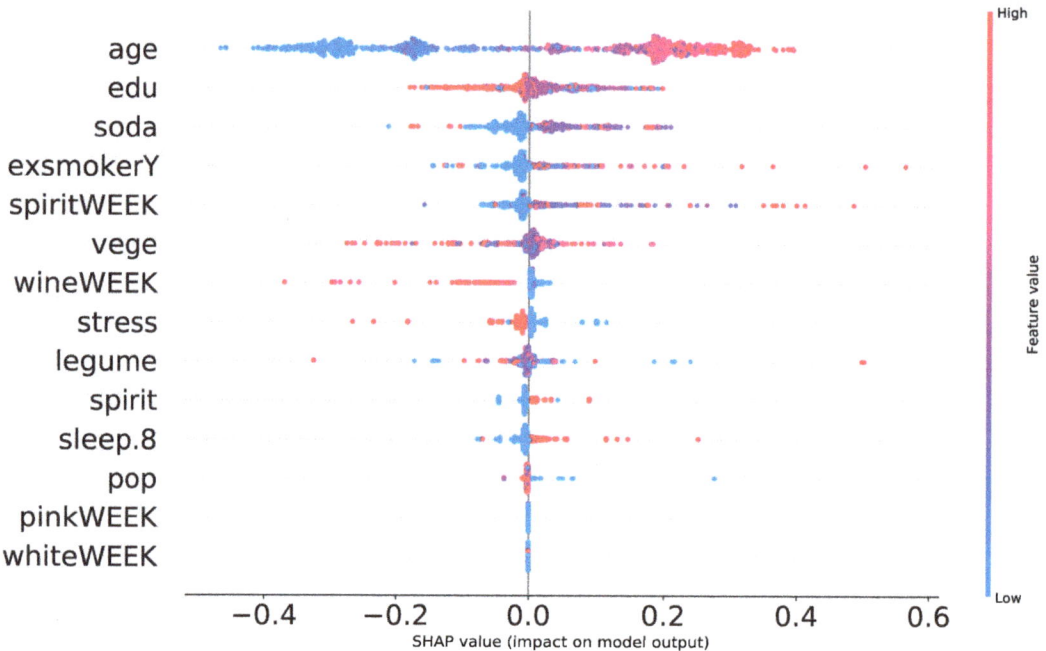

Figure 10. Impact of Variables with SHAP, Decision Tree with *Stud* Selection, without Pathologies.

GA with tournament selection for Decision Tree (DT-T): In this case, the number of variables used was 15, obtaining an average accuracy of 0.6862 and reaching 0.7525 in the best case, as shown in Table 3. Among the variables chosen are portions of legumes, fermented drinks, the time without smoking of an ex-smoker, and the IPAQ, the variable that reflects physical exercise. Other variables found are the most common, such as sex, age, soft drinks, and education, among others.

Figure 11 shows the heat map obtained by representing the frequency of variables selected from the dataset without pathologies using the FS models with genetic algorithms. There is a closer correlation between the variables selected with the GB and DT models, compared to the case of the dataset with pathologies. Again the most commonly selected variables are age, sex, and education followed by alcoholic beverages. As a new addition among the most selected variables, we find the variable soda.

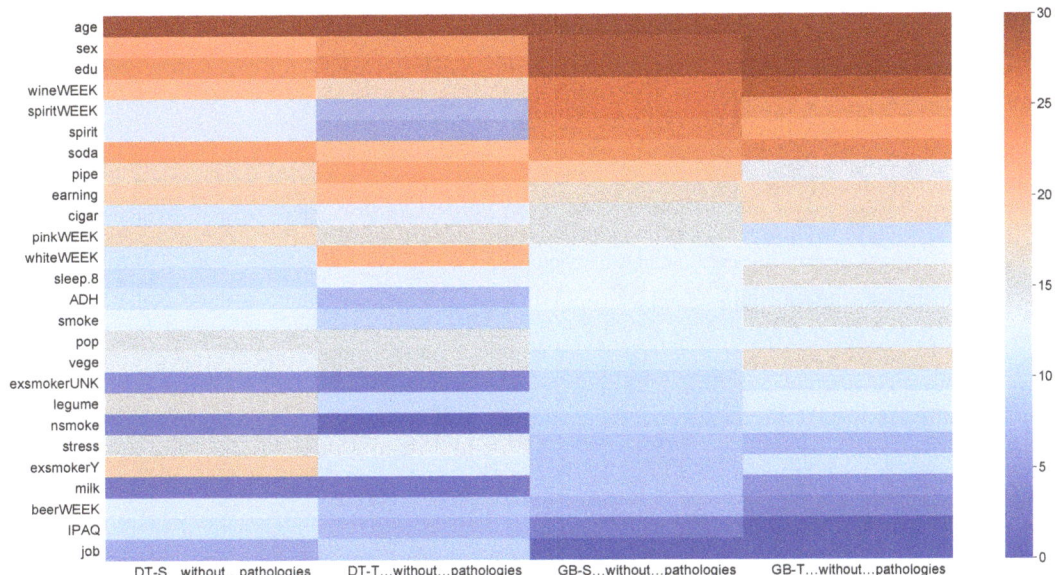

Figure 11. Heat map representation of the frequency of different variables obtained using FS with GA, dataset without pathologies.

5.2.6. Results for XGBoost with Classical Feature Selection (XGB-RFECV)

With only two variables, sex and age, an average accuracy ratio of 0.6850 and a maximum of 0.7186 were obtained, as shown in Table 2. The evolution of the accuracy rate as a function of the number of variables used is shown in Figure 3. From a medical point of view, the selection of only these two variables does not provide any information.

5.2.7. Results for XGBoost with Evolutionary Feature Selection

Results obtained with the XGBoost using the evolutionary GA were as follows.

GA with *Stud* selection for XGBoost (XGB-S): As shown in Table 2, a mean accuracy rate of 0.6989 and a maximum of 0.7424 using 12 variables were achieved. In this case, among the variables selected, we found some of the most common ones, such as sex, age, edu, or soda. In addition, there are variables such as vegetation or those related to alcoholic beverages.

GA with tournament selection for XGBoost (XGB-T): Using the 11 selected variables, an average precision ratio of 0.6948 and a maximum of 0.7288 were obtained, as can be seen in Table 2. In addition to the most common variables, such as sex, age, education, or soda, we can find variables related to alcoholic beverages, both distilled and fermented, vegetable intake, and earning.

Table 4 gives a sample of time in seconds taken by the GA for the different algorithms and datasets. One run of each type was carried out individually (no other program was running in parallel) on the same machine to obtain these values. The time is strongly related to the algorithm chosen.

Table 4. Measured times in seconds for each method and problem.

	GB-T	GB-S	DT-T	DT-S	XGB-T (GPU)	XGB-S (GPU)	XGB-T (CPU)	XGB-S (CPU)
With pathologies	4641.42	5203.35	178.31	189.13	13,140.59	13,492.46	2130.07	2085.69
Without pathologies	4311.17	4384.38	200.53	194.88	13,781.81	12,732.41	2220.16	2558.31

6. Statistical Analysis

We performed a nonparametric statistical analysis once the results were obtained for the two proposed problems, with and without pathologies. The Friedman test [32] is a nonparametric [33] alternative to the two-way parametric analysis of variance that tries to detect significant differences between the behavior of two or more algorithms. It can be used to identify if, in a set of k samples (where $k \geq 2$), at least two of the samples represent populations with different median values. The first step is to convert the original results, in Table 5, into ranks to produce Table 6. Table 5 shows the average accuracy rate for each of the algorithms on the two datasets.

Table 5. Algorithm results by problem (average).

Problem	DT-S	DT-T	DT-RFECV	DT	XGB-S	XGB-T	XGB-RFECV	XGB	GB-S	GB-T	GB-RFECV	GB	ADB	BG	BNB	ET	GNB	LR	RFC
With Pathologies	0.7150	0.7103	0.6962	0.6914	0.7216	0.7029	0.7085	0.6981	**0.7382**	0.7332	0.7324	0.7305	0.6424	0.7114	0.6675	0.7119	0.6603	0.7098	0.7292
Without Pathologies	0.6934	0.6862	0.6821	0.6799	0.6989	0.6948	0.6850	0.6803	0.7295	0.7307	0.7171	0.7169	0.6236	0.7086	0.6154	0.7125	0.6488	0.7060	**0.7331**

Table 6. Friedman Ranks for the different proposed algorithms.

Problem	With Feature Selection								Without Feature Selection								
	GB-S	GB-T	GB-RFECV	XGB-S	DT-S	DT-T	XGB-T	XGB-RFECV	DT-RFECV	RFC	GB	ET	BG	LR	XGB	DT	GNB	BNB	ADB
With Pathologies	1	2	3	6	7	10	13	12	15	5	4	8	9	11	14	16	18	17	19
Without Pathologies	3	2	4	9	11	12	10	13	14	1	5	6	7	8	15	16	17	19	18
Mean	2	2	3.5	7.5	9	11	11.5	12.5	14.5	3	4.5	7	8	9.5	14.5	16	17.5	18	18.5

- Gather observed results for each algorithm/problem pair.
- For each problem i, rank values from 1 (best result) to k (worst result). Denote these ranks as:

$$r_i^j = (1 \leq j \leq k) \quad (5)$$

where k is the number of algorithms.

- For each algorithm j, average the ranks obtained in all problems to obtain the final rank:

$$R_j = \frac{1}{n} \sum_{i=1}^{n} r_i^j \quad (6)$$

where n is the number of datasets.

In Table 6, the best algorithms are Gradient Boosting with *Stud* selection method, Gradient Boosting with tournament method, and Random Forest without feature selection. In both problems, algorithms with feature selection outperform their traditional version.

Under the null hypothesis (H0), which states that all algorithms behave similarly, so their ranks R_j should be equal, the Friedman statistic F_f can be calculated as:

$$F_f = \frac{12n}{k(k+1)} \left[\sum_j R_j^2 - \frac{k(k+1)^2}{4} \right] \quad (7)$$

Distributed according to a χ^2 distribution with $k-1$ degrees ($k = 19$) of freedom [33]. As usual, we can define the *p*-value as the probability of obtaining a result as extreme or even more extreme than the observed one, provided that the null hypothesis is true. As usual, we have chosen a *p*-value of 0.05. For this *p*-value, the critical value according to the distribution of χ^2 with 18 degrees of freedom is 28.8693. To reject H0, it is necessary to exceed this value. Applying the formula of Friedman's statistic, Equation (7), a value of 34.6421 is obtained, so the critical value is exceeded and the hypothesis H0 is rejected, confirming that there are significant differences between the models.

7. Conclusions and Future Work

In this work, a set of classification systems of persons at risk of suffering from overweight/obesity have been developed. Four different FS strategies have been employed for the two experiment datasets, one FS from the literature, two evolutionary FS methods, and no FS. For this work, ten machine-learning models have been employed. For the application of the feature-selection methods, we chose three of these 10 methods, based on reasons of performance and possible application to medical clinical practice. Thus, the FS methods were successfully applied in Gradient Boosting, XGBoost, and Decision Tree algorithms. The most important conclusions of this work are:

- Although not a surprising finding, we have found GA to be a very competent tool to perform feature selection and thus improve the training of classification models.
- The *Stud* selection, which uses an elitist set that is always part of the crossover process, achieves promising results.
- If we look at the fitness of the best individual in each of the final populations, they all maintain a similar standard deviation, perhaps indicating that we might expect to obtain similar results in the future with new data sets.
- About the variables related to pathologies, it is necessary to identify what is the cause and what is the effect. A good example would be sleep apnea. On several occasions, models have been based on apnea when classifying, but this disorder is caused by overweight or obesity in many cases. A person may have apnea due to being overweight or obese but will not necessarily be overweight or obese due to suffering from apnea.
- Finally, significant differences were found among the algorithms, with Gradient Boosting with feature selection being the one obtaining the best results.

The models developed in this work will be the basis of a recommendation system. This system will be able to warn people about behavioral tendencies that will end up producing overweight and obesity and recommend healthy habits to replace them.

Future Work

Although the overall accuracy of the model is important, from a medical point of view, classifying an individual at risk of overweight/obesity as not at risk is more detrimental than classifying a healthy person as at risk. The model must take this into account, so it would be interesting to find different fitness functions or develop a multi-objective evolutionary algorithm that could increase accuracy and reduce the number of false negatives for at-risk individuals. Therefore, we intend to test with precision and recall as fitness functions instead of accuracy.

In this work, and for space reasons, we have only used accuracy as a fitness function. There remains, therefore, as future work, the investigation of other fitness functions as well as the heuristic search or the dynamic selection as they have already been tested in the literature of other fields [34,35].

It is also necessary to study carefully the parameters of the evolutionary algorithm, testing with different combinations of population size and number of generations.

It would be highly recommended to increase the volume of the dataset as it could improve the accuracy of the models. As part of the project, genetic information of individuals will be incorporated, so it will be necessary to perform a study on the impact of these variables and their possible interaction with the current ones.

In this study we achieved an accuracy close to 0.8. Can these results be considered good enough, can they be improved, are they unacceptable? Although the members of our research team who are specialists in medicine consider these results as good, we lack having a context with a clear metric to determine where the minimum level is and how high we can achieve. Establishing this scale remains ambitious future work.

Author Contributions: Conceptualization: J.I.H., J.M.V. and J.J.Z.-L.; methodology: J.J.Z.-L., D.P., O.G. and J.I.H.; software: D.P. and A.G.-G.; validation: O.G., J.J.Z.-L. and J.M.V.; formal analysis: J.J.Z.-L., N.d.l.H. and J.I.H.; investigation: D.P. and K.Z.-N.; resources: K.Z.-N., J.J.Z.-L., N.d.l.H. and J.I.H.; data curation: K.Z.-N., D.P., A.G.-G. and N.d.l.H.; writing—original draft preparation: D.P. and A.G.-G.; writing—review and editing: J.M.V., O.G. and J.I.H.; visualization: D.P., A.G.-G. and J.M.V.; supervision: J.I.H.; project administration: J.I.H. and J.J.Z.-L.; funding acquisition: J.J.Z.-L. and J.I.H. All authors have read and agreed to the published version of the manuscript.

Funding: Work financed by the regional government of Madrid and co-financed by the EU Structural Funds through the Community of Madrid projects B2017/BMD3773 (GenObIA-CM) and Spanish Ministry of Economy and Competitiveness with number RTI2018-095180-B-I00 and PID2021-125549OB-I00.

Institutional Review Board Statement: The study was conducted according to the guidelines of the Declaration of Helsinki, and approved by the Clinical Research Ethics Committee of the Community of Madrid (Comité Ético de Investigación Clínica de la Comunidad de Madrid (CEIC-R)) and the Clinical Research Ethics Regional Committee (CEIC-R) (Comité Ético de Investigación Clínica-Regional (CEIC-R)). Genetic analyses have always been carried out in compliance with the provisions specified in the Biomedical Research Law (14/2007) and in the Personal Data Protection Law (Law 15/1999) from Spain.

Informed Consent Statement: Informed consent was obtained from all subjects involved in the study.

Data Availability Statement: The data that support the findings of this study are available on reasonable request from the corresponding author [J.I. Hidalgo]. The data are not publicly available due to legal restrictions.

Acknowledgments: We would also like to thank the centers that provided the data and made this work possible. 1. Atención Primaria. 2. Hospital Clínico San Carlos. 3. Hospital Universitario 12 de Octubre. 4. Hospital Universitario La Paz. 5. Hospital General Universitario Gregorio Marañón. 6. Hospital Universitario Ramón y Cajal. 7. Hospital Universitario Infanta Leonor.

Conflicts of Interest: The authors declare no conflict of interest.

Appendix A

Appendix A.1. Representation of Project Variables

Table A1. Representation of project variables.

ID	Variable	Description	Type
1	sex	Sex of the person	General information
2	age	Age of the person in years	General information
3	pop	Volume of the population where the person resides	General information
4	edu	Academic level attained by the person	General information
5	earning	Income level of the person	General information
6	job	Work type performed by the person	General information
7	stress	person self-perceived stress	General information
8	sleep.8	The person sleeps more than eight hours	General information
9	spirit	The person drinks spirits	General information
10	spiritWEEK	Units of spirit drinks per week	Alcoholic drinks

Table A1. *Cont.*

ID	Variable	Description	Type
11	wine_beer	The person drinks beer or wine	Alcoholic drinks
12	beerWEEK	Units of beer per week	Alcoholic drinks
13	wineWEEK	Units of red wine per week	Alcoholic drinks
14	whiteWEEK	Units of white wine per week	Alcoholic drinks
15	pinkWEEK	Units of rosé wine per week	Alcoholic drinks
16	smoke	The person smokes	Tobacco
17	nsmoke	Sigarettes consumed per day	Tobacco
18	pipe	Pipe tobacco consumed per day	Tobacco
19	cigar	Cigars consumed per day	Tobacco
20	exsmokerY	Time since a smoker quit smoking in years	Tobacco
21	exsmokerUNK	The person has given up smoking but does not remember how long ago.	Tobacco
22	cancer	The person has suffered or suffers from cancer	Pathologies
23	cancer_mam	The person has suffered or suffers from breast cancer.	Pathologies
24	cancer_col	The person has suffered or suffers from colon cancer	Pathologies
25	cancer_pros	The person has suffered or suffers from prostate cancer.	Pathologies
26	cancer_lung	The person has suffered or suffers from lung cancer.	Pathologies
27	cancer_other	The person has suffered or suffers from another type of cancer.	Pathologies
28	heart_attack	The person has suffered an acute myocardial infarction.	Pathologies
29	heart_angina	The person has suffered angina pectoris.	Pathologies
30	heart_failure	The person has suffered heart failure	Pathologies
31	diabetes	The person has type 2 diabetes mellitus	Pathologies
32	metabolic_syn	The person suffers from metabolic syndrome	Pathologies
33	apnea	The person suffers from sleep apnea	Pathologies
34	asthma	The person has asthma	Pathologies
35	COPD	The person suffers from chronic obstructive pulmonary disease.	Pathologies
36	ADH	The person has adherence to the Mediterranean diet.	Nutritional habits
37	vege	Servings of vegetables consumed by the individual per day	Nutritional habits
38	soda	Servings of carbonated and/or sweetened drinks consumed by the subject per day	Nutritional habits
39	legume	Servings of legumes consumed by the subject per week	Nutritional habits
40	milk	Servings of milk or dairy products consumed by the subject per day	Nutritional habits
41	IPAQ	Subject scores on the International Physical Activity Questionnaire (IPAQ)	Physical exercise

Appendix A.2. Representation of All Variables in the Original Data Set

Table A2. Representation of all variables in the original data set.

ID	Variable	Description	Type
1	n	Inclusion number	General information
2	center	Center	General information
3	sex	Sex of the person	General information
4	age	Age of the person in years	General information
5	height	Height (m)	General information
6	weight	Weight (Kg)	General information
7	IMC	BMI	General information
8	waist	Waist circumference (cm)	General information
9	pop	Volume of the population where the person resides	General information
10	edu	Academic level attained by the person	General information

Table A2. *Cont.*

ID	Variable	Description	Type
11	earning	Income level of the person	General information
12	job	Work type performed by the person	General information
13	stress	Person self-perceived stress	General information
14	sleep.8	The person sleeps more than eight hours	General information
15	spirit	The person drinks spirits	Alcoholic drinks
16	spiritWEEK	Units of spirit drinks per week	Alcoholic drinks
17	wine_beer	The person drinks beer or wine	Alcoholic drinks
18	beerWEEK	Units of beer per week	Alcoholic drinks
19	wineWEEK	Units of red wine per week	Alcoholic drinks
20	whiteWEEK	Units of white wine per week	Alcoholic drinks
21	pinkWEEK	Units of rosé wine per week	Alcoholic drinks
22	smoke	The person smokes	Tobacco
23	nsmoke	Cigarettes consumed per day	Tobacco
24	pipe	Pipe tobacco consumed per day	Tobacco
25	cigar	Cigars consumed per day	Tobacco
26	exsmokerY	Time since a smoker quit smoking (years)	Tobacco
27	exsmokerM	Time since a smoker quit smoking (months)	Tobacco
28	exsmokerUNK	The person has given up smoking but does not remember how long ago	Tobacco
29	cancer	The person has suffered or suffers from cancer	Pathologies
30	cancer_mam	The person has suffered or suffers from breast cancer.	Pathologies
31	cancer_col	The person has suffered or suffers from colon cancer	Pathologies
32	cancer_pros	The person has suffered or suffers from prostate cancer.	Pathologies
33	cancer_lung	The person has suffered or suffers from lung cancer	Pathologies
34	cancer_other	The person has suffered or suffers from another type of cancer.	Pathologies
35	heart_attack	The person has suffered an acute myocardial infarction.	Pathologies
36	heart_angina	The person has suffered angina pectoris.	Pathologies
37	heart_failure	The person has suffered heart failure	Pathologies
38	diabetes	The person has type 2 diabetes mellitus	Pathologies
39	hemo	Glycosylated hemoglobin (%)	Pathologies
40	metabolic_syn	The person suffers from metabolic syndrome	Pathologies
41	apnea	The person suffers from sleep apnea	Pathologies
42	asthma	The person has asthma	Pathologies
43	COPD	The person suffers from chronic obstructive pulmonary disease.	Pathologies
44	ADH	The person has adherence to the Mediterranean diet.	Nutritional habits
45	ADH_tot	Total ADH points	Nutritional habits
46	olive	Use olive oil	Nutritional habits
47	n_olive	Use olive oil (POINTS)	Nutritional habits
48	tot_olive	Tablespoons of olive oil consumed in total per day	Nutritional habits
49	ntot_olive	Tablespoons of olive oil consumed in total per day (POINTS)	Nutritional habits
50	vege	Servings of vegetables consumed per day	Nutritional habits
51	n_vege	Servings of vegetables consumed per day (POINTS)	Nutritional habits
52	fruit	Pieces of fruit (including natural juice) consumed per day	Nutritional habits
53	n_fruit	Pieces of fruit (including natural juice) consumed per day (POINTS)	Nutritional habits
54	burger	Red meat portions	Nutritional habits
55	n_burger	Red meat portions (POINTS)	Nutritional habits
56	cream	Servings of butter, margarine or cream consumed per day	Nutritional habits
57	n_cream	Servings of butter, margarine or cream consumed per day (POINTS)	Nutritional habits
58	soda	Glasses of carbonated and/or sweetened beverages per day	Nutritional habits
59	n_soda	Glasses of carbonated and/or sweetened beverages per day (POINTS)	Nutritional habits
60	wine_week	Wine consumed per week	Nutritional habits

Table A2. *Cont.*

ID	Variable	Description	Type
61	n_wine_week	Wine consumed per week (POINTS)	Nutritional habits
62	legume	Servings of legumes per week	Nutritional habits
63	n_legume	Servings of legumes per week (POINTS)	Nutritional habits
64	fish	Servings of fish or seafood consumed per week	Nutritional habits
65	n_fish	Servings of fish or seafood consumed per week (POINTS)	Nutritional habits
66	cake	Times per week consuming commercial bakery products	Nutritional habits
67	n_cake	Times per week consuming commercial bakery products (POINTS)	Nutritional habits
68	nuts	Servings of nuts and dried fruit consumed per week	Nutritional habits
69	n_nuts	Servings of nuts and dried fruit consumed per week (POINTS)	Nutritional habits
70	chicken	Preferably consume chicken, turkey or rabbit meat instead of beef, pork, hamburgers or sausages	Nutritional habits
71	n_chicken	Preferably consume chicken, turkey or rabbit meat instead of beef, pork, hamburgers or sausages (POINTS)	Nutritional habits
72	sauce	Times a week eat cooked vegetables, pasta, rice or other dishes seasoned with a tomato, garlic, onion or leek sauce simmered with olive oil	Nutritional habits
73	n_sauce	Times a week eat cooked vegetables, pasta, rice or other dishes seasoned with a tomato, garlic, onion or leek sauce simmered with olive oil (POINTS)	Nutritional habits
74	milk	Milk or dairy products (yogurts, cheese) consumed per day	Nutritional habits
75	n_milk	Milk or dairy products (yogurts, cheese) consumed per day (POINTS)	Nutritional habits
76	milk_light	The person takes skimmed dairy products	Nutritional habits
77	n_milk_light	The person takes skimmed dairy products (POINTS)	Nutritional habits
78	IPAQ	IPAQ Points	Physical exercise
79	cal_IPAQ	IPAQ Calories	Physical exercise
80	exercise_H	Days of intense physical exercise	Physical exercise
81	exercise_H_mets	Days of intense physical exercise (Mets)	Physical exercise
82	exercise_H_min	Intense physical exercise in one day (minutes)	Physical exercise
83	exercise_H_tot	Not sure about the time of intense physical exercise in one day	Physical exercise
84	exercise_L	Days of moderate physical exercise	Physical exercise
85	exercise_L_mets	Days of moderate physical exercise (Mets)	Physical exercise
86	exercise_L_min	Moderate physical exercise in one day (minutes)	Physical exercise
87	exercise_L_tot	Not sure about the time of moderate physical exercise in one day	Physical exercise
88	exercise_walk	Days of sedentary physical exercise	Physical exercise
89	exercise_walk_mets	Days of sedentary physical exercise (Mets)	Physical exercise
90	exercise_walk_min	Sedentary physical exercise in one day (minutes)	Physical exercise
91	exercise_walk_tot	Not sure about the time of sedentary physical exercise in one day	Physical exercise
92	exercise_sit_min	Time spent sitting during a day (minutes)	Physical exercise
93	exercise_sit	The person is not sure how much time was spent sitting during a day (minutes)	Physical exercise

Appendix A.3. Models Parameters

Table A3. Details about models parameters.

MODEL	Objective	ccp_Alpha	Class_Weight	Criterion	Learning_Rate	Loss	Max_Depth	Max_Features	n_estimators	Splitter	Bootstrap	Algorithm	Base_Estimator	Max_iter	Solver	Tol	Penalty	Prios	Var_Smoothing	Binarize	Fit_Prior
XGB	binary:logistic	-	-	-	None	-	None	-	100	-	-	-	-	-	-	-	-	-	-	-	-
GB	-	0.0	-	friedman_mse	0.1	deviance	3	None	100	-	-	-	-	-	-	-	-	-	-	-	-
DT	-	0.0	None	gini	-	-	6	None	-	best	-	-	-	-	-	-	-	-	-	-	-
RFC	-	0.0	None	gini	-	-	None	auto	100	-	True	-	-	-	-	-	-	-	-	-	-
ADB	-	0.0	None	gini	1.0	-	None	None	500	best	-	SAMME	DecisionTreeClassifier	-	-	-	-	-	-	-	-
LR	-	-	balanced	-	-	-	-	-	-	-	-	-	-	100	lbfgs	0.0001	l2	-	-	-	-
ET	-	0.0	None	gini	-	-	None	auto	250	-	False	-	-	-	-	-	-	-	-	-	-
GNB	-	-	-	-	-	-	-	-	-	-	-	-	-	-	-	-	-	None	1×10^{-9}	-	-
BNB	-	1.0	-	-	-	-	-	-	-	-	-	-	-	-	-	-	-	-	-	0.0	True
BG	-	-	-	-	-	-	-	auto	250	-	True	-	-	-	-	-	-	-	-	-	-

References

1. Keys, A.; Fidanza, F.; Karvonen, M.J.; Kimura, N.; Taylor, H.L. Indices of relative weight and obesity. *J. Chronic Dis.* **1972**, *25*, 329–343. [CrossRef]
2. Spanish Ministry of Health (Ministerio de Sanidad, Consumo y Bienestar Social). Encuesta Nacional de Salud. España 2017. Available online: https://www.mscbs.gob.es/estadEstudios/estadisticas/encuestaNacional/encuestaNac2017/ENSE2017_notatecnica.pdf (accessed on 15 January 2021).
3. World Health Organization. *Obesity: Preventing and Managing the Global Epidemic*; World Health Organization: Geneva, Switzerland, 2000; 252p.
4. Khatib, W.; Fleming, P.J. The Stud GA: A mini revolution? In Proceedings of the Parallel Problem Solving from Nature—PPSN V, Amsterdam, The Netherlands, 27–30 September 1998; Eiben, A.E., Bäck, T., Schoenauer, M., Schwefel, H.P., Eds.; Springer: Berlin/Heidelberg, Germany, 1998; pp. 683–691.
5. El Naqa, I.; Murphy, M.J. What is machine learning? In *Machine Learning in Radiation Oncology*; Springer: Berlin/Heidelberg, Germany, 2015; pp. 3–11.
6. De Prado, M.L. *Advances in Financial Machine Learning*; John Wiley & Sons: Hoboken, NJ, USA, 2018.
7. Bishop, C.M.; Nasrabadi, N.M. *Pattern Recognition and Machine Learning*; Springer: Berlin/Heidelberg, Germany, 2006; Volume 4.
8. Braga-Neto, U. *Fundamentals of Pattern Recognition and Machine Learning*; Springer: Berlin/Heidelberg, Germany, 2020.
9. Kononenko, I. Machine learning for medical diagnosis: History, state of the art and perspective. *Artif. Intell. Med.* **2001**, *23*, 89–109. [CrossRef]
10. Ahsan, M.M.; Luna, S.A.; Siddique, Z. Machine-Learning-Based Disease Diagnosis: A Comprehensive Review. *Healthcare* **2022**, *10*, 541. [CrossRef] [PubMed]
11. Pirgazi, J.; Alimoradi, M.; Abharian, T.E.; Olyaee, M.H. An Efficient hybrid filter-wrapper metaheuristic-based gene selection method for high dimensional datasets. *Sci. Rep.* **2019**, *9*, 18580. [CrossRef] [PubMed]
12. Chandrashekar, G.; Sahin, F. A survey on feature-selection methods. *Comput. Electr. Eng.* **2014**, *40*, 16–28. [CrossRef]
13. Guyon, I.; Elisseeff, A. An introduction to variable and feature selection. *J. Mach. Learn. Res.* **2003**, *3*, 1157–1182.
14. Reunanen, J. Overfitting in making comparisons between variable selection methods. *J. Mach. Learn. Res.* **2003**, *3*, 1371–1382.
15. Pudil, P.; Novovičová, J.; Kittler, J. Floating search methods in feature selection. *Pattern Recognit. Lett.* **1994**, *15*, 1119–1125. [CrossRef]
16. Misra, P.; Yadav, A.S. Improving the classification accuracy using recursive feature elimination with cross-validation. *Int. J. Emerg. Technol.* **2020**, *11*, 659–665.
17. Kumar, G.R.; Ramachandra, G.; Nagamani, K. An efficient feature selection system to integrating SVM with genetic algorithm for large medical datasets. *Int. J.* **2014**, *4*, 272–277.
18. Reddon, H.; Gerstein, H.C.; Engert, J.C.; Mohan, V.; Bosch, J.; Desai, D.; Bailey, S.D.; Diaz, R.; Yusuf, S.; Anand, S.S.; et al. Physical activity and genetic predisposition to obesity in a multiethnic longitudinal study. *Sci. Rep.* **2016**, *6*, 18672. [CrossRef] [PubMed]
19. Chatterjee, A.; Gerdes, M.W.; Martinez, S.G. Identification of Risk Factors Associated with Obesity and Overweight—A Machine Learning Overview. *Sensors* **2020**, *20*, 2734. [CrossRef] [PubMed]
20. Muhamad Adnan, M.H.B.; Husain, W.; Abdul Rashid, N. A hybrid approach using Naïve Bayes and Genetic Algorithm for childhood obesity prediction. In Proceedings of the 2012 International Conference on Computer Information Science (ICCIS), Chongqing, China, 17–19 August 2012; Volume 1, pp. 281–285. [CrossRef]
21. Mirjalili, S. Genetic algorithm. In *Evolutionary Algorithms and Neural Networks*; Springer: Berlin/Heidelberg, Germany, 2019; pp. 43–55.
22. Affenzeller, M.; Winkler, S.; Wagner, S.; Beham, A. *Genetic Algorithms and Genetic Programming: Modern Concepts and Practical Applications*; Chapman and Hall/CRC Publishers: London, UK, 2009.
23. Kohavi, R. A Study of Cross-Validation and Bootstrap for Accuracy Estimation and Model Selection. In Proceedings of the 14th International Joint Conference on Artificial Intelligence, IJCAI'95, Montreal, QC, Canada, 20–25 August 1995; Morgan Kaufmann Publishers Inc.: San Francisco, CA, USA, 1995; Volume 2; pp. 1137–1143.
24. Rao, R.; Fung, G. On the Dangers of Cross-Validation. An Experimental Evaluation. In Proceedings of the 2008 SIAM International Conference on Data Mining, Atlanta, GA, USA, 24–26 April 2008; pp. 588–596. [CrossRef]
25. Miller, B.L.; Goldberg, D.E. Genetic algorithms, tournament selection, and the effects of noise. *Complex Syst.* **1995**, *9*, 193–212.
26. Bäck, T. Selective Pressure in Evolutionary Algorithms: A Characterization of Selection Mechanisms. In Proceedings of the First IEEE Conference on Evolutionary Computation, Orlando, FL, USA, 27–29 June 1994; pp. 57–62.
27. Jolly, K. *Machine Learning with scikit-learn Quick Start Guide: Classification, Regression, and Clustering Techniques in Python*; Packt Publishing Ltd.: Birmingham, UK, 2018.
28. Friedman, J.H. Stochastic Gradient Boosting. *Comput. Stat. Data Anal.* **2002**, *38*, 367–378. [CrossRef]
29. Chen, T.; Guestrin, C. Xgboost: A scalable tree boosting system. In Proceedings of the 22nd Acm Sigkdd International Conference on Knowledge Discovery and Data Mining, San Francisco, CA, USA, 13–17 August 2016; pp. 785–794.
30. Myles, A.J.; Feudale, R.N.; Liu, Y.; Woody, N.A.; Brown, S.D. An introduction to Decision Tree modeling. *J. Chemom. J. Chemom. Soc.* **2004**, *18*, 275–285. [CrossRef]

31. Lundberg, S.M.; Lee, S.I. A Unified Approach to Interpreting Model Predictions. In *Advances in Neural Information Processing Systems 30*; Guyon, I., Luxburg, U.V., Bengio, S., Wallach, H., Fergus, R., Vishwanathan, S., Garnett, R., Eds.; Curran Associates, Inc.: Red Hook, NY, USA, 2017; pp. 4765–4774.
32. Derrac, J.; García, S.; Molina, D.; Herrera, F. A practical tutorial on the use of nonparametric statistical tests as a methodology for comparing evolutionary and swarm intelligence algorithms. *Swarm Evol. Comput.* **2011**, *1*, 3–18. [CrossRef]
33. Eisinga, R.; Heskes, T.; Pelzer, B.; Te Grotenhuis, M. Exact p-values for pairwise comparison of Friedman rank sums, with application to comparing classifiers. *BMC Bioinform.* **2017**, *18*, 68. [CrossRef] [PubMed]
34. Chen, H.; Jiang, W.; Li, C.; Li, R. A Heuristic Feature Selection Approach for Text Categorization by Using Chaos Optimization and Genetic Algorithm. *Math. Probl. Eng.* **2013**, *2013*, 1–6. [CrossRef]
35. Malhotra, R.; Khanna, M. Dynamic selection of fitness function for software change prediction using Particle Swarm Optimization. *Inf. Softw. Technol.* **2019**, *112*, 51–67. [CrossRef]

Article

Hybrid Discrete Particle Swarm Optimization Algorithm with Genetic Operators for Target Coverage Problem in Directional Wireless Sensor Networks

Yu-An Fan and Chiu-Kuo Liang *

Department of Computer Science and Information Engineering, Chung Hua University, Hsinchu 30012, Taiwan
* Correspondence: ckliang@chu.edu.tw

Abstract: For a sensing network comprising multiple directional sensors, maximizing the number of covered targets but minimizing sensor energy use is a challenging problem. Directional sensors that can rotate to modify their sensing directions can be used to increase coverage and decrease the number of activated sensors. Solving this target coverage problem requires creating an optimized schedule where (1) the number of covered targets is maximized and (2) the number of activated directional sensors is minimized. Herein, we used a discrete particle swarm optimization algorithm (DPSO) combined with genetic operators of the genetic algorithm (GA) to compute feasible and quasioptimal schedules for directional sensors and to determine the sensing orientations among the directional sensors. We simulated the hybrid DPSO with GA operators and compared its performance to a conventional greedy algorithm and two evolutionary algorithms, GA and DPSO. Our findings show that the hybrid scheme outperforms the greedy, GA, and DPSO algorithms up to 45%, 5%, and 9%, respectively, in terms of maximization of covered targets and minimization of active sensors under different perspectives. Finally, the simulation results revealed that the hybrid DPSO with GA produced schedules and orientations consistently superior to those produced when only DPSO was used, those produced when only GA was used, and those produced when the conventional greedy algorithm was used.

Keywords: target coverage problem; directional sensor networks; genetic algorithm; discrete particle swarm optimization

Citation: Fan, Y.-A.; Liang, C.-K. Hybrid Discrete Particle Swarm Optimization Algorithm with Genetic Operators for Target Coverage Problem in Directional Wireless Sensor Networks. *Appl. Sci.* **2022**, *12*, 8503. https://doi.org/10.3390/app12178503

Academic Editor: Vincent A. Cicirello

Received: 14 July 2022
Accepted: 22 August 2022
Published: 25 August 2022

Publisher's Note: MDPI stays neutral with regard to jurisdictional claims in published maps and institutional affiliations.

Copyright: © 2022 by the authors. Licensee MDPI, Basel, Switzerland. This article is an open access article distributed under the terms and conditions of the Creative Commons Attribution (CC BY) license (https://creativecommons.org/licenses/by/4.0/).

1. Introduction

A wireless sensor network (WSN) comprises many tiny sensors that can both acquire data and communicate. These sensors can monitor their environment and transmit the acquired data to other sensors or base stations to coordinate specific tasks [1]. WSNs are widely used in many fields, such as smart homes, military operations, environmental monitoring, animal conservation, medicine, and poaching prevention. Most studies have assumed that sensors are omnidirectional; however, this assumption may not hold for other types of sensors with directional sensing ranges, such as video, ultrasonic, or infrared [2]. Such sensors are called "directional sensors". The sensing range of an omnidirectional sensor node is a circular disk, whereas that of a directional sensor is a smaller, sector-like sensing area determined by its sensing angle. Research results for omnidirectional sensor networks cannot be directly applied to directional sensor networks (DSNs), which comprise numerous directional sensors. Many challenges specific to DSNs still require investigation.

Several methods of amplifying the sensing capabilities of sensor nodes have been proposed. One involves placing several of the same type of directional sensors on one sensor node, with each sensor facing a different direction. In [3], four pairs of ultrasonic sensors on a single node were used for omnidirectional detection of ultrasonic signals. Another technique is placing the sensor node on a mobile device to enable it to move and change direction. Third, sensor nodes can be equipped with devices that enable them to switch directions or to rotate. This third technique was adopted in the DSN investigated in this study.

A key challenge in DSN coverage is gathering data from a predetermined area. Coverage problems can be categorized as area coverage or target coverage problems. In area coverage, the purpose is to monitor a predetermined region; that of target coverage is to determine a set of sensors that can provide coverage for a set of targets [2]. Another challenge in DSNs is energy-consumption management; reducing energy usage can extend the network lifetime because sensor nodes are typically battery powered and, thus, can only continuously operate for a limited period of time [4].

In this study, we focused the target coverage problem for a DSN with a set of directional sensors that can rotate to any direction. Specifically, the full coverage of a set of targets in a predetermined region is required, but the limited sensing range of directional sensors prevents the targets from being entirely covered by the deployed sensors. Thus, rotating the directional sensors may be necessary to cover as many targets as possible. To increase the network lifetime, a scheduling technique in which some sensor nodes were active and others were inactivated was used to reduce energy use while maximizing the target coverage. This problem has increased complexity because of its multiple objectives. For such complex problems, the exact solution can be obtained by exhausting a lot of computational effort which increases exponentially with the problem size. The solutions obtained by traditional heuristic algorithms are only near to the optimal; thus, these algorithms should only be used to solve smaller or simple problems. As the problem size increases, the metaheuristic algorithms can be used to improve the solutions because of their high performance in a global search. Many metaheuristic algorithms, such as genetic algorithm (GA) [5–7], particle swarm optimization (PSO) [8], differential evolution (DE) [9], simulated annealing (SA) [10,11], and ant-colony optimization (ACO) [12,13], have been introduced for solving complex optimization problems that cannot be solved using traditional deterministic algorithms in the last two decades and some of them may lead to real improvement in optimization algorithms [14,15]. However, they will easily fall into the local optimal and the cost of computation iteration will increase explosively when the problem scale increases. Therefore, instead of proposing a new metaheuristic algorithm, this paper focuses on presenting a hybrid method for optimization problems. In our previous work [16], we focused on using a genetic algorithm to achieve optimal, or near-optimal, solutions to the target coverage problem. However, the accuracy of the solutions obtained by GA is affected by the selection and mutation operators and the convergence rate is slow. This paper first presents a discrete particle swarm optimization (DPSO) algorithm for the target coverage problem. Particle swarm optimization has gained increasing popularity in solving complex optimization problems due to its simplicity and high convergence speed. This paper also presents a hybrid metaheuristic algorithm based on the DPSO method combined with GA operators, namely DPSO_GA, to solve the target coverage problem. The use of GA operators aims to achieve the goal of intelligent exploration-exploitation. The hybridization of PSO and GA involves using two approaches sequentially or in parallel or using GA operators within the PSO framework [17–23]. In this study, the hybridization consists of a two-phase mechanism where the evolution process is accelerated by using DPSO and diversity is maintained by using GA. Our presented algorithm can achieve more precise solutions than conventional heuristic algorithms can. The main contributions of this paper are as follows:

1. We defined and proved the complexity of the target coverage problem in directional sensor networks with rotatable sensors, and we also mathematically formulated the problem;
2. We proposed a greedy algorithm according to the targets' maximally being covered by a number of directional sensors—the proposed greedy solutions are used as base for comparison;
3. We firstly proposed a metaheuristic algorithm, namely DPSO, for the target coverage problem to determine the schedule of the covered sector for each directional sensor

4. We secondly proposed a hybrid metaheuristic algorithm based on the DPSO method and combined with GA operators, namely DPSO_GA, to solve the target coverage problem;
5. We performed experiments to evaluate the performance of the proposed algorithms, described the results, and compared them with the results of a greedy algorithm;
6. We discussed the quality of the solutions generated by proposed DPSO and DPSO_GA algorithms along with the solutions generated by previous proposed GA, highlighting the most suitable algorithm for the target coverage problem in various environments.

The remainder of this paper is organized as follows. In Section 2, we present a review of the related work of target coverage problem. In Section 3, we define the target coverage problem in the DSN. Section 4 presents the details of our proposed algorithms. Section 5 presents the experimental results. In Section 6, we present a discussion of the experimental results. Finally, we conclude the paper and propose some future research topics in Section 7.

2. Related Works

The coverage problem is regarded as a key task in both WSNs or DSNs that are intended to collect information from an area of interests or to monitor targets in a predetermined region. Numerous studies have presented solutions for the coverage problem in omnidirectional WSNs [24–28]. However, the algorithms proposed in these WSNs cannot be directly applied to DSNs because of the limited angle of view of directional sensors. Therefore, new solutions specific to DSNs are necessary. The remainder of this section reviews studies conducted on the target coverage problem in DSNs.

Ma and Liu [29] presented a model of DSNs in which the orientations of the sensor nodes are static. They analyzed the probability of full coverage (i.e., each target is covered by at least one sensor). Ai and Abouzeid [30] presented a model of a sensor network in which the orientations of the sensors' nodes are adjustable. They defined the maximum-coverage-with-minimum-sensor (MCMS) problem, in which the coverage rate is maximized while minimizing the number of active sensors. They also demonstrated that the MCMS problem is NP-complete and presented two greedy algorithms, namely, the centralized greedy algorithm (CGA) and distributed greedy algorithm (DGA) for MCMS problems. Chen et al. [31] developed a weighted centralized greedy algorithm by modifying the CGA with adjustable weight functions to obtain a higher coverage rate. Cai et al. [4] addressed the multiple directional cover set (MDCS) problem of organizing sensor directions into a group of nondisjoint sets to prolong the network lifetime. Only one cover set is activated at a time, in which all targets are covered. They also demonstrated that the MDCS problem is NP-complete and proposed several algorithms for obtaining solutions to this problem. Han et al. [32] addressed the maximum set cover for DSN (MSCD) problems, which are NP-complete, and proposed an algorithm to provide energy-efficient cover sets that could cover all targets. The network lifetime was maximized by activating each cover set for various durations on a schedule. Gil and Han [33] presented a greedy target coverage scheduling algorithm for the MSCD problem and a GA that could identify optimal cover sets that could maximize the network lifetime while monitoring all targets by using the evolutionary global search technique. Li et al. [34] used a bounded service delay constraint to extend the network lifetimes on the target Q-coverage problem in DSNs. They proposed an algorithm to identify a collection of coverage sets that the bounded service delay constraint and the coverage quality requirement are satisfied; the target in each coverage set was not required to be served continuously but could be served with a tolerable service delay. Mohamadi et al. [35] reduced sensor energy consumption by partitioning the DSN into several cover sets, each of which could cover all targets, and activating these covers successively. They presented an irregular cellular learning automata-based distributed algorithm to identify a near-optimal solution for selecting an appropriate working direction for each sensor. Zannat et al. [36] addressed the target coverage problem in a visual sensor network

(VSN) that comprised a number of self-configurable visual sensors with adjustable spherical sectors with limited angles. Razali et al. [37] investigated priority-based target coverage with adjustable sensing ranges. They proposed two scheduling algorithms—greedy-based and learning-automata-based algorithms—to organize the sensors into a few cover sets that were successively activated to maximize network lifetime. The simulation revealed that the learning automata-based scheduling algorithm was superior to the greedy-based algorithm in terms of extending network lifetime. Zishan et al. [38] addressed heterogeneous coverage in VSNs, in which the coverage requirements of targets vary. Their main goal was to maximize the coverage of all targets (on the basis of their coverage requirements) by activating a minimal number of sensors. They solved this heterogeneous coverage problem by modifying the formulation of an existing integer linear programming method for the single and k-coverage MCMS problems. They also presented a sensor-oriented greedy algorithm to obtain an approximate solution of the formulated problem. Bakht et al. [39] successfully addressed the k-coverage problem by determining the orientation of a minimum number of directional sensors in which each target is monitored at least k times, with a learning automata-based algorithm.

The sensor nodes used in these studies comprise a fixed number of sensors with fixed orientations. However, many sensor nodes can rotate. Liang and Chen [40] addressed the maximum coverage with rotatable angles (MCRA) problem; they maximized the coverage rate of targets and minimized the total rotation (in degrees) of the sensors. They took advantage of the ability to adjust the working direction and presented two centralized greedy algorithms for the MCRA problem. In both greedy methods, the weights of sectors were used to select the appropriate working direction to which the sensors should rotate to cover more targets. Wu and Lu [41] redefined the MCMS problem for a DSN comprising rotatable sensors. They proposed a greedy algorithm that used a minimum number of sectors to cover targets within the sensing range of the sensors. Lo and Liang [16] also addressed the MCMS problem in a DSN with rotatable sensors. They proposed a GA for scheduling the orientation of active sensors to obtain a higher coverage rate of targets compared with that of other greedy algorithms.

We considered the target coverage problem with rotatable sensors. The goal aims to determine an appropriate orientation schedule in which each active sensor rotates to the scheduled orientation such that (1) the total number of covered targets is maximized and (2) the total number of active sensors is minimized. We focused on presenting different metaheuristic algorithms, including GA, DPSO, and the hybridization of DPSO and GA, for the problem due to its complexity. The hybridization of PSO and GA approaches has been applied to many applications in the last two decades [17–24]. Robinson et al. [17] proposed a hybrid PSO and GA algorithm by taking the population of one algorithm when the algorithm fails to improve, and using it as the new population of the other algorithm. Shi et al. [18] proposed a hybrid algorithm, namely PSO–GA-series-hybrid evolutionary algorithm (PGSHEA), which is to integrate PSO and GA methods in series. The PSO algorithm is terminated after a certain number of iterations and the best particles are selected and encoded into chromosomes to constitute the population for GA algorithm. The GA algorithm is terminated after a specific number of iterations and the best solutions of GA are transmitted back to the PSO populations. Yang et al. [19] proposed a PSO–GA-based hybrid evolutionary algorithm (HEA) for solving unconstrained and constrained optimization problems. Their proposed evolution strategy is divided into two stages in which the evolution process is accelerated by using PSO and diversity is maintained by using GA. Valdez et al. [20] proposed a fuzzy approach in the PSO-GA hybridization. They used several fuzzy rules to determine whether to consider GA or PSO particles and changes their parameters or to take a decision. Ghamisi and Benediktsson [21] proposed a feature selection method by hybridizing GA and PSO. The proposed method was confirmed to be able to automatically select the most informative features within an acceptable processing time without requiring the users to set the number of desired features beforehand. Moussa and Azar [22] introduced a combination of PSO and GA approach to classify software modules as fault-prone or not using object-oriented metrics. Nik et al. [23] presented various combinations of the hybrid GA and PSO algorithms to find an optimal arrangement of surveyed pavement inspection units (SIUs) in massive networks.

In a preliminary work [16], we focused on using a GA to achieve optimal, or near-optimal, solutions to the target coverage problem with a small problem size. In this study, we extend the previous work by formulating the target coverage problem with rotatable sensors mathematically; proposing a greedy algorithm and two swarm algorithms (a DPSO and a hybrid DPSO with GA operators, namely DPSO_GA); and conducting experiments to verify the performance of our proposed algorithms compared with that of the greedy algorithm and previous GA.

3. Preliminaries and Problem Statement

3.1. Network Model and Problem Definition

We studied the target coverage problem with the following model. All of the networking nodes are located in an obstacle-free predetermined region, including sensors and targets. $T = \{t_1, t_2,..., t_M\}$ is a set of M targets distributed randomly; the position of each is known. $S = \{s_1, s_2,..., s_N\}$ is a set of N homogeneous directional sensors with the same field of view θ and the same sensing radius r; these sensors should be scheduled to cover all targets. For a directional sensor s with field of view θ and sensing radius r located at location (x, y) in the region (Figure 1), the sensing area for sensing orientation α is a sector of a circle with center (x, y) and radius r; the sector is bounded by two radii between $\alpha - \theta/2$ and $\alpha + \theta/2$. This sector is represented by D_α. Target t is covered by sensor s if its location is within D_α. The directional sensors can rotate to any direction to monitor a specific target.

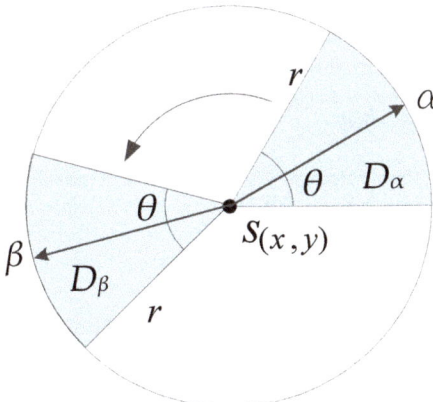

Figure 1. Directional sensing for a rotatable sensor.

Because of the limited field of view of each directional sensor, a target may not be covered by a sensor, even if it is located within the sensor's sensing radius. In this case, the directional sensor can rotate to cover the targets (Figure 2). In Figure 2a, three targets, namely, t_1, t_2, and t_3, are located within the sensing range of deployed sensor s with original orientation α. Targets t_1 and t_2 are covered by the sensor; however, target t_3 can also be covered if sensor s rotates its sensing orientation clockwise to β.

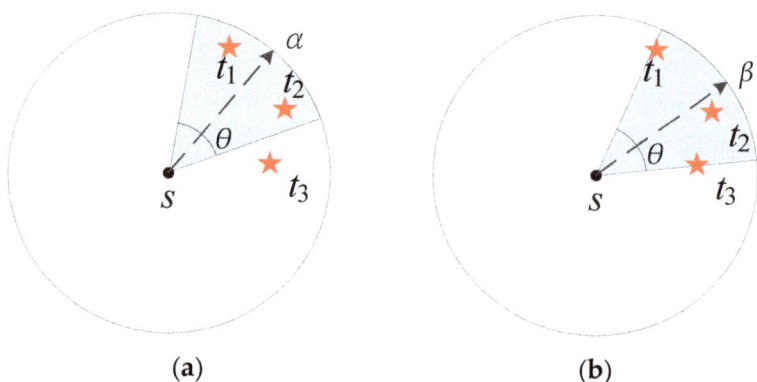

Figure 2. Rotation of a directional sensor s with three targets (indicated as red stars) located inside its sensing range. (**a**) Target t_3 is not covered by sensor s with sensing orientation α. (**b**) Target t_3 is covered by sensor s with sensing orientation β.

Definition 1. *For sector D_α, $|T(D_\alpha)|$ denotes the number of targets in this sector, where $T(D_\alpha)$ is the set of targets in T covered by sector D_α. Then, D_α is called a cover sector of T if $|T(D_\alpha)| > 0$.*

Definition 2. *Let D_α and D_β be two different cover sectors of T for directional sensor s. Then, D_α is equivalent to D_β and vice versa if $T(D_\alpha) = T(D_\beta)$. Furthermore, D_α is dominated by D_β or D_β dominates D_α if $T(D_\alpha) \subset T(D_\beta)$.*

Definition 3. *Let D_α be a cover sector of T for a directional sensor s. Then, D_α is called a maximal cover sector if D_α is not dominated by any other cover sector in sensor s.*

Proposition 1. *Let k be the number of targets within the sensing radius r of sensor s. Then, at most k different (or nonequivalent) maximal cover sectors exist for sensor s.*

For the target coverage problem, after the initial deployment of sensors, not all targets are covered under the random deployment fashion. Our goal is to rotate the initial orientations of the sensor to maximize the number of covered targets while minimizing the number of activated sensors. Therefore, the target coverage problem with rotatable sensors is defined as follows:

Definition 4. *Maximum Coverage with Minimum Rotatable Sensors (MCMRS) Problem: Assume a set of targets $T = \{t_1, t_2, \ldots, t_M\}$ to be covered, a set of homogeneous rotatable directional sensors $S = \{s_1, s_2, \ldots, s_N\}$ with p_i possible orientations each, and a set of cover sectors $D = \{D_{ij}, 1 \leq i \leq n, 1 \leq j \leq p_i\}$, where D_{ij} is the jth maximal cover sector for sensor s_i, p_i denotes the maximum number of cover sectors for sensor s_i, and $T(D_{ij})$ is a subset of T. Find a subset Z of D with the constraint that at most one D_{ij} can be selected for each i to maximize the cardinality of the union of chosen $\cup_{(i,j)} T(D_{ij})$ while minimizing the cardinality of $Z = \{D_{ij}, (i,j) \text{ chosen}\}$.*

Let x_{ij}^k and y_{ij} be binary variables defined as follows:

$$x_{ij}^k = \begin{cases} 1 & \text{if target } t_k \text{ is covered by } D_{ij}, \\ 0 & \text{otherwise.} \end{cases}$$

$$y_{ij} = \begin{cases} 1 & \text{if } D_{ij} \text{ is active,} \\ 0 & \text{otherwise.} \end{cases}$$

$$\varphi_k = \begin{cases} 1 & \text{if target } t_k \text{ is covered by any directional sensor,} \\ 0 & \text{otherwise.} \end{cases}$$

$$\delta_k^i = \begin{cases} 1 & \text{if target } t_k \text{ is located within the sensing radius of sensor } i, \\ 0 & \text{otherwise.} \end{cases}$$

The MCMRS problem can be formulated mathematically as follows:

$$\max \sum_{k=1}^{m} \varphi_k, \tag{1}$$

$$\min \sum_{i=1}^{n} \sum_{j=1}^{p_i} y_{ij} \tag{2}$$

Subject to:

$$\frac{\sum_{i=1}^{n} \sum_{j=1}^{p_i} x_{ij}^k \cdot y_{ij}}{n} \le \varphi_k \le \sum_{i=1}^{n} \sum_{j=1}^{p_i} x_{ij}^k \cdot y_{ij} \text{ for every target } t_k k \in T, \tag{3}$$

$$\sum_{j=1}^{p_i} y_{ij} \le 1 \text{ for every sensor } s_i i \in S, \tag{4}$$

$$\frac{\sum_{k=1}^{m} \delta_k^i}{m} \le p_i \le \sum_{k=1}^{m} \delta_k^i \text{ for every sensor } s_i i \in S, \tag{5}$$

The objective in (1) represents the number of covered targets to be maximized while the objective in (2) represents the number of activated sensors to be minimized. Equation (3) reveals that the value of φ_k can only be 0 or 1, indicating that each target must be covered or not covered. If target t_k is covered by any directional sensor, then $\varphi_k = 1$ to conform with the left inequality; otherwise, $\varphi_k = 0$ to conform with the right inequality. For each sensor, at most one cover sector that can be activated exists, as indicated by (4), and the number of maximal cover sectors is bounded by the number of targets located within that sensor, as indicated in (5).

3.2. Hardness of the Problem

We demonstrate that the MCMRS problem is NP-complete.

Theorem 1. *The MCMRS problem is NP-complete.*

Proof. Consider the special case of the MCMRS problem where sensors are omnidirectional (i.e., $\theta = 360°$). In this special case, maximizing the number of covered targets for a given k which represents the upper bound of the number of activated sensors can be viewed as a known NP-complete problem, the classic Maximum Coverage Problem [42]. Thus, the result is as follows. □

3.3. Greedy Algorithm

In this subsection, we propose a heuristic algorithm in which, at each step, the available maximal cover sector of an unselected sensor that covers the uncovered targets with greatest importance is identified. In our algorithm, we assign each target a weight indicating its importance. We first consider the maximally covered number (MCN) [31] for each target t_k, where $MCN(t_k) = |\{D_{ij} | t_k \in T(D_{ij}), 1 \le i \le n, 1 \le j \le p_i\}|$ indicates the number of maximal sectors that cover t_k. The weight of target t_k is $1/MCN(t_k)$. Therefore, targets with a smaller MCN value have greater weight (i.e., coverage of targets covered by fewer sectors should be considered more critical). We determine the weight of each maximal cover sector by summing the weights of all targets covered by the maximal cover sector. We then begin scheduling by selecting the sector with the highest weight among all unselected sectors. When a sector is selected, the corresponding sensor is scheduled as active with the orientation of the corresponding selected sector, and all remaining cover sets are eliminated from the unselected sectors for that sensor. The weights of all targets covered by the selected sector are set to 0, and the weights of the remaining unselected sectors that cover those targets are updated. This scheduling process continues until the set of unselected sectors is empty. The pseudocode of our proposed greedy algorithm is shown in Algorithm 1.

Algorithm 1: Greedy Algorithm (GREEDY)

1:     Find the set of cover sectors $D = \{D_{ij}, 1 \leq i \leq N, 1 \leq j \leq p_i\}$, where D_{ij} is the jth maximal cover sector of sensor i. Find the weight of each target $t_k, 1 \leq k \leq M$, and the weight of each sector in D.
2:     $Z = \{\}$
3:     **while** (D is not empty) **do**
4:         Choose the cover sector in D with maximum weight, denoted D_{nq}
5:         $Z = Z \cup D_{nq}$
6:         Delete D_{nj} from D where $1 \leq j \leq p_n$
7:         Set the weight all targets covered by D_{nq} to 0
8:         Update the weight of all sectors in D that contain the targets covered by D_{nq}
9:     **end while**
10:    **return** Z

The time complexity of the greedy target coverage algorithm, denoted as GREEDY, is as follows. Determining the MCN values of all targets requires mn steps. The running time for identifying all cover sectors is thus bounded by $O(mn)$. In the worst case, D has $O(mn)$ cover sectors. Therefore, we can identify the cover sector in D with maximum weight in $O(mn\log(mn))$. Because at most n loops are required to select the cover sectors and each update takes at most $O(m\log(mn))$, the time complexity of GREEDY is $O(mn\log(mn))$ in the worst case.

4. Proposed Metaheuristic Schemes

In this section, we propose two metaheuristic algorithms for the MCMRS problem: DPSO and DPSO_GA, to obtain better solutions than that obtained by the greedy algorithm.

4.1. Particle Swarm Optimization (PSO)

Swarm intelligence is an innovative distributed intelligence paradigm for solving optimization problems. The paradigm was inspired by biological phenomena, such as flocking, swarming, and herding in vertebrates. Particle swarm optimization (PSO) incorporates swarming behaviors observed in schools of fish, flocks of birds, swarms of bees, or even human social behavior [43,44]. PSO is a population-based optimization technique that can be easily applied to solve various optimization problems. The major strength of a PSO algorithm is its fast convergence, which has advantages over other global optimization algorithms, such as GAs and simulated annealing.

In PSO, a swarm of particles explores the solution space of an optimization problem to identify an optimal or quasioptimal solution. Each particle represents a candidate solution and is identified with specific position in the D-dimensional search space. The position and velocity of ith particle is represented as $X_i = (x_{i1}, x_{i2}, \ldots, x_{iD})$ and $V_i = (v_{i1}, v_{i2}, \ldots, v_{iD})$, respectively. The fitness function is evaluated for each particle in the swarm, and the result is compared with the best previous fitness obtained by that particle and to the best fitness ever achieved by any particles of the swarm. Then, each particle updates its velocity and position in accordance with the following equations:

$$V_i^{t+1} = \omega * V_i^t + c_1 * r_1 * \left(P_i^t - X_i^t\right) + c_2 * r_2 * \left(G^t - X_i^t\right) \quad (6)$$

$$X_i^{t+1} = X_i^t + V_i^{t+1} \quad (7)$$

where $1 \leq i \leq P$; P is the size of the swarm; V_i^t and X_i^t are the velocity and position of the ith particle at iteration t, respectively; P_i^t is the current best solution for the ith particle at iteration t; G^t is the global best solution at iteration t; ω is the inertia weight; and c_1

and c_2 are acceleration coefficients. Here, r_1 and r_2 denote uniform random numbers between $(0, 1)$.

4.1.1. Particle Encoding

In target coverage problem, the coverage of targets depends on determining the appropriate sector of each directional sensor. Therefore, each particle in a population can be encoded as the selection of a cover sector to represent a candidate solution to the problem. The length of each particle is the same as the number of directional sensors. Thus, a possible schedule of directional sensors or particle can be expressed as follows:

$$(d_1, d_2, \ldots d_n),$$

where $0 \leq d_i \leq p_i$ represents the selected cover sector of the ith directional sensor. In this representation, $d_i = 0$ indicates that sensor s_i is inactive.

4.1.2. Fitness

The proposed particle encoding representation was evaluated to determine an optimal schedule of directional sensors with a fitness function. To determine the optimal solution for the problem, the evolutionary process of our proposed PSO should achieve two objectives: (1) maximizing the target coverage rate and (2) minimizing the active sensor rate. The second condition is equivalent to maximizing the number of inactive sensors. These two objectives suggest the following. If two schedules of directional sensors or particles have the same total number of covered targets, a chromosome with fewer active sensors is preferable because of its reduced energy consumption. To achieve these two objectives, an appropriate fitness function is proposed which is used in the metasearch algorithm for searching for a schedule that maximizes the fitness function. In general, a schedule of directional sensors with a high fitness value would satisfy all of the objectives well. Therefore, the fitness function for the target coverage problem is defined as follows:

$$F = w \cdot R_T + (1 - w) \cdot (1 - R_S),$$

where R_T is the target coverage rate and R_S is the active sensor rate. The target coverage rate is the ratio of the total number of covered targets to M, and the active sensor rate is the ratio of the total number of active sensors to N. In addition, w is a predefined weight where $0 \leq w \leq 1$.

4.2. DPSO Algorithm

Because the MCMRS problem defined in Equations (1)–(5) is a discrete optimization problem, traditional PSO algorithm cannot be used and, thus, discrete particle swarm optimization algorithm will be developed for the MCMRS problem. Therefore, the evolutionary process of particle velocity and position represented by (6) and (7) in the original PSO algorithm, respectively, must be modified such that they can span the discrete search domain. For our proposed MCMRS problem, we developed a DPSO algorithm which is an adaptation of the method established in [45] for updating the particle positions. According to [45], the position of the ith particle at iteration t can be updated as follows:

$$X_i^{t+1} = c_2 \bigoplus F_3 \left(c_1 \bigoplus F_2 \left(\omega \bigoplus F_1(X_i^t), P_i^t \right), G^t \right) \tag{8}$$

where F_1, F_2, and F_3 are operations with the probabilities ω, c_1, and c_2, respectively.

In the DPSO algorithm, the particle velocities and positions are updated by three operations. The first operation is $\tau_i^{t+1} = \omega \oplus F_1(X_i^t)$, which represents the velocity operation for a particle and generates a temporary particle τ_i^{t+1}. F_1 denotes the mutation operator, which is applied with a probability of ω. Generate a random number $r \in (0, 1)$. A new particle $\tau_i^{t+1} = F_1(X_i^t)$ is generated by applying the mutation operator to the current particle if $r < \omega$. Otherwise, set the new particle to be the current particle, i.e., $\tau_i^{t+1} = X_i^t$. The second operation is $\delta_i^{t+1} = c_1 \oplus F_2\left(\tau_i^{t+1}, P_i^t\right)$, which is the cognitive part of the particle. F_2 represents the crossover operator, which is applied with a probability of c_1. The outcome is, thus, either $\delta_i^{t+1} = F_2\left(\tau_i^{t+1}, P_i^t\right)$ or $\delta_i^{t+1} = \tau_i^{t+1}$, where δ_i^{t+1} is a temporary particle. The third operation is $X_i^{t+1} = c_2 \oplus F_3\left(\delta_i^{t+1}, G^t\right)$, which is the social part of the particle. F_3 represents the crossover operator, which is applied with a probability of c_2. The outcome is either $X_i^{t+1} = F_3\left(\delta_i^{t+1}, G^t\right)$ or $X_i^{t+1} = \delta_i^{t+1}$.

Mutation operator F_1 for position X_i^t represents the replacement of the original cover sector for a sensor by a randomly selected cover sector. For example, with $X_i^t = (1, 3, 0, 2, 3, 1, 4, 0, 2)$ and mutations occurring in the second, third, and seventh positions, we might obtain $\tau_i^{t+1} = (1, 2, 1, 2, 3, 1, 3, 0, 2)$. Operator F_2 represents selecting τ_i^{t+1} and P_i^t to be the first and second parents for a crossover operation, respectively. Each of the cover sectors in τ_i^{t+1} and P_i^t are exchanged, from left to right, if a random number $r \in (0, 1) < c_1$. For example, with $\tau_i^{t+1} = (1, 3, 0, 2, 3, 1, 4, 0, 2)$ and $P_i^t = (1, 3, 1, 2, 4, 1, 3, 0, 1)$, we obtain $\delta_i^{t+1} = (1, 3, 1, 2, 4, 1, 3, 0, 1)$ if all generated uniform random numbers are less than c_1. The crossover operator F_3 represents selecting δ_i^{t+1} and G^t as the first and second parents, respectively, for a crossover operator in a similar manner for that of F_2; each of the different cover sectors in δ_i^{t+1} and G^t, from left to right, is exchanged if a random number $r \in (0, 1) < c_2$.

The pseudocode of the procedure of our proposed DPSO algorithm for the MCMRS problem is shown in Algorithm 2.

Algorithm 2: Discrete Particle Swarm Optimization (DPSO) Algorithm

1: Initialize the parameters: P, ω, c_1, and c_2. Initialize the *position* and *velocity* for each particle randomly. Set t_{max} to be the maximum generations, and set t to 1
2: Evaluate the *fitness* of each particle, and set *particle_best* to the particle itself
3: Set *global_best* to the particle with the highest *fitness*
4:     **while** $t < t_{max}$ **do**
5:         **for each** *particle* **do**
6: Update *particle.position* and *particle.velocity* by applying the operations F_1, F_2, and F_3 in (8) with their respective probabilities
7:            Compute new *fitness*
8:            **if** new *fitness* > *particle_best.fitness*
9:                *particle_best.fitness* ← new *fitness*
10:                *particle_best* ← *particle*
11:            **if** new *fitness* > *global_best.fitness*
12:                *global_best.fitness* ← new *fitness*
13:                *global_best* ← *particle*
14:     **end**
15:     $t \leftarrow t + 1$
16: **end**
17: **return** *global_best*

4.3. Hybrid DPSO Algorithm with GA Operators

The proposed hybrid DPSO algorithm with GA operators comprises two stages in each iteration of the evolutionary process. The first stage is the same process as in the DPSO algorithm; the new position and velocity of each particle are generated in accordance with the particle's best solution and the global swarm's best solution. In the second stage, the GA operators, namely, selection, crossover, and mutation operations, are applied to create new particles for the next iteration. We present the GA operators used in the hybrid algorithm in the following subsections.

4.3.1. Selection

The selection operation selects candidate particles, based on their fitness, from the population of the current generation. Roulette-wheel selection was employed in our proposed DPSO_GA scheme. Specifically, a slot on a biased roulette wheel was assigned to each particle with the slot size proportional to its fitness value. As a result, particles with higher fitness values were more likely to be chosen into the crossover operation queue.

4.3.2. Crossover

In the crossover operation, two individuals were selected and used to produce two new individuals by exchanging genes or chromosomes in accordance with a probability denoted the crossover rate. A uniform crossover mask was used to produce two children by exchanging two parent genes from the corresponding mask positions. An example of a crossover operation is presented in Figure 3a:

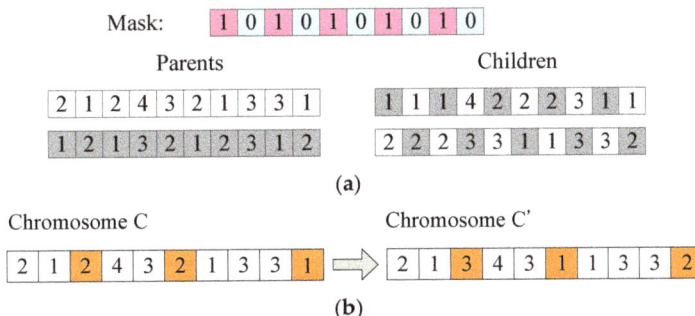

Figure 3. Genetic operators. (**a**) Crossover operation with mask. (**b**) Mutation operation (colored box numbers show the mutation of chromosome C to obtain chromosome C').

4.3.3. Mutation

For the MCMRS problem, the mutation operator is applied to each particle or chromosome to prevent premature convergence. It is used on each gene of a particle with a probability. A cover sector of the sensor is randomly selected to be the new value each chosen gene. An example of a mutation operation is presented in Figure 3b.

4.3.4. Pseudocode of the Hybrid DPSO_GA Algorithm

Algorithm 3 shows the pseudocode of the procedure of our hybrid DPSO_GA algorithm for the MCMRS problem.

Algorithm 3: Discrete PSO with GA operators (DPSO_GA) Algorithm

1: Initialize the parameters: P, ω, c_1, and c_2. Initilize the *position* and *velocity* for each particle randomly. Set t_{max} to be the maximum generations, and set t to 1
2: Evaluate the *fitness* of each particle, and set *particle_best* to the particle itself
3: Set *global_best* to the particle with the highest *fitness*
4:    **while** $t < t_{max}$ **do**
5:       **for each** *particle* **do**
6: Update *particle.position* and *particle.velocity* by applying the operations F_1, F_2, and F_3 in (8) with their respective probabilities
7:          Compute new *fitness*
8:          **if** new *fitness* > *particle_best.fitness*
9:             *particle_best.fitness* ← new *fitness*
10:            *particle_best* ← *particle*
11:          **if** new *fitness* > *global_best.fitness*
12:             *global_best.fitness* ← new *fitness*
13:             *global_best* ← *particle*
14:       **end**
15:       **while** *crossover_queue_length* < P **do**
16: Randomly select two particles and compare their *fitness*
17: Add *particle* with higher *fitness* into *crossover_queue*
18:       **end**
19:       **while** *crossover_queue* != ∅ **do**
20: Randomly select and remove two particles from *crossover_queue*
21: Crossover the selected particles with a certain probability
22: Insert these two particles into new population
23:       **end**
24:       **for each** *particle* in new population **do**
25:          Mutate *particle* with a certain probability
26:          Compute *fitness*
27:          *particle_best.fitness* ← *fitness*
28:          *particle_best* ← *particle*
29:       **end**
30: Descending sort of particles in new population according to their *fitness*
31: Select the particles with higher *fitness* as the swarm
32: Set *global_best* to the particle with the best *fitness*
33:       $t \leftarrow t + 1$
34:    **end**
35: **return** *global_best*

5. Performance Evaluation

We evaluated the results obtained with the DPSO and hybrid DPSO_GA algorithms in a set of experiments and compared the results with the results obtained with the greedy algorithm and GA. We performed all experiments with a program in C# on a Windows 10 computer. We investigated the effects of four network parameters on the number of covered targets and the number of activated sensors: the number of sensors N, the number of targets M, the field of view θ, and the sensing radius R. In the experiments, all targets and sensors were randomly scattered in a region of size 800 m × 800 m. Specifically, for each combination of network parameters, we randomly generated 40 instances of the network and reported the mean results. The simulation parameters are shown in Table 1.

Our simulations focused on determining the algorithm's performance for various network parameters in terms of the total number of covered targets, the total number of activated sensors, and the quality of the solutions. The quality of a solution was represented by its fitness value.

Table 1. Simulation parameters.

Parameter	Value	Meaning
N	50, 75, 100, 125, 150, 175, 200	Number of directional sensors
M	50, 100, 150, 200	Number of targets
R	60, 80, 100, 120	Sensing radius (m)
θ	30°, 60°, 90°, 120°	Field of view
w	0.5	Weight of fitness function
P	100	Population size
P_c	0.8	Probability of GA crossover operator
P_m	0.1	Probability of GA mutation operator
ω	0.1	Probability of DPSO mutation (F_1)
c_1	0.5	Probability of DPSO crossover (F_2)
c_2	0.5	Probability of DPSO crossover (F_3)
I_c	1000	Number of generations or iterations

Figure 4 shows how the number of sensors has an impact on the target coverage rate, active sensor rate, and fitness values under the GA, DPSO, and DPSO_GA algorithms, compared with the solutions obtained by the greedy algorithm. This experiment had 200 targets, and number of sensors N was 50, 75, 100, 125, 150, 175, or 200; the sensor and targets were randomly generated in the specified region. The sensors' coverage radius $R = 80$ m, the field of view $\theta = 60°$, and the weight of the fitness function $w = 0.5$. Figure 4a reveals that the greedy algorithm has at least a 17% higher target coverage rate than do the other three evolutionary algorithms. However, the active sensor rate of the greedy algorithm is also higher than others by at least 15% (Figure 4b). Therefore, in terms of the solution quality, the evolutionary algorithms achieve better performance than does the greedy algorithm by 0.01% to 40.08% (Figure 4c). Furthermore, among the proposed metaheuristic algorithms, the DPSO_GA algorithm outperforms both the GA and DPSO algorithms in terms of target coverage rate and fitness values; however, the DPSO_GA algorithm requires the greatest number of active sensors. The DPSO_GA algorithm outperforms the GA and DPSO algorithms by 0.4% to 2.9% in terms of the fitness value. Figure 4d plots the fitness values versus the number of generations for the GA, DPSO, and DPSO_GA schemes with $N = 100$, $M = 200$, $R = 80$, $\theta = 60°$, $w = 0.5$, and $Ic = 1000$. The DPSO_GA and DPSO approaches had higher fitness than GA did after 10 generations, and the DPSO_GA approach achieved solutions that were 0.42% and 0.82% better than those of DPSO and GA, respectively. Therefore, the proposed hybrid DPSO_GA algorithm can better identify high-fitness particles and transmit them to subsequent generations than the other algorithms. As a result, the hybrid scheme was more likely to achieve the global optimum solutions for the MCMRS problem in terms of the number of sensors.

Figure 5 shows how the number of targets has an impact on the target coverage rate, active sensor rate, and fitness values for the GA, DPSO, and DPSO_GA algorithms compared with the solutions obtained with the greedy algorithm. In the experiment, the number of targets varied from 50 to 200, and the number of sensors $N = 100$. All targets and sensors are randomly generated in the specified region. The sensor coverage radius $R = 80$ m, the field of view $\theta = 60°$, and the weight of fitness function $w = 0.5$. Figure 5a reveals that as the number of targets increases, the number of targets covered increases, but the target coverage rate decreases because increasing the number of targets increases both the number of uncovered and covered targets. The number of uncovered targets increases more quickly; thus, the total target coverage rate decreases. The greedy algorithm achieves 6% to 35% higher target coverage rates than do the other three evolutionary algorithms as the number of targets increases. However, Figure 5b reveals that the active sensor rate of the greedy algorithm is 4–41% higher than the other three algorithms. Therefore, the quality of the solutions of the evolutionary algorithms is higher than that of greedy algorithm for larger numbers of targets (Figure 5c). Moreover, among the proposed metaheuristic algorithms, the DPSO_GA algorithm outperforms the GA and DPSO algorithms in terms of the target coverage rate and the fitness values despite requiring the most active sensors.

The fitness of the DPSO_GA algorithm is 0.17–1.23% higher than that of the GA and DPSO algorithm. Figure 5d plots the fitness values versus the number of generations for the GA, DPSO, and DPSO_GA schemes with $N = 100$, $M = 150$, $R = 80$, $\theta = 60°$, $w = 0.5$, and $Ic = 1000$. The fitness values of the DPSO_GA and DPSO approaches were greater than that of the GA approach after the first three generations, that of the DPSO_GA approach was greater than the DPSO approach after 110 generations, and DPSO_GA obtained solutions that were 0.58% and 1.23% better than those of DPSO and GA, respectively, after 1000 generations. As a result, the hybrid scheme was more likely to achieve the global optimum solutions for the MCMRS problem in terms of the number of targets.

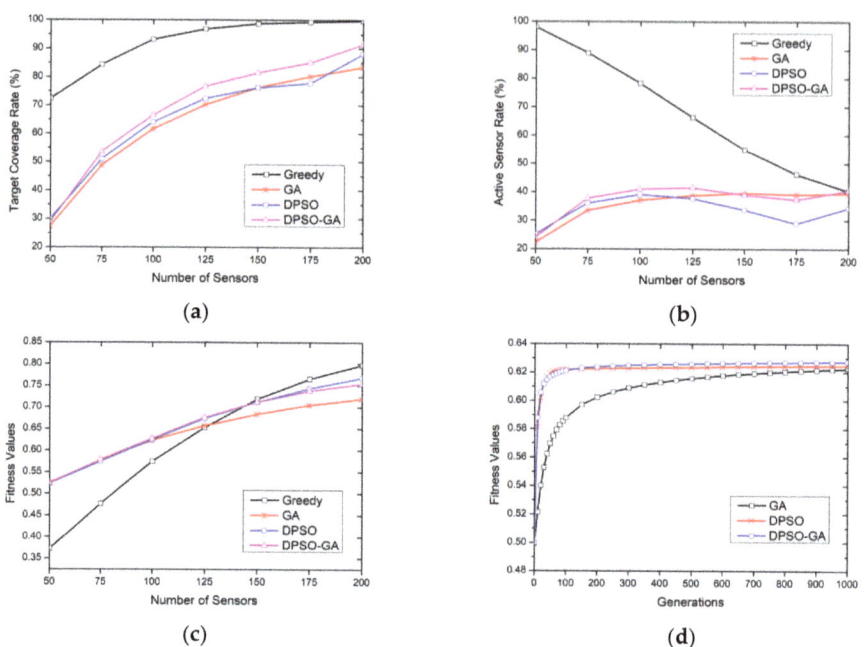

Figure 4. Effect of the number of sensors on the target coverage rate, active sensor rate, and fitness values for the greedy, GA, DPSO, and DPSO_GA algorithms. (**a**) Target coverage rate, (**b**) active sensor rate, (**c**) fitness values, and (**d**) fitness values versus the number of generations for GA, DPSO, and DPSO_GA.

Figure 6 shows how the field of view has an impact on the target coverage rate, active sensor rate, and fitness values in the GA, DPSO, and DPSO_GA algorithms compared with the solutions obtained by the greedy algorithm. In this experiment, $N = 200$ targets and $M = 100$ sensors were randomly generated in the specified region. The sensors' coverage radius $R = 80$ m and the weight of fitness function $w = 0.5$. The field of view varied from 30° to 120° in increments of 30°. Figure 6a,b reveals that, as in the other experiments, the greedy algorithm achieves a > 20% higher target coverage rate but requires a > 25% higher active sensor rate and thus is outperformed by the evolutionary algorithms by 2.53–45.61% (Figure 6c). The DPSO_GA algorithm again outperforms the GA and DPSO algorithms by 4.73% to 8.89% but requires more active sensors. Figure 6d plots the fitness values versus the number of generations for GA, DPSO, and DPSO_GA with $N = 100$, $M = 200$, $R = 80$, $\theta = 90°$, $w = 0.5$, and $Ic = 1000$. DPSO_GA and DPSO outperformed the GA approach after 12 generations, and DPSO_GA ultimately obtained solutions that were 3% and 5% better than those of DPSO and GA, respectively. As a result, the hybrid scheme was more likely to achieve the global optimum solutions for the MCMRS problem in terms of the field of view.

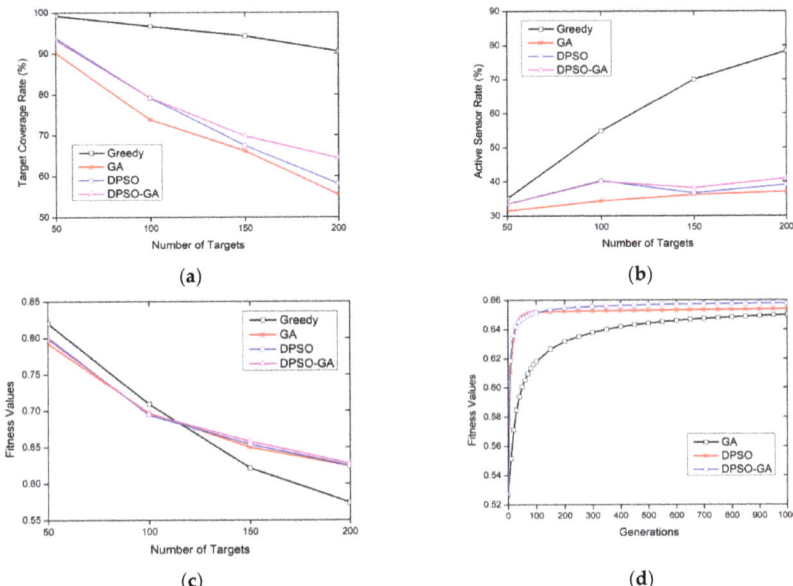

Figure 5. Effect of the number of targets on the target coverage rate, active sensor rate, and fitness values under the greedy, GA, DPSO, and DPSO_GA algorithms. (**a**) Target coverage rate, (**b**) active sensor rate, (**c**) fitness values, and (**d**) fitness values versus number of generations for the GA, DPSO, and DPSO_GA algorithms.

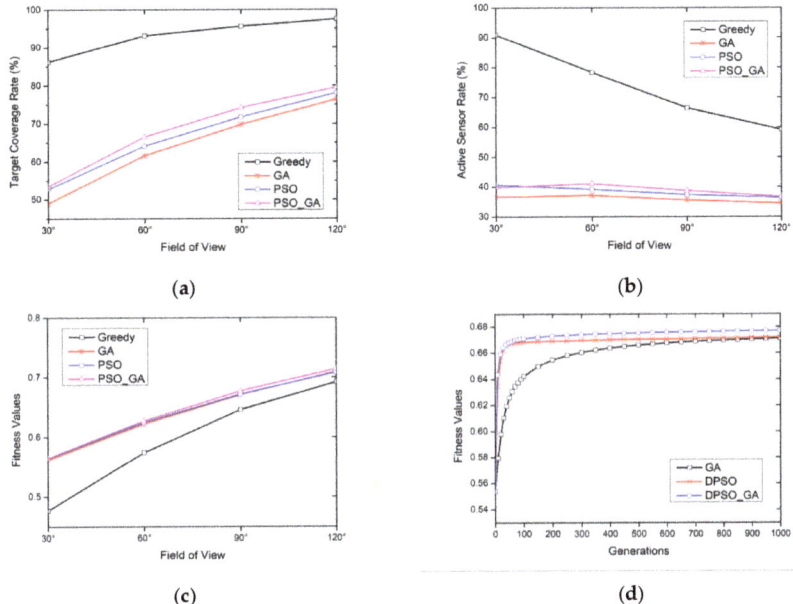

Figure 6. Effect of the field of view on the target coverage rate, active sensor rate, and fitness values in greedy, GA, DPSO, and DPSO_GA algorithms. (**a**) Target coverage rate, (**b**) active sensor rate, (**c**) fitness values, and (**d**) fitness values versus number of generations for the GA, DPSO, and DPSO_GA algorithms.

Figure 7 presents the effect of the sensing radius on the target coverage rate, active sensor rate, and fitness values for the GA, DPSO, and DPSO_GA algorithms compared with the solutions of the greedy algorithm. In this experiment, $N = 200$ targets and $M = 100$ sensors were randomly generated in the specified region. The field of view $\theta = 60°$ and the weight of fitness function $w = 0.5$. The sensing radius varied from 60 to 120 m in increments of 20 m. Figure 7a reveals that the number of covered targets increases as the sensing radius increases because the cover sector area increases. Again, Figure 7a,b reveals that the greedy algorithm achieved a 23–35% higher target coverage rate but required 19–51% more activated sensors than did the evolutionary algorithms and thus was outperformed by them by 1.2–15% (Figure 7c). Notably, the greedy algorithm outperforms the GA and DPSO algorithms if the sensing radius is 120 m because both the GA and DPSO algorithms become trapped in local optima. However, the proposed DPSO_GA operators' algorithm still outperforms the greedy algorithm. Figure 7d plots the fitness values versus the number of generations for GA, DPSO, and DPSO_GA with $N = 100$, $M = 200$, $R = 100$, $\theta = 60°$, $w = 0.5$, and $Ic = 1000$. DPSO_GA and DPSO had higher fitness than did the GA approach after five generations, and DPSO_GA, ultimately, achieved solutions 1.1% and 3.3% better than those of DPSO and GA, respectively. As a result, the hybrid scheme was more likely to achieve the global optimum solutions for the MCMRS problem in terms of the sensing radius.

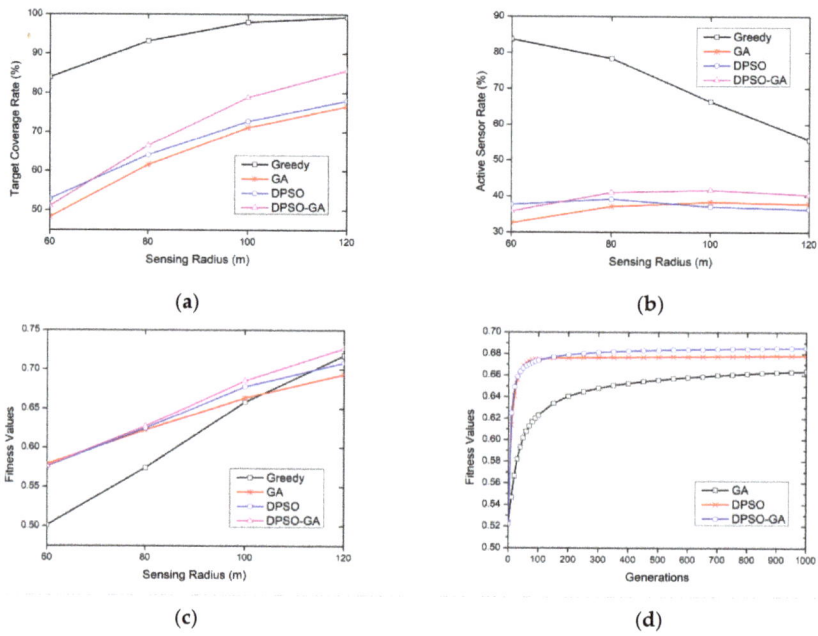

Figure 7. Effect of the sensing radius on the target coverage rate, active sensor rate, and fitness values under greedy, GA, DPSO, and DPSO_GA algorithms. (**a**) Target coverage rate, (**b**) active sensor rate, (**c**) fitness values, and (**d**) fitness values versus the number of generations for the GA, DPSO, and DPSO_GA algorithms.

6. Discussion

In this study, we presented a DPSO algorithm and a hybrid DPSO with GA operators' algorithm for solving the MCMRS problem in a DSN and compared their performance in terms of solution optimality and system efficiency. Directional sensors are increasingly used in traditional WSNs for target monitoring and tracking. Therefore, determining an effective schedule of active sensors to maximize target coverage rates is a crucial research topic. This paper provides effective algorithms to obtain high coverage and quality of solutions and contributes to the work of research topics related to the target coverage problem.

Although the proposed metaheuristic algorithms can effectively resolve the target coverage problem, this work has some limitations. (1) All targets were assumed to be static to reduce the complexity of the problem. (2) The coverage requirements of all targets were assumed to be identical; priority of target coverage was not considered. (3) All directional sensors were assumed to have identical parameters. (4) The proposed metaheuristic schemes have disadvantages, such as the premature convergence of the DPSO algorithm, and no guarantee of achieving the optimal solution.

Despite these limitations, the simulation results reveal that these algorithms outperform a conventional heuristic algorithm for solving complex optimization problems, such as the MCMRS problem in a DSN.

Our simulation results also revealed that the hybrid DPSO_GA algorithm outperformed both GA and DPSO algorithms in all of the evaluation metrics, especially for solving the largest and most complex problems. Thus, the hybrid algorithm effectively achieved the advantages of both the GA and DPSO algorithms; combining these algorithms resulted in a new metaheuristic algorithm with practical utility and high performance. Inspired by the satisfactory achievements obtained by the hybrid algorithm, it is worthy of discussing the applicability of the proposed mechanism to related nature-inspired and evolutionary algorithms to solve MCMRS problem. Finally, additional factors affecting the energy consumption of active sensors must be considered in practical applications, such as the energy required to rotate a sensor.

7. Concluding Remarks and Future Work

We investigated the MCMRS problem for a DSN. We required full coverage of a set of targets in a predetermined region. To conserve energy, we determined an arrangement of cover sectors such that the number of covered targets was maximized but the number of active sensors was minimized. Because the MCMRS problem is NP-complete, it will take a long computational time to determine the optimal schedule of active sensors. We employed two nondeterministic algorithms, the DPSO algorithm and the DPSO_GA algorithm, which is a hybrid DPSO scheme with GA operators, to manage this complexity and produce solutions in a relatively short computation time. Our proposed algorithms quickly determined high-quality sensor schedules and could be applied in situations with more numerous targets and sensors. The hybrid DPSO_GA algorithm produced solutions that were consistently superior to those produced by both the GA and the DPSO algorithms.

In the future, our proposed algorithms could be applied for targets with different coverage requirements (i.e., different coverage weights or priorities). It is also worth focusing on combining other metaheuristic methods to solve MCMRS problem. Moreover, energy is required to rotate the sensors in practical applications; this cost of reorientation could be investigated in future research.

Author Contributions: Conceptualization, C.-K.L.; software, Y.-A.F.; investigation, Y.-A.F.; methodology, C.-K.L.; validation, C.-K.L.; writing—original draft preparation, Y.-A.F. and C.-K.L.; and writing—review and editing, C.-K.L. All authors have read and agreed to the published version of the manuscript.

Funding: This research received no external funding.

Institutional Review Board Statement: Not applicable.

Informed Consent Statement: Not applicable.

Conflicts of Interest: The authors declare no conflict of interest.

References

1. Yick, J.; Mukherjee, B.; Ghosal, D. Wireless sensor network survey. *Comput. Netw.* **2008**, *52*, 2292–2330. [CrossRef]
2. Amac Guvensan, M.; Yavuz, G. On coverage issues in directional sensor networks: A survey. *Ad Hoc Netw.* **2011**, *9*, 1238–1255. [CrossRef]
3. Djugash, J.; Singh, S.; Kantor, G.; Zhang, W. Range-only slam for robots operating cooperatively with sensor networks. In Proceedings of the IEEE International Conference on Robotics and Automation, Orlando, FL, USA, 15–19 May 2006. [CrossRef]
4. Cai, Y.; Lou, W.; Li, M.; Li, X.Y. Energy efficient target-oriented scheduling in directional sensor networks. *IEEE Trans. Comput.* **2009**, *58*, 1259–1274. [CrossRef]
5. Goldberg, D.E. *Genetic Algorithms*; Pearson Education: New Delhi, India, 2006.
6. Deb, K.; Pratap, A.; Agarwal, S.; Meyarivan, T. A fast and elitist multiobjective genetic algorithm: NSGA-II. *IEEE Trans. Evol. Comput.* **2002**, *6*, 182–197. [CrossRef]
7. Gen, M.; Cheng, R. *Genetic Algorithms and Engineering Optimization*; Wiley: Toronto, ON, Canada, 2000.
8. Kennedy, J.; Eberhart, R. Particle swarm optimization. In Proceedings of the International Conference on Neural Networks, Perth, WA, Australia, 27 November–1 December 1995; pp. 1942–1948. [CrossRef]
9. Storn, R.; Price, K. Differential evolution—A simple and efficient heuristic for global optimization over continuous spaces. *J. Glob. Optim.* **1997**, *11*, 341–359. [CrossRef]
10. Chiang, W.C.; Russell, R.A. Simulated annealing meta-heuristics for the vehicle routing problem with time windows. *Ann. Oper. Res.* **1996**, *63*, 3–27. [CrossRef]
11. Tavares, R.S.; Martins, T.C.; Tsuzuki, M.S.G. Simulated annealing with adaptive neighborhood: A case study in off-line robot path planning. *Expert Syst. Appl.* **2011**, *38*, 2951–2965. [CrossRef]
12. Dorigo, M.; Maniezzi, V.; Colorni, A. Ant system: Optimization by a colony of cooperating agents. *IEEE Trans. Syst. Man Cybern. Part B Cybern.* **1996**, *26*, 29–41. [CrossRef]
13. Blum, C. Ant colony optimization: Introduction and recent trends. *Phys. Life Rev.* **2005**, *2*, 353–373. [CrossRef]
14. Jha, S.K.; Eyong, E.M. An energy optimization in wireless sensor networks by using genetic algorithm. *Telecommun. Syst.* **2018**, *67*, 113–121. [CrossRef]
15. Parvin, J.R.; Vasanthanayaki, C. Particle swarm optimization-based energy efficient target tracking in wireless sensor network. *Measurement* **2019**, *147*, 106882. [CrossRef]
16. Lo, Y.S.; Liang, C.K. A genetic algorithm for target coverage problem in directional sensor networks. In Proceedings of the 2nd Eurasian Conference on Educational Innovation (ECEI), Singapore, 27–29 January 2019; pp. 333–336. [CrossRef]
17. Robinson, J.; Sinton, S.; Rahmat-Samli, Y. Particle swarm, genetic algorithm, and their hybrids: Optimization of a profiled corrugated horn antenna. In Proceedings of the IEEE Antennas and Propagation Society International Symposium, San Antonio, TX, USA, 16–21 June 2002; pp. 16–21. [CrossRef]
18. Shi, X.H.; Lu, Y.H.; Lee, H.P.; Lin, W.Z.; Liang, Y.C. Hybrid evolutionary algorithms based on PSO and GA. In Proceedings of the 2003 Congress on Evolutionary Computation, Canberra, Australia, 8–12 December 2003; pp. 2393–2399. [CrossRef]
19. Yang, B.; Chen, Y.; Zhao, Z. A hybrid evolutionary algorithm by combination of PSO and GA for unconstrained and constrained optimization problems. In Proceedings of the 2007 IEEE International Conference on Control and Automation, Guangzhou, China, 30 May–1 June 2007; pp. 166–170. [CrossRef]
20. Valdez, F.; Melin, P.; Castillo, O. Evolutionary method combining particle swarm optimization and genetic algorithms using fuzzy logic for decision making. In Proceedings of the 2009 IEEE International Conference on Fuzzy Systems, Jeju, Korea, 20–24 August 2009; pp. 2114–2119. [CrossRef]
21. Ghamisi, P.; Benediktsson, J.A. Feature selection based on hybridization of genetic algorithm and particle swarm optimization. *IEEE Geosci. Remote Sens. Lett.* **2015**, *12*, 309–313. [CrossRef]
22. Moussa, R.; Azar, D. A PSO-GA approach targeting fault-prone software modules. *J. Syst. Softw.* **2017**, *132*, 41–49. [CrossRef]
23. Nik, A.A.; Nejad, F.M.; Zakeri, H. Hybrid PSO and GA approach for optimizing surveyed asphalt pavement inspection units in massive network. *Autom. Constr.* **2016**, *71*, 325–345. [CrossRef]
24. Cardei, M.; Du, D.Z. Improving wireless sensor network lifetime through power aware organization. *Wirel. Netw.* **2005**, *11*, 333–340. [CrossRef]
25. Cardei, M.; Thai, M.T.; Li, Y.; Wu, W. Energy-efficient target coverage in wireless sensor networks. In Proceedings of the IEEE 24th Annual Joint Conference of the IEEE Computer and Communications Societies, Miami, FL, USA, 13–17 March 2005. [CrossRef]
26. Zorbas, D.; Glynos, D.; Kotzanikolaou, P.; Douligeris, C. Solving coverage problems in wireless sensor networks using cover sets. *Ad Hoc Netw.* **2010**, *8*, 400–415. [CrossRef]
27. Ting, C.K.; Liao, C.C. A memetic algorithm for extending wireless sensor network lifetime. *Inf. Sci.* **2010**, *180*, 4818–4833. [CrossRef]
28. Mostafaei, H.; Shojafar, M. A new meta-heuristic algorithm for maximizing lifetime of wireless sensor networks. *Wirel. Pers. Commun.* **2015**, *82*, 723–742. [CrossRef]
29. Ma, H.; Liu, Y. On coverage problems of directional sensor networks. In Proceedings of the International Conference on Mobile Ad-Hoc and Sensor Networks (MSN 2005), Wuhan, China, 13–15 December 2005; pp. 721–731. [CrossRef]
30. Ai, J.; Abouzeid, A. Coverage by directional sensors in randomly deployed wireless sensor networks. *J. Comb. Optim.* **2006**, *11*, 21–41. [CrossRef]

31. Chen, U.R.; Chiou, B.S.; Chen, J.M.; Lin, W. An adjustable target coverage method in directional sensor networks. In Proceedings of the IEEE Asia-Pacific Services Computing Conference (APSCC 08), Yilan, Taiwan, 9–12 December 2008; pp. 174–180. [CrossRef]
32. Han, Y.H.; Kim, C.M.; Gil, J.M. A greedy algorithm for target coverage scheduling in directional sensor networks. *J. Wirel. Mob. Netw. Ubiquitous Comput. Dependable Appl.* **2010**, *9*, 96–106. [CrossRef]
33. Gil, J.M.; Han, Y.H. A target coverage scheduling scheme based on genetic algorithms in directional sensor networks. *Sensors* **2011**, *11*, 1888–1906. [CrossRef] [PubMed]
34. Li, D.; Liu, H.; Lu, X. Target Q-coverage problem with bounded service delay in directional sensor networks. *Int. J. Distrib. Sens. Netw.* **2012**, *2012*, 1–10. [CrossRef]
35. Mohamadi, H.; Ismail, A.S.B.H.; Salleh, S. A learning automata-based algorithm for solving coverage problem in directional sensor networks. *Computing* **2013**, *95*, 1–24. [CrossRef]
36. Zannat, H.; Akter, T.; Tasim, M.; Rahman, A. The coverage problem in visual sensor networks: A target oriented approach. *J. Netw. Comput. Appl.* **2016**, *75*, 1–15. [CrossRef]
37. Razali, M.N.; Salleh, S.; Mohamadi, H. Solving priority-based target coverage problem in directional sensor networks with adjustable sensing ranges. *Wirel. Pers. Commun.* **2017**, *95*, 847–872. [CrossRef]
38. Zishan, A.A.; Karim, I.; Shubha, S.S.; Rahman, A. Maximizing heterogeneous coverage in over and under provisioned visual sensor networks. *J. Netw. Comput. Appl.* **2018**, *124*, 44–62. [CrossRef]
39. Bakht, A.J.; Motameni, H.; Mohamadi, H. A learning automata-based algorithm to solve imbalanced k-coverage in visual sensor networks. *J. Intell. Fuzzy Syst.* **2020**, *39*, 2817–2829. [CrossRef]
40. Liang, C.K.; Chen, Y.T. The target coverage problem in directional sensor networks with rotatable angles. In Proceedings of the International Conference on Advances in Grid and Pervasive Computing (GPC 2011), Oulu, Finland, 11–13 May 2011; pp. 264–273. [CrossRef]
41. Wu, M.C.; Lu, W.F. On target coverage problem of angle rotatable directional sensor networks. In Proceedings of the 7th International Conference on Innovative Mobile and Internet Services in Ubiquitous Computing, Taichung, Taiwan, 3–5 July 2013; pp. 605–610. [CrossRef]
42. Hochbaum, D.S. Approximating covering and packing problems: Set cover, vertex cover, independent set, and related problems. In *Approximation Algorithms for NP-hard Problems*; Hochbaum, D.S., Ed.; PWS Publishing Co.: Boston, MA, USA, 1997; pp. 94–143.
43. Clerc, M.; Kennedy, J. The paarticle swarm-explosion, stability, and convergence in a multidirmensional complex space. *IEEE Trans. Evol. Comput.* **2002**, *6*, 58–73. [CrossRef]
44. Van den Bergh, F.; Engelbrecht, A.P. A cooperative approach to particle swarm optimization. *IEEE Trans. Evol. Comput.* **2004**, *8*, 225–239. [CrossRef]
45. Pan, Q.; Tasgetiren, M.F.; Liang, Y.C. A discrete particle swarm optimization algorithm for the permutation flowshop sequencing problem with makespan criterion. In *Research and Development in Intelligent Systems XXIII, SGAI 2006*; Bramer, M., Coenen, F., Tuson, A., Eds.; Springer: London, UK, 2006; pp. 19–31. [CrossRef]

Article

A Hybrid of Fully Informed Particle Swarm and Self-Adaptive Differential Evolution for Global Optimization

Shir Li Wang [1,*], Sarah Hazwani Adnan [2], Haidi Ibrahim [3], Theam Foo Ng [2,*] and Parvathy Rajendran [4,5]

1. Faculty of Art, Computing and Creative Industry, Universiti Pendidikan Sultan Idris (UPSI), Tanjong Malim 35900, Perak, Malaysia
2. Centre of Global Sustainability Studies (CGSS), Level 5, Hamzah Sendut Library, Universiti Sains Malaysia, Minden 11800, Pulau Pinang, Malaysia
3. School of Electrical and Electronic Engineering, Engineering Campus, Universiti Sains Malaysia, Nibong Tebal 14300, Pulau Pinang, Malaysia
4. School of Aerospace Engineering, Universiti Sains Malaysia, Engineering Campus, Nibong Tebal 14300, Pulau Pinang, Malaysia
5. Faculty of Engineering and Computing, First City University College, Bandar Utama, Petaling Jaya 47800 Selangor, Malaysia
* Correspondence: shirli_wang@fskik.upsi.edu.my (S.L.W.); tfng@usm.my (T.F.N.)

Citation: Wang, S.L.; Adnan, S.H.; Ibrahim, H.; Ng, T.F.; Rajendran, P. A Hybrid of Fully Informed Particle Swarm and Self-Adaptive Differential Evolution for Global Optimization. *Appl. Sci.* 2022, *12*, 11367. https://doi.org/10.3390/app122211367

Academic Editor: Vincent A. Cicirello

Received: 12 September 2022
Accepted: 5 November 2022
Published: 9 November 2022

Publisher's Note: MDPI stays neutral with regard to jurisdictional claims in published maps and institutional affiliations.

Copyright: © 2022 by the authors. Licensee MDPI, Basel, Switzerland. This article is an open access article distributed under the terms and conditions of the Creative Commons Attribution (CC BY) license (https://creativecommons.org/licenses/by/4.0/).

Abstract: Evolutionary computation algorithms (EC) and swarm intelligence have been widely used to solve global optimization problems. The optimal solution for an optimization problem is called by different terms in EC and swarm intelligence. It is called individual in EC and particle in swarm intelligence. Self-adaptive differential evolution (SaDE) is one of the promising variants of EC for solving global optimization problems. Adapting self-manipulating parameter values into SaDE can overcome the burden of choosing suitable parameter values to create the next best generation's individuals to achieve optimal convergence. In this paper, a fully informed particle swarm (FIPS) is hybridized with SaDE to enhance SaDE's exploitation capability while maintaining its exploration power so that it is not trapped in stagnation. The proposed hybrid is called FIPSaDE. FIPS, a variant of particle swarm optimization (PSO), aims to help solutions jump out of stagnation by gathering knowledge about its neighborhood's solutions. Each solution in the FIPS swarm is influenced by a group of solutions in its neighborhood, rather than by the best position it has visited. Indirectly, FIPS increases the diversity of the swarm. The proposed algorithm is tested on benchmark test functions from "CEC 2005 Special Session on Real-Parameter Optimization" with various properties. Experimental results show that the FIPSaDE is more effective and reasonably competent than its standalone variants, FIPS and SaDE, in solving the test functions, considering the solutions' quality.

Keywords: differential evolution; fully informed particle swarm; self-adaptive differential evolution; particle swarm optimization

1. Introduction

The real-life optimization problems are often unpredictable and dynamic, which means that the optimization problems are facing many time-varying and unimaginable constraints, such as routing problems, scheduling problems, prediction and more variants of broad application problems. The primary purpose of the optimization problems is to find the most optimum minimum or maximum value in the search space where often the problems have more than one maximum or minimum point. This type of problem is considered an NP-hard (nondeterministic polynomial time) problem, requiring high computational processing to solve it. Currently, evolutionary computation (EC) is widely used in numerous studies to solve this type of problem. The EC technique has proven to provide an effective search in a complex search to achieve optimal or near-optimal solutions and has been proven to have strong global search ability [1].

As a classic heuristic method, Particle Swarm Optimization (PSO) is one of the most used and reliable swarm intelligence techniques [2]. It has been successfully adopted into many practical applications due to its efficiency, fast optimization speed and simplicity of implementation. In the traditional PSO, each particle of the population is learning from its nearest neighbors, and this may cause the particle swarm to stagnate in the local region due to rapid convergence [1]. Unlike the traditional PSO, one of its variants, the Fully Informed Particle Swarm Optimization, FIPS, introduced by Mendes, Kennedy and Neves in 2004, has the weight contributions of all particles to have the same value. The particles are influenced by their whole neighborhood in a specific way according to different neighborhood topologies. This technique enables each particle to access the most successful solutions from the whole swarm, not necessarily the best from its nearest neighbor. Accordingly, the performance of FIPS algorithm types is also generally more dependent on the neighborhood topology [3].

Unlike PSO and its variants, Differential Evolution (DE) is also a widely used algorithm with a remarkable ability to find the optimal solution. DE relies upon its strength in handling starting initial points where multiple starting points are randomly chosen during sampling of potential solutions [4,5]. Several versions of the proposed adaptive DE focus on manipulating the parameter of DE with the introduction of a new way of controlling the value of existing parameters. The adaptive differential evolution (jDE) [6] implemented several DE strategies to control the diversity of the population. Additionally, it can self-adapt the scaling factor F and the crossover rate C_r. The parameter adaptive differential evolution (JADE) [7] relies on greedy mutation strategy (DE/current-to-$p$best) with optimal external achieve and utilizing the previously explored inferior solutions. In SaDE [8], suitable learning strategies and parameter settings are gradually self-adapted according to the previous learning experiences. Even more, the choice of learning strategy and the two essential control parameters of DE, F and C_r, are not required to be specified before the evolution phase.

In the past few years, the development of adaptive mechanisms has emerged to be one of the critical issues in the branch of the EC algorithm. Adaptive mechanism refers to the ability of the algorithm to change its behavior according to information available during its running phase. The number of works in which adaptive mechanisms are being successfully used has increased enormously over the past few decades. The applications include a wide variety of areas, such as routing problems, signal processing, optimization problem and medical fields. Furthermore, the efficiency of the adaptive mechanism mainly depends on the algorithm design and the algorithm used for adaptation.

One of the common drawbacks EC algorithms face in solving optimization problems is the lack of diversity in solutions causing a suboptimal solution. A potential approach to overcoming this drawback is a hybrid of EC and swarm intelligence. FIPS and SaDE emerge as promising EC and swarm intelligence algorithms in solving optimization problems. Therefore, they are chosen to form a hybrid in our study. FIPS and SaDE are hybridized owing to the solutions' diversity in FIPS and the optimal solution quality in SaDE. FIPS improves the solutions' diversity by gathering knowledge about its neighborhood's solutions. Each solution in FIPS receives information from all its neighbors rather than just the best one. SaDE is a simple, powerful and self-adaptive type of DE which finds an optimal solution through exploration and exploitation. SaDE minimally relies on user-specified parameters because it can self-adapt its parameters to solve optimization problems. FIPS improves the solutions through information gathering from its neighbors and maneuvering the group of solutions to move toward the optimal region, which is explored and exploited by SaDE. Therefore, in the hybrid of FIPS and SaDE, they complement each other by balancing the exploration of neighbors and exploitation of the optimal solutions.

2. Related Works
2.1. Improved Differential Evolution

DE is one of the most popular evolutionary algorithms used to solve optimization problems because it is simple, robust and computationally efficient. However, DE faces issues in convergence rate and local exploitation rate [9]. Therefore, various research efforts for its improvement are continuously carried out even though the algorithm has been introduced 25 years ago by Storn and Price [10]. Besides, surveys related DE variants are updated after a certain years to accommodate various modifications of DEs, as shown in [5,9,11–15]. The surveys show that most of the research efforts focus on improving the performance of DE through parameter settings and modifications in genetic operations consisting of initialization, differential mutation, crossover and selection. Another potential direction towards improving DE is the use of hybridization, and it has gained research attention [9,11,12].

Most of the optimization algorithms are not able to solve a variety of problems [9]. However, continuous research efforts are required to improve the algorithm through the complementary of their advantages. Hybridization is one the main research directions in improving the performance of DE because different optimization algorithms have different search behaviours and advantages [16].

It is implemented to enhance its performance and overcome its limitations such as convergence speed [11], premature convergence [9] and local minima [9]. The types of algorithms used for hybridization in DE can either belong to the same or different categories of algorithms. The work in [12] categorizes the algorithm in hybridization of DE into statistical techniques and algorithms. On the other hand, the work in [11] shows that most of the algorithms in a hybridization of DE are from the swarm intelligence algorithms. The commonly used swarm intelligence algorithms to hybridize with DE include PSO, ant colony algorithm (ACA) and artificial bee colony (ABC).

DE variants produce robust solutions owing to their exploration ability. However, they have issues related to premature convergence and local exploitation. On the other hand, one of the drawbacks of PSO is premature convergence [17]. Premature convergence is caused by improper velocity adjustment when PSO is configured by inappropriate acceleration coefficient and inertia weights [18]. Consequently, the particles move in undesired directions, causing stagnation around or being trapped in suboptimal [18]. A review of the modifications, extensions and hybridization of the PSO algorithm and their applications is presented in [19].

Most of the hybrid variants of DE is formed with PSO [11] and an extensive survey about the hybrid can be found in [9]. A hybrid of advanced DE (ADE) and (PSO) (APSO), namely, AHDEPSO, was proposed to solve unconstrained optimization problems in [9]. Given that a hybrid of DE and PSO offers complementary properties and has the effect of balancing the exploration and exploitation phases, the hybrid of their advanced variants has attracted more research attention. The research focusing on how to combine PSO and DE is still an open problem [16].

The premature convergence of PSO in the optimization process can be improved based on four different methods: parameter settings, neighborhood topology, learning strategy and hybridization [20]. Neighbourhood topology enhances PSO's exploration capability and different topologies affected solve different optimization problems [20]. On the other hand, hybridization is used to complement the weakness of intelligent algorithms by combining their helpful features. Therefore, FIPS, instead of the original PSO, is hybridized with DE to improve the performance of the optimization algorithms.

The self-adaptive mutation differential evolution algorithm based on particle swarm optimization (DEPSO) [21] uses balance to improve the optimization. The hybrid uses the selection probability of mutation strategy to decide between the modified DE/rand/1 and the PSO's mutation strategy. The modified mutation strategy with the elite archive strategy, called DE/e-rand/1, is proposed in DEPSO. The framework of DEPSO is still based on DE. Unlike DEPSO, our proposed FIPSaDE is based on the framework of PSO. The hybridization of DEPSO focuses on selecting mutation strategies from DE and PSO to generate the mutant vector. Therefore, its hybridization involves the mutation phase only. On the other hand, the hybridization in FIPSaDE involves integrating DE's algorithmic operations, from the mutation to selection phases, into the framework of PSO before the velocity and position updating.

Dash et al. proposed HDEPSO [22] to solve various benchmark functions and an optimization problem focusing on the effectiveness of the sharp edge FIR filter (SEFIRF). The proposed framework is similar to FIPSaDE, whereby, DE's mutation, crossover and selection are integrated with the best particles of PSO to enhance global searching ability. However, the settings of F and C_r were not adaptive. F is fixed as a constant and C_r refers to an arbitrary number between [0, 1]. In contrast, the settings of F and C_r in FIPSaDE are adaptive.

For the case of the hybrid approach, it may cause more uncertainties in parameter setting in view of the fact that a user needs to realize how to set parameters for at least two algorithms [23]. The task of parameter setting in solving optimization problem is problem-dependent, user-dependent and algorithm-dependent. The task becomes more complex when the algorithms of the hybrid are from different branches of machine learning. Therefore, researchers start to focus on the possibility of applying adaptation methods for users that have minimum knowledge about the algorithms, parameters settings and the problems. DE variants with varying adaptation levels have shown promising results in solving optimization problems and its popularity in setting DE's parameters can be found in the aforementioned reviews.

For example, the work in [24] showed the use of adaptation to tune the configurations of F, C_r, and mutation strategy at different stages of the evolution. Another work demonstrating the use of adaptation in controlling parameters F, C_r is the Success-History-based adaptive DE (SHADE) algorithm [25]. SHADE uses the nearest spatial neighborhood-based modification to the adaptation process of the parameters described above. Do et al. [26] also proposed an adaptive mechanism to determine the parameters F and F, C_r with the mutation and selection processes determined by the best individual-based mutation and elitist selection techniques. The adaptation in population sizing is summarized in the review by Piotrowski [27]. These research studies have shown the use of adaptation to control DE's various parameters.

The adaptability trend to set DE's parameters, from partial adaptiveness to full adaptiveness, is increasing. Since the adaptive DE variants have shown promising results, its use to form the hybrid can reduce the complexity of parameter setting because the number of algorithms that require user-specified parameters has declined. Therefore, we use a adaptive DE called SaDE in [8] to form the hybrid of DE and PSO in the current work. SaDE adapts the control parameter F and C_r based on the mutation strategy $DE/rand/1/bin strategy$.

2.2. Particle Swarm Optimization

PSO is another branch of EC that operates based on swarm-based intelligence. Its implementation is simple and easy, but it also suffers from premature convergence [18,28]. Therefore, various modifications are applied to the standard PSO to overcome its drawback. In [18], the researchers categorize the modification of PSO into four strategies: Modification of the PSO's controlling parameters (1); (2) Hybridization with other meta-heuristic algorithms such as GA and DE; (3) Cooperation; (4) Multi-swarm techniques. The work in [9,18] reflects that a hybrid of PSO and DE has gained popularity in recent years as the strategy to improve EC's performance in solving optimization problems. The performance of PSO is affected by its neighbourhood topology, where the particles within the neighbour topology communicate with each other and share information. FIPS is the PSO that relies on the particle's best positions of its neighbors in updating its velocity to prevent it from being stuck at local optimum. The hybrid variants of DE and PSO have shown better quality of solutions and computation efficiency than the original PSO [18]. Therefore, a hybrid of the adaptive SaDE and FIPS is formed to complement their weakness in solving optimization problems in our current work. Brief descriptions of SaDE and FIPS are provided in Sections 2.3 and 2.4.

2.3. Self-Adaptive Differential Evolution (SaDE)

SaDE is a variant of DE that can produce better results than the traditional DE algorithms [2]. This algorithm has been applied in various optimization problems [2–4,6,7]. In SaDE, two out of three critical control parameters of DE (i.e., F and C_r) are manipulated for DE improvement. Since these parameters are used as the external fixed parameters, these control parameters are not deemed to be evolving entities in the earliest EA algorithm. Later on, it was determined that certain parameters could be changed during the evolution process in order to reach the desired level of convergence [8].

The population size N_p is not favored as a chosen parameter because its value is not sensitive to the efficiency and robustness of the DE algorithm [8]. N_p is held as a user-specified parameter, whereas the F value is set to be between (0, 2) for different individuals, with a normal distribution of mean 0.5 and standard deviation of 0.3 [29]. Instead of the commonly used (0, 1] for F values, the range between (0, 2] was chosen to maintain both small F values for local search and large F values for global search. C_r would initially be naturally distributed in a range with a mean of C_{r_m} and a standard deviation, std, of 0.1. C_{r_m} value is set to 0.5 as a starting point where the values will be held for several generations before being replaced by a new value with a similar normal distribution for the next generation.

Throughout each generation, better C_r values associated with trial vectors that are able to reach the next generation will be registered. The mean value for the normal distribution of C_r is recalculated based on all observed values corresponding to the effective trial vectors during the cycle. In order to prevent potentially inappropriate long-term accumulated results, successful C_r values captured during the recalculation of the normal distribution mean are not saved.

2.4. Fully Informed Particle Swarm (FIPS)

The FIPS algorithm that was developed by Mendes et al. in 2004 [1] is a variant of PSO in which each particle is influenced by all of its K-neighbors. This idea is in contrast to the standard PSO algorithm, in which the particle is drawn to the best location it has visited as well as the best position found by the particle in its neighborhood that manages to produce the best result. The algorithm is motivated by how individuals in human society are influenced by a statistical summary of their world [30].

Each particle in the FIPS swarm is affected by a group of particles from its surrounding neighborhood, which is not necessarily to the best particle's location [1]. The velocity equation for FIPS is modified based on canonical PSO. Each particle inside the swarm is affected by the achievements of all its neighbors, rather than pointing to just one best particle performance from its neighbors [30]. FIPS's velocity and position updates are defined by Equations (1) and (2), respectively.

$$v_i(t+1) = w v_i + \sum_{m=1}^{|N_i|} \gamma_m(t) \frac{y_m(t) - x_i(t)}{|N_i|} \tag{1}$$

$$x_i(t+1) = x_i(t) + v_i(t+1) \tag{2}$$

where N_i refers to the set of particles in particle i's neighborhood. $\gamma_m(t)$ is in the range of $U(0, c_1 + c_2)^D$ and D is dimension of the problem. $y_m(t)$ refer to the best position previously visited by m. Coefficients c_1, c_2 and w represent cognitive, social and inertia weight, respectively. There are five variants of the algorithm, as stated by Mendes et al. [1], which are FIPS, wFIPS, wdFIPS, Self and wSelf:

- FIPS refers to a fully informed particle swarm optimization algorithm with w returning a constant value and the number of neighbourhood contributions being the same;
- wFIPS refers to a type of FIPS algorithm in which each neighbor's input is weighted by the goodness of its previous best particle;
- wdFIPS refers to a type of FIPS algorithm where the contribution of each neighbor is weighted by its distance in the search space from the target particle;
- Self refers to a FIPS algorithm in which the previous best particle received half the weight;
- wSelf refers to a FIPS algorithm where the previous best particle received half the weight and the contribution of each neighbor was weighted by the goodness of its previous best particle.

The swarm optimization algorithm is a type of algorithm that is greatly influenced by the neighborhood. As a result, the efficiency of the FIPS algorithm is generally even more dependent on its neighborhood topology. Ten topology types selected by Cleghorn and Engelbrecht [30] to be implemented on FIPS include All, Ring, Four Clusters, Pyramid, Square, UAll, UFour Clusters, UPyramid and USquare. Topology with the "U" prefix refers to a similar topology without the prefix, but with the particle's own index omitted from the neighborhood.

3. Methodology

FIPS and SaDE are hybridized owing to the solutions' diversity in FIPS and the quality of optimal solution in SaDE. One of the common drawbacks of using optimization algorithms is the lack of diversity leading to a suboptimal solution [31]. The use of FIPS could improve the drawback. In the FIPS, the neighbors around a solution become the source of influence to improve the solution's diversity. DE is remarkable in finding the optimal solution in a group of solutions. However, choosing suitable control parameter values is tricky as the values need to vary according to each problem. SaDE has the advantage that the user does not need to determine the optimal parameters settings, and time complexity does not increase as the rules for applying SaDE are simple [29]. SaDE performs well in finding the optimal solution in a group of solutions through self-adapting its parameters. Each particle in FIPS receives information from all of its neighbors rather than just the best one. The particle may sometimes trap in local optima, and the FIPS topology network may help to monitor and maneuver the particle's position. Therefore, the operation in FIPS improves the solutions through the exploration of its neighbors and guides the group of solutions moving toward the optimal region.

A hybridization between SaDE and FIPS, called FIPSaDE, could improve the performance in solving complex optimization problems. The main process of FIPSaDE is structured according to the usual structure of the PSO algorithm. The solution found in FIPSaDE is called *individual* or *solution* interchangeably. The FIPS parameter is updated each time before the mutation process of the SaDE algorithm and again during the selection process in controlling the quality of individuals chosen for the next generation. The process is repeated iteratively until the algorithm reaches the optimum value. The presence of the FIPS process creates a disturbance in the population in which the FIPS is focused on moving the solution to explore its surrounding. This helps in maintaining the diversity of the population created and producing an excellent optimal solution. The pseudocode of FIPSaDE is given in Algorithm 1.

In FIPSaDE, a swarm of individuals flies in a D-dimensional search space seek an optimal solution. Each individual i possesses a current velocity vector $v_i = [v_{i1}, v_{i2}, \ldots, v_{iD}]$ and a current position vector $x_i = [x_{i1}, x_{i2}, \ldots, x_{iD}]$, where D is the number of dimensions. The FIPSaDE process starts by randomly initializing v_i and x_i.

Algorithm 1: The Pseudocode of the FIPSaDE for Solving a Minimization Problem

**********************************Start of FIPS**********************************

Initialization
Define the swarm size N_p
for each individual $i \in [1 \ldots N_p]$ **do**
   Randomly generate x_i and v_i.
   Evaluate the fitness of x_i denoting it as $f(x_i)$.
   Set $Pbest_i = x_i$ and $f(Pbest_i) = f(x_i)$.
end for
Set $Gbest = Pbest_1$ and $f(Gbest) = f(Pbest_1)$.
for each individual $i \in [1 \ldots N_p]$ **do**
  if $f(Pbest_i) < f(Gbest)$ **then**
    $f(Gbest) = f(Pbest_i)$
  end if
end for
while $t <$ maximum iterations **do**
  for each individual $i \in [1 \ldots N_p]$ **do**
    -.-.-.-.-.-.-.-.-.-.-.-.-.-.-.-.-.-.**Start of SaDE**-.-.-.-.-.-.-.-.-.-.-.-.-.-.-.-.-.-.-
    Generate vector $[x_i, F_i, C_{r_i}]$.
    Update the parameters F_i and C_{r_i} based on Equations (3) and (4).
    Mutate x_i based on Equation (5).
    Crossover x_i based on Equation (6).
    Select x_i based on Equation (7).
    -.-.-.-.-.-.-.-.-.-.-.-.-.-.-.-.-.-.**End of SaDE**-.-.-.-.-.-.-.-.-.-.-.-.-.-.-.-.-.-.-
    Evaluate its velocity $v_i(t+1)$ based on Equation (1).
    Update the position $x_i(t+1)$ of the particle based on Equation (2).
    if $f(x_i(t+1)) < f(Pbest_i)$ **then**
      $Pbest_i = x_i(t+1)$
      $f(Pbest_i) = f(x_i(t+1))$
    end if
    if $f(Pbest_i) < f(Gbest)$ **then**
      $Gbest = Pbest_i$
      $f(Gbest) = f(Pbest_i)$
    end if
  end for
  $t = t + 1$
end while
return $Gbest$

**********************************End of FIPS**********************************

In each iteration t, FIPSaDE uses a self-adapting mechanism of two control parameters, i.e., F and C_r. Each individual i is extended with control parameter values F and C_r as a vector $[x_i, F_i, C_{r_i}]$. New control parameters $F_{i,(t+1)}$ and $C_{r_i,(t+1)}$ are calculated before the mutation operator based on Equations (3) and (4). Therefore, the parameters influence the mutation, crossover and selection operations of the new vector $x_{i,(t+1)}$.

$$F_{i,(t+1)} = \begin{cases} F_l + rand_1 * F_u & \text{if } rand_2 < \tau_1 \\ F_{i,t} & \text{otherwise} \end{cases} \quad (3)$$

$$C_{r_i,(t+1)} = \begin{cases} rand_3 & \text{if } rand_4 < \tau_2 \\ C_{r_i,t} & \text{otherwise} \end{cases} \quad (4)$$

where $rand_j$, for $j \in \{1, 2, 3, 4\}$ are uniform random values within the range [0, 1]. The parameters τ_1, τ_2, F_l, F_u are fixed to values 0.1, 0.1, 0.1, 0.9, respectively in [8].

In the mutation process, for each target vector x_i, a mutant vector q_i is generated according to Equation (5)

$$q_{i,(t+1)} = x_{r_1,t} + F_{i,t} * (x_{r_2,t} - x_{r_3,t}) \tag{5}$$

with randomly chosen indexes $r_1, r_2, r_3 \in [1, N_p]$.

Then, a crossover operator forms a trial vector $p_i = (p_{i1}, p_{i2}, \ldots p_{iD})$, with the target vector is mixed with the mutated vector using Equation (6).

$$p_{ij,(t+1)} = \begin{cases} q_{ij,(t+1)} & \text{if } (rand_j(0,1) \leq C_{r_i,t}) \text{ or } (j = j_{rand}) \\ x_{ij,t} & \text{otherwise} \end{cases} \tag{6}$$

for $i = 1, 2, \ldots N_p$ and $j = 1, 2, \ldots D$. Index $j_{rand} \in \{1, N_p\}$ is a randomly chosen integer that is responsible for the trial vector containing at least one component from the mutant vector.

In the selection process, a greedy selection scheme for a minimization is used and it is shown in Equation (7).

$$x_{i,(t+1)} = \begin{cases} p_{i,(t+1)} & \text{if } f(p_{i,(t+1)}) \leq f(x_{i,t}) \\ x_{i,t} & \text{otherwise} \end{cases} \tag{7}$$

In each iteration t, the best position that has been found by individual i, $Pbest_i = [Pbest_{i1}, Pbest_{i2}, \ldots, Pbest_{iD}]$ and the best position that has been found by the whole swarm $Gbest = [Gbest_1, Gbest_2, \ldots, Gbest_D]$ guide individual i to update its velocity and position by Equations (1) and (2).

The FIPSaDE is effective in traversing through swarm spaces while avoiding getting stuck inside local search using the information collected from individuals' neighborhood. The individuals of each population will contain a control parameter adapted from SaDE and also a traversing parameter from FIPS in order to exploit the swarm space that is searching for the best global point, while avoiding getting trapped inside a local point. If some population suffers from slow and premature convergence or if the population becomes stagnant, the velocity parameter and position parameter from each individual will be able to help the population to evolve out into a better point. Experimental results validated the effectiveness and strength of our proposed model.

4. Experimental Setup

The performance of FIPSaDE is evaluated against the 25 benchmark functions from "CEC 2005 Special Session on Real-Parameter Optimization" (CEC 2005) [32], namely, F1–F25. The benchmark functions consist of 5 unimodal functions, 7 basic multimodal functions, 2 expanded multimodal functions and 11 composition functions. The details of the functions are described in Table 1. The number of variables for each function is represented by D and the ranges of variable search are represented by S. All functions are a minimization of problems with their best minimum solutions can refer to [32]. All experiments were conducted in the Eclipse software by using the Java programming language and on a laptop featuring an Intel(R) Core (TM) i5-7200U CPU @ 2.50GHz processor with 4GB of RAM.

Table 1. Details on the benchmark functions.

Denotation	Test Function	S	Modality
F1	Shifted Sphere Function	$[-100, 100]^D$	Unimodal
F2	Shifted Schwefel's Problem 1.2	$[-100, 100]^D$	Unimodal
F3	Shifted Rotated High Conditioned Elliptic Function	$[-100, 100]^D$	Unimodal
F4	Shifted Schwefel's Problem 1.2 with Noise in Fitness	$[-100, 100]^D$	Unimodal
F5	Schwefel's Problem 2.6 with Global Optimum on Bounds	$[-100, 100]^D$	Unimodal
F6	Shifted Rosenbrock's Function	$[-100, 100]^D$	Basic Multimodal
F7	Shifted Rotated Griewank's Function without Bounds	$[0, 600]^D$	Basic Multimodal
F8	Shifted Rotated Ackley's Function with Global Optimum on Bounds	$[-32, 32]^D$	Basic Multimodal
F9	Shifted Rastrigin's Function	$[-5, 5]^D$	Basic Multimodal
F10	Shifted Rotated Rastrigin's Function	$[-5, 5]^D$	Basic Multimodal
F11	Shifted Rotated Weierstrass Function	$[-0.5, 0.5]^D$	Basic Multimodal
F12	Schwefel's Problem 2.13	$[-\Pi, \Pi]^D$	Basic Multimodal
F13	Shifted Expanded Griewank's plus Rosenbrock's Function	$[-5, 5]^D$	Expanded Multimodal
F14	Shifted Rotated Expanded Scaffer's $f6$ Function	$[-100, 100]^D$	Expanded Multimodal
F15	Hybrid Composition Function	$[-5, 5]^D$	Hybrid Composition
F16	Rotated Version of Hybrid Composition Function $f15$	$[-5, 5]^D$	Hybrid Composition
F17	Rotated Version of Hybrid Composition Function $f15$ with Noise in Fitness	$[-5, 5]^D$	Hybrid Composition
F18	Rotated Hybrid Composition Function	$[-5, 5]^D$	Hybrid Composition
F19	Rotated Hybrid Composition Function with Narrow Basin Global Optimum	$[-5, 5]^D$	Hybrid Composition
F20	Rotated Hybrid Composition Function with Global Optimum on the Bounds	$[-5, 5]^D$	Hybrid Composition
F21	Rotated Hybrid Composition Function	$[-5, 5]^D$	Hybrid Composition
F22	Rotated Hybrid Composition Function with High Condition Number Matrix	$[-5, 5]^D$	Hybrid Composition
F23	Non-continuous Rotated Hybrid Composition Function	$[-5, 5]^D$	Hybrid Composition
F24	Rotated Hybrid Composition Function	$[-5, 5]^D$	Hybrid Composition
F25	Rotated Hybrid Composition Function without Bounds	Initialize population in $[2, 5]^D$ but no exact search range set	Hybrid Composition

In this experiment, the performance of FIPSaDE is studied by comparing the algorithm with four known algorithms, which are FIPS by Mendes et al. [1], DE by Storn and Price [10], SaDE by Brest et al. [8] and DE-PSO by Pant et al. [31]. FIPS, DE and SaDE were selected as these algorithms can be considered as the parent or the ancestor of FIPSaDE. As for DE-PSO, it was selected as the algorithm is a hybrid algorithm that features a similar structure and concept as FIPSaDE. The parameters for the algorithms are shown in Tables 2 and 3. The optimization test for all algorithms is restricted to a maximum number of function evaluation (MAX_FEs), which are 5×10^4 for both 10D and 50D problem, and 3×10^4 for 30D problem. A success-threshold of 10^{-14} was administered for the experiments. This means the evolutionary processes are terminated if the best-fitness, $f_{best} < f_{target}$ is reached, with $f_{target} = f(x*) + 10^{-14}$. Otherwise, the processes continue until they reach MAX_FEs. The evolutions for all models of DEs are run until they meet the stopping criterion and repeated by using the k-th seed numbers, $k = 1, 2, \ldots, K$, with K referring to the maximum number of runs.

Table 2. Parameter settings of DE, PSO, DE-PSO, SaDE and FIPS.

Parameter	DE	PSO	DE-PSO	SaDE	FIPS
Population size, N_p ($D = 10, D = 30, D = 50$)	(25, 30, 50)	(25, 30, 50)	(25, 30, 50)	(25, 30, 50)	(25, 30, 50)
MAX_FEs ($D = 10, D = 30, D = 50$)	($5 \times 10^4, 3 \times 10^4,$ 5×10^4)	($5 \times 10^4, 3 \times 10^4,$ 5×10^4)	($5 \times 10^4, 3 \times 10^4,$ 5×10^4)	($5 \times 10^4, 3 \times 10^4,$ 5×10^4)	($5 \times 10^4, 3 \times 10^4,$ 5×10^4)
Crossover rate, Cr	0.1	-	0.95	$N(0.5 \pm 0.1)$	-
Scale factor, F	0.5	-	0.9	$F = (0, 2]$ with $N(0.5 \pm 0.3)$	-
Mutation strategy	de/rand/1/bin	-	de/rand/1/bin	de/rand/1/bin,	-
Acceleration rate (c_1, c_2)	-	(2.05, 2.05)	(2.0, 2.0)	-	(2.05, 2.05)
Inertial weight, w	-	0.7298	0.9	-	0.7298

Table 3. Parameter settings of FIPSaDE.

Parameter	Dimension		
	10	30	50
Population size, N_p	25	30	50
MAX_FEs	5×10^4	3×10^4	5×10^4
K runs per case	30	30	20
Crossover rate, Cr		$N(0.5 \pm 0.1)$	
Scale factor, F		$F = (0, 2]$ with $N(0.5 \pm 0.3)$	
Learning period		50	
Acceleration rate, c_1		2.05	
Acceleration rate, c_2		2.05	
Inertial weight, w		0.7298	
Neighborhood Type		Self	
Neighborhood Topology		Four cluster	
Symmetric		Asymmetric	

5. Results and Analysis

The difference between current fitness value and optimum value, also known as error value, is used to compare the algorithms' performance. The average errors of the independent runs of all algorithms for 10D, 30D and 50D problems are summarized in Tables 4–6. The last rows in Tables 4–6 show the frequency of having the best result for the algorithms, h_{win}. Based on the results in Tables 4–6, DE cannot find the solutions for functions F15–F17 because its solutions have undefined numeric results for the functions that are divided by zero.

When the settings are 10D and N is 25 (10D25N), SaDE, DE-PSO and FIPSaDE have the highest h_{win}, which is 6. This means the three algorithms have compatible performances in solving the problems. An interesting finding from the comparison of h_{win} for SaDE, DE-PSO and FIPSaDE for the setting of 10D25N is that they have the lowest error for different functions. This finding may indicate that they perform well for different characteristics of problems. As the combinations of D and N increase to 30D30N and 50D50N, FIPSaDE shows distinctive performances compared to the other algorithms by having the highest h_{win} in both settings. FIPSaDE has h_{win} = 9 and h_{win} = 12 in 30D30N and 50D50N, respectively. In other words, FIPSaDE's performances improves with the increase of problem dimensionality and population size. In contrast, the performances of FIPS are the worst among the algorithms for the three settings. FIPS, DE and SaDE show deterioration when the settings increase from 10D25N to 50D50N, with the highest deterioration being observed in FIPS. DE-PSO, which has similar characteristics to FIPSaDE, shows moderate performances regardless of the settings.

Table 4. Results of average error values for $10D$ test problem and $N = 25$ (10D25N).

Function	FIPS	DE	SaDE	DE-PSO	FIPSaDE
F1	1.6913E+00	5.6843E-14	**1.8948E-15**	5.6843E-14	7.7548E-11
F2	1.7168E+00	2.5199E-01	**5.6843E-14**	1.8182E+01	9.2022E-11
F3	4.8917E+03	6.3587E+05	3.3801E+04	5.8865E+04	**3.0103E+01**
F4	1.0672E+01	2.0546E+02	**7.0025E-11**	2.1796E+01	9.2033E-11
F5	**0.0000E+00**	9.1538E+02	6.6696E-13	2.3041E-12	8.9070E-11
F6	5.0970E+01	2.6962E+00	1.2781E+00	1.3425E+05	**5.3154E-01**
F7	1.2671E+03	**7.6374E-02**	1.2670E+03	1.8818E-01	1.2670E+03
F8	2.0217E+01	2.0227E+01	2.0384E+01	**2.0081E+01**	2.0110E+01
F9	8.0601E+00	**5.6843E-14**	2.6532E-01	5.6843E-14	4.9748E-01
F10	8.6778E+00	1.6527E+01	6.9076E+00	1.0315E+01	**5.8371E+00**
F11	**1.7263E+00**	6.9768E+00	3.0350E+00	1.7695E+00	2.3054E+00
F12	2.8017E+02	**2.4992E+01**	5.6407E+02	1.5613E+03	3.0828E+02
F13	5.6892E-01	**2.6653E-02**	8.8579E-01	4.2952E-01	3.2816E-01
F14	2.1172E+00	3.0997E+00	3.1387E+00	2.2248E+00	**1.9601E+00**
F15	**1.8014E+02**	-	2.8100E+02	1.9472E+02	3.1617E+02
F16	**1.0019E+02**	-	1.0218E+02	1.1402E+02	1.0514E+02
F17	1.0719E+02	1.0722E+03	1.1966E+02	1.4561E+02	**1.0136E+02**
F18	7.6687E+02	1.6996E+03	**7.0154E+02**	7.8021E+02	7.8756E+02
F19	7.7130E+02	1.8139E+03	**6.5847E+02**	7.5198E+02	7.2585E+02
F20	7.4034E+02	1.7606E+03	6.8165E+02	7.6896E+02	**6.4081E+02**
F21	6.9422E+02	1.2334E+03	7.0185E+02	**5.4388E+02**	7.3634E+02
F22	7.1363E+02	1.2253E+03	7.7433E+02	**7.4833E+02**	7.5392E+02
F23	7.8007E+02	1.2661E+03	8.5139E+02	**7.9163E+02**	7.9806E+02
F24	2.8126E+02	7.4243E+02	**2.0000E+02**	5.3977E+02	2.2654E+02
F25	1.7506E+03	7.4292E+02	1.7507E+03	**5.4948E+02**	1.7495E+03
h_{win}	4	4	6	6	6

Table 5. Results of average error values for $30D$ test problem and $N = 30$ (30D30N).

Function	FIPS	DE	SaDE	DE-PSO	FIPSaDE
F1	1.0003E+03	**1.3264E-14**	1.0232E-13	1.3873E+02	6.6317E-14
F2	7.4895E+03	8.1760E+03	5.9137E+02	3.3581E+02	**2.3457E-12**
F3	9.6860E+06	1.2073E+07	3.0102E+06	1.4827E+06	**8.9114E+04**
F4	1.3059E+04	3.2880E+04	8.8330E+03	4.2988E+02	**3.2211E+01**
F5	5.0737E+03	1.0061E+04	4.8601E+03	**1.9268E+03**	3.1928E+03
F6	4.0001E+07	9.1524E+00	1.4404E+02	4.5495E+06	**1.1960E+00**
F7	4.6963E+03	**1.5056E-02**	4.6963E+03	2.4008E-02	4.6963E+03
F8	2.0904E+01	2.0861E+01	2.1049E+01	**2.0349E+01**	2.0887E+01
F9	9.1200E+01	**1.5158E-14**	1.8278E+01	2.2971E+01	1.0267E+01
F10	1.1387E+02	1.8811E+02	1.1745E+02	6.2888E+01	**4.1795E+01**
F11	2.2143E+01	3.5289E+01	3.5379E+01	2.1214E+01	**2.0232E+01**
F12	9.3414E+04	**5.5912E+03**	1.6091E+04	1.8567E+04	6.4080E+03
F13	6.8397E+00	**2.5628E-01**	1.2032E+01	2.1437E+00	2.3846E+00
F14	**1.1821E+01**	1.2741E+01	1.3538E+01	1.2031E+01	1.1989E+01
F15	4.8635E+02	-	**3.2749E+02**	3.9304E+02	3.3594E+02
F16	2.4964E+02	-	1.8945E+02	2.5133E+02	**1.2952E+02**
F17	2.4655E+02	-	2.9869E+02	2.9779E+02	**1.4873E+02**
F18	9.1085E+02	1.2349E+03	**9.0925E+02**	9.0925E+02	9.1217E+02
F19	9.0797E+02	1.2258E+03	**9.0760E+02**	9.1250E+02	9.1158E+02
F20	9.1196E+02	1.2834E+03	9.1304E+02	9.0907E+02	**9.0881E+02**
F21	9.3243E+02	1.4813E+03	**5.0000E+02**	6.6047E+02	5.3210E+02
F22	8.9686E+02	1.6756E+03	9.4560E+02	**8.5655E+02**	8.9115E+02
F23	**5.9064E+02**	1.5085E+03	6.1697E+02	8.2535E+02	5.9916E+02
F24	4.1547E+02	1.4310E+03	**2.0000E+02**	6.4398E+02	2.2529E+02
F25	1.6313E+03	1.4382E+03	1.6448E+03	**2.8872E+02**	1.6171E+03
h_{win}	2	5	5	4	9

Table 6. Results of average error values for 50D test problem and $N = 50$ (50D50N).

Function	FIPS	DE	SaDE	DE-PSO	FIPSaDE
F1	2.9123E+03	**5.6843E-14**	5.1044E-10	6.6018E+02	8.2423E-14
F2	1.9747E+04	2.6854E+04	4.0747E+03	1.8025E+03	**2.7853E-12**
F3	4.6307E+07	2.7198E+07	7.2181E+06	4.5704E+06	**9.4374E+04**
F4	2.9802E+04	9.4951E+04	2.7156E+04	2.5763E+03	**8.5264E+01**
F5	1.2415E+04	2.2436E+04	9.7093E+03	6.1975E+03	**5.5761E+03**
F6	1.1081E+08	1.7165E+01	3.4873E+02	4.1990E+07	**1.7940E+00**
F7	6.1953E+03	**3.5798E-04**	6.1953E+03	1.1065E-02	6.1953E+03
F8	2.1079E+01	2.1056E+01	2.1194E+01	**2.1009E+01**	2.1084E+01
F9	1.9239E+02	**5.6843E-14**	9.1163E+01	8.3410E+01	1.5024E+01
F10	2.5784E+02	4.8818E+02	2.8407E+02	1.3299E+02	**9.3924E+01**
F11	4.4393E+01	5.7573E+03	7.0201E+01	4.5260E+01	**3.8998E+01**
F12	4.2248E+05	2.3101E+04	7.0532E+04	1.0593E+05	**1.6649E+04**
F13	1.5238E+01	**5.4804E-01**	2.7075E+01	4.4169E+00	4.7292E+00
F14	2.1285E+01	2.2126E+01	2.3325E+01	**2.0844E+01**	2.1229E+01
F15	4.5708E+02	-	**2.8346E+02**	3.8232E+02	3.0640E+02
F16	2.3026E+02	-	1.9536E+02	1.9984E+02	**7.8014E+01**
F17	2.5320E+02	-	3.1450E+02	1.5794E+02	**7.1639E+01**
F18	1.0033E+03	1.4072E+03	9.3804E+02	**9.2262E+02**	9.5502E+02
F19	9.7666E+02	1.3165E+03	9.4219E+02	**9.1941E+02**	9.4735E+02
F20	9.7761E+02	1.3865E+03	9.4763E+02	**9.1692E+02**	9.4149E+02
F21	1.2067E+03	7.9999E+02	6.6371E+02	8.8581E+02	**6.3703E+02**
F22	9.5024E+02	1.2092E+03	9.5214E+02	**9.0646E+02**	9.2008E+02
F23	**6.3049E+02**	7.5279E+02	6.7595E+02	9.2719E+02	6.9500E+02
F24	8.5524E+02	1.2819E+03	2.0000E+02	5.6974E+02	**2.0000E+02**
F25	1.6662E+03	1.2397E+03	1.6795E+03	**2.1600E+02**	1.6539E+03
h_{win}	1	4	1	7	**12**

FIPSaDE with its original variants, that is, FIPS and SaDE are further analyzed based on the best, the mean and standard deviation of f_{best} for different configurations of problem dimensionality and population size for all functions. The f_{best} obtained by FIPSaDE, FIPS and SaDE for all functions are shown in Tables 7–9, summarizing the best, the mean and standard deviation for f_{best} over a number of runs, which can be either 20 runs or 30 runs. The best, mean and standard deviation results of f_{best} are denoted as best-f_{best}, mean-f_{best} and std-f_{best}, respectively. For each table, the algorithm producing the best results for the same function is bolded and its frequency, h_{win} is calculated at the bottom.

For the setting of 10D25N, FIPSaDE has the highest h_{win} for best-f_{best}, mean-f_{best} and std-f_{best}. FIPS and SaDE have similar h_{win} values best-f_{best}, mean-f_{best} for the same setting. When the setting increases to 30D30N and 50D50N, FIPSaDE is still associated with the highest h_{win} for best-f_{best}, mean-f_{best} and std-f_{best}. Therefore, FIPSaDE shows better results for best-f_{best}, mean-f_{best} and std-f_{best} with the increase of problem dimensionality and population size.

FIPS's h_{win} values for best-f_{best}, mean-f_{best} deteriorate from 10D25N to 50D50N. Based on h_{win} for std-f_{best}, FIPS has consistent values that are 6 to 7, regardless of the problem dimensionality and population size. The finding indicates that the frequency of FIPS producing the highest best-f_{best} and mean-f_{best} among the algorithms deteriorates with the increase of problem dimensionality and population size. Additionally, FIPS's frequency in producing the lowest variants of solutions is consistent across the problem dimensionality and population size.

SaDE shows a decreasing trend as FIPS for best-f_{best}, mean-f_{best} when the settings change from 10D25N to 50D50N. However, the decreasing trend is not as drastic as FIPS. Based on the comparisons of std-f_{best}, SaDE and FIPS have similar values for the settings, except 10D25N. SaDE's std-f_{best} = 10 is higher than FIPS for the lower problem dimensionality and population size, 10D25N. The finding indicates that SaDE's frequency in producing lowest variations of solutions is slightly lower at high problem dimensionality and population size.

Table 7. Best-f_{best}, mean-f_{best} and std-f_{best} obtained by FIPS, SaDE and FIPSaDE algorithms for the settings of 10D25N.

Function	Best-f_{best}			Mean-f_{best}			Std-f_{best}		
	FIPS	SaDE	FIPSaDE	FIPS	SaDE	FIPSaDE	FIPS	SaDE	FIPSaDE
F1	−450.00	−450.00	−450.00	−448.31	−450.00	−450.00	2.70	0.00	0.00
F2	−450.00	−450.00	−450.00	−448.28	−450.00	−450.00	2.71	0.00	0.00
F3	−449.99	3085.63	−450.00	−439.33	33350.51	−450.00	15.16	27,313.82	0.00
F4	−449.99	−450.00	−450.00	−439.33	−450.00	−450.00	15.16	0.00	0.00
F5	−310.00	−310.00	−310.00	−310.00	−310.00	−310.00	0.00	0.00	0.00
F6	390.30	390.00	390.00	440.97	391.28	390.53	86.73	1.66	1.38
F7	1087.05	1087.05	1087.05	1087.05	1087.05	1087.05	0.03	0.00	0.00
F8	−119.88	−119.78	−120.00	−119.78	−119.62	−119.89	0.05	0.09	0.10
F9	−329.01	−330.00	−330.00	−321.94	−329.73	−329.50	3.60	0.52	0.51
F10	−327.02	−328.01	−328.01	−321.13	−323.09	−324.16	3.47	2.57	3.21
F11	90.81	90.67	90.54	91.73	93.03	92.31	0.68	1.82	1.30
F12	−460.00	−460.00	−460.00	−179.83	104.07	−151.72	678.16	1064.47	702.17
F13	−129.72	−129.46	−129.82	−129.43	−129.11	−129.67	0.15	0.21	0.08
F14	−298.72	−297.77	−299.74	−297.88	−296.86	−298.04	0.45	0.38	0.61
F15	121.83	178.14	160.45	300.14	401.00	436.17	129.47	148.90	140.78
F16	120.00	209.94	211.25	220.19	222.18	225.14	20.73	7.33	6.98
F17	207.88	214.00	201.17	227.19	239.66	221.36	8.66	17.61	8.77
F18	310.00	310.00	310.00	776.87	711.54	7979.56	239.17	226.16	169.84
F19	310.00	310.00	310.00	781.30	668.47	735.85	242.09	219.37	219.42
F20	310.00	310.00	310.00	750.34	913.31	650.81	277.37	346.97	230.60
F21	660.00	660.00	660.00	1054.22	1061.85	1096.34	275.44	276.19	220.53
F22	660.91	1109.44	1084.87	1073.63	1134.33	1113.92	133.25	22.27	29.15
F23	914.00	919.47	919.47	1140.07	1211.39	1158.06	178.64	254.86	271.52
F24	460.00	460.00	460.00	533.32	460.00	486.54	130.48	0.00	145.39
F25	92.99	1991.62	1995.92	1948.74	1020.66	2009.47	344.45	6.99	5.36
h_{win}	15	14	20	10	9	13	6	10	14

Table 8. Best-f_{best}, mean-f_{best} and std-f_{best} obtained by FIPS, SaDE and FIPSaDE algorithms for the settings of 30D30N.

Function	Best-f_{best}			Mean-f_{best}			Std-f_{best}		
	FIPS	SaDE	FIPSaDE	FIPS	SaDE	FIPSaDE	FIPS	SaDE	FIPSaDE
F1	−297.49	−450.00	−450.00	550.32	−450.00	−450.00	474.17	0.00	0.00
F2	2925.32	−334.80	−450.00	7039.46	141.37	−450.00	2090.36	394.74	0.00
F3	4,662,113.70	807,037.41	19,291.79	9,685,580.44	3,009,720.45	88,664.15	2,937,448.59	1,504,001.37	63,037.56
F4	3522.98	3339.36	−450.00	12,609.08	8382.96	−417.79	3289.58	3329.53	68.06
F5	2752.59	2851.22	1828.14	4763.71	4550.14	2882.76	1143.85	1175.35	534.97
F6	1,311,610.25	393.12	390.00	40,001,127.07	534.04	391.20	41,151,416.49	195.57	1.86
F7	4516.29	4516.29	4516.29	4516.29	4516.29	4516.29	0.00	0.00	0.00
F8	−119.18	−119.24	−119.21	−119.10	−118.95	−119.11	0.05	0.06	0.05
F9	−274.12	−325.88	−329.01	−238.80	−311.72	−319.73	15.94	7.87	4.57
F10	−257.68	−303.46	−307.12	−216.13	−212.55	−288.20	19.61	52.41	9.80
F11	108.32	110.59	104.48	112.14	125.38	110.23	1.49	5.28	2.56
F12	44,816.64	1924.29	−457.54	92,954.31	15,631.41	5948.00	35,286.01	13,703.43	9314.81
F13	−125.85	−120.51	−128.55	−123.16	−117.97	−127.62	1.59	1.15	0.59
F14	−288.87	−286.68	−289.13	−288.18	−286.46	−288.01	0.33	0.13	0.50
F15	518.41	230.34	320.00	606.35	447.49	455.94	45.94	106.58	97.05
F16	227.57	179.27	165.78	369.64	309.45	249.52	151.59	138.42	125.34
F17	238.75	297.00	172.70	366.55	418.69	268.73	141.11	92.36	109.60
F18	916.04	810.00	915.42	920.85	918.09	922.17	6.09	21.18	5.81
F19	851.22	810.00	915.96	917.97	917.60	921.58	19.54	20.83	4.63
F20	915.69	915.39	810.00	921.96	923.04	918.81	7.84	4.70	20.93
F21	953.46	860.00	860.00	1292.43	860.00	892.10	260.10	0.00	131.12
F22	1213.63	1245.75	1209.77	1256.86	1305.60	1251.15	22.58	41.09	23.76
F23	901.56	896.74	894.16	950.64	976.97	959.16	97.57	133.89	158.86
F24	550.61	460.00	460.00	675.47	460.00	485.29	76.41	0.00	138.53
F25	1880.42	1891.12	1868.00	1891.27	1904.77	1877.14	5.85	6.17	4.40
h_{win}	1	7	22	3	7	18	6	6	16

Table 9. Best-f_{best}, mean-f_{best} and std-f_{best} obtained by FIPS, SaDE and FIPSaDE algorithms for the settings of 50D50N.

Function	Best-f_{best}			Mean-f_{best}			Std-f_{best}		
	FIPS	SaDE	FIPSaDE	FIPS	SaDE	FIPSaDE	FIPS	SaDE	FIPSaDE
F1	1153.25	450.00	**−450.00**	2462.33	450.00	**−450.00**	1044.25	0.00	0.00
F2	12,312.68	967.61	**−450.00**	19,296.90	3624.73	**−450.00**	2864.28	1944.03	0.00
F3	19,742,966.18	**1.55**	26,213.65	46,306,427.65	6,442,716.38	**93,924.17**	18,623,890.22	2,656,381.58	54,148.58
F4	18,353.52	11,796.54	**−448.82**	29,351.56	26,706.41	**−364.74**	7366.24	7776.96	94.60
F5	8344.21	7515.12	**3823.55**	12,104.89	9399.31	**5266.12**	1713.64	1378.09	924.29
F6	13,540,381.90	490.65	**390.00**	110,811,270.60	738.73	**391.79**	55,887,725.95	277.30	2.03
F7	6015.32	6015.32	6015.32	6015.32	6015.32	6015.32	0.00	0.00	0.00
F8	**−118.99**	118.75	−118.97	**−118.92**	118.81	−118.92	0.03	0.04	0.03
F9	−174.99	208.21	**−326.02**	−137.61	238.84	**−314.98**	24.75	15.11	4.92
F10	−136.98	0.90	**−261.35**	−72.16	60.49	**−236.08**	36.26	73.04	16.82
F11	129.40	156.93	**120.76**	134.39	160.20	**129.00**	2.63	1.75	3.64
F12	241,193.33	19,055.20	**−154.39**	422,016.01	70,072.02	**16,188.50**	115,690.78	40,227.17	11,162.45
F13	−116.34	99.05	**−126.80**	−114.76	102.93	**−125.27**	1.39	2.25	0.94
F14	−279.28	276.38	**−280.33**	−278.71	276.68	**−278.77**	0.30	0.16	0.53
F15	426.38	**320.00**	323.00	577.08	**403.46**	426.40	38.77	81.82	86.84
F16	244.32	191.77	**160.24**	350.26	315.36	**198.01**	105.18	78.92	37.68
F17	283.90	378.02	**170.74**	373.20	434.50	**191.64**	102.40	56.01	11.83
F18	935.54	**810.00**	938.12	1013.28	**984.04**	965.02	41.81	34.89	17.86
F19	942.46	935.18	**929.32**	986.66	**952.19**	957.35	28.70	11.17	15.08
F20	917.00	939.19	**916.08**	987.61	957.63	**951.49**	37.23	15.20	14.64
F21	1412.69	**860.00**	860.00	1566.71	1023.71	**997.03**	37.54	258.81	246.13
F22	1288.45	1268.20	**1243.66**	1310.24	1312.14	**1280.08**	9.76	29.34	24.74
F23	928.47	**899.18**	899.12	**990.49**	1035.95	1055.00	105.67	203.07	216.18
F24	728.10	**460.00**	460.00	1115.24	**460.00**	460.00	301.48	0.00	0.00
F25	1912.16	1915.96	**1902.75**	1926.19	1939.45	**1913.91**	5.80	9.62	7.03
h_{win}	2	6	21	3	5	21	7	6	17

The overall findings from the comparisons show that FIPSaDE has the highest probability of producing a better solution compared to its original variants, which are FIPS and SaDE, with an increase in problem dimensionality and population size. An EC algorithm commonly uses a large population size to solve an optimization problem with high dimensionality. However, the approach may be more effective for FIPSaDE than FIPS and SaDE. Based on the comparisons of h_{win} for best-f_{best} and mean-f_{best}, FIPSaDE has the highest values, followed by FIPS and SaDE. Therefore, FIPSaDE consistently produces a group of better-quality solutions than its original variants when problem dimensionality and population size increase.

For most tested functions, FIPSaDE proves to produce better best-f_{best} and mean-f_{best} in the swarm space compared to the other algorithms as the problem space increases. FIPSaDE, a hybrid of both FIPS and SaDE, can maneuver and manage the swarm solutions more effectively. Therefore, FIPSaDE improves performance in finding the best fitness point as the problem space increases.

The variable mean-f_{best} for FIPSaDE, FIPS and SaDE is denoted as $\mu_{FIPSaDE}$, μ_{FIPS} and μ_{SaDE}, respectively. A t-test is performed with a significance level, α of 0.05 to evaluate the hypothesis whether $\mu_{FIPSaDE}$ of the proposed algorithm is better than μ_{FIPS} and μ_{SaDE} or not, as shown below.

$$H_0 : \mu_{FIPSaDE} \geq \mu_{FIPS}$$
$$H_1 : \mu_{FIPSaDE} < \mu_{FIPS}$$

$$H_0 : \mu_{FIPSaDE} \geq \mu_{SaDE}$$
$$H_1 : \mu_{FIPSaDE} < \mu_{SaDE}$$

The results of hypothesis tests for the paired FIPSaDE-FIPS and FIPS-SaDE based on different settings of problem dimensionality and population size are shown in Table 10. If the p-values are less α, they are bolded, indicating that the associated H_0 is rejected. As a result, there is enough evidence to accept H_1. The p-values associated with rejecting H_0 are bolded and its frequency is denoted as $h_{(-H_0)}$. At the bottom of the table, $h_{(-H_0)}$ shows the frequency H_0 being rejected for different settings. For the setting 10D25N, $\mu_{FIPSaDE}$ is significantly better than μ_{FIPS} and μ_{SaDE} in 10 and 9 functions, respectively. When the setting increases from 10D25N to 50D50N, the values $h_{(-H_0)}$ for FIPSaDE-FIPS increase from 10 to 18 and 22. Therefore, the frequency that FIPSaDE is significantly better than FIPS increases. When similar comparisons are made for FIPSaDE-SaDE, an increasing trend is observed, but it is less obvious. The findings about $h_{(-H_0)}$ show that the performances of FIPSaDE are significantly better than FIPS and SaDE when the problem dimensionality and population size increase.

Table 10. Hypothesis test between $\mu_{FIPSaDE}$, μ_{FIPS} and μ_{SaDE} for the settings of 10D25N, 30D30N and 50D50N.

Function	10D25N				30D30N				50D50N			
	t-Value		p-Value		t-Value		p-Value		t-Value		p-Value	
	FIPSaDE-FIPS	FIPSaDE-SaDE	FIPSaDE-FIPS	FIPSaDE-SaDE	FIPSaDE-FIPS	FIPSaDE-SaDE	FIPSaDE-FIPS	FIPSaDE-SaDE	FIPSaDE-FIPS	FIPSaDE-SaDE	FIPSaDE-FIPS	FIPSaDE-SaDE
F1	−3.44	28.07	**0.00**	1.00	−11.55	5.60	**0.00**	1.00	−12.47	−7.20E+12	**0.00**	**0.00**
F2	−3.47	42.75	**0.00**	1.00	−19.62	−8.21	**0.00**	**0.00**	−30.83	−9.37	**0.00**	**0.00**
F3	−3.86	−6.78	**0.00**	**0.00**	−17.89	−10.63	**0.00**	**0.00**	−11.10	−10.69	**0.00**	**0.00**
F4	−3.86	0.31	**0.00**	0.62	−21.69	−14.47	**0.00**	**0.00**	−18.04	−15.57	**0.00**	**0.00**
F5	66.38	63.86	1.00	1.00	−8.16	−7.07	**0.00**	**0.00**	−15.71	−11.14	**0.00**	**0.00**
F6	−3.18	−1.90	**0.00**	**0.03**	−5.32	−4.00	**0.00**	**0.00**	−8.87	−5.60	**0.00**	**0.00**
F7	−1.00	0.00	0.16	0.50	−18.08	−18.08	**0.00**	**0.00**	−1.76	−4.18	**0.04**	**0.00**
F8	−5.46	−11.41	**0.00**	**0.00**	−1.37	−11.75	0.09	**0.00**	0.51	−2.05E+04	0.69	**0.00**
F9	−11.41	1.76	**0.00**	0.96	−26.73	−4.82	**0.00**	**0.00**	−31.44	−155.85	**0.00**	**0.00**
F10	−3.49	−1.43	**0.00**	0.08	−18.01	−7.77	**0.00**	**0.00**	−18.34	−17.69	**0.00**	**0.00**
F11	2.16	−1.79	0.98	**0.04**	−3.53	−14.13	**0.00**	**0.00**	−5.37	−34.53	**0.00**	**0.00**
F12	0.16	−1.10	0.56	0.14	−13.06	−3.20	**0.00**	**0.00**	−15.62	−5.77	**0.00**	**0.00**
F13	−7.76	−13.78	**0.00**	**0.00**	−14.38	−40.84	**0.00**	**0.00**	−27.94	−417.89	**0.00**	**0.00**
F14	−1.14	−8.98	0.13	**0.00**	1.52	−16.31	0.93	**0.00**	−0.41	−4.44E+03	0.34	**0.00**
F15	3.90	0.94	1.00	0.82	−6.45	0.32	**0.00**	0.63	−7.09	0.86	**0.00**	0.80
F16	1.24	1.60	0.89	0.94	−3.34	−1.76	**0.00**	**0.04**	−6.09	−6.00	**0.00**	**0.00**
F17	−2.59	−5.09	**0.01**	**0.00**	−3.00	−5.73	**0.00**	**0.00**	−7.88	−18.97	**0.00**	**0.00**
F18	0.39	1.67	0.65	0.95	0.86	1.02	0.80	0.84	−4.65	1.94	**0.00**	0.97
F19	−0.76	1.19	0.22	0.88	0.98	1.02	0.84	0.84	−4.04	1.23	**0.00**	0.89
F20	−1.51	−3.45	0.07	**0.00**	−0.77	−1.08	0.22	0.14	−4.04	−1.30	**0.00**	0.10
F21	0.65	0.53	0.74	0.70	−7.53	1.34	**0.00**	0.91	−10.23	−0.33	**0.00**	0.37
F22	1.62	−3.05	0.94	**0.00**	−0.96	−6.28	0.17	**0.00**	−5.07	−3.74	**0.00**	**0.00**
F23	0.30	−0.78	0.62	0.62	0.25	−0.47	0.60	0.32	1.20	0.29	0.88	0.61
F24	−1.30	1.00	0.10	0.84	−6.58	1.00	**0.00**	0.84	−9.72	−1.01	**0.00**	0.16
F25	0.97	−0.74	0.83	0.23	−10.57	−19.98	**0.00**	**0.00**	−6.02	−9.59	**0.00**	**0.00**
$h_{(-H_0)}$	-	-	10	9	-	-	18	17	-	-	22	18

6. Conclusions

In this study, a hybrid of FIPS and SaDE called FIPSaDE is proposed, and its performance is validated by running the algorithm against the benchmark functions while also being compared to their respective original version, the SaDE and FIPS as well as it variants, such as DE and DE-PSO. The self-adaptation strategy of SaDE is adapted and maneuvered by the FIPS particle swarm, preventing the solutions from being trapped in the local region. Each algorithm can adaptively adjust the parameter values while the swarm is searching for the best solution needed. Based on different configurations of problem dimensionality and population sizes, the FIPSaDE algorithm consistently has the highest frequency of having lowest average errors than FIPS, DE, SaDE and DE-PSO. The frequency analysis of h_{win} and the hypothesis test show that FIPSaDE performs better than its respective original versions in terms of the best and mean of f_{best} as the problems' dimensionality increases. Future research will investigate the strength of proposed algorithms with other benchmark test functions. Since the current FIPSaDE is only tested on one topology, that is, Four Clusters, the impact of the other four other FIPS topologies, namely, All, Ring, Pyramid, Square, should be investigated in the future for their effects on the hybrid's performances. Moreover, various performance metrics could be applied to strengthen the comparison among the algorithms. The study of the convergence profiles and coverage curve of the proposed algorithm as compared to other algorithms could also be investigated in the future. Furthermore, investigation should be conducted to broaden the comparison among the hybridization methods.

Author Contributions: Conceptualization, S.L.W., S.H.A. and T.F.N.; Data curation, S.L.W., S.H.A., H.I. and P.R.; Formal analysis, S.L.W., S.H.A. and P.R.; Funding acquisition, T.F.N.; Investigation, S.L.W. and S.H.A.; Methodology, S.L.W., S.H.A., H.I. and T.F.N.; Project administration, T.F.N.; Resources, S.H.A., H.I. and P.R.; Software, S.H.A., H.I. and P.R.; Supervision, S.L.W., H.I. and T.F.N.; Validation, S.L.W. and T.F.N.; Visualization, S.L.W., S.H.A. and P.R.; Writing—Original draft, S.L.W., S.H.A., H.I. and T.F.N.; Writing—Review and editing, S.L.W., H.I. and T.F.N. All authors have read and agreed to the published version of the manuscript.

Funding: This research was funded by the Ministry of Higher Education Malaysia under the Fundamental Research Grant Scheme (FRGS) with grant number FRGS/1/2022/ICT02/UPSI/02/1.

Institutional Review Board Statement: Not applicable.

Informed Consent Statement: Not applicable.

Data Availability Statement: The original contributions presented in the study are included in the article. Further inquires can be directed to the corresponding author.

Conflicts of Interest: The authors declare no conflict of interest.

Abbreviations

The following abbreviations are used in this manuscript:

ABC	Artificial bee colony
ACA	Ant colony algorithm
CGSS	Centre of Global Sustainability Studies
DE	Differential evolution
EC	Evolutionary computation
FIPS	Fully informed particle swarm
FIPSaDE	FIPS-SaDE
GUI	Graphical user interface
NP	nondeterministic polynomial
PSO	Particle swarm optimization
UPSI	Universiti Pendidikan Sultan Idris
USM	Universiti Sains Malaysia
SaDE	Self-adaptive differential evolution

References

1. Mendes, R.; Kennedy, J.; Neves, J. The fully informed particle swarm: Simpler, maybe better. *IEEE Trans. Evol. Comput.* **2004**, *8*, 204–210. [CrossRef]
2. Zhalechian, M.; Tavakkoli-Moghaddam, R.; Rahimi, Y. A self-adaptive evolutionary algorithm for a fuzzy multi-objective hub location problem: An integration of responsiveness and social responsibility. *Eng. Appl. Artif. Intell.* **2017**, *62*, 1–16. [CrossRef]
3. Surekha, P.; Archana, N.; Sumathi, S. Unit commitment and economic load dispatch using self adaptive differential evolution. *Wseas Trans. Power Syst.* **2012**, *7*, 159–171.
4. Chen, Y.; Mahalex, V.; Chen, Y.; He, R.; Liu, X. Optimal satellite orbit design for prioritized multiple targets with threshold observation time using self-adaptive differential evolution. *J. Aerosp. Eng.* **2015**, *28*, 04014066. [CrossRef]
5. Das, S.; Mullick, S.S.; Suganthan, P.N. Recent advances in differential evolution–an updated survey. *Swarm Evol. Comput.* **2016**, *27*, 1–30. [CrossRef]
6. Al-Anzi, F.S.; Allahverdi, A. A self-adaptive differential evolution heuristic for two-stage assembly scheduling problem to minimize maximum lateness with setup times. *Eur. J. Oper. Res.* **2007**, *182*, 80–94. [CrossRef]
7. Ali, M.; Ahn, C.W. An optimized watermarking technique based on self-adaptive DE in DWT-SVD transform domain. *Signal Process.* **2014**, *94*, 545–556. [CrossRef]
8. Brest, J.; Greiner, S.; Boskovic, B.; Mernik, M.; Zumer, V. Self-adapting control parameters in differential evolution: A comparative study on numerical benchmark problems. *IEEE Trans. Evol. Comput.* **2006**, *10*, 646–657. [CrossRef]
9. Parouha, R.P.; Verma, P. A systematic overview of developments in differential evolution and particle swarm optimization with their advanced suggestion. *Appl. Intell.* **2022**, *52*, 10448–10492. [CrossRef]

10. Storn, R.; Price, K. Differential evolution–A simple and efficient heuristic for global optimization over continuous spaces. *J. Glob. Optim.* **1997**, *11*, 341–359. [CrossRef]
11. Bilal; Pant, M.; Zaheer, H.; Garcia-Hernandez, L.; Abraham, A. Differential Evolution: A review of more than two decades of research. *Eng. Appl. Artif. Intell.* **2020**, *90*, 103479. [CrossRef]
12. Wang, S.L.; Morsidi, F.; Ng, T.F.; Budiman, H.; Neoh, S.C. Insights into the effects of control parameters and mutation strategy on self-adaptive ensemble-based differential evolution. *Inf. Sci.* **2020**, *514*, 203–233. [CrossRef]
13. Al-Dabbagh, R.D.; Neri, F.; Idris, N.; Baba, M.S. Algorithmic design issues in adaptive differential evolution schemes: Review and taxonomy. *Swarm Evol. Comput.* **2018**, *43*, 284–311. [CrossRef]
14. Das, S.; Suganthan, P.N. Differential evolution: A survey of the state-of-the-art. *IEEE Trans. Evol. Comput.* **2011**, *15*, 4–31. [CrossRef]
15. Neri, F.; Tirronen, V. Recent advances in differential evolution: A survey and experimental analysis. *Artif. Intell. Rev.* **2010**, *33*, 61–106. [CrossRef]
16. Parouha, R.P.; Verma, P. An innovative hybrid algorithm for bound-unconstrained optimization problems and applications. *J. Intell. Manuf.* **2022**, *33*, 1273–1336. [CrossRef]
17. Tao, X.; Guo, W.; Li, Q.; Ren, C.; Liu, R. Multiple scale self-adaptive cooperation mutation strategy-based particle swarm optimization. *Appl. Soft Comput.* **2020**, *89*, 106124. [CrossRef]
18. Shami, T.M.; El-Saleh, A.A.; Alswaitti, M.; Al-Tashi, Q.; Summakieh, M.A.; Mirjalili, S. Particle Swarm Optimization: A Comprehensive Survey. *IEEE Access* **2022**, *10*, 10031–10061. [CrossRef]
19. Jain, M.; Saihjpal, V.; Singh, N.; Singh, S.B. An Overview of Variants and Advancements of PSO Algorithm. *Appl. Sci.* **2022**, *12*, 8392. [CrossRef]
20. Chen, Y.; Liang, J.; Wu, Y.; He, B.; Lin, L.; Wang, Y. Self-Regulating and Self-Perception Particle Swarm Optimization with Mutation Mechanism. *J. Intell. Robot. Syst.* **2022**, *105*, 1–21. [CrossRef]
21. Wang, S.; Li, Y.; Yang, H. Self-adaptive mutation differential evolution algorithm based on particle swarm optimization. *Appl. Soft Comput.* **2019**, *81*, 105496. [CrossRef]
22. Dash, J.; Dam, B.; Swain, R. Design and implementation of sharp edge FIR filters using hybrid differential evolution particle swarm optimization. *Aeu-Int. J. Electron. Commun.* **2020**, *114*, 153019. [CrossRef]
23. Yang, X.S. Nature-inspired optimization algorithms: Challenges and open problems. *J. Comput. Sci.* **2020**, *46*, 101104. [CrossRef]
24. Mallipeddi, R.; Suganthan, P.N.; Pan, Q.K.; Tasgetiren, M.F. Differential evolution algorithm with ensemble of parameters and mutation strategies. *Appl. Soft Comput.* **2011**, *11*, 1679–1696. [CrossRef]
25. Ghosh, A.; Das, S.; Das, A.K.; Senkerik, R.; Viktorin, A.; Zelinka, I.; Masegosa, A.D. Using spatial neighborhoods for parameter adaptation: An improved success history based differential evolution. *Swarm Evol. Comput.* **2022**, *71*, 101057. [CrossRef]
26. Do, D.T.; Lee, S.; Lee, J. A modified differential evolution algorithm for tensegrity structures. *Compos. Struct.* **2016**, *158*, 11–19. [CrossRef]
27. Piotrowski, A.P. Review of differential evolution population size. *Swarm Evol. Comput.* **2017**, *32*, 1–24. [CrossRef]
28. Borowska, B. Learning Competitive Swarm Optimization. *Entropy* **2022**, *24*, 283. [CrossRef]
29. Qin, A.K.; Huang, V.L.; Suganthan, P.N. Differential evolution algorithm with strategy adaptation for global numerical optimization. *IEEE Trans. Evol. Comput.* **2009**, *13*, 398–417. [CrossRef]
30. Cleghorn, C.W.; Engelbrecht, A. Fully informed particle swarm optimizer: Convergence analysis. In Proceedings of the 2015 IEEE Congress on Evolutionary Computation (CEC), Sendai, Japan, 25–28 May 2015; pp. 164–170. [CrossRef]
31. Pant, M.; Thangaraj, R.; Grosan, C.; Abraham, A. Hybrid differential evolution-particle swarm optimization algorithm for solving global optimization problems. In Proceedings of the 2008 Third International Conference on Digital Information Management, London, UK, 13–16 November 2008; pp. 18–24.
32. Suganthan, P.N.; Hansen, N.; Liang, J.J.; Deb, K.; Chen, Y.P.; Auger, A.; Tiwari, S. Problem definitions and evaluation criteria for the CEC 2005 special session on real-parameter optimization. *KanGAL Rep.* **2005**, *2005005*, 2005.

Article

Optimization on Linkage System for Vehicle Wipers by the Method of Differential Evolution

Tsai-Jung Chen [1], Ying-Ji Hong [2,*], Chia-Han Lin [1] and Jing-Yuan Wang [1]

[1] Department of Vehicle Engineering, National Pingtung University of Science and Technology, Pingtung 912, Taiwan
[2] Department of Mathematics, National Cheng-Kung University, Tainan City 701, Taiwan
* Correspondence: yjhong@mail.ncku.edu.tw

Abstract: We consider an optimization problem on the maximal magnitude of angular acceleration of the output-links of a commercially available center-driven linkage system (CDLS) for vehicle wipers on windshield. The purpose of this optimization is to improve the steadiness of a linkage system without weakening its normal function. Thus this optimization problem is considered under the assumptions that the frame of the fixed links of linkage system is unchanged and that the input-link rotates at the same constant angular speed with its length unchanged. To meet the usual requirements for vehicle wipers on windshield, this optimization problem must be solved subject to 10 specific constraints. We expect that optimizing the maximal magnitude of angular acceleration of the output-links of a linkage system would also be helpful for reducing the amplitudes of sound waves of wiper noise. We establish the motion model of CDLS and then justify this model with ADAMS. We use a "Differential Evolution" type method to search for the minimum of an objective function subject to 10 constraints for this optimization problem. Our optimization computation shows that the maximal magnitude of angular acceleration of both output-links of this linkage system can be reduced by more than 10%.

Keywords: intelligence optimization algorithm; differential evolution; linkage system; vehicle wipers

1. Introduction

Linkage systems are widely used on vehicles. Research on the design of linkage systems is important for vehicle industry development. In this article, we consider the linkage systems for vehicle wipers. Our goal is to improve the *steadiness* of a linkage system for vehicle wipers without weakening its normal function. This improvement on the *steadiness* of a linkage system for vehicle wipers could reduce the possible material fatigue of a linkage system. Besides, this improvement could also be helpful for reducing the unpleasant rubber wiper noise on vehicle windshield.

The linkage systems considered in this article are the center driven linkage systems (CDLS). A center driven linkage system is a composition of two crank-rocker linkage systems driven by a single central input-link. A crank-rocker linkage system is shown in the following Figure 1.

In Figure 1, the link OB is a fixed link with length q. The link OA, with length s, may rotates full 360 degrees and is called the "crank" of this linkage system. Usually the crank is driven by a motor. The link BC, with length p, may only moves on a limited circular sector and is called the "rocker" of this linkage system. The link AC usually has the longest length l and acts as the transfer link of this linkage system. Generally the Grashof law

$$s + l < p + q$$

for crank-rocker system must be satisfied. See [1] for more discussions on linkage systems.

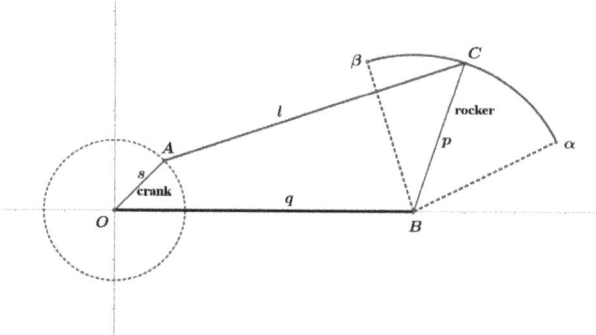

Figure 1. Crank-rocker linkage system.

To improve the *steadiness* of a linkage system without weakening its normal function, we will consider an optimization problem on the maximal magnitude of *angular acceleration* of the output-links of a CDLS under the assumptions that the frame of the fixed links of linkage system is unchanged and that the input link rotates at the same (constant) angular speed with its length unchanged. To meet the usual requirements for vehicle wipers on windshield, this optimization problem must be solved subject to 10 specific constraints.

To tackle this optimization problem, we will use a Differential Evolution type search method. The usual stages of a DE search algorithm is shown in the following Figure 2.

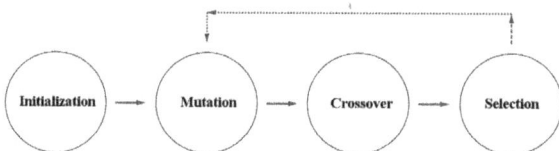

Figure 2. Usual stages of DE search algorithm.

The method of Differential Evolution (DE) was introduced by Storn and Price in 1995 [2–4]. In this DE search method, two important parameters F (the *scale factor*) and CR (the *crossover ratio*), for the mutation and crossover operations of DE, can be selected respectively from the interval $[0, 1]$. These parameters are important both for controlling the *diversity* of "generation population" and for controlling the *convergence rate* of DE.

The space complexity of DE is usually relatively low, when compared with the highly efficient "restart covariance matrix adaptation evolution strategies (restart CMA-ES)". Besides, the control parameters F (the *scale factor*) and CR (the *crossover ratio*) can be adjusted to improve the performance of the DE algorithm easily *without* causing serious *computational burden*. Moreover, the introduction of mutation operator by "difference vectors" in DE obviously speed up the computational convergence of DE.

Because of its simplicity and its experimental efficiency, this DE search method soon has become very popular. Since then, various variants of DE were presented. For examples, there are DE using "Trigonometric Mutation" [5]; "opposition-based learning" DE [6–9]; DEGL with Neighborhood-Based Mutation [10]; SADE (DE with parameters adjusted through self-adaptive learning) [11–14]. See [15–17] for more information on the various variants of DE.

Though various complicated types of DE methods were presented, research articles on the theoretical analysis of DE are rare. Most of the variants of DE mentioned above were presented only with experimental studies. For classical DE algorithms, "weak convergence" for most objective functions were proved under reasonable assumptions in [18–20]. The theoretical analysis of the DE search methods in [18–20] provides the foundation for the convergence of the DE type search method adopted in this article.

Our objective function for this optimization problem is defined to be the sum of the maximal absolute values of *angular acceleration* of the output-links of this linkage system:

$$f(r_3, r_4, r_5, r_6) = \max_{0 \leq \theta_2 \leq 2\pi} |\alpha_4(r_3, r_4, \theta_2)| + \max_{0 \leq \theta_2 \leq 2\pi} |\alpha_6(r_5, r_6, \theta_2)|.$$

This optimization problem must be solved subject to 10 specific constraints. These constraints are from the usual requirements for vehicle wipers on windshield. This objective function is a function of the connecting-links and the output-links respectively on the driver side and on the passenger side of CDLS. Details of this objective function will be discussed in Section 2.

We assume that the input link of this CDLS rotates at a *constant speed* with angular velocity of 1 rad/s. Our optimization computation for this linkage system shows that the maximal magnitude of the angular acceleration of the output-link on the driver side of this CDLS can be reduced by 10.69%. Besides, our optimization computation for this linkage system shows that the maximal magnitude of the angular acceleration of the output-link on the passenger side of this CDLS can be reduced by 12.10%. Interestingly, our algorithm takes only 30 seconds to finish its searching process for the minimum of the objective function for this optimization problem subject to 10 specific constraints.

This improvement on the *steadiness* of a linkage system for vehicle wipers is related to our earlier work [21] on the vibration frequencies of rubber wiper on *convex* windshield. We expect that improvement on the *steadiness* of a linkage system for vehicle wipers would be helpful for reducing the amplitudes of sound waves of wiper noise around each reversal of the motion of rubber wipers on windshield.

Optimization problems on the "expectation output function" of a linkage system were considered before by some people using Genetic Algorithms on MATLAB. See, for example [22].

2. Materials and Methods

A flowchart of our method is shown in Figure 3. In this section, we will discuss a kinematic model of CDLS. On the basis of this model, mathematical formulas for the angles, the angular velocities, the values of angular acceleration, and the transmission angle of various parts of a CDLS will be established. Numerical simulations of the motion of a CDLS will be performed on MATLAB based on these mathematical formulas. The objective function is defined as follows:

$$f(r_3, r_4, r_5, r_6) = \max_{0 \leq \theta_2 \leq 2\pi} |\alpha_4(r_3, r_4, \theta_2)| + \max_{0 \leq \theta_2 \leq 2\pi} |\alpha_6(r_5, r_6, \theta_2)|$$

in which α_4 and α_6 are the values of angular acceleration of the output-links respectively on the driver side and on the passenger side of a CDLS. We will discuss the mathematical formulas for α_4 and for α_6 in Section 2.1, where we discuss the motion model of CDLS. We assume that the input link of CDLS rotates at a *constant speed* with angular velocity of 1 rad/s.

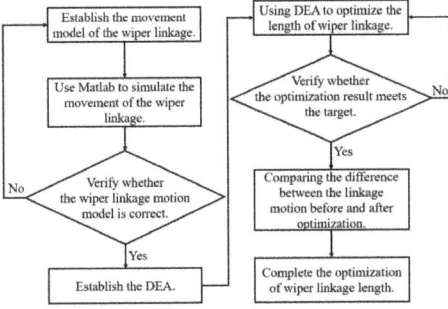

Figure 3. Flowchart of our research method.

We will discuss a method of Differential Evolution (DE) in Section 2.2. Then, in Section 2.3, we will use the method of DE to search for the minimum of the objective function $f = \max\limits_{0 \leq \theta_2 \leq 2\pi} |\alpha_4| + \max\limits_{0 \leq \theta_2 \leq 2\pi} |\alpha_6|$ subject to 10 specific constraints.

2.1. Establishing the Wiper Linkage Motion Model

In Figure 4, the structure of a Center Driven Linkage System (CDLS) is illustrated. A simplified planar model of CDLS is shown in Figure 5. On this planar model, we choose A and AD respectively as the origin and the *x*-axis of our Cartesian (*x*, *y*) coordinate system.

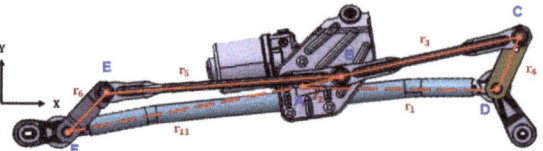

Figure 4. Structure of a center driven linkage system (CDLS).

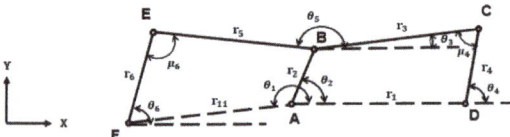

Figure 5. Simplified planar model of the CDLS.

We will use mm as the length unit for the link components of a CDLS. We will use radian as the unit for the angles in the following discussions on CDLS.

The linkage, consisting of Link AB, Link BC, Link DC and Link AD, is responsible for driving the wiper on the driver side of CDLS. Link AD is a fixed link. Link AB is the input-link that receives power to drive the links BC and DC. Link BC and Link DC are respectively the connecting link and the output link on the driver side of CDLS. The angle μ_4 between CB and CD is usually called the *transmission angle* on the driver side of CDLS [23]. The transmission angle μ_4 changes as the input-link AB rotates. The mechanism is most efficient (with the effective torque maximized) when the transmission angle is $\pi/2$ (rad). Usually the transmission angle is required to move in the range from 40° to 140° [23].

The linkage, consisting of Link AB, Link BE, Link FE and Link AF is responsible for driving the wiper on the passenger side of CDLS. Link AF is a fixed link. Link AB is the input-link that receives power to drive the links BE and FE. Link BE and Link FE are respectively the connecting link and the output link on the passenger side of CDLS. The angle μ_6 between EB and EF is usually called the *transmission angle* on the passenger side of CDLS. The mechanism is most efficient (with the effective torque maximized) when the transmission angle is $\pi/2$ (rad).

The lengths of the links AB, BC, CD and AD are respectively denoted by r_2, r_3, r_4 and r_1. The angle between AB and AD, the angle between BC and AD, and the angle between DC and AD are respectively denoted by θ_2, θ_3, and θ_4. See Figure 5. The lengths of the links BE, FE and AF are respectively denoted by r_5, r_6 and r_{11}. The angle between BE and AD, the angle between FE and AD, and the angle between AF and AD are respectively denoted by θ_5, θ_6, and θ_1. See Figure 5.

In the following paragraphs, we will discuss some mathematical formulas for this planar model.

2.1.1. Motion of the Driver Side Linkage of CDLS

Using the vector loop method, we have, in view of Figure 5, the following equality:

$$\vec{AB} + \vec{BC} + \vec{CD} + \vec{DA} = \vec{0} \tag{1}$$

Expressing this vector equality in terms of (x, y) coordinates, we have the following relations:

$$r_2 \cdot \cos\theta_2 + r_3 \cdot \cos\theta_3 - r_4 \cdot \cos\theta_4 - r_1 = 0 \quad (2)$$

and

$$r_2 \cdot \sin\theta_2 + r_3 \cdot \sin\theta_3 - r_4 \cdot \sin\theta_4 = 0 \quad (3)$$

in which θ_2 is the input variable (of the input link AB) and θ_4 is the unknown output (of the output link DC) to be determined. Using the Equations (2) and (3), we may obtain the following relation without θ_3:

$$A \cdot \cos\theta_4 + B \cdot \sin\theta_4 + C = 0 \quad (4)$$

in which $A = -2 \cdot r_2 \cdot r_4 \cdot \cos\theta_2 + 2 \cdot r_1 \cdot r_4$, $B = -2 \cdot r_2 \cdot r_4 \cdot \sin\theta_2$ and $C = r_2^2 + r_4^2 + r_1^2 - r_3^2 - 2 \cdot r_1 \cdot r_2 \cdot \cos\theta_2$.

Let

$$\beta = \tan\left(\frac{\theta_4}{2}\right)$$

We may use the half-angle formula for tan to express (4) in the following form

$$(C - A) \cdot \beta^2 + 2B \cdot \beta + (C + A) = 0 \quad (5)$$

Solving (5) for β, we find that

$$\beta = \frac{-B \pm \sqrt{B^2 - (C^2 - A^2)}}{C - A}$$

and so

$$\theta_4 = 2 \cdot \arctan\left[\frac{-B \pm \sqrt{B^2 - (C^2 - A^2)}}{C - A}\right] \quad (6)$$

In (6), one solution of θ_4 is for the motion of linkage *above* the x-axis AD, and the other solution of θ_4 is for the motion of linkage *below* the x-axis AD. Thus the solution of θ_4 we adopt is the following

$$\theta_4 = 2 \cdot \arctan\left[\frac{-B + \sqrt{B^2 - (C^2 - A^2)}}{C - A}\right] \quad (7)$$

Substituting the result (7) into the relations (2) and (3), we find that

$$\theta_3 = \arctan\left[\frac{-r_2 \cdot \sin\theta_2 + r_4 \cdot \sin\theta_4}{-r_2 \cdot \cos\theta_2 + r_4 \cdot \cos\theta_4 + r_1}\right] \quad (8)$$

Note that the angular velocity ω_2 (the derivative of θ_2 with respect to time) of Link AB is the *constant* 1 rad/s. Differentiating the Equations (2) and (3) with respect to the time variable t, we may find the angular velocities ω_3 (the derivative of θ_3 with respect to time) and ω_4 (the derivative of θ_4 with respect to time) respectively of Link BC and of Link DC:

$$\omega_3 = \frac{r_2 \cdot \omega_2 \cdot \sin(\theta_2 - \theta_4)}{r_3 \cdot \sin(\theta_4 - \theta_3)} \quad (9)$$

and

$$\omega_4 = \frac{r_2 \cdot \omega_2 \cdot \sin(\theta_2 - \theta_3)}{r_4 \cdot \sin(\theta_4 - \theta_3)} \quad (10)$$

Since the angular velocity ω_2 of Link AB is the *constant* 1 rad/s, we note that the *angular acceleration* α_2 of Link AB is 0. By differentiating the relations (2) and (3) twice with respect to the time variable t, we may find the angular acceleration α_3 and the angular acceleration α_4 respectively of Link BC and of Link DC:

$$\alpha_3 = \frac{r_2 \cdot [\omega_2^2 \cdot \cos(\theta_2 - \theta_4) + \alpha_2 \cdot \sin(\theta_2 - \theta_4)]}{r_3 \cdot \sin(\theta_4 - \theta_3)} + \frac{r_3 \cdot \omega_3^2 \cdot \cos(\theta_3 - \theta_4) - r_4 \cdot \omega_4^2}{r_3 \cdot \sin(\theta_4 - \theta_3)} \quad (11)$$

and

$$\alpha_4 = \frac{r_2 \cdot [\omega_2^2 \cdot \cos(\theta_2 - \theta_3) + \alpha_2 \cdot \sin(\theta_3 - \theta_2)]}{r_4 \cdot \sin(\theta_4 - \theta_3)} + \frac{r_3 \cdot \omega_3^2 - r_4 \cdot \omega_4^2 \cdot \cos(\theta_3 - \theta_4)}{r_4 \cdot \sin(\theta_4 - \theta_3)} \quad (12)$$

Using the Law of Cosines, we note that

$$r_1^2 + r_2^2 - 2r_1 \cdot r_2 \cdot \cos\theta_2 = r_3^2 + r_4^2 - 2r_3 \cdot r_4 \cdot \cos\mu_4 \quad (13)$$

Thus we find that the transmission angle μ_4 on the driver side of CDLS can be expressed as follows:

$$\mu_4 = \arccos\left[\frac{r_3^2 + r_4^2 - r_1^2 - r_2^2 + 2r_1 \cdot r_2 \cdot \cos\theta_2}{2r_3 \cdot r_4}\right] \quad (14)$$

2.1.2. Motion of the Passenger Side Linkage of CDLS

Similar mathematical formulas are valid for the motion of the passenger side linkage of CDLS. In this case, θ_2 is the input variable (of the input link AB) and θ_6 is the unknown output (of the output link FE) to be determined. Using the vector loop method and the half-angle formula for tan, we may show that:

$$\theta_1 - \theta_6 = 2 \cdot \arctan\left[\frac{-E \pm \sqrt{E^2 - (F^2 - D^2)}}{F - D}\right] \quad (15)$$

and

$$\theta_1 - \theta_5 = \arctan\left[\frac{-r_2 \cdot \sin(\theta_1 - \theta_2) + r_6 \cdot \sin(\theta_1 - \theta_6)}{-r_2 \cdot \cos(\theta_1 - \theta_2) + r_6 \cdot \cos(\theta_1 - \theta_6) + r_{11}}\right] \quad (16)$$

with $D = -2 \cdot r_2 \cdot r_6 \cdot \cos(\theta_1 - \theta_2) + 2 \cdot r_{11} \cdot r_6$, $E = -2 \cdot r_2 \cdot r_6 \cdot \sin(\theta_1 - \theta_2)$, and

$$F = r_2^2 + r_6^2 + r_{11}^2 - r_5^2 - 2 \cdot r_{11} \cdot r_2 \cdot \cos(\theta_1 - \theta_2)$$

The solution we adopt for θ_6 is the following

$$\theta_1 - \theta_6 = 2 \cdot \arctan\left[\frac{-E + \sqrt{E^2 - (F^2 - D^2)}}{F - D}\right] \quad (17)$$

Using the method introduced in Section 2.1.1, it can be shown that the angular velocities ω_5 (the derivative of θ_5 with respect to time) and ω_6 (the derivative of θ_6 with respect to time) respectively of Link BE and of Link FE can be expressed as:

$$\omega_5 = \frac{r_2 \cdot \omega_2 \cdot \sin(\theta_2 - \theta_6)}{r_5 \cdot \sin(\theta_6 - \theta_5)} \quad (18)$$

and

$$\omega_6 = \frac{r_2 \cdot \omega_2 \cdot \sin(\theta_2 - \theta_5)}{r_6 \cdot \sin(\theta_6 - \theta_5)} \quad (19)$$

Besides, the angular acceleration α_5 of Link BE and the angular acceleration α_6 of Link FE can be expressed as follows:

$$\alpha_5 = \frac{r_2 \cdot [\omega_2^2 \cdot \cos(\theta_2 - \theta_6) + \alpha_2 \cdot \sin(\theta_2 - \theta_6)]}{r_5 \cdot \sin(\theta_6 - \theta_5)} + \frac{r_5 \cdot \omega_5^2 \cdot \cos(\theta_6 - \theta_5) - r_6 \cdot \omega_6^2}{r_5 \cdot \sin(\theta_6 - \theta_5)} \quad (20)$$

and

$$\alpha_6 = \frac{r_2 \cdot [\omega_2^2 \cdot \cos(\theta_5 - \theta_2) + \alpha_2 \cdot \sin(\theta_5 - \theta_2)]}{r_6 \cdot \sin(\theta_6 - \theta_5)} + \frac{r_5 \cdot \omega_5^2 - r_6 \cdot \omega_6^2 \cdot \cos(\theta_6 - \theta_5)}{r_6 \cdot \sin(\theta_6 - \theta_5)} \quad (21)$$

Using the Law of Cosines as in Section 2.1.1, it can be shown that the *transmission angle* μ_6 on the passenger side of CDLS can be expressed as follows:

$$\mu_6 = \arccos\left[\frac{r_5^2 + r_6^2 - r_2^2 - r_{11}^2 + 2r_2 \cdot r_{11} \cdot \cos(\theta_1 - \theta_2)}{2r_5 \cdot r_6}\right] \quad (22)$$

2.2. Method of Differential Evolution

Solving an optimization problem is to find the minimum/maximum (points) of an objective function on a specific set. Solving optimization problems subject to *constraints* is usually very important for engineering sciences. Over the past few years, the method of Differential Evolution (DE) has been one of the most important population based search methods. The method of Differential Evolution is usually effective in solving global optimization problems subject to constraints [4,24–26].

We discuss the ideas of DE briefly. Assume that

$$g(x) = g(x_1, \cdots, x_d)$$

is a continuous function of d variables on a specific region Ω of $\mathbf{R}^d$ specified by certain *inequality conditions (constraints)*:

$$h_1(x) \leq 0, \cdots, h_q(x) \leq 0$$

in which $x = (x_1, \cdots, x_d) \in \mathbf{R}^d$. We want to find the minimum of g on Ω:

$$\min_{x \in \Omega} g(x) = \min_{(x_1, \cdots, x_d) \in \Omega} g(x_1, \cdots, x_d) \quad (23)$$

subject to the *constraints*

$$h_1(x) \leq 0, \cdots, h_q(x) \leq 0 \quad (24)$$

The approach of DE is to approximate the minimum of $g(x) = g(x_1, \cdots, x_d)$ on the region Ω specified by (24) through several generations of *estimation*. In each generation, certain points on the region Ω are selected for the estimation of the minimum of g on Ω.

In order to ensure that the *estimated minimum* of g on Ω found in the (G+1)-th generation is always better than the *estimated minimum* of g on Ω found in the G-th generation, some selected points used in the G-th generation must be *reconsidered* in the (G+1)-th generation. This is operated as follows. Assume that, in the G-th generation, the *estimated minimum* of g on Ω appears somewhere at the following selected points:

$$x_{i,G} = (x_{i,G,1}, \cdots, x_{i,G,d}) \Omega$$

in which $i = 1, \cdots, M$. Then, in the (G+1)-th generation, M extra points

$$Z_{i,G} = (Z_{i,G,1}, \cdots, Z_{i,G,d}) \Omega$$

found through the "*mutation* and *crossover* operations" (to be discussed later), will be introduced for comparison with the M points $x_{i,G}$ taken from the G-th generation. We define $x_{i,G+1}$ (for the next generation G+1) as follows:

$$x_{i,G+1} = \begin{cases} x_{i,G}, & \text{if } g(x_{i,G}) \leq g(Z_{i,G}); \\ Z_{i,G}, & \text{if } g(x_{i,G}) > g(Z_{i,G}). \end{cases}$$

This definition of the points $x_{i,G+1}$ on Ω will ensure that

$$g(x_{i,G+1}) \leq g(x_{i,G})$$

for each $i = 1, \cdots, M$. For detailed explanation of the selection process, see Section 2.2.4.

In the subsequent paragraphs, we will discuss the details of our DE method. In our DE method, there are some parameters. The values of these parameters will be fixed throughout the operation of our DE method. Different choices of the values of these parameters might lead to different search results for the estimated minimum of $g(x)$ on the region Ω specified by (24). These parameters are listed as follows: the *crossover ratio* CR, the *scale factor* F, the *population size* M (the number of selected points on Ω for each generation), and the maximum number of *iterations* $G_{\max}$ (so that $G = 0, 1, \cdots, G_{\max}$). The crossover ratio CR and the scale factor F are respectively real numbers in the interval $[0, 1]$. Usually we will use the term "population" to mean a collection of points in a generation.

2.2.1. Initialization of Population

Assume that the region Ω specified by (24) is included in the following bounded domain D of $\mathbf{R}^d$:

$$D = [x_1^{(L)}, x_1^{(U)}] \times \cdots [x_d^{(L)}, x_d^{(U)}] \tag{25}$$

in which $x_1^{(L)}$ and $x_1^{(U)}$ are respectively are the minimum and maximum of the k-th component of all points

$$x = (x_1, \cdots, x_d) \Omega$$

To initiate our DE method, we must choose M points in D for the *initial generation* indexed by $G = 0$:

$$x_{i,0} = (x_{i,0,1}, \cdots, x_{i,0,d})$$

in which $i = 1, \cdots, M$. To randomize the selection of these points, we may assume that

$$x_{i,0,j} = x_j^{(L)} + (rand_{i,j}[0,1]) \cdot (x_j^{(U)} - x_j^{(L)}) \tag{26}$$

in which $j = 1, \cdots, d$. Here, each $rand_{i,j}[0,1]$ is a randomly chosen real number in the interval $[0,1]$. Thus $0 \leq rand_{i,j}[0,1] \leq 1$.

2.2.2. Mutation

For each $i = 1, \cdots, M$, we consider an associated mutation point

$$v_{i,G} = (v_{i,G,1}, \cdots, v_{i,G,d}) \text{ in } \mathbf{R}^d$$

This point is constructed as follows:

$$v_{i,G} = x_{\alpha(i,G),G} + F \cdot [x_{\beta(i,G),G} - x_{\gamma(i,G),G}] \tag{27}$$

in which $\alpha(i,G)$, $\beta(i,G)$ and $\gamma(i,G)$ are three different integers chosen from the set $\{1, 2, 3, \cdots, M\}$. Usually it is required that the integers $\alpha(i,G)$, $\beta(i,G)$ and $\gamma(i,G)$ are all different from i.

In (25), the parameter F is a real number in the interval $[0, 1]$. We usually call F the *scale factor*. This parameter F is important both for controlling the *diversity* of generation population and for controlling the *convergence rate* of DE. If F is small, the process of DE method could be trapped around a local minimum of g due to poor diversity of generation population. If F is close to 1, the convergence rate of the DE method could be very small. Usually F is chosen in the range $[0.3, 0.6]$.

The mutation point $v_{i,G}$, associated with $x_{i,G}$, will be used to construct the point $Z_{i,G}$ through the *"crossover"* operation. We will discuss this in the next subsection.

2.2.3. Crossover

The crossover operation is a method for producing an extra point $Z_{i,G}$ using the coordinates of $v_{i,G}$ and of $x_{i,G}$ with randomness. The *crossover ratio CR* is a real number in the interval $[0,1]$. For each i of the population set $\{1, 2, 3, \cdots, M\}$, we choose d real numbers $w_{i,G,1}, \cdots, w_{i,G,d}$ in $[0,1]$ as *"weights"*. Besides we choose, for this i, an integer $s_{i,G}$ from the dimension set $\{1, \cdots, d\}$. We define the point

$$Z_{i,G} = (Z_{i,G,1}, \cdots, Z_{i,G,d})$$

As follows:

$$Z_{i,G,j} = \begin{cases} v_{i,G,j}, & \text{if } j = s_{i,G} \text{ or } w_{i,G,j} \leq CR; \\ x_{i,G,j}, & \text{if } j \neq s_{i,G} \text{ and } CR < w_{i,G,j}. \end{cases} \tag{28}$$

The definition (28) implies the $s_{i,G}$-th coordinate of the point $Z_{i,G}$ is definitely taken from that of $v_{i,G}$. However, for $j \neq s_{i,G}$, the j-th coordinate of $Z_{i,G}$ is *dependent on CR*. When CR is small, more of the the coordinates of $Z_{i,G}$ are taken from that of $x_{i,G}$. When CR is large, more of the the coordinates of $Z_{i,G}$ are taken from that of $v_{i,G}$. Thus the parameter CR is important for controlling the deviation of the population $\{Z_{1,G}, \cdots, Z_{M,G}\}$ from the population $\{x_{1,G}, \cdots, x_{M,G}\}$.

It could happen that the some of the points $\{Z_{1,G}, \cdots, Z_{M,G}\}$ are *not* in the bounded domain

$$D = [x_1^{(L)}, x_1^{(U)}] \times \cdots [x_d^{(L)}, x_d^{(U)}]$$

If $Z_{i,G,j} < x_j^{(L)}$ or $Z_{i,G,j} > x_j^{(U)}$ happens, we will *redefine* $Z_{i,G,j} < x_j^{(L)}$ as follows:

$$Z_{i,G,j} = x_j^{(L)} + (randZ_{i,G,j}[0,1]) \cdot (x_j^{(U)} - x_j^{(L)}) \tag{29}$$

in which $randZ_{i,G,j}[0,1]$ is a randomly chosen real number in the interval $[0,1]$. This *modification of crossover* ensures that all the points $\{Z_{1,G}, \cdots, Z_{M,G}\}$ are included in the bounded domain D.

2.2.4. Selection Process

For points $x = (x_1, \cdots, x_d)$ on the bounded domain D, it might happen that some of the constraints of (24) are not satisfied. Thus we introduce a function CV to measure the *"Constraint Violation"* of a point. This function is defined as follows:

$$CV(x) = \frac{1}{q} \cdot \sum_{j=1}^{q} \max[0, h_j(x)] \tag{30}$$

It can be inferred readily that $CV(x) = 0$ if and only if all the constraints of (24) are satisfied by x.

To define the population $\{x_{1,G+1}, \cdots, x_{M,G+1}\}$ for the next generation G+1, we will consider the CV values and the g values respectively of $\{x_{1,G}, \cdots, x_{M,G}\}$ and of $\{Z_{1,G}, \cdots, Z_{M,G}\}$. When $CV(x_{i,G}) \neq CV(Z_{i,G})$, we define

$$x_{i,G+1} = \begin{cases} x_{i,G}, & \text{if } CV(x_{i,G}) < CV(Z_{i,G}); \\ Z_{i,G}, & \text{if } CV(x_{i,G}) \geq CV(Z_{i,G}). \end{cases} \tag{31}$$

When $CV(x_{i,G}) = CV(Z_{i,G})$, we define

$$x_{i,G+1} = \begin{cases} x_{i,G}, & \text{if } g(x_{i,G}) < g(Z_{i,G}); \\ Z_{i,G}, & \text{if } g(x_{i,G}) \geq g(Z_{i,G}). \end{cases} \qquad (32)$$

2.3. Optimization on the Angular acceleration of Output Links of CDLS

Now we will use the method of Differential Evolution, discussed in Section 2.2, to solve an optimization problem on the maximal magnitude of *angular acceleration* of the output-links of a commercially available center driven linkage system (CDLS) for vehicle wipers. We consider the objective function f defined as follows:

$$f(r_3, r_4, r_5, r_6) = \max_{0 \leq \theta_2 \leq 2\pi} |\alpha_4(r_3, r_4, \theta_2)| + \max_{0 \leq \theta_2 \leq 2\pi} |\alpha_6(r_5, r_6, \theta_2)|$$

See Section 2.1 for explanations of our model.

We assume that the input link AB of this CDLS rotates at a *constant* angular velocity $\omega_2 = 1$ (rad/s). See Figure 5. We record/calculate the relevant items per $\pi/180$ second. The width of wiping region for the driver side of this CDLS is $85\pi/180$ (rad). The width of wiping region for the passenger side of this CDLS is $80\pi/180$ (rad). An error of $1\pi/1800$ (rad) is allowed for the width of the wiping region. Patterns of the wiping region are regulated in [27].

We will use the method of Section 2.2 to consider the minimum problem of f subject to 10 specific constraints. For this CDLS, $r_1 = 210.5$ (mm), $r_{11} = 206.8$ (mm) and $r_2 = 45$ (mm) are fixed constants. The constraints for the objective function f are listed as follows.

(1). $150 \leq r_3 \leq 250$ (mm).
(2). $50 \leq r_4 \leq 75$ (mm).
(3). $150 \leq r_5 \leq 250$ (mm).
(4). $50 \leq r_6 \leq 75$ (mm).
(5). $\left| \max_{0 \leq \theta_2 \leq 2\pi} \theta_4 - \min_{0 \leq \theta_2 \leq 2\pi} \theta_4 - \frac{85\pi}{180} \right| \leq \frac{\pi}{1800}$ (rad).
(6). $\frac{42\pi}{180} \leq \mu_4 \leq \frac{138\pi}{180}$ (rad).
(7). $|\omega_4| \leq 0.3\pi$ (rad/s).
(8). $\left| \max_{0 \leq \theta_2 \leq 2\pi} \theta_6 - \min_{0 \leq \theta_2 \leq 2\pi} \theta_6 - \frac{80\pi}{180} \right| \leq \frac{\pi}{1800}$ (rad).
(9). $\frac{42\pi}{180} \leq \mu_6 \leq \frac{138\pi}{180}$ (rad).
(10). $|\omega_6| \leq 0.3\pi$ (rad/s).

The values of the parameters for our DE method are chosen as follows.
(1). Crossover ratio $CR = 0.6$.
(2). Scale factor $F = 0.6$.
(3). Population size $M = 150$.
(4) Maximum number of iterations $G_{\max} = 100$.

3. Results

Section 3.1 is devoted to the justification of our modeling. We compare our calculation results based on the mathematical formulas derived in Section 2.1 with that by commercial ADAMS to justify our modeling. Section 3.2 is devoted to the computation results using the optimization method of Differential Evolution.

3.1. Comparison of Our Simulation Results with That by ADAMS

We calculate ω_3, ω_4, ω_5, ω_6, α_3, α_4, α_5 and α_6 on MATLAB using the mathematical formulas derived in Section 2.1. The values of the parameters of our calculations are listed in the following Table 1. The time interval and the sampling time for our computations on MATLAB are respectively $[0, 2\pi]$ and $\pi/180$. The initial value of θ_2 is 0.

Table 1. Parameter values used in simulations.

r_1 (mm)	r_{11} (mm)	r_3 (mm)	r_4 (mm)	r_5 (mm)	r_6 (mm)	θ_1 (mm)	r_2 (mm)	ω_2 (rad/s)	α_2 (rad/s^2)
210.5	206.8	209	66.8	206.0	69.9	$\frac{207\pi}{180}$	45	1	0

We input the same parameters into the commercial software ADAMS to justify our modeling. Comparison of the computation results from our mathematical formulas on MATLAB with that from the commercial software ADAMS is shown in Figures 6–13.

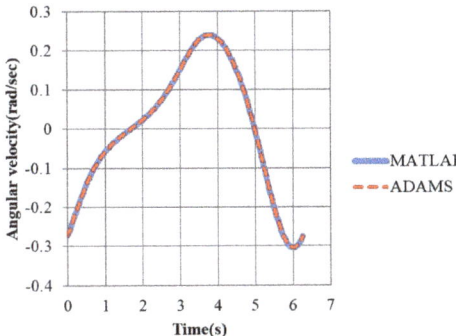

Figure 6. Comparison of our modeling computation with that of ADAMS on ω_3.

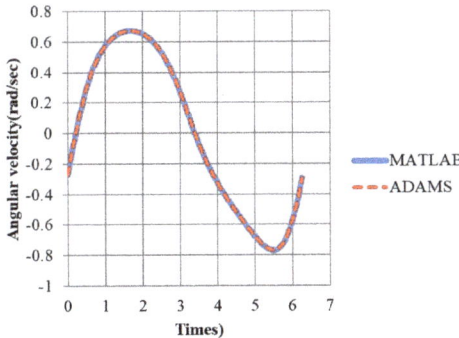

Figure 7. Comparison of our modeling computation with that of ADAMS on ω_4.

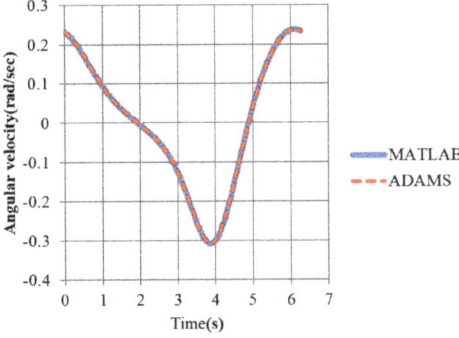

Figure 8. Comparison of our modeling computation with that of ADAMS on ω_5.

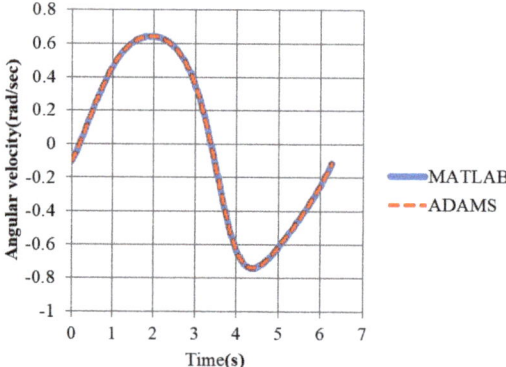

Figure 9. Comparison of our modeling computation with that of ADAMS on ω_6.

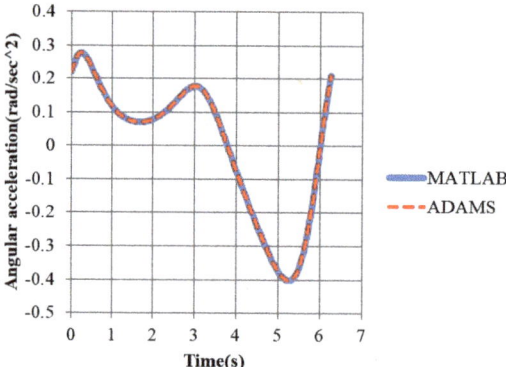

Figure 10. Comparison of our modeling computation with that of ADAMS on α_3.

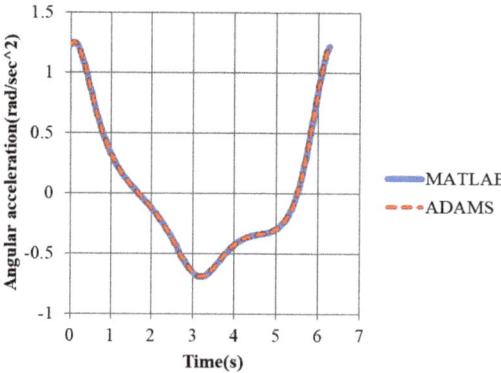

Figure 11. Comparison of our modeling computation with that of ADAMS on α_4.

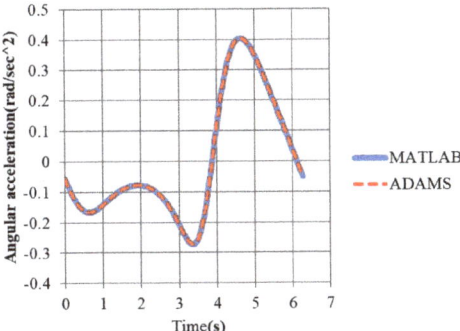

Figure 12. Comparison of our modeling computation with that of ADAMS on α_5.

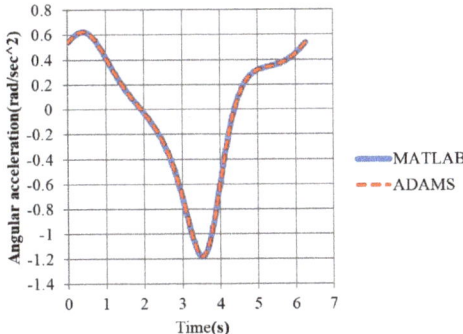

Figure 13. Comparison of our modeling computation with that of ADAMS on α_6.

Comparison of our calculation results based on the mathematical formulas derived in Section 2.1 with that by commercial ADAMS shows that our modeling correct and reliable.

3.2. Comparison of the Linkage Lengths before DE Optimization with That after DE Optimization

Our DE search results on the optimization problem of the objective function

$$f(r_3, r_4, r_5, r_6) = \max_{0 \leq \theta_2 \leq 2\pi} |\alpha_4(r_3, r_4, \theta_2)| + \max_{0 \leq \theta_2 \leq 2\pi} |\alpha_6(r_5, r_6, \theta_2)|$$

of the maximal absolute values $|\alpha_4|$ and $|\alpha_6|$ of angular acceleration of the output-links of a commercially available CDLS are shown in Table 2.

Table 2. Linkage lengths before/after DE optimization.

	r_3 (mm)	r_4 (mm)	r_5 (mm)	r_6 (mm)
Linkage length before DE	209.0	66.8	206.0	69.9
Linkage length after DE	202.0	66.7	196.4	69.5

We notice that, on Figures 14 and 15, the *maximal absolute values* $|\alpha_4|$ and $|\alpha_6|$ of angular acceleration of the output-links respectively on the driver side and on the passenger side of CDLS become smaller after our DE optimization.

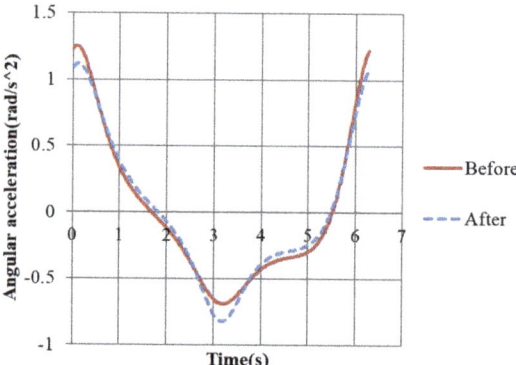

Figure 14. The angular acceleration α_4 of the driver side output-link of CDLS before/after DE optimization.

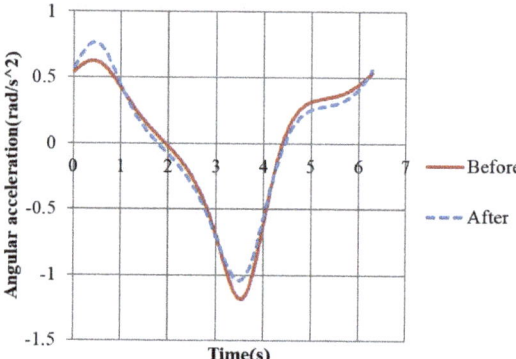

Figure 15. The angular acceleration α_6 of the passenger side output-link of CDLS before/after DE optimization.

We notice that, on Figures 16 and 17, both the angular velocities of the output-links respectively on the driver side and on the passenger side of CDLS do not change much after our DE optimization.

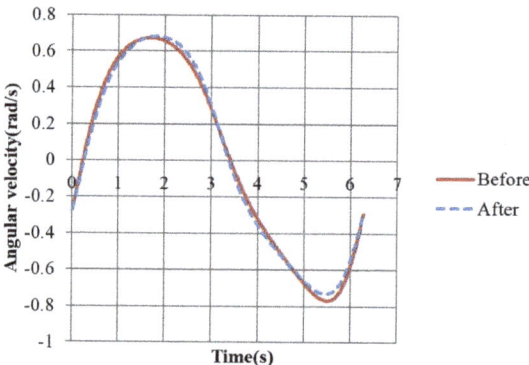

Figure 16. The angular velocity ω_4 of the driver side output-link of CDLS before/after DE optimization.

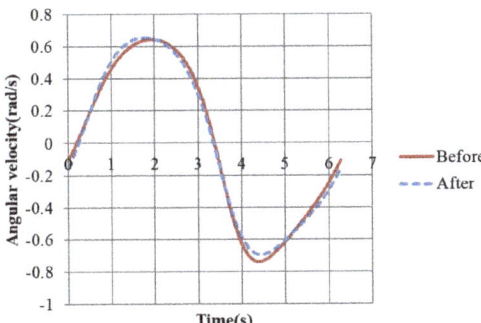

Figure 17. The angular velocity ω_6 of the driver side output-link of CDLS before/after DE optimization.

We notice that, on Figures 18 and 19, both *transmission angles* of CDLS become closer to $\pi/2$(rad) or $90°$ after our DE optimization. In fact, this optimization slightly improves the effective torques of the output links of CDLS. See Table 3.

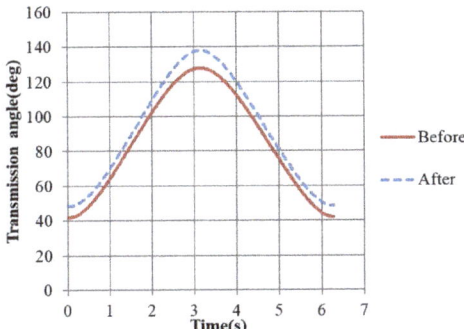

Figure 18. The transmission angle μ_4 of the driver side output-link of CDLS before/after DE optimization.

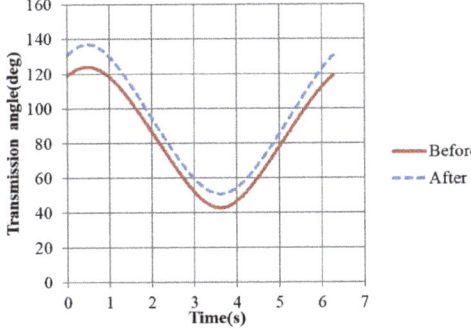

Figure 19. The transmission angle μ_6 of the passenger side output-link of CDLS before/after DE optimization.

Table 3. The deviation of μ_4 and μ_6 respectively from $\pi/2$ before/after DE optimization.

	$\max_{0\leq\theta_2\leq 2\pi}|\mu_4 - \frac{\pi}{2}|$	$\min_{0\leq\theta_2\leq 2\pi}|\mu_6 - \frac{\pi}{2}|$
Before DE	0.8379(rad)	0.8232(rad)
After DE	0.8377(rad)	0.8205(rad)

The resulting reduction of deviation of transmission angles from $\pi/2$ clearly shows us that the torque-efficiency by the input-link is improved. Thus improvement on the *steadiness* of a linkage system for vehicle wipers could also reduce the possible material fatigue of a linkage system.

Notice that, on Table 4, our optimization reduces the maximal magnitude $\max|\alpha_4|$ of angular acceleration of the output-link on the driver side of CDLS by 10.69%. Besides, notice that, on Table 5, our optimization reduces the maximal magnitude $\max|\alpha_6|$ of angular acceleration of the output-link on the passenger side of CDLS by 12.10%.

Table 4. α_4 and ω_4 of the driver side output-link of CDLS before/after DE optimization.

	$\max_{0\leq\theta_2\leq 2\pi}\alpha_4$	$\min_{0\leq\theta_2\leq 2\pi}\alpha_4$	$\max_{0\leq\theta_2\leq 2\pi}\omega_4$	$\min_{0\leq\theta_2\leq 2\pi}\omega_4$
Before DE	1.254 (rad/s^2)	-0.690 (rad/s^2)	0.674 (rad/s^2)	-0.769 (rad/s^2)
After DE	1.120 (rad/s^2)	-0.819 (rad/s^2)	0.683 (rad/s^2)	-0.730 (rad/s^2)
Change	-10.69%	18.70%	1.34%	-5.07%

Table 5. α_6 and ω_6 of the passenger side output-link of CDLS before/after DE optimization.

	$\max_{0\leq\theta_2\leq 2\pi}\alpha_6$	$\min_{0\leq\theta_2\leq 2\pi}\alpha_6$	$\max_{0\leq\theta_2\leq 2\pi}\omega_6$	$\min_{0\leq\theta_2\leq 2\pi}\omega_6$
Before DE	0.624(rad/s^2)	-1.182(rad/s^2)	0.644(rad/s)	-0.739(rad/s)
After DE	0.767(rad/s^2)	-1.039(rad/s^2)	0.657(rad/s)	-0.696(rad/s)
Change	22.92%	-12.10%	2.02%	-5.82%

4. Discussion

In this article, we have explained how to improve the *steadiness* of a linkage system for vehicle wipers without weakening its normal function. This improvement is based on an optimization process on the maximal absolute values of *angular acceleration* of the output-links of a linkage system. We have demonstrated how to optimize the maximal magnitude of *angular acceleration* of the output-links of a commercially available center driven linkage system (CDLS) for vehicle wipers using a Differential Evolution type search method. This improvement on the *steadiness* of a linkage system for vehicle wipers could reduce the possible material fatigue of a linkage system. It also improves the torque-efficiency of the input-link.

This optimization problem is related to another problem which we now discuss. Reducing the unpleasant rubber wiper noise on vehicle windshield has been an interesting research topic over the past few decades [28–42]. Some simple spring-mass models for the vibration of rubber wiper were presented for simulations before [32,39,41]. Improvement on the electrical motor of a linkage system for rubber wipers was discussed in [42].

Since rubber wipers are made of the almost *hyper-elastic* material "rubber" [43–45], it is natural to expect that the mathematical theories of *Elasticity* Mechanics [46–48], and the mathematical theories of Partial Differential Equations [49–52], should be helpful for us to understand the complicated vibration of rubber wiper. In our earlier research [21], we seriously analyze the *Elasticity* Mechanics and Physics of rubber wiper on *convex* windshield through mathematical theories of Partial Differential Equations and through physical aspects of vibration [46–53]. We discovered two important mathematical formulas

$$\sqrt{\frac{\lambda+2\mu}{\rho}} \cdot \frac{n}{2l} \text{ Hz (Class I) and } \sqrt{\frac{\mu}{\rho}} \cdot \frac{n}{2l} \text{ Hz (Class II)}$$

For two classes of "characteristic vibration frequencies" of rubber wiper on *convex* windshield through mathematical analysis of the partial differential system of Lame equations

$$\frac{\partial^2 v}{\partial t^2} = \frac{(\lambda+\mu)}{\rho} \cdot \nabla \left(\frac{\partial v_1}{\partial x_1} + \frac{\partial v_2}{\partial x_2} + \frac{\partial v_3}{\partial x_3} \right) + \frac{\mu}{\rho} \cdot \left(\frac{\partial^2 v}{\partial x_1^2} + \frac{\partial^2 v}{\partial x_2^2} + \frac{\partial^2 v}{\partial x_3^2} \right)$$

$$= \frac{(\lambda+\mu)}{\rho} \cdot \nabla (\text{div } v) + \frac{\mu}{\rho} \cdot \Delta v$$

here ρ kg/m^3 and l m are respectively the density and the length of the rubber wiper. The constants λ and μ are the "Lame coefficients" of the rubber wiper. These material constants are related to "the Young modulus E" and "the Poisson ratio (the coefficient of *transverse* contraction) σ" of rubber wiper as follows:

$$\lambda = \frac{\sigma \cdot E}{(1+\sigma) \cdot (1-2\sigma)} \text{ and } \mu = \frac{E}{2 \cdot (1+\sigma)}$$

When rubber wipers move back and forth on a vehicle windshield, sharp wiper noise is generated around each reversal of the motion of rubber wipers on windshield. Usually the maximal magnitude of angular acceleration of an output-link of a CDLS happens around each *reversal* of the motion of this output-link. See Figures 14 and 15.

Since larger angular acceleration of an output-link usually would lead to greater normal/shear *stress* on the rubber wiper associated with this output-link, we expect that reduction of the maximal magnitude of angular acceleration of an output-link would also be helpful for reducing the amplitudes of sound waves of wiper noise around each reversal of the motion of this output-link. Further studies on these topics should be interesting.

Author Contributions: Methodology, T.-J.C. and Y.-J.H.; Software, C.-H.L. and J.-Y.W. All authors have read and agreed to the published version of the manuscript.

Funding: This research was funded by the National Science and Technology Council of Taiwan Government, grants 110-2115-M-006-008- and 111-2115-M-006-008-.

Data Availability Statement: Not applicable.

Conflicts of Interest: The authors declare no conflict of interest.

References

1. McCarthy, J.M.; Soh, G.S. *Geometric Design of Linkages*, 2nd ed.; Springer: New York, NY, USA, 2010.
2. Storn, R.; Price, K.V. *Differential Evolution—A Simple and Efficient Adaptive Scheme for Global Optimization over Continuous Spaces*; Technical Report, TR-95-012; International Computer Science Institute (ICSI) of University of California: Berkeley, CA, USA, 1995.
3. Storn, R.; Price, K.V. Differential Evolution—A Simple and Efficient Heuristic for Global Optimization over Continuous Spaces. *J. Glob. Optim.* **1997**, *11*, 341–359. [CrossRef]
4. Storn, R.; Price, K.V. *Differential Evolution: A Practical Approach to Global Optimization*; Springer: Berlin/Heidelberg, Germany, 2005.
5. Fan, Y.-F.; Lampinen, J. A Trigonometric Mutation Operation to Differential Evolution. *J. Glob. Optim.* **2003**, *27*, 105–129. [CrossRef]
6. Tizhoosh, H.R. Opposition-Based Learning: A New Scheme for Machine Intelligence. In Proceedings of the International Conference on Computational Intelligence for Modeling Control and Automation—CIMCA, Vienna, Austria, 28–30 November 2005.
7. Tizhoosh, H.R. Reinforcement Learning based on Actions and Opposite Actions. In Proceedings of the ICGST International Conference on Artificial Intelligence and Machine Learning (AIML-05), Cairo, Egypt, 10–21 December 2005.
8. Tizhoosh, H.R. Opposition-based Reinforcement Learning. *J. Adv. Comput. Intell. Intell. Inform.* **2006**, *10*, 578–585. [CrossRef]

9. Rahnamayan, S.; Tizhoosh, H.R.; Salama, M.M.A. Opposition Based Differential Evolution. *IEEE Trans. Evol. Comput.* **2008**, *12*, 64–79. [CrossRef]
10. Das, S.; Abraham, A.; Chakraborty, U.K.; Konar, A. Differential Evolution using a Neighborhood Based Mutation Operator. *IEEE Trans. Evol. Comput.* **2009**, *13*, 526–553. [CrossRef]
11. Qin, A.K.; Huang, V.L.; Suganthan, P.N. Differential Evolution Algorithm with Strategy Adaptation for Global Numerical Optimization. *IEEE Trans. Evol. Comput.* **2009**, *13*, 398–417. [CrossRef]
12. Brest, J.; Sepesy Maučec, M. Self-Adaptive Differential Evolution Algorithm Using Population Size Reduction and Three Strategies. *Soft Comput.* **2011**, *15*, 2157–2174. [CrossRef]
13. Lou, Y.; Yuen, S.Y.; Chen, G. Non-Revisiting Stochastic Search Revisited: Results, Perspectives, and Future Directions. *Swarm Evol. Comput.* **2021**, *61*, 100828. [CrossRef]
14. Črepinšek, M.; Liu, S.-H.; Mernik, M.; Ravber, M. The Trap of Sisyphean Work in Differential Evolution and How to Avoid It. In *Differential Evolution: From Theory to Practice*, 1st ed.; Vinoth Kumar, B., Oliva, D., Suganthan, P.N., Eds.; Springer: Singapore, 2022; pp. 137–174.
15. Das, S.; Suganthan, P.N. Differential Evolution: A Survey of the State-of-the-Art. *IEEE Trans. Evol. Comput.* **2011**, *15*, 4–31. [CrossRef]
16. Das, S.; Mullick, S.S.; Suganthan, P.N. Recent Advances in Differential Evolution—An Updated Survey. *Swarm Evol. Comput.* **2016**, *27*, 1–30. [CrossRef]
17. Karol R.Opara, K.O.; Arabas, J. Differential Evolution: A Survey of Theoretical Analyses. *Swarm Evol. Comput.* **2019**, *44*, 546–558. [CrossRef]
18. Rudolph, G. Convergence of Evolutionary Algorithms in General Search Spaces. Proceedings of IEEE International Conference on Evolutionary Computation, Nagoya, Japan, 20–22 May 1996.
19. Hu, Z.; Su, Q.; Yang, X.; Xiong, Z. Not Guaranteeing Convergence of Differential Evolution on a Class of Multimodal Functions. *Appl. Soft Comput.* **2016**, *41*, 479–487. [CrossRef]
20. Hu, Z.; Xiong, S.; Su, Q.; Fang, Z. Finite Markov Chain Analysis of Classical Differential Evolution Algorithm. *J. Comput. Appl. Math.* **2014**, *268*, 121–134. [CrossRef]
21. Chen, T.-J.; Hong, Y.-J. Geometric Analysis of the Vibration of Rubber Wiper Blade. *Taiwan. J. Math.* **2021**, *25*, 491–516. [CrossRef]
22. Tang, Y.; Liu, Z.; He, B. Optimization Design for Crank-rocker Mechanism Based on Genetic Algorithm. In Proceedings of the 2012 Second International Conference on Intelligent System Design and Engineering Application, Sanya, China, 6–7 January 2012.
23. Balli, S.S.; Chand, S. Transmission angle in mechanisms. *Mech. Mach. Theory* **2002**, *37*, 175–195. [CrossRef]
24. Wang, Y.; Cai, Z.; Hang, Q. Differential Evolution with Composite Trial Vector Generation Strategies and Control Parameters. *IEEE Trans. Evol. Comput.* **2011**, *15*, 55–66. [CrossRef]
25. Ma, C.-F. *Optimization Methods and MATLAB Programming*; Science Press, Chinese Academy of Sciences: Beijing, China, 2010.
26. Arafa, M.; Sallam, E.A.; Fahmy, M.M. An enhanced differential evolution optimization algorithm. In Proceedings of the 2014 Fourth International Conference on Digital Information and Communication Technology and its Applications (DICTAP), Bangkok, Thailand, 6–8 May 2014.
27. Electronic Code of Federal Regulations. Available online: https://www.ecfr.gov/cgi-bin/text-idx?SID=7d443eb75ceba033fed91e90f816b574&node=se49.6.571_1104&rgn=div8 (accessed on 10 October 2020).
28. Begout, M. Les Problèmes Liés au Frottement Élastomère-Verre dans L'Automobile. Ph.D. Thesis, Université Paul Sabatier de Toulouse, Toulouse, France, 1979.
29. Okura, S.; Sekiguchi, T.; Oya, T. *Dynamic Analysis of Blade Reversal Behavior in a Windshield Wiper System*; SAE 2000 World Congress 2000, Technical Paper 2000-01-0127; SAE International: Warrendale, PA, USA, 2000.
30. Goto, S.; Takahashi, H.; Oya, T. Investigation of wiper blade squeal noise reduction measures. In *SAE 2001 Noise & Vibration Conference & Exposition*; Technical Paper 2001-01-1410; SAE International: Warrendale: Warrendale, PA, USA, 2001.
31. Goto, S.; Takahashi, H.; Oya, T. Clarification of the Mechanism of Wiper Blade Rubber Squeal Noise Generation. *JSAE Rev.* **2001**, *22*, 57–62. [CrossRef]
32. Grenouillat, R.; Leblanc, C. Simulation of chatter vibrations for wiper systems. In *SAE 2002 World Congress & Exhibition*; Technical Paper 2002-01-1239; SAE International: Warrendale, PA, USA, 2002.
33. Berger, S.; Sinou, J.-J.; Aubry, E. Model of Chatter Vibrations and Stability Analysis of a Non-linear Wiper System. *Int. Rev. Mech. Eng.* **2008**, *2*, 349–356.
34. Stein, G.J.; Zahoransky, R.; Múčka, P. On Dry Friction Modelling and Simulation in Kinematically Excited Oscillatory Systems. *J. Sound Vib.* **2008**, *311*, 74–96. [CrossRef]
35. Bódai, G.; Goda, T.J. Friction Force Measurement at Windscreen Wiper/Glass Contact. *Tribol. Lett.* **2012**, *45*, 515–523. [CrossRef]
36. Nechak, L.; Berger, S.; Aubry, E. Prediction of Random Self Friction-Induced Vibrations in Uncertain Dry Friction Systems Using a Multi-Element Generalized Polynomial Chaos Approach. *J. Vib. Acoust.* **2012**, *134*, 041015. [CrossRef]
37. Min, D.; Jeong, S.; Yoo, H.H.; Kang, H.; Park, J. Experimental Investigation of Vehicle Wiperblade's Squeal Noise Generation due to Windscreen Waviness. *Tribol. Int.* **2014**, *80*, 191–197. [CrossRef]
38. Bódai, G.; Goda, T.J. Sliding Friction of Wiper Blade: Measurement, FE Modeling and Mixed Friction Simulation. *Tribol. Int.* **2014**, *70*, 63–74. [CrossRef]

39. Lancioni, G.; Lenci, S.; Galvanetto, U. Dynamics of windscreen wiper blades: Squeal noise, reversal noise and chattering. *Int. J. Non-Linear Mech.* **2015**, *80*, 132–143. [CrossRef]
40. Reddyhoff, T.; Dober, O.; Rouzic, J.L.; Gotzen, N.-A.; Parton, H.; Dini, D. Friction Induced Vibration in Windscreen Wiper Contacts. *J. Vib. Acoust.* **2015**, *137*, 041009. [CrossRef]
41. Unno, M.; Shibata, A.; Yabuno, H.; Yanagisawa, D.; Nakano, T. Analysis of the Behavior of a Wiper Blade around the Reversal in Consideration of Dynamic and Static Friction. *J. Sound Vib.* **2017**, *393*, 76–91. [CrossRef]
42. Viscardi, M.; Di Leo, R.; Ciminello, M.; Brandizzi, M. Preliminary Experimental/Numerical Study for the Vibration Annoyance Control of a Windshield Wiper Mechanical System through a Synchronized Switch Shunt Resonator (SSSR) Technology. *J. Theor. Appl. Mech.* **2018**, *56*, 283–296. [CrossRef]
43. Muhr, A.H. Modeling the Stress-Strain Behavior of Rubber. *Rubber Chem. Technol.* **2005**, *78*, 391–425. [CrossRef]
44. Wang, W.; Chen, S. Hyperelasticity, Viscoelasticity, and Nonlocal Elasticity Govern Dynamic Fracture in Rubber. *Phys. Rev. Lett.* **2005**, *95*, 144301. [CrossRef]
45. Yang, C.; Persson, B.N.J. Molecular Dynamics Study of Contact Mechanics: Contact Area and Interfacial Separation from Small to Full Contact. *Phys. Rev. Lett.* **2008**, *100*, 024303. [CrossRef]
46. Frankel, T. *The Geometry of Physics*; Cambridge University Press: Cambridge, UK, 2004.
47. Chaichian, M.; Merches, I.; Tureanu, A. *Mechanics*; Springer: Berlin/Heidelberg, Germany, 2012.
48. Ernst, E. The Boussinesq Form of Saint-Venant's Principle. *Math. Model. Methods Appl. Sci.* **2000**, *10*, 863–875. [CrossRef]
49. Gilbarg, D.; Trudinger, N.S. *Elliptic Partial Differential Equations of Second Order*; Springer: Berlin/Heidelberg, Germany, 2001.
50. Alinhac, S. *Hyperbolic Partial Differential Equations*; Springer: New York, NY, USA, 2009.
51. Evans, L.C. *Partial Differential Equations*; American Mathematical Society: Providence, RI, USA, 2010.
52. Serov, V. *Fourier Transform and Their Applications to Mathematical Physics*; Springer International Publishing AG: New York, NY, USA, 2017.
53. Rossing, T.D.; Fletcher, N.H. *Principles of Vibration and Sound*; Springer: New York, NY, USA, 2004.

Disclaimer/Publisher's Note: The statements, opinions and data contained in all publications are solely those of the individual author(s) and contributor(s) and not of MDPI and/or the editor(s). MDPI and/or the editor(s) disclaim responsibility for any injury to people or property resulting from any ideas, methods, instructions or products referred to in the content.

Article

Random Orthogonal Search with Triangular and Quadratic Distributions (TROS and QROS): Parameterless Algorithms for Global Optimization

Bruce Kwong-Bun Tong [1,2,*], Chi Wan Sung [3] and Wing Shing Wong [2]

[1] Department of Electronic Engineering and Computer Science, Hong Kong Metropolitan University, Hong Kong, China
[2] Department of Information Engineering, The Chinese University of Hong Kong, Hong Kong, China
[3] Department of Electrical Engineering, City University of Hong Kong, Hong Kong, China
* Correspondence: kbtong@hkmu.edu.hk; Tel.: +852-3120-2607

Abstract: In this paper, the behavior and performance of Pure Random Orthogonal Search (PROS), a parameter-free evolutionary algorithm (EA) that outperforms many existing EAs on the well-known benchmark functions with finite-time budget, are analyzed. The sufficient conditions to converge to the global optimum are also determined. In addition, we propose two modifications to PROS, namely Triangular-Distributed Random Orthogonal Search (TROS) and Quadratic-Distributed Random Orthogonal Search (QROS). With our local search mechanism, both modified algorithms improve the convergence rates and the errors of the obtained solutions significantly on the benchmark functions while preserving the advantages of PROS: parameterless, excellent computational efficiency, ease of applying to all kinds of applications, and high performance with finite-time search budget. The experimental results show that both TROS and QROS are competitive in comparison to several classic metaheuristic optimization algorithms.

Keywords: global optimization; metaheuristics; orthogonal search; parameterless; TROS; QROS

Citation: Tong, B.K.-B.; Sung, C.W.; Wong, W.S. Random Orthogonal Search with Triangular and Quadratic Distributions (TROS and QROS): Parameterless Algorithms for Global Optimization. *Appl. Sci.* **2023**, *13*, 1391. https://doi.org/10.3390/app13031391

Academic Editor: Vincent A. Cicirello

Received: 13 December 2022
Revised: 6 January 2023
Accepted: 7 January 2023
Published: 20 January 2023

Copyright: © 2023 by the authors. Licensee MDPI, Basel, Switzerland. This article is an open access article distributed under the terms and conditions of the Creative Commons Attribution (CC BY) license (https://creativecommons.org/licenses/by/4.0/).

1. Introduction

Black-box optimization refers to optimizing an objective function in the absence of prior knowledge of the function. The only knowledge of the objective function is the observed outputs of the given inputs. Metaheuristics such as evolutionary computation (EC) techniques and evolutionary algorithms (EAs) are widely employed for black-box optimization of various fields across science and engineering [1–3] due to their certain primary advantages over the traditional techniques (such as Newton's method and gradient descent) including high robustness in solving complex real-world problems, and domain knowledge and regularity (e.g., convexity, continuity, differentiability) of the functions to be optimized are generally not required [4].

However, these algorithms often require a large number of function evaluations to evolve or update the candidate solutions in order for errors reduced to a satisfactory level [5]. For many real-world problems, evaluating the objective function is often costly. These function evaluations may involve conducting computational intensive numerical simulations or expensive physical experiments [6–9] When only a limited number of objective function evaluations is affordable, EC's trial and error approach becomes unattractive.

Moreover, EC techniques are commonly recognized as heuristic search algorithms because their theoretical analysis often lags behind the development of the algorithms [4,10]. Their performance is often sensitive to the problem-dependent hyper-parameters [11]. Various approaches have been suggested to overcome these drawbacks. For example, building an inexpensive surrogate model for evaluation so as to reduce the number of objective function evaluations [5,12], and automatic or self-adaptive tuning of the hyperparameters [11,13,14].

In this paper, we propose two algorithms for expensive black-box optimization. The two algorithms are modified from Pure Random Orthogonal Search (PROS) [15]. PROS is a (1 + 1) Evolutionary Strategy ((1 + 1) ES), a kind of EA, that has only one parent in the population and generates one child in each generation that competes with its parent. The finite-time performance of PROS is promising and outperforms several existing EC techniques on most benchmark functions [15]. Unlike those well known EAs, PROS is free of parameters and thus the tuning of parameters is not required. To the best of our knowledge, there were no mathematical analyses on the behaviour and performance of the algorithm reported in the literature. In this paper, we aim to fill this gap. The major contributions of this research work are summarized as follows:

1. The behavior and performance analysis of PROS, which are not available in [15], are provided here.
2. Two effective novel (1 + 1) ES based on PROS, namely TROS and QROS, are proposed and they outperform PROS on a set of benchmark problems.
3. The performance of TROS and QROS are found competitive with three well-known optimization algorithms (GA, PSO and DE) on a set of benchmark problems.

Problem Formulation

In this paper, we restrict attention to global optimization of expensive black-box functions. The goal of global optimization is to find $\mathbf{x}^*$ which is a global minimum of f,

$$\mathbf{x}^* = \arg\min_{\mathbf{x}\in\Omega} f(\mathbf{x})$$

where $f : \mathbb{R}^D \to \mathbb{R}$ is a scalar-valued objective function defined on the decision space $\Omega \subseteq \mathbb{R}^D$ and $\mathbf{x} = (x_1, x_2, ..., x_D)$ represents a vector of decision variables with D dimensions.

We have made the following assumptions.

Assumption 1. *f has a single global optimum $f^* = \min_{\mathbf{x}\in\Omega} f(\mathbf{x})$.*

Assumption 2. *f can be evaluated at all points of Ω in an arbitrary order.*

Assumption 3. *The slope of f is bounded with a Lipschitz constant L such that*

$$|f(\mathbf{x}_1) - f(\mathbf{x}_2)| \leq L\|\mathbf{x}_1 - \mathbf{x}_2\|, \quad \forall \mathbf{x}_1, \mathbf{x}_2 \in \Omega,$$

where $L > 0$ and $\|\cdot\|$ denotes the Euclidean norm.

It has to be noted that although f is assumed a Lipschitz function, the corresponding Libschitz constant L is unknown to an optimization algorithm.

2. Analysis of the Pure Random Orthogonal Search (PROS) Algorithm

The PROS algorithm was originally proposed by Plevris et al. in [15]. We describe it with our revised notation as follows:

Initially, a candidate solution vector $\mathbf{x}^{(0)}$ is generated randomly from Ω (lines 1 and 2). Then, for each iteration t, j is chosen randomly from 1 to D (line 4) and a real number r is drawn randomly with uniform distribution within the search space of the j-th decision variable (line 5). A new candidate solution vector $\mathbf{y}$ is obtained by replacing the value of the j-th decision variable of the current best solution vector $\mathbf{x}^{(t)}$ with r (lines 6 and 7). The new candidate solution vector $\mathbf{y}$ is evaluated and the new best solution vector $\mathbf{x}^{(t+1)}$ is then updated based on the comparison result between $f(\mathbf{y})$ and $f(\mathbf{x}^{(t)})$. If $f(\mathbf{y})$ is smaller, $\mathbf{y}$ is accepted as the new best solution vector. Otherwise, $\mathbf{y}$ is rejected (lines 8 to 12). The iteration counter t is updated and the search process is repeated until a termination criterion is met (lines 13 and 14). Finally, the best solution vector found by the algorithm is returned as the final result (line 15).

In the following analysis, we assume the algorithm runs continuously until the global optimum is found (defined below of this section). It is because we are interested in analyzing its ultimate performance. In practice, it may take an infinitely long time to reach the global optimum. Therefore, other termination criteria are usually adopted. For example, the algorithm stops when no improvement is made after a certain number of iterations or when a pre-defined maximum number of iterations is reached.

The error of the algorithm after running for t iterations is given by

$$e_t = f(\mathbf{x}^{(t)}) - f(\mathbf{x}^*).$$

Here, $\mathbf{x}^{(t)}$ is the solution vector found by the algorithm in the t-th iteration and is also the best solution vector found by the algorithm after t iterations. It is because the values of the solution vectors found by PROS is monotonically non-increasing. That is,

$$f(\mathbf{x}^{(t)}) \leq f(\mathbf{x}^{(t-1)}) \leq \cdots \leq f(\mathbf{x}^{(0)}),$$

Consider the pseudo code from line 8 to line 12 of Algorithm 1. In the t-th iteration, when a better solution $\mathbf{y}$ is found, it will be assigned to $\mathbf{x}^{(t+1)}$. In this case, $f(\mathbf{x}^{(t+1)}) < f(\mathbf{x}^{(t)})$. When no better solution is found in the t-th iteration, $\mathbf{x}^{(t)}$ will be assigned to $\mathbf{x}^{(t+1)}$. In this case, $f(\mathbf{x}^{(t+1)}) = f(\mathbf{x}^{(t)})$. Combining the two conditions, $f(\mathbf{x}^{(t+1)}) \leq f(\mathbf{x}^{(t)})$ and the inequalities then follow.

Algorithm 1: Pure Random Orthogonal Search (PROS)

input : nil
output: the best solution vector $\mathbf{x}^{(t)}$ found by the algorithm
$t \leftarrow 0$;
Initialize $\mathbf{x}^{(t)} = (x_1, x_2, \ldots, x_D)$ randomly from $\Omega = [a_1, b_1] \times [a_2, b_2] \times \cdots \times [a_D, b_D]$;
repeat
    Sample a random integer $j \in \{1, \ldots, D\}$;
    Sample a random real number $r \sim \mathcal{U}(a_j, b_j)$;
    $\mathbf{y} \leftarrow \mathbf{x}^{(t)}$;
    $y_j \leftarrow r$;
    if $f(\mathbf{y}) < f(\mathbf{x}^{(t)})$ **then**
        $\mathbf{x}^{(t+1)} \leftarrow \mathbf{y}$;
    else
        $\mathbf{x}^{(t+1)} \leftarrow \mathbf{x}^{(t)}$;
    end
    $t \leftarrow t + 1$;
until a termination criterion is met;
return $\mathbf{x}^{(t)}$

Definition 1 (Region of Global Optimum). *It is said that a candidate solution vector $\mathbf{x}$ is in the region of global optimum R_ϵ if the error of $\mathbf{x}$ is not larger than ϵ. That is,*

$$R_\epsilon = \{\mathbf{x} : \mathbf{x} \in \Omega \text{ and } f(\mathbf{x}) - f^* \leq \epsilon\},$$

where ϵ is a small positive real number denoting the tolerance of error.

Definition 2 (Convergence Time). *The convergence time T of an algorithm is defined to be the first time t when*

$$\mathbf{x}^{(t)} \in R_\epsilon.$$

2.1. One-Dimensional Functions

In this section, we study the performance of PROS on one-dimensional functions defined on the domain $\Omega = [a, b]$. Since we are interested in the case where ϵ is small, we assume that $\frac{2\epsilon}{L} < b - a$.

Lemma 1. *For any $\epsilon > 0$, the region of the global minimum contains an interval of minimum length $\frac{\epsilon}{L}$ for all one-dimensional functions.*

Proof. Let $\mathbf{x}^* = (x_1^*)$ be the global optimum, $\mathbf{x}^l = (x_1^l) \in R_\epsilon$ be a point on the left of $\mathbf{x}^*$ and $\mathbf{x}^r = (x_1^r) \in R_\epsilon$ be a point on the right of $\mathbf{x}^*$. We are going to find a sufficient condition to ensure that $[x_1^l, x_1^r]$ is in the region of global optimum. Since f is a Lipschitz function, the interval $[x_1^l, x_1^*]$ belongs to R_ϵ if x_1^l satisfies

$$f(\mathbf{x}^l) - f(\mathbf{x}^*) \leq L\|\mathbf{x}^l - \mathbf{x}^*\| = L(x_1^* - x_1^l) \leq \epsilon,$$

or equivalently, $x_1^l \geq x_1^* - \frac{\epsilon}{L}$. We want to make x_1^l as small as possible, so we let

$$x_1^l = \max\{a, x_1^* - \frac{\epsilon}{L}\}.$$

Similarly, the interval $[x_1^*, x_1^r]$ belongs to R_ϵ if

$$x_1^r = \min\{b, x_1^* + \frac{\epsilon}{L}\}.$$

Since we have assumed $\frac{2\epsilon}{L} < b - a$, it is impossible that $x_1^l = a$ and $x_1^r = b$ hold simultaneously. Therefore, at least one of the intervals has length $\frac{\epsilon}{L}$. Hence, the length of the region of global optimum is at least $\frac{\epsilon}{L}$. □

Theorem 1. *The expected convergence time of PROS for one-dimensional functions is bounded above by $\frac{L(b-a)}{\epsilon}$.*

Proof. Let l be the length of the region of global optimum, R_ϵ. The probability that a point chosen uniformly at random falls in R_ϵ is given by

$$p = \frac{l}{b-a} \geq \frac{\epsilon}{L(b-a)}.$$

The convergence time T is geometrically distributed with parameter p. Its expected value is given by

$$\mathbb{E}[T] = \frac{1}{p} \leq \frac{L(b-a)}{\epsilon}.$$

□

Corollary 1. *PROS converges in probability to the region of global optimum for all one-dimensional functions, i.e.,*

$$\lim_{t \to \infty} P\{\mathbf{x}^{(t)} \in R_\epsilon\} = 1, \quad \forall \epsilon > 0.$$

Proof. For any realization $\{\mathbf{x}^{(t)} : t = 0, 1, 2, \ldots\}$, $f(\mathbf{x}^{(t)})$ is monotonically non-increasing in t. If $\mathbf{x}^{(t)} \in R_\epsilon$, then $\mathbf{x}^{(t+\tau)} \in R_\epsilon$ for all non-negative integer τ. Therefore, $P\{\mathbf{x}^{(t)} \in R_\epsilon\}$ is monotonically non-decreasing in t. Since the sequence is bounded above by 1, it is convergent.

Given any $\epsilon > 0$, the expected convergence time is bounded according to Theorem 1. If $\lim_{t \to \infty} P\{\mathbf{x}^{(t)} \notin R_\epsilon\}$ was non-zero, then the expected convergence time would be unbounded, which leads to a contradiction. □

2.2. Multi-Dimensional Functions

In this section, we study the performance of PROS on D-dimensional functions with domain $\Omega = [a_1, b_1] \times [a_2, b_2] \times \cdots \times [a_D, b_D]$, where $D > 1$. In particular, we focus on the class of totally separable functions as defined below. For notation simplicity, define $\mathbf{x}_{-i} \triangleq (x_1, x_2, \ldots, x_{i-1}, x_{i+1}, \ldots, x_D)$.

Definition 3 (Partially Separable Function). *A function $f(\mathbf{x})$ is partially separable with coordinate i if $\arg\min_{x_i} f(x_i, \mathbf{x}_{-i})$ is independent of $\mathbf{x}_{-i}$.*

An example of partially separable functions with coordinate $i = 1$ is :

$$f(\mathbf{x}) = x_1^2 + \sum_{i=2}^{D-1}(x_i - x_{i+1})^2.$$

Definition 4 (Totally Separable Function). *A function $f(\mathbf{x})$ is totally separable if it is partially separable with every of the D coordinates.*

An example of totally separable functions is:

$$f(\mathbf{x}) = \prod_{i=1}^{D}(x_i^2 + 1).$$

In each iteration, PROS minimizes f in one of the random coordinate i. As f is totally separable, each coordinate i can be minimized independently. The following result shows that PROS converges to the region of global optimum in probability.

Theorem 2. *PROS converges in probability to the region of global optimum for all totally separable functions, i.e.,*

$$\lim_{t \to \infty} P\{\mathbf{x}^{(t)} \in R_\epsilon\} = 1, \quad \forall \epsilon > 0.$$

Proof. Since f is Lipschitz, given any $\mathbf{x}_{-i}$, we have

$$|f(x_i, \mathbf{x}_{-i}) - f(\tilde{x}_i, \mathbf{x}_{-i})| \leq L|x_i - \tilde{x}_i|.$$

Therefore, f can be regarded as a one-dimensional Lipschitz function in x_i with the same constant L. As shown in the previous subsection, there is an interval V_{ϵ_i} of minimum length $\frac{\epsilon_i}{L}$ such that for all $x_i \in V_{\epsilon_i}$,

$$|f(x_i, \mathbf{x}_{-i}) - f(x_i^*, \mathbf{x}_{-i})| \leq \epsilon_i.$$

Note that V_{ϵ_i} is independent of $\mathbf{x}_{-i}$. Under PROS, it is clear that

$$\lim_{t \to \infty} P\{x_i^{(t)} \in V_{\epsilon_i}\} = 1, \quad \forall \epsilon_i > 0.$$

Since f is Lipschitz, it is a continuous function. Given any $\epsilon > 0$, there exists sufficiently small positive ϵ_i's such that when $x_i^{(t)} \in V_{\epsilon_i}$ for all i, we must have $\mathbf{x}^{(t)} \in R_\epsilon$. The statement then follows from

$$P\{\mathbf{x}^{(t)} \in R_\epsilon\} \geq P\{x_i^{(t)} \in V_{\epsilon_i}, \forall i\} = \prod_{i=1}^{D} P\{x_i^{(t)} \in V_{\epsilon_i}\}.$$

□

The expected convergence time can be obtained for the following subclass of totally separable functions.

Definition 5 (Additively Separable Function). *A function f is additively separable if $f(\mathbf{x})$ can be written in the form of $f_1(x_1) + f_2(x_2) + \cdots + f_D(x_D)$, where $\mathbf{x} = (x_1, x_2, \ldots, x_D)$ and $f_1, f_2, \ldots, f_D$ are one-dimensional functions.*

An example of additively separable functions is the sum-of-spheres function:

$$f_{sphere}(\mathbf{x}) = \sum_{i=1}^{D} x_i^2.$$

For the optimization of D-dimensional additively separable functions, it is equivalent to optimize D one-dimensional functions independently, i.e.,

$$\min_{\mathbf{x} \in \Omega} f(\mathbf{x}) = \sum_{i=1}^{D} \min_{x_i \in [a_i, b_i]} f_i(x_i).$$

Theorem 3. *The expected convergence time of PROS for D-dimensional additively separable functions is bounded above by*

$$\mathbb{E}[T] \leq \frac{D^2 L}{\epsilon} \sum_{i=1}^{D} (b_i - a_i).$$

Proof. A point $\mathbf{x} = (x_1, x_2, \ldots, x_D)$ belongs to R_ϵ if

$$\sum_{i=1}^{D} (f_i(x_i) - f_i^*) \leq \epsilon,$$

where f_i^* is the global minimum of the i-th one-dimensional function. Then,

$$f_i(x_i) - f_i^* \leq \frac{\epsilon}{D}, \text{ for } i \in \{1, \ldots, D\} \quad (1)$$

is a sufficient condition for $\mathbf{x} \in R_\epsilon$.

Let S_i be the iteration time PROS enters the region for x_i as stated in (1). Under PROS, at each iteration t, the coordinate to be optimized is chosen uniformly at random. As in the proof of Theorem 1, S_i is a geometric random variable with parameter

$$p_i \geq \frac{1}{D} \left(\frac{\epsilon/D}{L(b_i - a_i)} \right) = \frac{\epsilon}{D^2 L (b_i - a_i)}.$$

Note that S_i's are not independent. We bound the convergence time T as follows:

$$T \leq \max\{S_1, \ldots, S_D\} \leq S_1 + \cdots + S_D.$$

Hence,

$$\mathbb{E}[T] \leq \sum_{i=1}^{D} \mathbb{E}[S_i] = \sum_{i=1}^{D} \frac{1}{p_i} \leq \frac{D^2 L}{\epsilon} \sum_{i=1}^{D} (b_i - a_i).$$

□

3. Modified PROS with Local Search Mechanism

Although the PROS algorithm converges to the global optimum, provided that the sufficient conditions are satisfied, it converges slowly when compared with other well-known EC algorithms [15]. The major reason is PROS simply performs uniform orthogonal search in every iteration with no local search mechanism. The probability of finding an improved solution (defined in the below subsection) diminishes as it moves closer to the global optimum. Consider the situation that in the t-th iteration, $\mathbf{x}^{(t)}$ is already very close to $\mathbf{x}^*$ but has not yet fallen into in the region of global optimum. There is a high chance that $\mathbf{x}^{(t+1)}$ would reach the region of the global optimum if a narrow-range local search is performed in the $(t+1)$-th iteration. However, with uniform orthogonal search, the chance to reach the global optimum is relatively low, thus making it converge slowly. One may consider using a sampling policy other than uniform to perform local search.

3.1. Triangular-Distributed Random Orthogonal Search (TROS)

In this section, we present our first proposed algorithm called Triangular-Distributed Random Orthogonal Search (TROS). The TROS algorithm (Algorithm 2) is presented as follows:

Algorithm 2: Triangular-Distributed Random Orthogonal Search (TROS)

input : nil
output: the best solution vector $\mathbf{x}^{(t)}$ found by the algorithm
$t \leftarrow 0$;
Initialize $\mathbf{x}^{(t)} = (x_1, x_2, \ldots, x_D)$ randomly from
$\Omega = [a_1, b_1] \times [a_2, b_2] \times \cdots \times [a_D, b_D]$;
repeat
    Sample a random integer $j \in \{1, \ldots, D\}$;
    Sample a random real number $r \sim \mathcal{T}(a_j, b_j, x_j^{(t)})$;
    $\mathbf{y} \leftarrow \mathbf{x}^{(t)}$;
    $y_j \leftarrow r$;
    if $f(\mathbf{y}) < f(\mathbf{x}^{(t)})$ **then**
        | $\mathbf{x}^{(t+1)} \leftarrow \mathbf{y}$;
    else
        | $\mathbf{x}^{(t+1)} \leftarrow \mathbf{x}^{(t)}$;
    end
    $t \leftarrow t + 1$;
until *a termination criterion is met*;
return $\mathbf{x}^{(t)}$

Compared with PROS, TROS has one change in line 5 of Algorithm 1. Instead of sampling the next point of the j-th decision variable using uniform distribution, the triangular distribution $\mathcal{T}$ is used in the TROS algorithm. The probability density function of the triangular distribution $\mathcal{T}(a, b, c)$ is

$$f_{\mathcal{T}}(x) = \begin{cases} \frac{2}{b-a} \frac{x-a}{c-a} & , a < x \leq c \\ \frac{2}{b-a} \frac{b-x}{b-c} & , c < x < b \\ 0 & , x \leq a \text{ or } x \geq b \end{cases} \quad (2)$$

where a, b and c are the parameters of the distribution that represents the lower limit of x, upper limit of x and mode of x, respectively. The triangular distribution is illustrated in Figure 1.

In each iteration of TROS, the next point is sampled using the triangular distribution with the settings $a = a_j, b = b_j, c = x_j^{(t)}$ where $j \in \{1, 2, \ldots, D\}$ is the randomly chosen decision variable for the current iteration, a_j and b_j are the lower bound and upper bound of the j-th decision variable, and $x_j^{(t)}$ is the value of the j-th decision variable of the current best solution vector. With this distribution, there is a higher chance to draw a sample that is near to the current position $x_j^{(t)}$ than is far from the current position. As a result, the algorithm performs exploitation (that is, encourages local search) on the j-th decision variable.

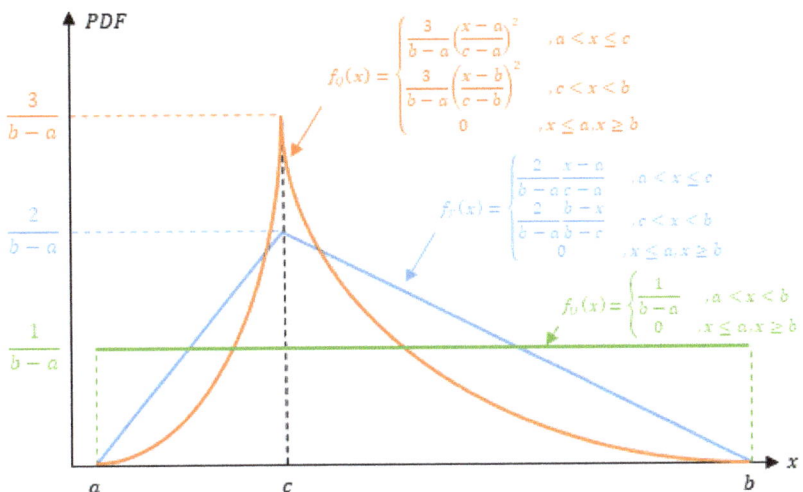

Figure 1. The probability density function of the triangular distribution $\mathcal{T}(a,b,c)$, the quadratic distribution $\mathcal{Q}(a,b,c)$, and uniform distribution $\mathcal{U}(a,b)$.

3.2. Quadratic-Distributed Random Orthogonal Search (QROS)

In this section, we present our second proposed algorithm called Quadratic-Distributed Random Orthogonal Search (QROS). The QROS algorithm (Algorithm 3) is presented as follows:

Algorithm 3: Quadratic-Distributed Random Orthogonal Search (QROS)
---
input : nil
output: the best solution vector $\mathbf{x}^{(t)}$ found by the algorithm
$t \leftarrow 0$;
Initialize $\mathbf{x}^{(t)} = (x_1, x_2, \ldots, x_D)$ randomly from
$\Omega = [a_1, b_1] \times [a_2, b_2] \times \cdots \times [a_D, b_D]$;
repeat
    Sample a random integer $j \in \{1, \ldots, D\}$;
    Sample a random real number $r \sim \mathcal{Q}(a_j, b_j, x_j^{(t)})$;
    $\mathbf{y} \leftarrow \mathbf{x}^{(t)}$;
    $y_j \leftarrow r$;
    if $f(\mathbf{y}) < f(\mathbf{x}^{(t)})$ **then**
        $\mathbf{x}^{(t+1)} \leftarrow \mathbf{y}$;
    else
        $\mathbf{x}^{(t+1)} \leftarrow \mathbf{x}^{(t)}$;
    end
    $t \leftarrow t + 1$;
until *a termination criterion is met*;
return $\mathbf{x}^{(t)}$

Compared with TROS, QROS has one change in line 5 of Algorithm 2. The value of the j-th decision variable is sampled using the quadratic distribution $\mathcal{Q}$. The probability density function of the quadratic distribution $\mathcal{Q}(a, b, c)$ is

$$f_\mathcal{Q}(x) = \begin{cases} \alpha_\mathcal{Q}(x-a)^2 & , a < x \leq c \\ \beta_\mathcal{Q}(x-b)^2 & , c < x < b \\ 0 & , x \leq a \text{ or } x \geq b \end{cases} \quad (3)$$

where $\alpha_\mathcal{Q} = \frac{3}{(b-a)(c-a)^2}$, $\beta_\mathcal{Q} = \frac{3}{(b-a)(c-b)^2}$, a, b and c are the lower limit of x, upper limit of x and mode of x, respectively. The quadratic distribution is illustrated in Figure 1.

In each iteration of QROS, the next point is sampled using the quadratic distribution with the settings $a = a_j, b = b_j, c = x_j^{(t)}$ where $j \in \{1, 2, \ldots, D\}$ is the randomly chosen decision variable for the current iteration, a_j and b_j are the lower bound and upper bound of the j-th decision variable, and $x_j^{(t)}$ is the value of the j-th decision variable of the current best solution vector. With this distribution, there is a higher chance to draw a sample that is near to the current position $x_j^{(t)}$ than is far from the current position. Compared with TROS, QROS encourages exploitation even more than TROS.

It has to be noted that both TROS and QROS are still "parameter-free" algorithms, as a and b are fixed boundaries, and c is determined based on the current best solution vector. With a simple modification, both TROS and QROS are able to improve the convergence speed of PROS and keep the major properties of PROS: parameterless, excellent computational efficiency, easy to apply to all kinds of applications, and high performance with finite-time search budget.

3.3. Analysis of the Modified Algorithms

In this subsection, we are going to explain the motivation and analyze the performance of the two algorithms. Sampling using uniform distribution can be considered as performing global search on a decision variable, while sampling using the triangular distribution and the quadratic distribution can be considered as performing local search on a decision variable.

Definition 6 (Improved Solution). *The candidate solution vector $\mathbf{x}^{(t+1)}$ found by the algorithm is said to be improved if its error is less than the error of the current best solution. That is,*

$$f(\mathbf{x}^{(t+1)}) - f(\mathbf{x}^*) < f(\mathbf{x}^{(t)}) - f(\mathbf{x}^*)$$

or simply

$$f(\mathbf{x}^{(t+1)}) < f(\mathbf{x}^{(t)})$$

where t denotes the current iteration.

For continuous (or Lipschitz) unimodal functions defined on a bounded domain, finding an improved solution vector always lead to the convergence to the global optimum. However, it may not be true for multimodal functions. Therefore, we are not interested in simply finding an improved solution vector. Here, we define a restrictive subset of improved solution vectors.

Definition 7 (Tame Solution). *The candidate solution vector $\mathbf{x}^{(t+1)}$ found by the algorithm is said to be a tame solution if the errors of all points in between the line containing the candidate solution and the optimum solution are all less than the error of the current best solution. That is,*

$$f(\alpha \mathbf{x}^{(t+1)} + (1-\alpha)\mathbf{x}^*) < f(\mathbf{x}^{(t)}), \forall \alpha \in [0,1]$$

where t denotes the current iteration.

A local optimum in a multimodal function is not considered as a *tame* solution if there exists a high hill in between the local optimum and the global optimum. The purpose of defining *tame* solution is when a *tame* solution is found, it does not only mean the solution is improved but also implies one could apply some simple techniques to converge to the global optimum. Therefore, we are interested in knowing the probability of finding a *tame* solution by each of the algorithms.

We define following terms for the analysis of the performance of the proposed algorithms. Let $\mathbf{x}^{(t)} = (x_1^{(t)}, x_2^{(t)}, \ldots, x_D^{(t)})$ be the current best solution vector, and $j \in \{1, \ldots, D\}$ be the chosen decision variable of the current iteration. Given $\mathbf{x}_{-j}^{(t)} = (x_1^{(t)}, \ldots, x_{j-1}^{(t)}, x_{j+1}^{(t)}, \ldots, x_D^{(t)})$, assume $f(\mathbf{x}_{-j}^{(t)}, x_j)$ has a unique global minimum and is denoted as x_j^*. That is,

$$f(\mathbf{x}_{-j}^{(t)}, x_j^*) \leq f(\mathbf{x}_{-j}^{(t)}, x_j), \forall x_j \in [a_j, b_j].$$

Let R_j be the set of values of the j-th decision variable that belong to improved solution vectors. That is,

$$R_j = \{x_j | f(\mathbf{x}_{-j}^{(t)}, x_j) < f(\mathbf{x}_{-j}^{(t)}, x_j^{(t)}), x_j \in [a_j, b_j]\}.$$

Let S_j be the set of values of the j-th decision variable that belong to *tame* solution vectors. That is,

$$S_j = \{x_j | (\alpha x_j + (1-\alpha) x_j^*) \in R_j, \forall \alpha \in [0,1], x_j \in [a_j, b_j]\}.$$

Let x_j^l be the smallest x_j value and x_j^r be the largest x_j value of the *tame* solutions, respectively. That is,

$$x_j^l = \inf S_j$$

and

$$x_j^r = \sup S_j.$$

Let l_j be the length of the interval S_j. That is,

$$l_j = x_j^r - x_j^l.$$

By Lemma 1, l_j is non-negative and is bounded below by

$$l_j \geq \frac{e}{L}$$

where e is the error of $f(\mathbf{x}_{-j}^{(t)}, x_j^{(t)})$ with respect to $f(\mathbf{x}_{-j}^{(t)}, x_j^*)$. That is,

$$e = f(\mathbf{x}_{-j}^{(t)}, x_j^{(t)}) - f(\mathbf{x}_{-j}^{(t)}, x_j^*).$$

Figure 2 shows two examples of multimodal functions with the corresponding set of *tame* solutions denoted by S_j.

Definition 8 (Probability of *Tame* Convergence). *The probability of tame convergence is defined as the probability of an algorithm with sampling policy π to find a tame solution in its next iteration, given the current best solution vector. That is,*

$$P\{x_j^{(t+1)} \in S_j | \mathbf{x}^{(t)}, x_j^{(t+1)} \sim \pi_{\mathbf{x}^{(t)}}\}.$$

Lemma 2. *The probability of tame convergence of PROS is $\frac{l_j}{b_j - a_j}$.*

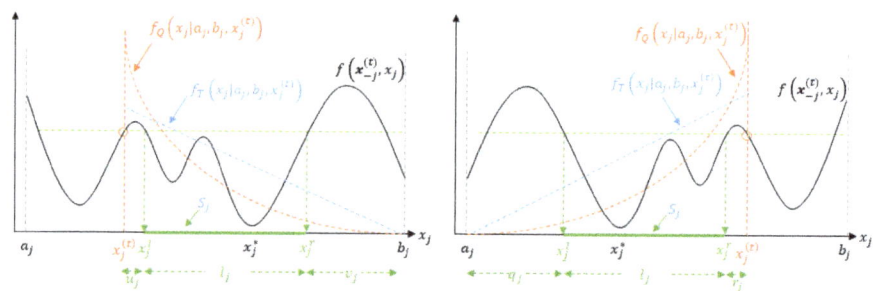

Figure 2. **Left**: an example function where $x^{(t)} < x_j^*$. **Right**: another example function where $x^{(t)} > x_j^*$. The set of *tame* solutions is denoted as S_j.

Proof. Let P_j^U be the probability that a *tame* solution is found by sampling using uniform distribution on the j-th decision variable in the $(t+1)$-th iteration, given the best solution vector in the t-th iteration. Then

$$P_j^U = P\{x_j^{(t+1)} \in S_j | \mathbf{x}_{-j}^{(t)}, x_j^{(t+1)} \sim \mathcal{U}(a_j, b_j)\} = \frac{l_j}{b_j - a_j}.$$

□

Theorem 4. *The probability of tame convergence of TROS is* $\frac{l_j}{b_j-a_j}(\frac{l_j+2v_j}{u_j+l_j+v_j})$ *when* $x_j^{(t)} < x_j^*$ *and is* $\frac{l_j}{b_j-a_j}(\frac{l_j+2q_j}{q_j+l_j+r_j})$ *when* $x_j^{(t)} > x_j^*$ *where* $u_j = x_j^l - x_j^{(t)}$, $v_j = b_j - x_j^r$, $q_j = x_j^l - a_j$ *and* $r_j = x_j^{(t)} - x_j^r$.

Proof. Let P_j^T be the probability that a *tame* solution is found by sampling using the triangular distribution on the j-th decision variable in the $(t+1)$-th iteration, given the best solution vector in the t-th iteration. That is,

$$P_j^T = P\{x_j^{(t+1)} \in S_j | \mathbf{x}_{-j}^{(t)}, x_j^{(t+1)} \sim \mathcal{T}(a_j, b_j, x_j^{(t)})\}.$$

For case 1, $x_j^{(t)} < x_j^*$:

$$P_j^T = \frac{l_j}{2}(\frac{2}{b_j - a_j}\frac{v_j}{u_j + l_j + v_j} + \frac{2}{b_j - a_j}\frac{l_j + v_j}{u_j + l_j + v_j})$$

$$= \frac{l_j}{b_j - a_j}(\frac{l_j + 2v_j}{u_j + l_j + v_j}).$$

Similarly, for case 2, $x_j^{(t)} > x_j^*$:

$$P_j^T = \frac{l_j}{2}(\frac{2}{b_j - a_j}\frac{q_j}{q_j + l_j + r_j} + \frac{2}{b_j - a_j}\frac{l_j + q_j}{q_j + l_j + r_j})$$

$$= \frac{l_j}{b_j - a_j}(\frac{l_j + 2q_j}{q_j + l_j + r_j}).$$

□

Corollary 2. *The conditions that TROS has a higher probability of tame convergence than that of PROS, the same probability as that of PROS, and a lower probability than that of PROS are as follows, respectively.*

For case 1, $x_j^{(t)} < x_j^*$:

$$P_j^T \begin{cases} > P_j^U & , v_j > u_j \\ = P_j^U & , v_j = u_j \\ < P_j^U & , v_j < u_j \end{cases}.$$

For case 2, $x_j^{(t)} > x_j^*$:

$$P_j^T \begin{cases} > P_j^U & , q_j > r_j \\ = P_j^U & , q_j = r_j \\ < P_j^U & , q_j < r_j \end{cases}.$$

Corollary 3. *The probability of tame convergence of TROS is higher than or equal to that of PROS for all convex functions.*

Proof. For convex functions, $u_j = 0$ (for case 1) and $r_j = 0$ (for case 2). An example convex function is shown in Figure 3.

For case 1, $x_j^{(t)} < x_j^*$:

$$P_j^T = \frac{l_j}{b_j - a_j}\left(\frac{l_j + 2v_j}{u_j + l_j + v_j}\right)$$

$$= P_j^U\left(1 + \frac{v_j}{l_j + v_j}\right)$$

$$\therefore P_j^T \begin{cases} > P_j^U & , v_j > 0 \\ = P_j^U & , v_j = 0 \end{cases}.$$

Similarly, for case 2, $x_j^{(t)} > x_j^*$:

$$P_j^T = \frac{l_j}{b_j - a_j}\left(\frac{l_j + 2q_j}{q_j + l_j + r_j}\right)$$

$$= P_j^U\left(1 + \frac{q_j}{q_j + l_j}\right)$$

$$\therefore P_j^T \begin{cases} > P_j^U & , q_j > 0 \\ = P_j^U & , q_j = 0 \end{cases}.$$

Therefore, for convex functions, P_j^T can never be less than P_j^U. □

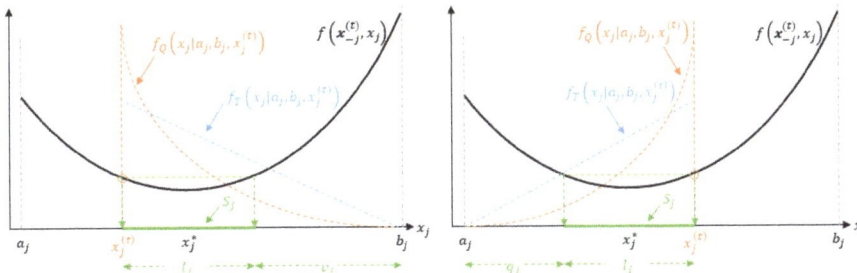

Figure 3. An example convex function. Left: for case 1, $x_j^{(t)} < x_j^*$, $u_j = 0$. Right: case 2, $x_j^{(t)} > x_j^*$, $r_j = 0$.

Theorem 5. *The probability of tame convergence of QROS is* $\frac{l_j}{b_j-a_j}\frac{(l_j+v_j)(l_j+2v_j)+v_j^2}{(u_j+l_j+v_j)^2}$ *when* $x_j^{(t)} < x_j^*$ *and is* $\frac{l_j}{b_j-a_j}\frac{(q_j+l_j)(2q_j+l_j)+q_j^2}{(q_j+l_j+r_j)^2}$ *when* $x_j^{(t)} > x_j^*$ *where* $u_j = x_j^l - x_j^{(t)}$, $v_j = b_j - x_j^r$, $q_j = x_j^l - a_j$ *and* $r_j = x_j^{(t)} - x_j^r$.

Proof. Let P_j^Q be the probability that a *tame* solution is found by sampling using the quadratic distribution on the j-th decision variable in the $(t+1)$-th iteration, given the best solution vector in the t-th iteration. That is,

$$P_j^Q = P\{x_j^{(t+1)} \in S_j | \mathbf{x}_{-j}^{(t)}, x_j^{(t+1)} \sim \mathcal{Q}(a_j, b_j, x_j^{(t)})\}.$$

For case 1, $x_j^{(t)} < x_j^*$:

$$P_j^Q = \int_{x_j^l}^{x_j^r} f_Q(x) dx$$

$$= \int_{x_j^l}^{x_j^r} \frac{3(x-b_j)^2}{(b_j-a_j)(x_j^{(t)}-b_j)^2} dx$$

$$= \frac{l_j}{b_j-a_j} \frac{(l_j+v_j)(l_j+2v_j)+v_j^2}{(u_j+l_j+v_j)^2}.$$

Similarly, for case 2, $x_j^{(t)} > x_j^*$:

$$P_j^Q = \int_{x_j^l}^{x_j^r} f_Q(x) dx$$

$$= \int_{x_j^l}^{x_j^r} \frac{3(x-a_j)^2}{(b_j-a_j)(x_j^{(t)}-a_j)^2} dx$$

$$= \frac{l_j}{b_j-a_j} \frac{(q_j+l_j)(2q_j+l_j)+q_j^2}{(q_j+l_j+r_j)^2}.$$

□

Corollary 4. *The probability of tame convergence of QROS is higher than or equal to that of TROS and PROS for all convex functions.*

Proof. For convex functions, $u_j = 0$ (for case 1) and $r_j = 0$ (for case 2).
For case 1, $x_j^{(t)} < x^*$:

$$P_j^Q = \frac{l_j}{b_j-a_j} \frac{(l_j+v_j)(l_j+2v_j)+v_j^2}{(u_j+l_j+v_j)^2}$$

$$= \underbrace{P_j^U \left(1 + \frac{v_j}{l_j+v_j}\right)}_{P_j^T} + \frac{v_j^2}{(l_j+v_j)^2}$$

$$\therefore \begin{cases} > P_j^T > P_j^U & , v_j > 0 \\ = P_j^T = P_j^U & , v_j = 0 \end{cases}.$$

For case 2, $x_j^{(t)} > x^*$:

$$P_j^Q = \frac{l_j}{b_j - a_j} \frac{(q_j + l_j)(2q_j + l_j) + q_j^2}{(q_j + l_j + r_j)^2}$$

$$= \underbrace{P_j^U \left(1 + \frac{q_j}{q_j + l_j}\right)}_{P_j^T} + \frac{q_j^2}{(q_j + l_j)^2}$$

$$\therefore \begin{cases} > P_j^T > P_j^U & , q_j > 0 \\ = P_j^T = P_j^U & , q_j = 0 \end{cases}.$$

Therefore, for convex functions, P_j^Q can never be less than P_j^T and P_j^U. □

The probability density function of distributions $\mathcal{U}$, $\mathcal{T}$ and $\mathcal{Q}$ are characterized by a degree-0, degree-1 and degree-2 polynomials, respectively. From Corollary 4, the increment in the probablity of *tame* convergence diminishes as the degree of the polynomial increases. We expect sampling using an even higher degree distribution (e.g., cubic distribution with the mode narrower than $\mathcal{Q}$ and $\mathcal{T}$) would give higher probability of *tame* convergence, but the gain diminishes as the degree grows.

4. Experiments

In order to evaluate the merits of the two modified algorithms (TROS and QROS), a set of benchmark test problems is selected from the literature [15–17] and shown in Table 1. The benchmark problems include unimodal and highly multimodal functions, convex and non-convex functions, and separable and non-separable functions. TROS and QROS were compared with the original PROS algorithm and three well-known and widely used EC algorithms for optimization: genetic algorithm (GA) [18], particle swarm optimization (PSO) [19] and differential evolution (DE) [20]. PROS, TROS and QROS were implemented in Java by the authors. GA, PSO and DE were run using pymoo, an open source framework for multi-objective optimization in Python, version 0.6.0 [21]. The code for the benchmark problems running on pymoo were revised by the authors (for experiments of shifted search space). The hyper-parameters of GA, PSO and DE are selected using the default values from the pymoo package (they are the typical values suggested in the literature).

Table 1. Benchmark test problems. (Unimodal functions: f_1–f_5; multimodal functions: f_6–f_{12}; convex functions: f_1–f_2, f_5; non-convex functions: f_3–f_4, f_6–f_{12}; separable functions: f_1–f_2, f_6–f_7, f_{12}; non-separable functions: f_3–f_5, f_8–f_{11})

No.	Function	Formulation and Global Optimum	Search Space
f_1	Sphere	$\sum_{i=1}^{D} x_i^2$ $\mathbf{x}^* = (0,0,\ldots,0), f(\mathbf{x}^*) = 0$	$[-10, 10]^D$
f_2	Ellipsoid	$\sum_{i=1}^{D} i x_i^2$ $\mathbf{x}^* = (0,0,\ldots,0), f(\mathbf{x}^*) = 0$	$[-10, 10]^D$
f_3	Schwefel 1.2	$\sum_{i=1}^{D} (\sum_{j=1}^{i} x_j)^2$ $\mathbf{x}^* = (0,0,\ldots,0), f(\mathbf{x}^*) = 0$	$[-5.12, 5.12]^D$
f_4	Rosenbrock	$\sum_{i=1}^{D-1} [100(x_{i+1} - x_i^2)^2 + (x_i - 1)^2]$ $\mathbf{x}^* = (1,1,\ldots,1), f(\mathbf{x}^*) = 0$	$[-2.048, 2.048]^D$
f_5	Zakharov	$\sum_{i=1}^{D} x_i^2 + (\frac{1}{2}\sum_{i=1}^{D} i x_i)^2 + (\frac{1}{2}\sum_{i=1}^{D} i x_i)^4$ $\mathbf{x}^* = (0,0,\ldots,0), f(\mathbf{x}^*) = 0$	$[-10, 10]^D$

Table 1. *Cont.*

No.	Function	Formulation and Global Optimum	Search Space
f_6	Alpine 1	$\sum_{i=1}^{D} \|x_i \sin(x_i) + 0.1x_i\|$ $\mathbf{x}^* = (0,0,\ldots,0), f(\mathbf{x}^*) = 0$	$[-10, 10]^D$
f_7	Rastrigin	$10D + \sum_{i=1}^{D}[x_i^2 - 10\cos(2\pi x_i)]$ $\mathbf{x}^* = (0,0,\ldots,0), f(\mathbf{x}^*) = 0$	$[-5.12, 5.12]^D$
f_8	Ackley	$20 + \exp(1) - 20\exp(-0.2\sqrt{\frac{1}{D}\sum_{i=1}^{D} x_i^2})$ $- \exp(\frac{1}{D}\sum_{i=1}^{D} \cos(2\pi x_i))$ $\mathbf{x}^* = (0,0,\ldots,0), f(\mathbf{x}^*) = 0$	$[-32.768, 32.768]^D$
f_9	Griewank	$\sum_{i=1}^{D} \frac{x_i^2}{4000} - \prod_{i=1}^{D} \cos(\frac{x_i}{\sqrt{i}}) + 1$ $\mathbf{x}^* = (0,0,\ldots,0), f(\mathbf{x}^*) = 0$	$[-600, 600]^D$
f_{10}	HGBat	$\|(\sum_{i=1}^{D} x_i^2)^2 - (\sum_{i=1}^{D} x_i)^2\|^{1/2} +$ $(0.5\sum_{i=1}^{D} x_i^2 + \sum_{i=1}^{D} x_i)/D + 0.5$ $\mathbf{x}^* = (-1,-1,\ldots,-1), f(\mathbf{x}^*) = 0$	$[-15, 15]^D$
f_{11}	HappyCat	$\|\sum_{i=1}^{D} x_i^2 - D\|^{1/4} +$ $(0.5\sum_{i=1}^{D} x_i^2 + \sum_{i=1}^{D} x_i)/D + 0.5$ $\mathbf{x}^* = (-1,-1,\ldots,-1), f(\mathbf{x}^*) = 0$	$[-20, 20]^D$
f_{12}	Weierstrass	$\sum_{i=1}^{D}[\sum_{k=0}^{20} 0.5^k \cos(2\pi 3^k(x_i + 0.5))$ $- D\sum_{k=0}^{20} 0.5^k \cos(\pi 3^k)]$ $\mathbf{x}^* = (0,0,\ldots,0), f(\mathbf{x}^*) = 0$	$[-0.5, 0.5]^D$

The experiments were carried out on a 3.20 GHz computer with 16GB RAM under a Windows 10 platform. Multiple runs were conducted for each problem by each algorithm with a different seed for the generation of random numbers for each run. For fair comparison, all algorithms end with the same maximum number of objective function evaluations. We follow the experimental settings of the PROS paper [15] for the population size and the maximum number of objective function evaluations. In order to increase the reliability of the statistical results, the number of runs was increased to 100, 100 and 30 for 5D, 10D, and 50D problems, respectively (instead of 10 for all dimensions in [15]). The settings are summarized in Table 2.

Table 2. Settings of the numerical experiments.

Settings		$D = 5$	$D = 10$	$D = 50$
Population Size	$10D$	50	100	500
Max. No. of Generations	$(20D - 50)$	50	150	950
Max. No. of Objective Function Evaluations	$10D(20D - 50)$	2500	15,000	475,000
No. of Runs		100	100	30

4.1. Experiment I: Basic Benchmark Problems

Figure 4 shows the experimental results of the mean error for 100 independent runs of each of the algorithms on each benchmark problem for dimension $D = 5$. The x-axes are the number of objective function evaluations (in log scale) and the y-axes are the average error of the algorithm after certain objective function evaluations. As shown in the figure, QROS converged faster than the other five algorithms except on f_3 and f_5 where PSO converged the fastest. TROS and PROS had similar trends as QROS on the benchmark functions: they converged faster than GA, PSO and DE on most benchmark functions except on f_3 and f_5. Figures 5 and 6 show the experimental results for dimension $D = 10$ and $D = 50$, respectively. Similar to the results for $D = 5$, QROS converged faster than the other five algorithms except on f_3 (for 10D) and f_5 (for 10D and 50D) where PSO converged the fastest.

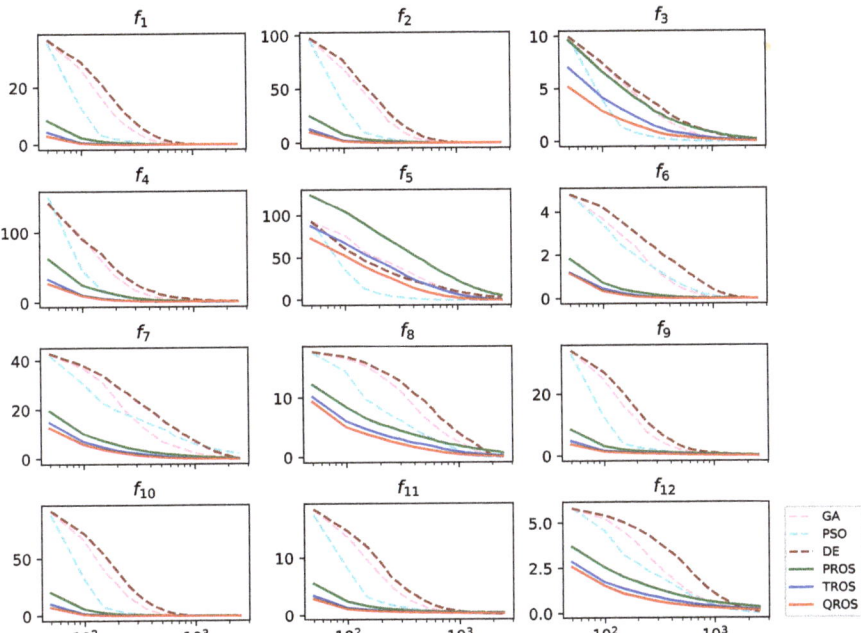

Figure 4. The convergence curve of 100 runs on 5D benchmark functions.

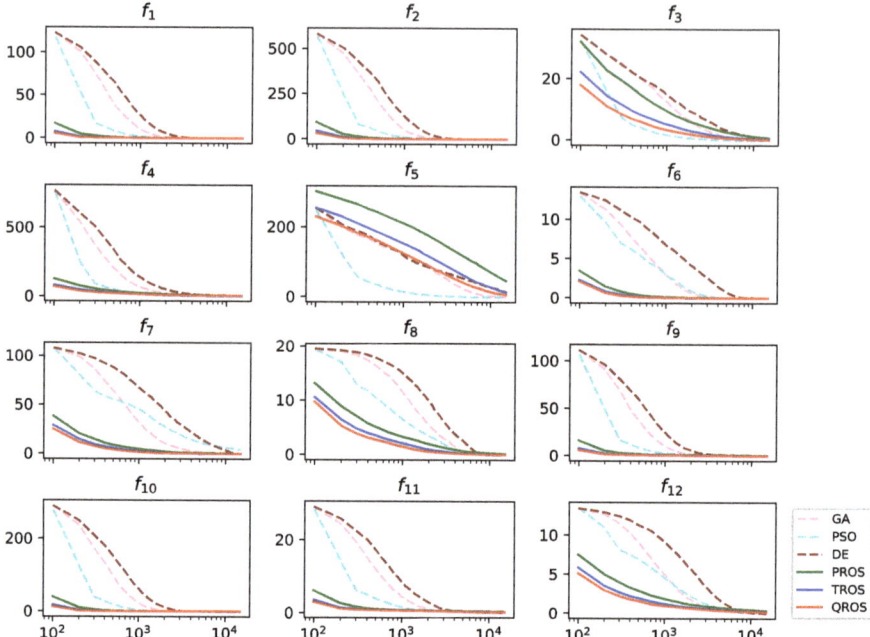

Figure 5. The convergence curve of 100 runs on 10D benchmark functions.

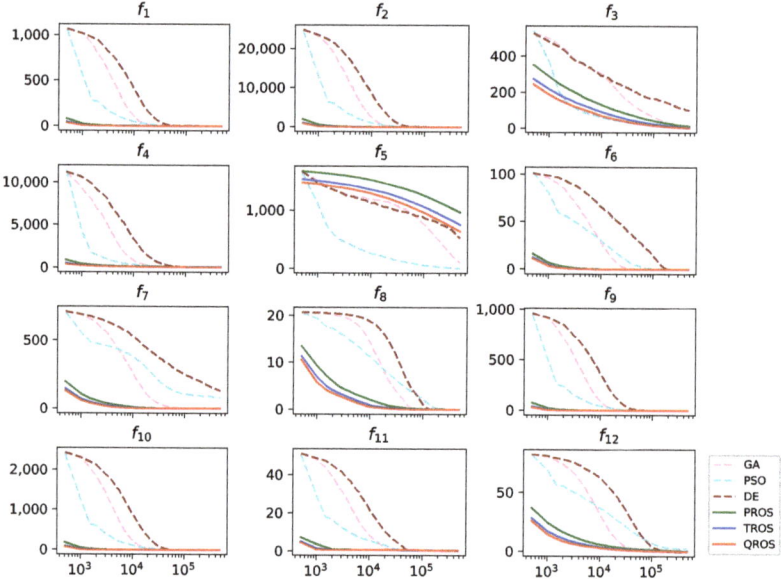

Figure 6. The convergence curve of 30 runs on 50D benchmark functions.

Table 3 shows the mean of the errors of the final solution vectors returned by PROS, TROS and QROS for 100 independent runs of on each benchmark problem for dimension $D = 5$. The mean errors represent the medium to long term performance of the algorithms. The best results (the smallest mean final errors) for each benchmark problem are highlighted in bold font. The corresponding standard deviations are placed next with parenthesis. Owing to limited space, the results of GA, PSO and DE are not shown here, as they have been reported in [15] already. As seen from the table, QROS had the smallest mean final errors on most of the benchmark problems. Similar results can be seen from Tables 4 and 5 which show the means of the final errors for dimension $D = 10$ and $D = 50$, respectively. In general, QROS and TROS are more efficient in reducing the mean final errors on most benchmark problems when compared with PROS.

Table 3. Statistical results of 100 runs on 5D benchmark functions.

f	PROS	TROS	QROS
f_1	4.65e-03 (4.85e-03)	1.06e-03 (8.88e-04)	**4.13e-04 (4.54e-04)**
f_2	1.38e-02 (1.68e-02)	3.13e-03 (2.65e-03)	**1.21e-03 (1.33e-03)**
f_3	2.17e-01 (1.80e-01)	7.61e-02 (5.05e-02)	**4.32e-02 (2.99e-02)**
f_4	**1.34e+00 (1.17e+00)**	1.48e+00 (1.16e+00)	1.55e+00 (1.17e+00)
f_5	5.13e+00 (6.46e+00)	5.33e-01 (8.45e-01)	**1.28e-01 (1.81e-01)**
f_6	4.81e-03 (3.68e-03)	2.38e-03 (9.69e-04)	**1.62e-03 (7.19e-04)**
f_7	2.41e-01 (2.49e-01)	5.30e-02 (4.49e-02)	**2.91e-02 (3.01e-02)**
f_8	7.48e-01 (4.80e-01)	2.81e-01 (1.65e-01)	**1.53e-01 (8.95e-02)**
f_9	2.69e-01 (1.41e-01)	1.33e-01 (6.76e-02)	**8.98e-02 (3.69e-02)**
f_{10}	4.01e-01 (1.40e-01)	3.34e-01 (1.24e-01)	**2.83e-01 (1.26e-01)**
f_{11}	4.67e-01 (1.34e-01)	4.05e-01 (1.20e-01)	**3.72e-01 (9.48e-02)**
f_{12}	3.49e-01 (1.04e-01)	2.22e-01 (6.44e-02)	**1.77e-01 (4.67e-02)**

Table 4. Statistical results of 100 runs on 10D benchmark functions.

f	PROS	TROS	QROS
f_1	8.62e-04 (7.26e-04)	2.38e-04 (1.85e-04)	**1.03e-04 (7.51e-05)**
f_2	4.45e-03 (3.33e-03)	1.34e-03 (1.23e-03)	**5.68e-04 (4.63e-04)**
f_3	6.84e-01 (3.62e-01)	2.52e-01 (1.35e-01)	**1.35e-01 (7.28e-02)**
f_4	**3.40e+00 (3.28e+00)**	3.72e+00 (2.99e+00)	4.53e+00 (2.91e+00)
f_5	4.69e+01 (2.47e+01)	1.40e+01 (9.29e+00)	**6.26e+00 (4.77e+00)**
f_6	3.10e-03 (9.52e-04)	1.70e-03 (5.30e-04)	**1.10e-03 (3.52e-04)**
f_7	4.48e-02 (3.77e-02)	1.24e-02 (9.46e-03)	**5.24e-03 (3.56e-03)**
f_8	1.61e-01 (8.23e-02)	7.13e-02 (3.00e-02)	**4.32e-02 (1.55e-02)**
f_9	1.78e-01 (6.53e-02)	9.15e-02 (3.68e-02)	**6.77e-02 (2.74e-02)**
f_{10}	4.73e-01 (2.46e-01)	4.35e-01 (1.97e-01)	**4.31e-01 (1.99e-01)**
f_{11}	4.90e-01 (1.65e-01)	4.30e-01 (1.49e-01)	**3.76e-01 (1.17e-01)**
f_{12}	3.40e-01 (8.16e-02)	2.20e-01 (4.97e-02)	**1.68e-01 (3.85e-02)**

Table 5. Statistical results of 30 runs on 50D benchmark functions.

f	PROS	TROS	QROS
f_1	1.12e-04 (3.30e-05)	2.82e-05 (7.60e-06)	**1.24e-05 (4.68e-06)**
f_2	2.86e-03 (9.68e-04)	7.25e-04 (2.22e-04)	**3.29e-04 (1.32e-04)**
f_3	1.34e+01 (3.13e+00)	6.45e+00 (1.57e+00)	**3.75e+00 (8.21e-01)**
f_4	8.06e+01 (3.80e+01)	6.05e+01 (3.41e+01)	**5.48e+01 (2.46e+01)**
f_5	9.66e+02 (1.63e+02)	7.58e+02 (1.51e+02)	**6.44e+02 (1.36e+02)**
f_6	2.46e-03 (3.48e-04)	1.32e-03 (1.93e-04)	**8.85e-04 (1.26e-04)**
f_7	5.81e-03 (1.72e-03)	1.32e-03 (2.79e-04)	**6.25e-04 (1.60e-04)**
f_8	2.06e-02 (3.28e-03)	9.90e-03 (1.13e-03)	**6.49e-03 (9.52e-04)**
f_9	2.63e-02 (1.51e-02)	**1.17e-02 (1.01e-02)**	1.39e-02 (1.77e-02)
f_{10}	5.99e-01 (2.48e-01)	5.80e-01 (2.25e-01)	**5.66e-01 (2.27e-01)**
f_{11}	6.51e-01 (1.28e-01)	**6.17e-01 (1.11e-01)**	6.21e-01 (1.08e-01)
f_{12}	5.34e-01 (4.80e-02)	3.43e-01 (2.93e-02)	**2.61e-01 (2.84e-02)**

4.2. Experiment II: Random Shifted Benchmark Problems

One may notice that both TROS and QROS are in favor of objective functions with the global optimum located at the center of the search space. For example, the global optimum of $f_1, f_2, f_3, f_5, f_6, f_7, f_8, f_9, f_{12}$ is $(0, 0, \ldots, 0)$. In order to test the effectiveness of TROS and QROS in a general case, another experiment was conducted. The same set of benchmark problems were used in this experiment, but the search space was shifted so that the global optimum may be located anywhere within the bounded search space. A random point $\mathbf{o} = (o_1, o_2, \ldots, o_D)$ is drawn uniformly from $\Omega' = [a_1 - x_1^*, b_1 - x_1^*] \times [a_2 - x_2^*, b_2 - x_2^*] \times \cdots \times [a_D - x_D^*, b_D - x_D^*]$ for each benchmark function for each run where a_i, b_i, x_i^* are the lower limit, upper limit and the optimum value of the i-th decision variable in the original search space of the benchmark function being optimized and $i \in \{1, 2, \ldots, D\}$. The shifted search space becomes $\Omega'' = [a_1 - o_1, b_1 - o_1] \times [a_2 - o_2, b_2 - o_2] \times \cdots \times [a_D - o_D, b_D - o_D]$. It has to be noted that the benchmark problems are carefully chosen in this experiment to ensure for each benchmark problem, x^* is still the global optimum in the shifted search space but x^* is not necessarily be located at the center of the new search space.

Figures 7–9 show the experimental results of the mean error of each of the algorithms on each benchmark problem for dimension $D = 5$, $D = 10$ and $D = 50$, respectively. Similar to the non-shifted experiments, PROS, TROS and QROS converged faster than GA, PSO and DE on most benchmark problems except on f_5 when $D = 50$. When only considering the three random orthogonal search algorithms, PROS usually converged quickly initially, then TROS surpassed PROS, followed by QROS surpassed TROS on most of the benchmark problems.

Tables 6–8 show the statistical results of the final error of each of the algorithms on each benchmark problem for dimension $D = 5$, $D = 10$ and $D = 50$, respectively. Similar to the non-shifted experiments, both QROS and TROS often improve the mean final errors. The performance of QROS and TROS are quite promising on the shifted benchmark functions.

Table 6. Statistical results of 100 runs on 5D benchmark functions with shifted global optimum.

f	PROS	TROS	QROS
f_1	4.31e-03 (4.38e-03)	1.67e-03 (3.92e-03)	**1.15e-03 (5.90e-03)**
f_2	1.24e-02 (1.20e-02)	4.41e-03 (7.05e-03)	**2.60e-03 (9.36e-03)**
f_3	2.64e-01 (2.23e-01)	9.43e-02 (8.36e-02)	**5.54e-02 (4.39e-02)**
f_4	1.38e+00 (3.51e+00)	**9.47e-01 (3.44e+00)**	1.82e+00 (1.12e+01)
f_5	7.76e+00 (1.24e+01)	8.39e-01 (1.16e+00)	**2.60e-01 (4.88e-01)**
f_6	**4.62e-03 (2.28e-03)**	6.57e-03 (9.53e-03)	1.21e-02 (1.96e-02)
f_7	2.23e-01 (2.23e-01)	**1.39e-01 (3.07e-01)**	2.97e-01 (5.00e-01)
f_8	7.34e-01 (4.52e-01)	3.27e-01 (2.86e-01)	**2.34e-01 (3.01e-01)**
f_9	2.94e-01 (1.97e-01)	1.54e-01 (7.26e-02)	**1.18e-01 (1.06e-01)**
f_{10}	3.90e-01 (1.68e-01)	3.18e-01 (1.22e-01)	**3.13e-01 (1.38e-01)**
f_{11}	4.82e-01 (1.21e-01)	4.28e-01 (1.43e-01)	**4.15e-01 (1.93e-01)**
f_{12}	3.47e-01 (1.03e-01)	**2.64e-01 (1.33e-01)**	2.79e-01 (2.32e-01)

Table 7. Statistical results of 100 runs on 10D benchmark functions with shifted global optimum.

f	PROS	TROS	QROS
f_1	8.71e-04 (5.89e-04)	2.55e-04 (1.99e-04)	**1.64e-04 (2.97e-04)**
f_2	4.64e-03 (3.58e-03)	1.33e-03 (1.18e-03)	**9.91e-04 (2.60e-03)**
f_3	7.99e-01 (4.64e-01)	2.78e-01 (1.76e-01)	**1.62e-01 (9.67e-02)**
f_4	1.35e+00 (6.26e+00)	8.60e-01 (2.81e+00)	**8.20e-01 (2.68e+00)**
f_5	6.23e+01 (4.86e+01)	1.80e+01 (1.70e+01)	**7.34e+00 (7.04e+00)**
f_6	3.12e-03 (9.80e-04)	**2.38e-03 (2.74e-03)**	6.05e-03 (9.72e-03)
f_7	**4.53e-02 (3.06e-02)**	6.38e-02 (2.58e-01)	2.65e-01 (5.19e-01)
f_8	1.64e-01 (7.02e-02)	8.10e-02 (3.39e-02)	**6.33e-02 (1.06e-01)**
f_9	1.85e-01 (6.35e-02)	9.56e-02 (3.76e-02)	**7.97e-02 (3.49e-02)**
f_{10}	3.84e-01 (2.06e-01)	3.61e-01 (1.80e-01)	**3.37e-01 (1.96e-01)**
f_{11}	4.50e-01 (1.46e-01)	4.51e-01 (2.05e-01)	**4.45e-01 (1.92e-01)**
f_{12}	3.31e-01 (7.39e-02)	**2.52e-01 (9.44e-02)**	3.03e-01 (2.26e-01)

Table 8. Statistical results of 30 runs on 50D benchmark functions with shifted global optimum.

f	PROS	TROS	QROS
f_1	1.13e-04 (2.63e-05)	2.88e-05 (1.00e-05)	**1.30e-05 (4.43e-06)**
f_2	2.90e-03 (7.79e-04)	7.43e-04 (3.28e-04)	**3.31e-04 (1.23e-04)**
f_3	1.78e+01 (4.20e+00)	8.55e+00 (1.75e+00)	**5.46e+00 (1.20e+00)**
f_4	2.02e+01 (3.98e+01)	**2.59e+00 (8.90e+00)**	3.84e+00 (1.35e+01)
f_5	1.93e+03 (4.23e+02)	1.55e+03 (3.24e+02)	**1.30e+03 (2.85e+02)**
f_6	2.62e-03 (3.48e-04)	**1.69e-03 (5.51e-04)**	3.19e-03 (2.83e-03)
f_7	5.89e-03 (1.37e-03)	**1.58e-03 (4.07e-04)**	2.99e-01 (5.23e-01)
f_8	2.09e-02 (2.66e-03)	9.85e-03 (1.31e-03)	**6.68e-03 (1.06e-03)**
f_9	2.77e-02 (1.27e-02)	**1.92e-02 (1.70e-02)**	5.13e-02 (4.70e-02)
f_{10}	3.68e-01 (1.11e-01)	3.70e-01 (1.55e-01)	**3.32e-01 (9.46e-02)**
f_{11}	5.82e-01 (1.11e-01)	**5.74e-01 (9.44e-02)**	6.06e-01 (1.34e-01)
f_{12}	5.51e-01 (4.15e-02)	**3.93e-01 (1.21e-01)**	5.14e-01 (2.79e-01)

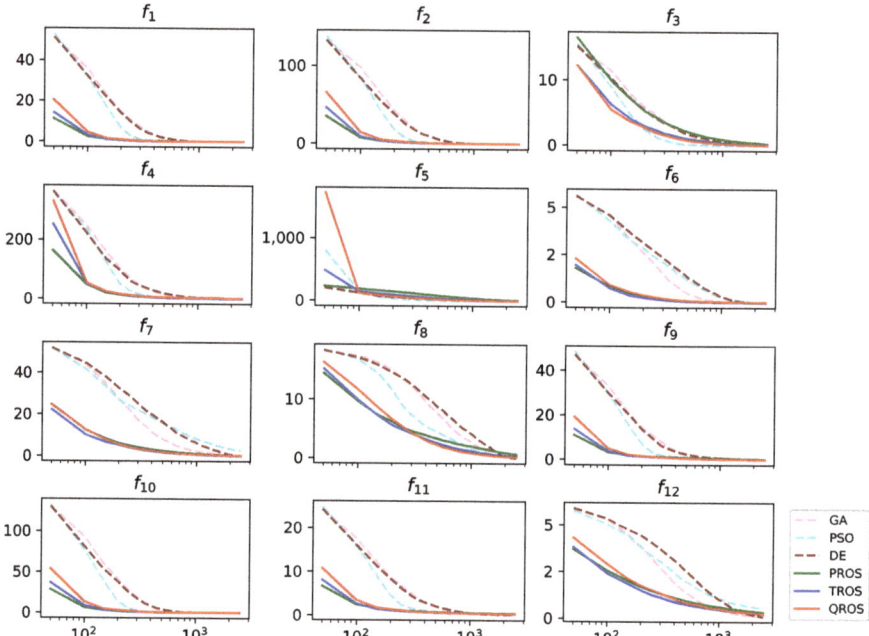

Figure 7. The convergence curve of 100 runs on 5D benchmark functions with shifted global optimum.

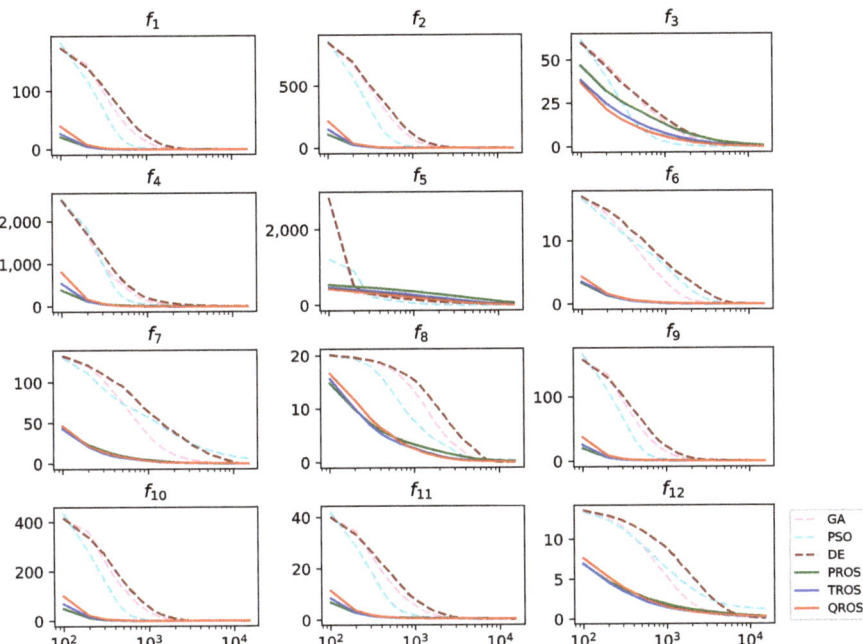

Figure 8. The convergence curve of 100 runs on 10D benchmark functions with shifted global optimum.

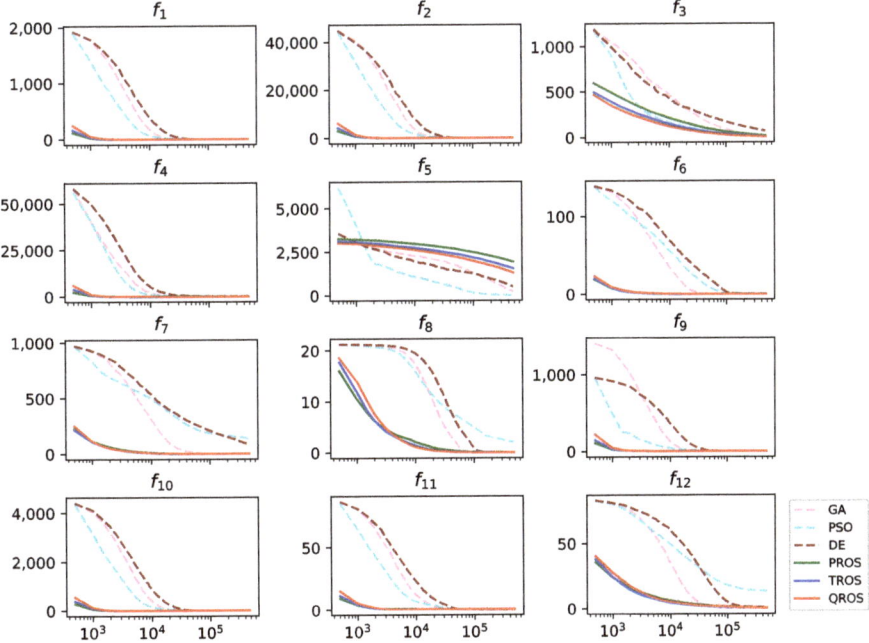

Figure 9. The convergence curve of 30 runs on 50D benchmark functions with shifted global optimum.

5. Future Work

In our future work, we plan to extend the algorithms in two directions. The first one is to allow switching between various sampling polices (e.g., uniform, triangular, quadratic) in an adaptive way. This is particular useful for black-box optimization where the form of the function is unknown to the optimization algorithm. One could learn the best sampling policy for that particular function and adapt gradually from the results of previous function evaluations. The second direction is to gradually rotate the objective function based on the current best solution set [22,23]. By doing so, a complex function would be transformed to a simpler convex function in the small area of concern. Any simple techniques such as TROS and QROS would quickly converge to the global optimum if it is in the area of concern. We are interested in investigating the convergence of TROS and QROS with the addition of transformation techniques.

6. Conclusions

In this paper, we perform an analysis of PROS. PROS is a parameterless EA that outperforms several well-known optimization algorithms on benchmark functions when the search budget is limited. It is attractive for expensive optimizing problems. However, to the best of our knowledge, there is no mathematical analysis of the performance and behavior of PROS reported in the literature. In this paper, we fill this gap.

Moreover, we propose TROS and QROS, which preserve the advantages of PROS and outperform it. We perform an analysis of *tame* convergence for PROS, TROS and QROS. We conduct two sets of experiments to evaluate the performance of TROS and QROS. We increase the number of runs compared to those conducted in [15] in order to increase the reliability of the statistical results. Experimental results illustrate that TROS and QROS outperform PROS to a large extent. Both TROS and QROS converge quickly on the benchmark problems. Although GA, PSO and DE usually have better performance in terms of smaller error in the final solution when a considerable amount of objective function evaluation is given, the performances of QROS and TROS are not much inferior to that of GA, PSO and DE. In fact, QROS and TROS are even sensible choices when we consider the computational complexity.

Author Contributions: Conceptualization, B.K.-B.T., C.W.S. and W.S.W.; methodology, B.K.-B.T. and W.S.W.; software, B.K.-B.T.; validation, B.K.-B.T., C.W.S. and W.S.W.; formal analysis, B.K.-B.T., C.W.S. and W.S.W.; investigation, B.K.-B.T.; resources, B.K.-B.T.; data curation, B.K.-B.T.; writing—original draft preparation, B.K.-B.T.; writing—review and editing, B.K.-B.T., C.W.S. and W.S.W.; visualization, B.K.-B.T.; supervision, W.S.W.; project adminstration, B.K.-B.T.; funding acquisition, B.K.-B.T. All authors have read and agreed to the published version of the manuscript.

Funding: The APC was partially funded by Hong Kong Metropolitan University.

Institutional Review Board Statement: Not applicable.

Informed Consent Statement: Not applicable.

Data Availability Statement: Data are available from the authors upon reasonable request.

Conflicts of Interest: The authors declare no conflict of interest.

References

1. Cao, Y.; Zhang, H.; Li, W.; Zhou, M.; Zhang, Y.; Chaovalitwongse, W.A. Comprehensive Learning Particle Swarm Optimization Algorithm With Local Search for Multimodal Functions. *IEEE Trans. Evol. Comput.* **2019**, *23*, 718–731. [CrossRef]
2. Kang, Q.; Song, X.; Zhou, M.; Li, L. A Collaborative Resource Allocation Strategy for Decomposition-Based Multiobjective Evolutionary Algorithms. *IEEE Trans. Syst. Man Cybern. Syst.* **2019**, *49*, 2416–2423. [CrossRef]
3. Liu, J.; Liu, Y.; Jin, Y.; Li, F. A Decision Variable Assortment-Based Evolutionary Algorithm for Dominance Robust Multiobjective Optimization. *IEEE Trans. Syst. Man Cybern. Syst.* **2022**, *52*, 3360–3375. [CrossRef]
4. Sarker, R.A.; Kamruzzaman, J.; Newton, C.S. Evolutionary Optimization (Evopt): A Brief Review And Analysis. *Int. J. Comput. Intell. Appl.* **2003**, *3*, 311–330. [CrossRef]
5. Tian, J.; Tan, Y.; Zeng, J.; Sun, C.; Jin, Y. Multiobjective Infill Criterion Driven Gaussian Process-Assisted Particle Swarm Optimization of High-Dimensional Expensive Problems. *IEEE Trans. Evol. Comput.* **2019**, *23*, 459–472. [CrossRef]
6. Jin, Y.; Wang, H.; Chugh, T.; Guo, D.; Miettinen, K. Data-Driven Evolutionary Optimization: An Overview and Case Studies. *IEEE Trans. Evol. Comput.* **2019**, *23*, 442–458. [CrossRef]

7. Yang, C.; Ding, J.; Jin, Y.; Chai, T. Offline Data-Driven Multiobjective Optimization: Knowledge Transfer Between Surrogates and Generation of Final Solutions. *IEEE Trans. Evol. Comput.* **2020**, *24*, 409–423. [CrossRef]
8. Liu, Y.; Liu, J.; Jin, Y. Surrogate-Assisted Multipopulation Particle Swarm Optimizer for High-Dimensional Expensive Optimization. *IEEE Trans. Syst. Man Cybern. Syst.* **2022**, *52*, 4671–4684. [CrossRef]
9. Zhou, Y.; He, X.; Chen, Z.; Jiang, S. A Neighborhood Regression Optimization Algorithm for Computationally Expensive Optimization Problems. *IEEE Trans. Cybern.* **2022**, *52*, 3018–3031. [CrossRef] [PubMed]
10. Gutjahr, W.J. Convergence Analysis of Metaheuristics. In *Matheuristics: Hybridizing Metaheuristics and Mathematical Programming*; Maniezzo, V., Stützle, T., Voß, S., Eds.; Springer: Boston, MA, USA, 2010; pp. 159–187. [CrossRef]
11. Zamani, S.; Hemmati, H. A Cost-Effective Approach for Hyper-Parameter Tuning in Search-based Test Case Generation. In Proceedings of the 2020 IEEE International Conference on Software Maintenance and Evolution (ICSME), Adelaide, Australia, 28 September–2 October 2020; pp. 418–429. [CrossRef]
12. Gu, Q.; Wang, Q.; Xiong, N.N.; Jiang, S.; Chen, L. Surrogate-assisted Evolutionary Algorithm for Expensive Constrained Multi-objective Discrete Optimization Problems. *Complex Intell. Syst.* **2022**, *8*, 2699–2718. [CrossRef]
13. Sarker, R.A.; Elsayed, S.M.; Ray, T. Differential Evolution With Dynamic Parameters Selection for Optimization Problems. *IEEE Trans. Evol. Comput.* **2014**, *18*, 689–707. [CrossRef]
14. Karafotias, G.; Hoogendoorn, M.; Eiben, A.E. Parameter Control in Evolutionary Algorithms: Trends and Challenges. *IEEE Trans. Evol. Comput.* **2015**, *19*, 167–187. [CrossRef]
15. Plevris, V.; Bakas, N.P.; Solorzano, G. Pure Random Orthogonal Search (PROS): A Plain and Elegant Parameterless Algorithm for Global Optimization. *Appl. Sci.* **2021**, *11*, 5053. [CrossRef]
16. Vesterstrom, J.; Thomsen, R. A Comparative Study of Differential Evolution, Particle Swarm Optimization, and Evolutionary Algorithms on Numerical Benchmark Problems. In Proceedings of the 2004 Congress on Evolutionary Computation (IEEE Cat. No.04TH8753), Portland, OR, USA, 19–23 June 2004; Volume 2, pp. 1980–1987. [CrossRef]
17. Omidvar, M.N.; Li, X.; Tang, K. Designing Benchmark Problems for Large-scale Continuous Optimization. *Inf. Sci.* **2015**, *316*, 419–436. [CrossRef]
18. Holland, J.H. *Adaptation in Natural and Artificial Systems*; The University of Michigan Press: Ann Arbor, MI, USA, 1975.
19. Kennedy, J.; Eberhart, R. Particle Swarm Optimization. In Proceedings of the ICNN'95—International Conference on Neural Networks, Perth, Australia, 27 November–1 December 1995; Volume 4, pp. 1942–1948. [CrossRef]
20. Storn, R.; Price, K. Differential Evolution—A Simple and Efficient Heuristic for global Optimization over Continuous Spaces. *J. Glob. Optim.* **1997**, *11*, 341–359. [CrossRef]
21. Blank, J.; Deb, K. Pymoo: Multi-Objective Optimization in Python. *IEEE Access* **2020**, *8*, 89497–89509. [CrossRef]
22. Hansen, N. Adaptive Encoding: How to Render Search Coordinate System Invariant. In *Parallel Problem Solving from Nature—PPSN X*; Rudolph, G., Jansen, T., Beume, N., Lucas, S., Poloni, C., Eds.; Springer: Berlin/Heidelberg, Germany, 2008; pp. 205–214. [CrossRef]
23. Loshchilov, I.; Schoenauer, M.; Sebag, M. Adaptive Coordinate Descent. In Proceedings of the 13th Annual Conference on Genetic and Evolutionary Computation, GECCO '11, Dublin, Ireland, 12–16 July 2011; pp. 885–892. [CrossRef]

Disclaimer/Publisher's Note: The statements, opinions and data contained in all publications are solely those of the individual author(s) and contributor(s) and not of MDPI and/or the editor(s). MDPI and/or the editor(s) disclaim responsibility for any injury to people or property resulting from any ideas, methods, instructions or products referred to in the content.

Article

Estimation of Interaction Locations in Super Cryogenic Dark Matter Search Detectors Using Genetic Programming-Symbolic Regression Method

Nikola Anđelić *,†, Sandi Baressi Šegota †, Matko Glučina and Zlatan Car

Department of Automation and Electronics, Faculty of Engineering, University of Rijeka, Vukovarska 58, 51000 Rijeka, Croatia
* Correspondence: nandelic@riteh.hr
† These authors contributed equally to this work.

Abstract: The Super Cryogenic Dark Matter Search (SuperCDMS) experiment is used to search for Weakly Interacting Massive Particles (WIMPs)—candidates for dark matter particles. In this experiment, the WIMPs interact with nuclei in the detector; however, there are many other interactions (background interactions). To separate background interactions from the signal, it is necessary to measure the interaction energy and to reconstruct the location of the interaction between WIMPs and the nuclei. In recent years, some research papers have been investigating the reconstruction of interaction locations using artificial intelligence (AI) methods. In this paper, a genetic programming-symbolic regression (GPSR), with randomly tuned hyperparameters cross-validated via a five-fold procedure, was applied to the SuperCDMS experiment to estimate the interaction locations with high accuracy. To measure the estimation accuracy of obtaining the SEs, the mean and standard deviation (σ) values of R^2, the root-mean-squared error ($RMSE$), and finally, the mean absolute error (MAE) were used. The investigation showed that using GPSR, SEs can be obtained that estimate the interaction locations with high accuracy. To improve the solution, the five best SEs were combined from the three best cases. The results demonstrated that a very high estimation accuracy can be achieved with the proposed methodology.

Keywords: cross-validation; genetic programming; interaction location; SuperCDMS; symbolic regression

Citation: Anđelić, N.; Baressi Šegota, S.; Glučina, M.; Car, Z. Estimation of Interaction Locations in Super Cryogenic Dark Matter Search Detectors Using Genetic Programming-Symbolic Regression Method. *Appl. Sci.* **2023**, *13*, 2059. https://doi.org/10.3390/app13042059

Academic Editor: Vincent A. Cicirello

Received: 23 January 2023
Revised: 1 February 2023
Accepted: 2 February 2023
Published: 5 February 2023

Copyright: © 2023 by the authors. Licensee MDPI, Basel, Switzerland. This article is an open access article distributed under the terms and conditions of the Creative Commons Attribution (CC BY) license (https://creativecommons.org/licenses/by/4.0/).

1. Introduction

The Cryogenic Dark Matter Search (CDMS) can be described as a series of specially designed experiments that are used to detect Weakly Interacting Massive Particles (WIMPs), i.e., dark matter, using an array of semiconductor detectors at extremely low temperatures (mK). The exact definition of WIMPs does not exist; however, they are described as hypothetical new elementary particles (candidates for dark matter) that interact via gravity and other forces and are not part of the Standard Model. The product of every particle interaction in the germanium and silicon substrate produces ionization and phonons, which are measured using CDMS detectors [1]. The measurement of ionization and phonons determines the energy deposited in the crystal for each interaction and provides information on which kind of particle caused the event. Every particle's interaction with atomic electrons (electron recoils) and atomic nuclei (nuclear recoils) results in different ratios of ionization and phonon signals. The majority of particle interactions are electron recoils, while WIMPs are expected to be nuclear recoils. Although the WIMP scattering events are unique, they are rare when compared to the vast majority of unwanted interactions.

Two types of experiments exist, i.e., CDMS and SuperCDMS. The CDMS experiment provided the most-sensitive test of potential WIMP–nucleon interactions, as reported in [2,3]. The XENON experiment [4] later surpassed the CDMS experiment. The Super-CDMS experiment [5] is the successor of the CDMS, which uses new and improved detectors and an increased target mass to improve sensitivity by a factor of 5–8 factor and is limited by the residual cosmogenic background at the current location.

One of the challenges in CDMS and SuperCDMS experiments is the detection accuracy of particle interactions and the reconstruction of particle interaction locations, which could be greatly improved with the use of machine learning (ML) algorithms. In recent years, scientists have implemented various ML algorithms to estimate or detect dark matter using data obtained from CDMS/SuperCDMS experiments. Some of these research papers are briefly described in the following subsection.

1.1. Application of ML Algorithms in CDSM/SuperCDSM Experiments

A semi-supervised ML approach to dark matter search was proposed in [6]. This paper used a convectional auto-encoder and semi-supervised convolutional neural network (CNN) to directly detect dark matter. A deep convolutional neural network (DCNN) was used in [7] to solve the tasks of retrieving Lagrangian patches from which dark matter halos will condense. The results of the investigation showed that, if the proposed methods were properly tuned, they can outperform likelihood-based methods. The recurrent neural network (RNN) was implemented on the trigger FPGA to maximize the sensitivity to the low-mass dark matter of the SuperCDMS SNOLAB experiment [8]. By doing so, the energy estimator based on the combined information of filtered traces from individual detector channels was improved. With the performed modifications, the trigger threshold was lowered by 22%. The deep learning method was proposed in [9] to map the 3D galaxy distribution in hydrodynamic simulations and the underlying dark matter distribution. In this research, as two-phase CNN was used to generate fast galaxy catalogs, and the results were compared with traditional cosmological techniques. The proposed method outperformed the traditional techniques. Gradient-boosted trees were used in [10] to model dark matter halo formation. In [11], the Bayesian optimization for likelihood-free inference (BOLFI) algorithm was used to reconstruct the two-dimensional position and to determine the size of the interaction charge signal. The results showed that the BOLFI algorithm provides improved accuracy of 15% in reconstruction when compared in the case of events at large radii (R > 30 cm, the outer 37% of the detector). The investigation also showed that the proposed algorithm provided smaller uncertainties compared to other methods.

1.2. Definition of Novelty and the Research Hypotheses

From the previous short literature overview, complex ML models such as DNN have been used in CDMS detectors' investigation. Although the estimation results were high, the problem is that these algorithms require substantial computational resources. Another problem is that these types of ML methods cannot be easily expressed as an equation.

The authors present a novel approach through the use of genetic programming-symbolic regression (GPSR), which was applied to determine equations that can accurately reconstruct the particle interaction locations in the SuperCDMS. A public dataset [12] provided by a team from the University of Minnesota was used in the research. In comparison to the reviewed research, the research questions can be posed as follows:

- Can an SE be obtained using GPSR that can accurately reconstruct the locations of the interactions in SuperCDMS detectors?
- Is it possible to obtain a set of robust SEs using GPSR with randomly tested hyperparameters and validated through k-fold cross-validation that can reconstruct the locations of the interactions in the SuperCDMS with high accuracy?
- Is it possible to achieve even higher estimation accuracy in the reconstruction of the locations' interactions by combining multiple SEs that were obtained from different GPSR executions?
- Are all input variables required as model inputs to accurately reconstruct the interaction locations?

The presented paper consists of the following sections, i.e., the Materials and Methods, Results, Discussion, and Conclusions. In the Materials and Methods, the detailed dataset description is provided with statistical analysis following the description of the research methodology. In this section, the used method is described. In the Results Section, the best set of SEs obtained after five-fold cross-validation (5-CV) is shown with GPRS hyperparameters used to obtain them. Within the Discussion, the obtained models and results of the statistical analysis are further presented and discussed. Based on the hypotheses, presented results, and discussion, the Conclusions are given in the final section. Besides that, the Conclusions Section provides the pros and cons of the proposed method and possible directions for future work.

2. Materials and Methods

The materials, namely the dataset, are given a short description with a basic statistical analysis. Based on the dataset description, a research methodology is provided in which detailed steps are given. Then, the GPSR algorithm is described, as well as the process of developing a random hyperparameter search (RHS) method and the process of obtaining the SEs through the use of the 5-CV. Finally, the computational resources used in this research are described.

2.1. Dataset Description

For this investigation, the publicly available dataset from Kaggle [12] was used. The dataset was provided by the team from the University of Minnesota, and their research [13] was also focused on addressing the problem of accurately reconstructing the locations of interactions in the SuperCDMS detectors using machine learning methods.

In [13], the prototype SuperCDMS germanium detector was tested with a radioactive source positioned on a movable stage that can perform scanning from the center of the detector up to the near edge. The SuperCDMS germanium detector is a disk-shaped object that is 10 cm in diameter and 3 cm in height, with phonon sensors placed on the top and bottom surfaces to detect particles from a radioactive source. The sensors were used to measure phonons, i.e., quantized vibrations of the crystal lattice, which are produced from interacting particles and travel from the interaction location to the sensors. The number of phonons and relative arrival time to the particular sensors depends on the interaction and the sensor positions. The output of each sensor channel is a waveform for every interaction. In this experiment, the sensors were grouped into six regions labeled A, B, C, D, E, and F on both sides of the detector, as shown in Figure 1.

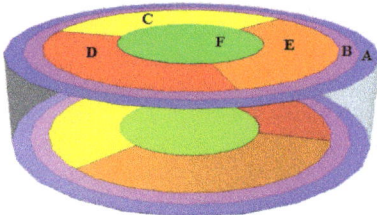

Figure 1. The detector regions of the SuperCDMS detector.

To produce interactions at 13 different locations on the detector along a radial path from the center up to the detector's outer edge, a movable radioactive source was used. The 13 different locations are shown in Figure 2, while the numeric values of the locations are listed in Table 1.

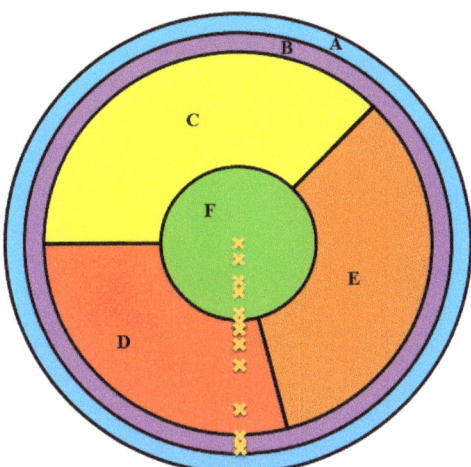

Figure 2. Top view showing detector regions (A, B, C, D, E, and F) and interaction locations indicated with the "x" symbol in orange color.

Table 1. The measured coordinates of the interaction locations shown in Figure 2.

Location	1	2	3	4	5	6	7	8	9	10	11	12	13
Coordinate x	0	0	0	0	0	0	0	0	0	0	0	0	0
Coordinate y	0	−3.96	−9.98	−12.502	−17.992	−19.7	−21.034	−24.077	−29.5	−36.116	−39.4	−41.01	−41.9

The entire dataset consisted of 21 parameters and 7151 samples. However, the first parameter row number ("Row") was omitted from this research. The dataset was obtained by extracting a set of parameters for each interaction from the signals of five sensors. The extracted parameters are sensitive to interaction location, relative timing between pulses in different channels, and features of the pulse shape. The relative amplitudes are also relevant, but were not included in the dataset due to amplification instabilities in the experiment. For each interaction, the following parameters were recorded:

- B, C, D, and F start—the time at which the signal pulse rises to 20% of the signal peak to Channel A. The variables in the original dataset are labeled as PBstart, PCstart, PDstart, and PDstart, respectively.
- A, B, C, D, and F rise—the time required for the signal to rise from 20 % to 80% of its peak. These variables in the original dataset are labeled as PArise, PBrise, PCrise, PDrise, and PFrise,
- A, B, C, D, and F width—the width of the pulse at 80% of the pulse height. The height was measured in seconds. These variables in the original dataset are labeled as PAwidth, PBwidth, PCwidth, PDwidth, and PFwidth.
- A, B, C, D, and F fall—the time required for a pulse to fall from 40% to 20 % of its peak. These variables in the original dataset are labeled as PAfall, PBfall, PCfall, PDfall, and PFfall.

As seen from the minimum and maximum values shown in Table 2, all the input variables labeled X_0 to X_{18} are all very small values in the 10^{-6} to 10^{-4} range. The target (output) variable, which is y, is in the −41.9 to 0 range.

Table 2. The results of the statistical investigation of dataset variables including the GSPR variable representation.

Dataset Variable	Data Points	Mean Value	σ	Minimum Value	Maximum Value	GPSR Variable Symbol
PBstart		-2.484×10^{-6}	5.346×10^{-6}	-1.800×10^{-5}	1.613×10^{-5}	X_0
PCstart		2.491×10^{-5}	2.596×10^{-5}	-2.160×10^{-5}	6.359×10^{-5}	X_1
PDstart		-1.316×10^{-5}	1.014×10^{-5}	-2.990×10^{-5}	5.149×10^{-5}	X_2
PFstart		-5.240×10^{-6}	2.580×10^{-5}	-3.520×10^{-5}	4.211×10^{-5}	X_3
PArise		1.710×10^{-5}	4.501×10^{-6}	8.464×10^{-6}	2.682×10^{-5}	X_4
PBrise		1.973×10^{-5}	3.843×10^{-6}	9.338×10^{-6}	3.221×10^{-5}	X_5
PCrise		3.012×10^{-5}	7.718×10^{-6}	1.157×10^{-5}	5.354×10^{-5}	X_6
PDrise		1.298×10^{-5}	2.078×10^{-6}	8.705×10^{-6}	3.444×10^{-5}	X_7
PFrise		1.675×10^{-5}	7.026×10^{-6}	8.060×10^{-6}	3.158×10^{-5}	X_8
PAfall	7151	2.148×10^{-5}	1.826×10^{-5}	1.521×10^{-5}	6.814×10^{-5}	X_9
PBfall		2.516×10^{-5}	3.781×10^{-5}	1.596×10^{-5}	9.416×10^{-5}	X_{10}
PCfall		2.712×10^{-5}	4.878×10^{-5}	1.388×10^{-5}	9.884×10^{-5}	X_{11}
PDfall		2.588×10^{-5}	4.621×10^{-5}	1.415×10^{-5}	9.806×10^{-5}	X_{12}
PFfall		2.472×10^{-5}	5.361×10^{-5}	1.163×10^{-5}	9.355×10^{-5}	X_{13}
PAwidth		1.696×10^{-5}	3.659×10^{-5}	7.105×10^{-5}	2.271×10^{-5}	X_{14}
PBwidth		2.071×10^{-5}	3.343×10^{-5}	7.133×10^{-5}	2.983×10^{-5}	X_{15}
PCwidth		2.467×10^{-5}	2.453×10^{-5}	1.648×10^{-5}	3.443×10^{-5}	X_{16}
PDwidth		1.494×10^{-5}	5.623×10^{-5}	3.603×10^{-5}	3.033×10^{-5}	X_{17}
PFwidth		1.765×10^{-5}	7.636×10^{-5}	3.321×10^{-5}	2.665×10^{-5}	X_{18}
y		-21.758	14.091	-41.9	0	y

The extremely low values of the input variables in the GPSR can lead to poor performance, i.e., the low estimation accuracy of the obtained SEs. To improve the performance of GPSR, the StandardScaler technique [14,15] was applied to the input dataset variables. The standard scaling technique is a technique that standardizes dataset variables and scales them to the unit variance. The standard manner of scaling the variable values is calculated per:

$$z = \frac{x - \sigma}{s}, \tag{1}$$

with m and s being the mean and σ of the dataset variable. The idea is to apply the StandardScaler method to all input variables before using the dataset in the GPSR algorithm.

Besides the statistical analysis in the dataset analysis, it is important to perform the correlation analysis. In this paper, Pearson's correlation analysis [16] was performed to determine the correlations between the output and individual inputs. The obtained correlation value between two dataset variables is calculated in the range of $<-1.0, 1.0>$. The absolute value of the correlation result is proportional to the rate at which two variables change together—with a positive value indicating a mutual increase and a negative value indicating that one variable decreases while the other increases. A heat map of the calculated coefficients is given in Figure 3.

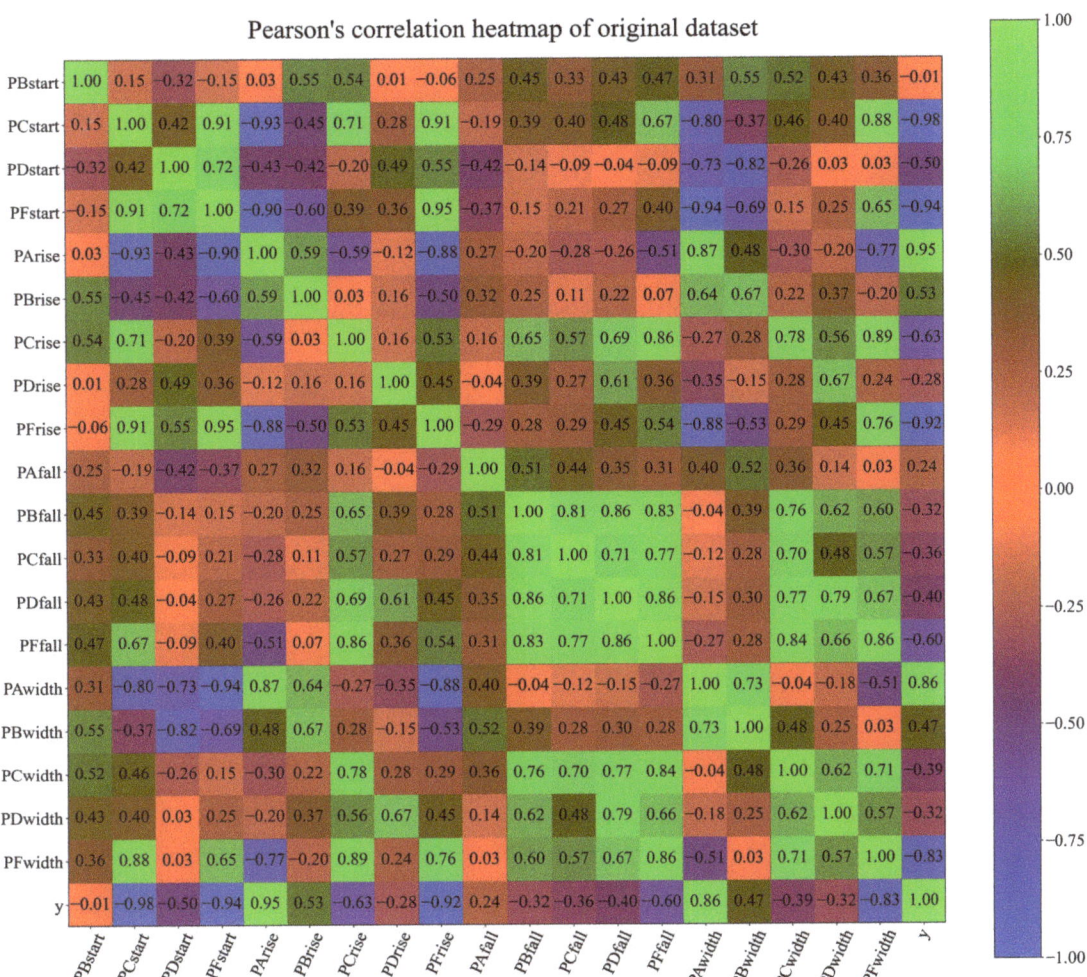

Figure 3. Pearson's correlation heat map.

As seen from Figure 3, the highest negative correlation (−0.98–−0.83) was achieved between PCStart, PFStart, PFrise, PFwidth, and the output variable y. The highest positive correlation (0.86–0.95) was achieved between PAwidth, PArise, and output variable y, respectively. The poorest correlation coefficient was achieved in the cases of PBstart-y (−0.01), PBfall-y (−0.32), PCfall-y (−0.36), PCwidth-y (−0.39), and PCwidth-y (−0.32). From the correlation heat map, it can be noticed that PBFall, PCfall, PDfall, and PFfall are mutually highly correlated variables, while showing a poor correlation in regard to the output. However, since GPSR is not a computationally intensive algorithm, it is good practice to include all input variables in the investigation. The reason why all input variables were included was to compare the variables used in the final models with their correlations.

2.2. Research Methodology

Figure 4 demonstrates the flow of the performed research methodology.

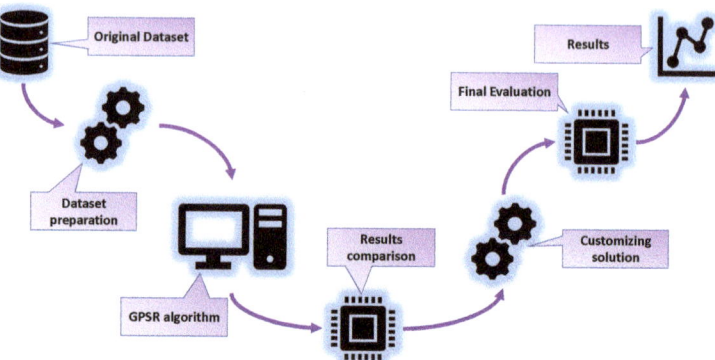

Figure 4. The flow chart of the research methodology.

Based on the previous flowchart, the research methodology can be summarized in the following steps:

- Dataset preparation—Checking the dataset for null values and deleting the first column ("Row"), which is not relevant for the analysis. Separating the dataset into input and output variables and applying of StandardScaler method on the input dataset variables. Dividing the dataset into training and testing datasets in a 70:30 ratio.
- GPSR algorithm—The GPSR algorithm is combined with a RHS to find the hyperparameters that yield the best-performing models. Perform training of GPSR on the training dataset using cross-validation, for the evaluation of the testing dataset.
- Result comparison—Perform a comparison of the best sets of SEs in terms of the estimation accuracy.
- Customizing solution—Combining the five best SEs to obtain a robust estimator for the reconstruction of the interaction locations.
- Final evaluation—Perform the final evaluation of the customized solution on the entire dataset.

2.3. Genetic Programming-Symbolic Regression

Genetic programming-symbolic regression (GPSR) begins its execution by creating the unit population. The population members' quality initially is very low. Then, through a consecutive number of generations, they are fit for a specific task with the application of genetic operations.

To build the initial population in GPSR, the dataset is required to have labeled input variables, the target output variable, the range of constant values, and mathematical functions. It should be noted that, in this paper, the following mathematical functions were used: addition, subtraction, multiplication, division, square root, cube root, absolute value, sine, cosine, tangent, minimum value, maximum value, natural logarithm, logarithm with base 2 and 10, respectively. It should be noted that mathematical functions such as division, natural logarithm, logarithm with base 2 and 10, and square root are specifically defined to avoid zero division errors and infinite or complex values during GPSR execution. The definition of these functions is given in Appendix A. The constant range hyperparameter is defined in Table 3. From these three sets, the GPSR randomly selects components to create population members, i.e., initial population. The hyperparameters required for the development of the initial population are population size, the maximum number of generations, and the constant range.

According to [17], there are three commonly used methods for creating population members, and these are full, grow, and ramped-half-and-half. The ramped-half-and-half method is a combination of the full and grow methods, i.e., both methods are used to create half of the dataset. To ensure a higher population diversity in the ramped-half-and-half method, the population members' depth is set in the range of hyperparameter init_depth. For example, if the init_depth value is set to (6,20), this means that the population members will be created with the full and grow method, where the population member's depth will be in the range from 6 to a maximum of 20.

The population members (SEs) and obtained SEs from each GPSR execution can be measured in terms of length. The length is counted as the sum of all mathematical functions, variables, and constants contained in a single SE. In the Results Section, besides the depth of the SEs, the length of the SEs will be also given.

In this paper, the fitness function used in all investigations is the mean absolute error (MAE), which can be written in the following form:

$$MAE = \frac{\sum_{i=1}^{n}(y_i - x_i)}{n}, \qquad (2)$$

where y_i, x_i, and n are predicted values by the population member, the true (output) dataset value, and the number of dataset samples [18]. To calculate the MAE, first, the population member must be evaluated, i.e., the values of the input variables must be provided to calculate the output. This output is the predicted y_i variable in Equation (2).

The tournament selection method was used to select the parents to which the genetic operations will be applied. When this selection is used, the population members are selected randomly for comparison. The member with the best fitness value is chosen as the winner of the process. The tournament_size hyperparameter defines the number of units used in this process.

Sometimes, during the execution of GPSR and due to the low correlation between dataset variables, the size of candidate solutions can rapidly grow through the generations. The size of the population members can grow so large that this can result in the bloat phenomenon. Various methods can be used to prevent this phenomenon, such as size fair crossover [19], size fair mutation [20], the Tarpeian method [21], and the parsimony pressure method [22]. The latter method is the most-commonly used and will be used in this paper. This method penalizes large programs during tournament selections by increasing the fitness value score, so they are not selected as the winners of tournament selection. This hyperparameter used for preventing the bloat phenomenon is parsimony_coefficient, and it is one of the most-sensitive coefficients to adjust, so when defining its range, extensive initial testing is required.

The execution of GPSR like in the majority of evolutionary algorithms can go indefinitely if the termination criteria are not defined. In GPSR, this is achieved with the stopping_criteria hyperparameter and the generations hyperparameter. The stopping_criteria is the fitness function value, which will stop the execution if achieved. If this fitness is not achieved, the hyperparameter generations will stop the execution when the number of iterations is equal to it. In all the performed investigations, the execution was stopped due to the generations parameter, due to the small value of the stopping_criteria. After each tournament selection, the best population member is obtained, and in that population member, one of the genetic operations is performed. Four genetic operators were selected in GPSR. The first of them is a crossover, the second a subtree mutation, the third a hoist mutation, and the fourth a point mutation. Each hyperparameter represents the probability and the possible sum of all genetic operations that must be equal to 1 to prevent tournament selection winners from being cloned and entering in the next generation. In this paper, the sum of all genetic operation probabilities was lower than one, so some winners enter the next generation unchanged.

In the case of a crossover operation, two winners are required, which determine the parent and donor. On the parent and the donor, random subtrees are selected, and the subtree from the donor replaces the subtree of the parent to form the member of the next generation. However, in the case of subtree mutation, the selection is a little different.

To select a subtree mutation, only one tournament winner is needed, which is later replaced with a randomly generated subtree created using mathematical functions, the constants from a constant range, and the input variables from the dataset. A similar situation occurs with the hoist mutation, which also requires only one tournament winner on which the random tree is defined. Then, a random node is selected on that subtree, which replaces the entire subtree. The point mutation demands only one tournament winner, then afterward, random nodes are selected. The constants are then replaced by randomly selected constants from the constant range, and also, the variables are replaced by the randomly selected input variables and the other functions by other randomly selected functions. However, in the case of the functions, the arity of randomly selected functions must be the same as the function in the tournament winner.

2.4. Random Hyperparameter Search

To find the optimal combination of hyperparameters, the RHS method was used. To develop this method for GPSR, the initial testing of each hyperparameter was required.

The GPSR hyperparameter ranges defined after initial testing are listed in Table 3.

Table 3. The range of GPSR hyperparameters.

Hyperparameter Name	Range
population_size	1000–2000
number of generations	100–200
tournament_selection	10–500
init_depth	3–15
crossover	0.001–1.0
subtree_mutation	0.001–1.0
hoist_mutation	0.001–1.0
point_mutation	0.001–1.0
stopping_criteria	$0-1 \times 10^{-8}$
maximum_samples	0.99–1
constant_range	−10,000–10,000
parsimony_coefficient	$0-1 \times 10^{-4}$

In the case of GPSR, there were four parameters that were most-influential: the size of the population (population_size), the number of possible generations (number_of_generations), the tree depth (init_depth), and finally, the parsimony coefficient [23]. The population size and the number of possible generations are highly correlated hyperparameters. If the size of the population and number of generations is too large, it can lead to a very long execution time. The ranges shown in Table 3 proved to be optimal. Initially, init_depth was increased, but the GPSR execution showed that higher values of the init_depth hyperparameter can lead to longer execution times without any benefit to the estimation accuracy of the obtained SE. As the stated parsimony_coefficient is the most-crucial and most-sensitive hyperparameter to define, the range is defined by trial and error, so the range shown in Table 3 will prevent the occurrence of bloat phenomena, but will allow the stable growth of the population members.

The initial assumption was that crossover or subtree mutation will be the most-influential genetic operation in the evolution process. However, an initial investigation showed that very high values of these two genetic operations can lead to local minimums, i.e., the fitness function value is constant, while the size of the population members grows rapidly. Therefore, the value of all four genetic operations was set to the 0.001–1 range. In this way, the sum of all possible genetic operations was nowhere near one, so some tournament selection winners entered the next generation unchanged.

It should be noted that each GPSR execution is terminated after reaching the defined number of maximum generations. As described, the GPSR algorithm has two termination criteria, i.e., the maximum number of generations and the stopping criteria. However, the stopping criterion, i.e., the value of the fitness function, was set to an extremely small value to ensure that the execution of the GPSR algorithm is terminated after reaching the maximum number of generations.

The maximum samples were set in the 0.99 to 1 range, so each time the fitness function was computed, almost the entire training dataset was used. If the value of the maximum samples was set to 1, then during the GPSR execution, the out-of-bag raw (OOB) fitness would not be shown. OOB fitness refers to the fitness of each population member in held-out samples. To see OOB fitness during execution, the maximum samples must be less than 1. To ensure that the population's growth is stable, a relatively large range of constants was provided to ensure that the small range of constants can lead to the large growth of members.

2.5. Training and Evaluation Process with the GPSR Algorithm

The entire flowchart of the training process with GPSR is shown in Figure 5.

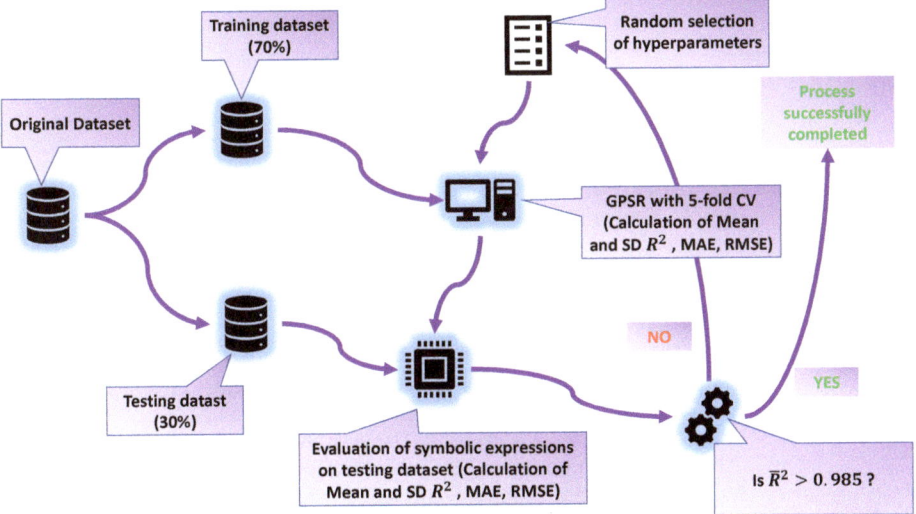

Figure 5. The flowchart of obtaining symbolic expressions using the GPSR algorithm.

Due to a large number of samples (7157), the original dataset after preprocessing with StandardScaler (only input variables) was divided into a 70:30 ratio: 70% of the dataset was used in the 5-CV and the remaining 30% for the final evaluation.

The investigated process begins with randomly selecting GPSR hyperparameters (range values defined in Table 3), then the execution of GPSR algorithm with the 5-CV on the training dataset. In the 5-CV [24], there are 5 splits, i.e., the GPSR execution is trained 5 times, and for each time, a new SE is obtained. After each split, the performance metrics are calculated on the training and validation folds. After the process of training is performed, the mean and σ values are calculated and stored. When this step is performed, the obtained SEs are evaluated on the testing dataset, and the performance metrics (mean and σ) values are obtained. For the termination criteria of the entire process, the mean value of R^2 must be greater than 0.985. If the value is above 0.985, the process is terminated; otherwise, the process begins from the beginning with a random selection of the GPSR hyperparameters.

2.6. Performance Metrics and Methodology

In this paper, to evaluate the SEs, three performance metrics were used, i.e., the coefficient of determination (R^2) [25], the mean absolute error MAE [18] (Equation (2)), and the root-mean-squared error ($RMSE$) [26]. The R^2 is in the 0 to 1 range, where 0 represents the worst-possible value, while the value of 1 represents the best-possible value and is aimed for. Regarding the MAE and $RMSE$ values, the goal is to obtain as low a value as possible.

As stated in the description of the GPSR algorithm, the MAE metric was used to evaluate the population members (fitness function). After each SE is obtained, it is evaluated to determine the R^2, MAE, and $RMSE$ values. To calculate the mean and σ values of the performance metrics in GPSR with RHS and the 5-CV, the following steps are performed:

- During the 5-CV, calculate the performance metric values on the training and validation sets.
- After the 5-CV process is complete, calculate the performance metrics (mean values obtained on the training dataset).
- Perform the last evaluation of the trained model (five SEs) on the testing dataset, and calculate the performance (evaluation) metric values (values achieved on the testing dataset),
- Calculate the mean and σ values of the performance metrics from the obtained values on the training and testing datasets.

2.7. Computational Resources

The entire research was performed on a desktop computer consisting of an Intel I7-4770 processor supported by 16 GB of DDR3 RAM. All scripts were written and made in the Python Programming language (Version 3.9.12). The statistical analysis was conducted using pandas [27] and the matplotlib [28] library. The datasets were scaled using the StandardScaler method from the scikit-learn library (Version 1.2.0) [14]. GPSR was used from the gplearn library (Version 0.4.1.) [29].

3. Results

The Results Section is divided into two subsections entitled "Results obtained using GPSR algorithm with RHS and 5-CV" and "Custom solution and final evaluation". In Section 3.1, the estimation accuracy of the three best cases of the SEs is presented, obtained during the 5-CV (training dataset) and final evaluation (test dataset). In Section 3.2, the custom set of the five SEs was created by picking the SEs with the highest estimation accuracy from the previously presented subsection. Finally, the performance evaluation of the modified (customized) set of SEs on the entire dataset is presented.

3.1. Results Acquired Using GPSR with RHS and 5-CV

The GPSR algorithm with the 5-CV was executed multiple times. Before each execution, the hyperparameters were randomly selected from the predefined ranges shown in Table 3. From the obtained results, the three best cases were selected that achieved the highest estimation accuracy. The combination of the hyperparameters that were used to obtain the highest estimation accuracies is listed in Table 4.

From Table 4, it can be seen that, for Cases 1 and 3, the population size was very large. In Case 1, the point mutation (0.43) was the dominating genetic operation, while in Cases 2 and 3, the crossover (0.41) and subtree mutation (0.469) dominated the genetic operations. However, in Case 3, both crossover and subtree mutation had higher probabilities than the other two types of mutations. The stopping criteria, as planned, were never reached by any population member in all three cases, and each execution stopped after the maximum possible number of generations was attained. The parsimony coefficient (Pcoef) value was much higher in Cases 2 and 3 than in Case 1. The results of these three cases are shown in Figure 6 and Table 5.

Table 4. The combination of the GPSR hyperparameter values for which the highest estimation accuracy was achieved.

Case No.	GPSR Hyperparameters (Population Size, Number of Generations, Tournament Selection, Init_Depth, Crossover, Subtree_Mutation, Hoist_Mutation, Point_Mutation, Stopping_Criteria, Max_Samples, Constant_Range, Parsimony_Coefficient)
1	1978, 233, 254, (6, 15), 0.043, 0.22, 0.22, 0.43, 5.49×10^{-7}, 0.99, $(-9793.08, 1402.72)$, 9.15×10^{-6}
2	1075, 196, 461, (3, 15), 0.41, 0.4, 0.05, 0.097, 9.95×10^{-7}, 0.99, $(-8821.29, 3713.89)$, 9.2×10^{-4}
3	1500, 171, 108, (7, 11), 0.45, 0.469, 0.043, 0.0071, 8.17×10^{-7}, 0.99, $(-4076.61, 4272.87)$, 7.16×10^{-4}

The Splits 1–5 shown in Figure 6 and Table 5 indicate the GPSR algorithm execution on the different training sets (4 folds) and the evaluation on the validation set (1 fold) in the 5-CV. After each split, the performance metric values were calculated on the training and validation set and the mean and σ values were calculated. When all five splits were performed, the total mean and σ of all assessment (evaluation) metrics were calculated. These "Total" values represent the final values of the 5-CV process. It should be noted that, after each split, the SE was obtained, so each case consisted of five SEs.

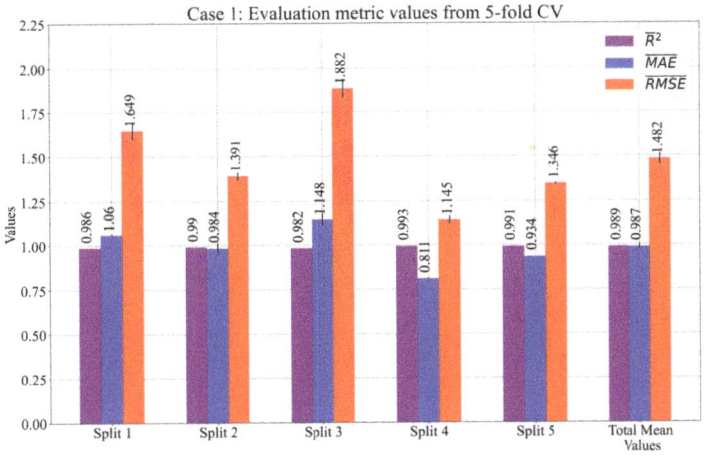

Figure 6. Cont.

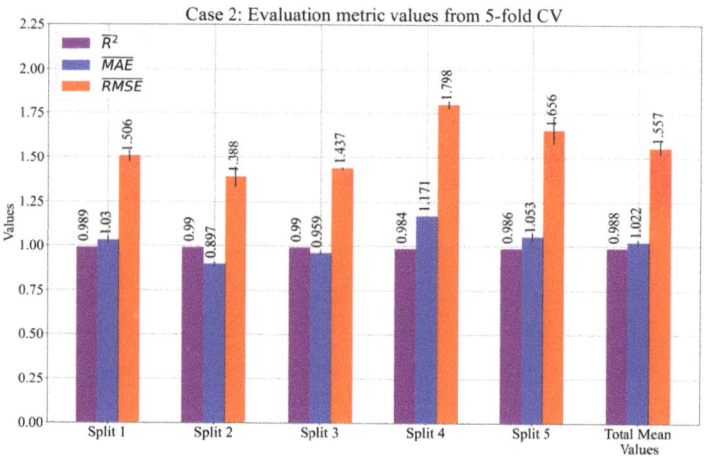

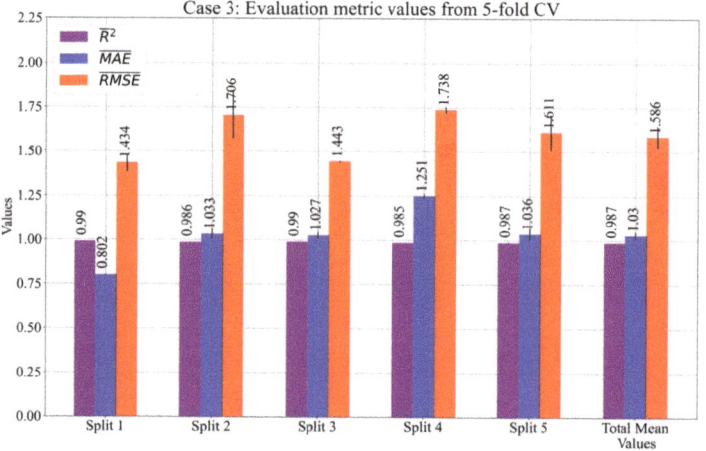

Figure 6. The mean values of the obtained performance metric during GPSR 5-CV. The σ values are presented as error bars.

As seen from Figure 6 and Table 5, in all three cases, the mean performance metric values obtained with the 5-CV on the training data were very high with low σ values. Table 6 shows the depth and length of each SE with the average depth and length of the SEs in each case.

To clarify the terms length and depth in Table 6, the length is a representation of several elements (constants, functions, and possible variables) in the SEs, while the depth is measured when the SE is shown in tree form. The depth is measured from the required root node to the deepest leaf of the SE. Since the tree form of the SE is not important for this investigation, the length of the SE will be investigated. The highest SE lengths and average length were achieved in Case 1 followed by Case 3 and Case 2. From the length and the estimation accuracy in Table 5, the best case is Case 2, since the estimation accuracy is slightly lower than Case 1, while the average length of the SEs is the lowest.

Table 5. The mean and σ values of R^2, MAE, and $RMSE$ were achieved in a 5-fold cross-validation process.

	Performance metric	Split 1	Split 2	Split 3	Split 4	Split 5	Total
Case 1	$\overline{R^2}$	0.9864	0.9903	0.9821	0.9935	0.9909	0.9886
	$\sigma(R^2)$	0.0009	0.0003	0.0013	0.0002	0.0001	0.0006
	$\overline{MAE}$	1.0599	0.9837	1.1479	0.8113	0.9342	0.9874
	$\sigma(MAE)$	0.0057	0.0291	0.0354	0.0035	0.0003	0.0148
	$\overline{RMSE}$	1.6494	1.3905	1.8818	1.1445	1.3456	1.4824
	$\sigma(RMSE)$	0.0501	0.0222	0.0542	0.0230	0.0057	0.0311
	Performance metric	Split 1	Split 2	Split 3	Split 4	Split 5	Total
Case 2	$\overline{R^2}$	0.9889	0.9905	0.9896	0.9835	0.9860	0.9877
	$\sigma(R^2)$	0.0001	0.0006	0.0001	0.0001	0.0009	0.0004
	$\overline{MAE}$	1.0304	0.8973	0.9593	1.1713	1.0530	1.0223
	$\sigma(MAE)$	0.0233	0.0115	0.0108	0.0006	0.0238	0.0140
	$\overline{RMSE}$	1.5057	1.3883	1.4369	1.7984	1.6564	1.5571
	$\sigma(RMSE)$	0.0331	0.0562	0.0062	0.0218	0.0772	0.0389
	Performance metric	Split 1	Split 2	Split 3	Split 4	Split 5	Total
Case 3	$\overline{R^2}$	0.9899	0.9856	0.9896	0.9846	0.9868	0.9873
	$\sigma(R^2)$	0.0004	0.0019	0.0001	0.0001	0.0013	0.0008
	$\overline{MAE}$	0.8020	1.0330	1.0266	1.2512	1.0365	1.0299
	$\sigma(MAE)$	0.0041	0.0302	0.0170	0.0128	0.0328	0.0194
	$\overline{RMSE}$	1.4336	1.7058	1.4429	1.7375	1.6111	1.5862
	$\sigma(RMSE)$	0.0489	0.1311	0.0051	0.0170	0.1016	0.0608

The performance metric mean and σ values obtained for each case on the testing dataset are shown in Figure 7 and Table 7.

Table 6. The symbolic expressions length and depth of each case with average length and depth.

Case No.	Symbolic Expression No.	Length	Depth	Average Length	Average Depth
1	1	220	33	206.6	28.4
	2	257	41		
	3	106	13		
	4	351	36		
	5	99	19		
2	1	75	17	81.4	16.2
	2	87	16		
	3	109	18		
	4	58	15		
	5	78	15		

Table 6. Cont.

Case No.	Symbolic Expression No.	Length	Depth	Average Length	Average Depth
3	1	145	31	161.6	24.8
	2	156	19		
	3	127	27		
	4	230	17		
	5	150	30		

Table 7. The performance metric values obtained on the test dataset.

Case Number	Evaluation Metric	Symbolic Expressions					Mean	σ
		1	2	3	4	5		
1	R^2	0.9827	0.9892	0.9811	0.9934	0.9905	0.9876	0.0048
	MAE	1.1180	0.9919	1.1645	0.8225	0.9619	1.0096	0.1215
	$RMSE$	1.8649	1.4646	1.9488	1.1537	1.3810	1.5506	0.3050
2	R^2	0.9878	0.9896	0.9887	0.9836	0.9837	0.9867	0.0025
	MAE	1.0615	0.9393	0.9894	1.1815	1.1450	1.0634	0.0911
	$RMSE$	1.5664	1.4490	1.5057	1.8173	1.8116	1.6300	0.1551
3	R^2	0.9909	0.9865	0.9889	0.9828	0.9873	0.9873	0.0027
	MAE	0.8032	1.0518	1.0513	1.3162	1.0666	1.0578	0.1623
	$RMSE$	1.3529	1.6464	1.4958	1.8588	1.5978	1.5903	0.1677

As seen from Figure 7 and Table 7, the mean values of the performance metric are almost the same as those obtained on the training dataset. The σ values are slightly higher when compared to those obtained on the training dataset. Although Case 2 showed better performance on the training dataset, the results of the testing dataset showed that Case 3 slightly outperformed Case 2.

It can be noticed that the highest scores in Table 7 are marked with red color. The marked scores represent the chosen SE of three cases that will be used for a customized solution.

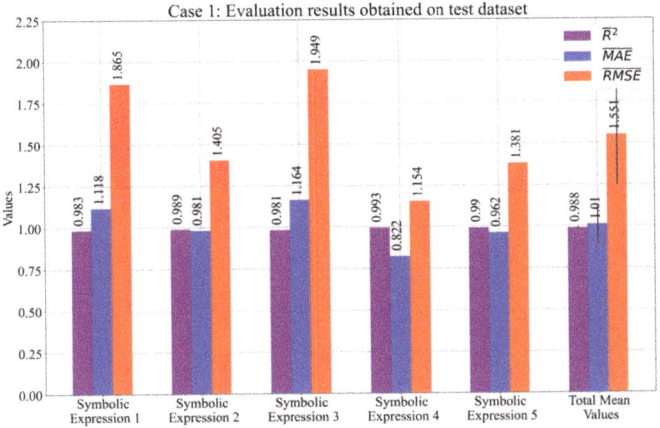

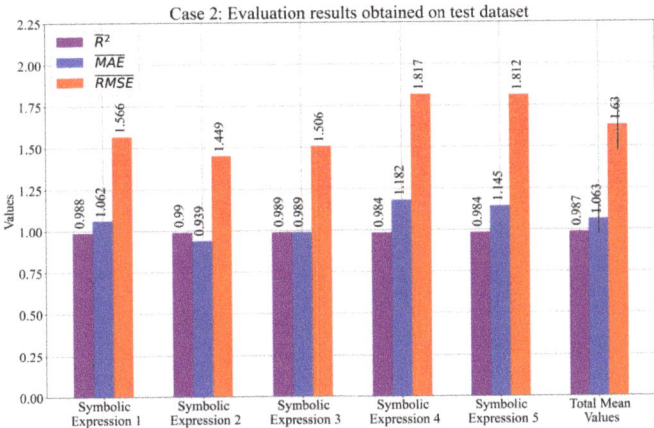

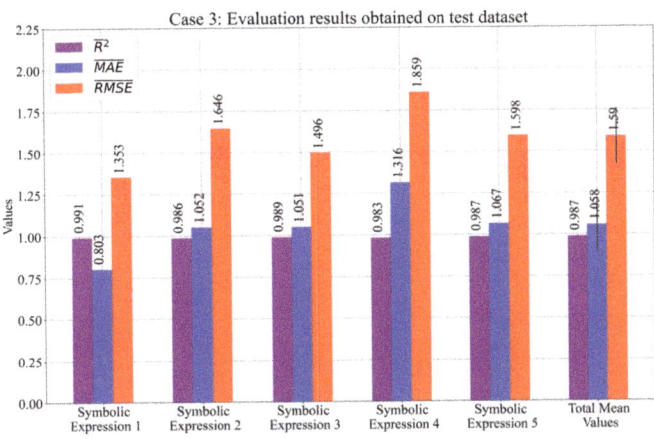

Figure 7. The mean values of the performance metric obtained on the testing dataset. The σ values are presented as error bars.

3.2. Combination of the Best SE

Based on the results obtained on the test dataset, the five SEs with the best estimation accuracy were selected. In Table 7, the results of the best SEs that were selected are marked with red color. The main criterion for choosing the SEs was that the $\overline{R}^2$ score was higher than 0.989. The chosen SEs can be written in the following form:

$$
\begin{aligned}
y_1 =\ & -14.061 \Bigg(\Big| \sin\Big(\sqrt{\sqrt[3]{\tfrac{\min(\sin(X_1), \sqrt[3]{X_0})}{\sqrt{X_9}}}} \Big) - \min(-X_{18} + X_4 - |\min(\sqrt{X_{18}}, -X_{18} - X_3 \\
& + X_4 + X_6 - |X_2 + X_5 - \min(-X_3 + X_6 + \cos(\cos(X_5)), -X_{18} - X_3 + X_4 + X_6 \\
& - |-|\max(X_2 + X_5, 1.44 \log(\tan(X_3))) - \min(X_5 + \max(X_{17}, X_{18}), \sqrt[3]{X_{15}}, -X_{18} - X_3 \\
& + 2X_4 - |X_7| - 0.43 \log(X_1) - 0.43 \log(X_2) + \max(X_{17}, X_{18}) - \sqrt[3]{X_1} + \sqrt[3]{X_{15}} - \tan(\sqrt{X_4}))| \\
& - \cos(\cos(X_1 + X_{17})) + \cos(0.43 \log(X_1))| - 0.43 \log(X_1))| - \cos(\sin(X_8)))| - \cos(\min(X_3, X_6)) \\
& + 0.43 \log(\sqrt{\cos(\cos(X_5))}), \sin(\sqrt{\sqrt[3]{\tfrac{\sin(X_1)}{\sqrt{X_9}}}}))\Big| \Bigg)^{\tfrac{1}{3}},
\end{aligned}
\qquad (3)
$$

$$
\begin{aligned}
y_2 =\ & 0.43 \log(X_5 - 1.44 \log(\sin(\max(X_1, X_{14})))) + 0.43 \log(X_5 - \sin(\log(\max(X_1, X_{14})))) \\
& + 0.43 \log(\max(X_5, X_5 - X_6) - \log(\max(X_1, X_5))) - 4X_1 + 4X_{14} + 0.43 \log(0.43 \log(X_{14})) \\
& + X_{15} - 3X_{18} + \sin(\sin(1.44 \log(X_{18}))) + \cos(X_2(X_2 - X_6)) - X_3 + X_4 + \sin(X_5) - X_6 \\
& + 1.129 \sqrt[3]{\log(X_6)} + X_8 - 20.4537,
\end{aligned}
\qquad (4)
$$

$$
\begin{aligned}
y_3 =\ & - \max(X_0 X_{14} + X_4, X_2 + 2X_3 + 2X_5, 1.44 \log(X_7), \max(X_6, |X_2|) \\
& + \min(X_{18}, X_4) \min(X_4, X_6)) - \min(X_{18}, X_4) + \min(-3X_3 - 2X_5, 0.43 \log(\cos(\min(X_4, X_6))) \\
& + 2X_8)) \log(\sqrt{X_1}) - X_{16} - X_{18} - 4X_3 + 6X_4 + 2X_5 - \cos(X_6) - 19.2567,
\end{aligned}
\qquad (5)
$$

$$
\begin{aligned}
y_4 =\ & \min(X_5, \Big(-\max(\sqrt[3]{|X_{17}| - X_1}, \sqrt[3]{\log(X_{12})}) - |X_7 + \log(1.44 \log(\max(X_1, X_{17})))| \\
& + 1.44 \log(\cos(\max(|X_2|, \tan(X_1)))) - \min(|X_5|, \Big(\min(X_4 - X_0, 1.44 \log(X_{12}(X_{18} - X_4))) \\
& + \sqrt[3]{\min(X_0, X_2)} - |\log(X_{17})| - \sin(X_{17}) \Big)^{\tfrac{1}{3}} + \sqrt[3]{\min(X_0, X_2)} - 1.44 \log(\sin(|X_3|)) - X_{10} \\
& - 1.44 \log(\cos(\min(X_2, X_3))) - X_0) - \Big(\sqrt[3]{\min(X_0, X_2)} - 1.44 \log(\cos(|X_2|)) \\
& + 0.43 \log(\log(X_{13})) - \sin(X_{17}) \Big) \tfrac{1}{3} + \sqrt[3]{\min(X_0, X_2)} - 1.44 \log(\cos(|X_2|)) \\
& - \Big(\sqrt[3]{\sqrt[3]{\min(X_0, X_2)}} - 1.44 \log(\cos(|X_2|)) + 0.43 \log(\log(X_{13})) - \sin(X_{17}) + \sqrt[3]{\min(X_0, X_2)} \\
& - 1.44 \log(\cos(\min(X_0, X_2))) \Big)^{\tfrac{1}{3}} - \Big(\min(X_4 - X_0, |X_5|) + \sqrt[3]{\min(X_0, X_2)} - 1.44 \log(\cos(|X_2|)) \\
& - \sin(X_{17}) \Big)^{\tfrac{1}{3}} - \min(X_5, 1.44 \log(X_0) - |X_0|) + \max(\min(X_{12}, X_{15}) + X_{13}, X_0 X_6 + \sin(X_7)) \\
& - \min(0.43 \log(X_{16}), \max(-1255.06, X_0, X_{18})) + 1.44 \log(\min(X_4 - X_0, \max(-1255.06, X_0, \\
& X_{18}))) + 1.44 \log(\cos(\max(\sin(\cos(X_{13})), \sqrt[3]{X_{15}} - \sqrt[3]{X_1}))) + 1.44 \log(\cos(\max(-9072.8, \\
& \tan(X_1)))) - \tan(\min(X_0, \cos(X_{14} + X_4))) - 3\sqrt[3]{\min(X_0, X_2)} - \min(X_{14}, X_6) \\
& + 1.44 \log(\cos(\min(X_2, X_3))) + 1.44 \log(\cos(\min(X_2, \log(X_2)))) - |X_2 + X_7 + \log(X_0)| \\
& - |X_7 + X_8 + \log(X_0)| - 1.44 \log(\log(X_0) + X_7) - X_0 + 1.44 \log(\cos(X_0)) + X_1 + X_{10} - X_{11} \\
& + 2X_{12} + \sqrt[3]{X_{12} - 3012.83} - X_{13} + \cos(X_{17}) + \sin(\cos(X_{17})) - X_2 + X_4 + X_7 \\
& - 1.44 \log(X_7) + X_8 - X_9 \Big)^{\tfrac{1}{3}} + \min(X_0, X_1) + \min(-19.0691, \sin(X_{18}))
\end{aligned}
\qquad (6)
$$

$$
\begin{aligned}
&+ \quad X_1(\sqrt[3]{X_0 - 2345.71} - X_2) + \sqrt{\cos(X_2)} + X_4) + X_0, \\
y_5 =\ &- \max(7.64(X_1 + X_3), 0.43 \log(\min(\left(X_2 | \min(221.844 X_2 \cot(\cos(X_2)), -9574.83 \right. \\
&\quad \min(8546.04(X_1 + X_3), X_8(X_0 + X_1 + \min((2X_0 + X_1)X_8, 1529.14(1529.14(X_1 + X_{17}) \\
&- \quad (2X_0 + X_1)X_8)))))|\Big) \Big/ \Big(|X_1 + \sqrt[3]{|\cos(X_2)|}|\Big), 0.43 \log(\sqrt[9]{\sin(\max(\sqrt{X_0}, \sqrt[3]{X_4}))})))) \\
&+ \quad |\min(7.33(X_1 + X_3), X_8(X_0 + X_1 + |\cos(X_2)|))| - 21.09.
\end{aligned}
\tag{7}
$$

However, it should be noted that, to use Equations (3)–(7) the values of input variables must be scaled using the StandardScaler method. In Equations (3)–(7), the mathematical functions such as the square root, natural logarithm, logarithm with base 2 and 10, and division function are specifically defined to avoid complex, infinite values or zero division errors. The definition of these functions is given in Appendix A.

The analysis of Equations (3)–(7) showed that, using all five SEs together to compute the output all 19 variables is required. However, if Equations (3)–(7) are used individually, not all input variables are required. To compute the output in Equation (3), 13 out of 19 input variables are required. The variables that are required to compute the output are $X_0,...,X_9$, X_{15}, X_{17}, and X_{18}. It is evident from Table 2 that these variables are PBstart, PCstart, PDstart, PFstart, PArise, PBrise, PCrise, PDrise, PFrise, PAfall, PBwidth, PCwidth, and PFwidth, respectively. The total number of variables required to compute the output using Equation (4) is 10 out of 19, and these variables are $X_1,...,X_6$, X_8, X_{14}, X_{15}, and X_{18}. From Table 2, these input variables are PCstart, PDstart, PFstart, PArise, PBrise, PCrise, PFrise, PAwidth, PBwidth, and PFwidth, respectively. In Equation (5), to compute the output, the 12 out of 19 input variables required are $X_0,..., X_8$, X_{14}, X_{16}, and X_{18}. These variables are PBstart, PCstart, PDstart, PFstart, PArise, PBrise, PCrise, PDrise, PFrise, PAwidth, PCwdith, and PFwidth, respectively. All input variables are required to compute the output using Equation (6). The lowest number of input variables (7) is required to compute the output when Equation (7) is used. Equation (7) consist of $X_0,...,X_4$, X_8, and X_{17}, and from Table 2, these variables are PBstart, PCstart, PDstart, PFstart, PArise, PFrise, and PDwidth.

The set of previously shown SEs was evaluated on the entire dataset, and the results are presented in Figure 8 and Table 8.

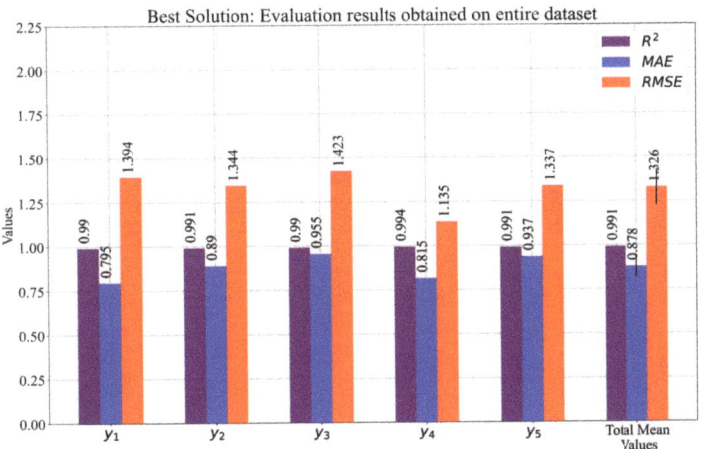

Figure 8. The graphical representation of the performance metric values obtained with the best combination of SEs on the entire dataset. σ is shown in the case of the total mean values in the form of an error bar.

Table 8. The performance metric values achieved with the best combination of SEs on the entire dataset.

Evaluation Metric	y_1	y_2	y_3	y_4	y_5	Mean	σ
R^2	0.99022	0.990905	0.989798	0.993511	0.991003	0.991087	0.001291
MAE	0.795499	0.889602	0.954781	0.815055	0.936876	0.878362	0.063662
$RMSE$	1.393535	1.343819	1.423247	1.135059	1.336586	1.326449	0.100901

The results shown in Table 8 showed that, using a customized combination of the SEs, even higher estimation accuracy was achieved. The mean values were better than those shown in Table 7, and the σs were smaller.

The reconstruction of the interaction locations was performed in [13], where the authors used the DNN. For the evaluation of the results, the authors used the $RMSE$, and the lowest achieved value was 1.53. If these results are compared to the results shown in Table 8, it can be seen that the proposed method outperformed the results in [13].

4. Discussion

The initial statistical analysis of the dataset variables showed that all the input variables had a really small value range when compared to the output (target) variable. With this in mind, scaling actions were taken, i.e., all input variables were scaled using the StandardScaler method. Based on Pearson's correlation analysis, it was possible to see that the (6 out of 18) input variables had a very good correlation with the output variable y. Only one (PBStart) dataset variable did not correlate with the output variable.

Although the initial investigation of the GPSR algorithm to define each GPSR hyperparameter range for the development of the RHS method was a painstaking and time-consuming process, eventually, it resulted in the faster search of multiple hyperparameter combinations, for which the highest estimation accuracies were achieved. The larger population size and smaller tournament selection size proved to be a good combination in obtaining the SEs with high evaluation accuracy; however, these two were not only responsible. An initial investigation showed that, initially, a small range of any genetic operation would lead to a local minimum value of the fitness value, which would eventually result in SEs with lower estimation accuracy. To improve the range of all genetic operations, it was set to the 0.001–1 range. This wide range produced some interesting hyperparameter combinations (Table 4). In Case 1, the dominant genetic operation was point mutation (0.43). Cases 2 and 3 had high values of crossover and subtree mutation, i.e., they were the main genetic operations in these two cases. As already stated, the idea was to enable the GPSR algorithm to reach the lowest value of the fitness function possible, so the stopping criteria were preset to an extremely low value, and the GPSR execution ended after a maximum number of generations was reached.

When Case 1's hyperparameters and the SEs' lengths were compared to those of Cases 2 and 3, it can be noticed that the initial tree depth size and Pcoef had a huge influence on the length and depth of the SEs. Almost all the SEs obtained in Case 1 were large in length and depth. The lowest SE size was obtained on "Split 3" with a length of 106 and a depth of 13. However, these SEs had the lowest estimation performance on the test dataset ($R^2 = 0.9811$, $MAE = 1.1654$, and $RMSE = 1.9488$) when compared to the other four SEs of the same case. Generally, the Pcoef in Case 1 was set to 9.15×10^{-6}, which generated SEs of average length 206.6, as seen from Table 6. The Pcoef in Cases 2 and 3 was much larger (9.2×10^{-4}, 7.16×10^{-4}), which generated average lengths of 81.4 and 161.6, respectively. From these results, the high influence of the Pcoef on the size of the obtained SEs can be noticed.

From the results obtained on the training and testing datasets (Figures 6 and 7 and Tables 5 and 7), it can be seen that the mean values of the performance metrics were similar (values of R^2 high and near 1, while MAE and $RMSE$ values of 1 and 1.5, respectively). However, in Case 2, the performance metric values were slightly higher on the training compared to the testing dataset. Therefore, Case 3 slightly outperformed Case 2 on the testing dataset. From the obtained results, the highest evaluation metric values ($\overline{R^2} \pm \sigma(R^2)$, $\overline{MAE} \pm \sigma(MAE)$, $\overline{RMSE} \pm \sigma(RMSE)$) on the testing dataset were achieved in Case 1 (0.9876 ± 0.0048, 1.0096 ± 0.1215, 1.5506 ± 0.3050.)

The combined set of the five best SEs selected from Cases 1, 2, and 3 achieved the highest estimation accuracy when evaluated on the entire dataset. The results of the evaluation metrics ($\overline{R^2} \pm \sigma(R^2)$ and $\overline{MAE} \pm \sigma(MAE)$, $\overline{RMSE} \pm \sigma(RMSE)$) in this case were equal to 0.991087 ± 0.001291, 0.878362 ± 0.063662, and 1.326449 ± 0.100901. Therefore, the results showed that the custom combination of the best SEs in terms of the evaluation metric values outperformed Cases 1–3.

The analysis of the number of variables required to compute the output if a custom solution (five SEs) is used showed that all input variables were required. Even if only Equation (5) was used, all input variables were required. However, the high estimation accuracy can be achieved only using Equation (7), and this equation required only seven input variables. However, 4 (PCstart, PFstart, PArise, and PFrise) out of those 7 variables had a high correlation with the output variable, as seen from Figure 3.

5. Conclusions

In the conducted research, the GPSR algorithm with RHS and five-fold CV was used to obtain a system of robust SEs for the estimation of the interaction locations in Super Cryogenic Matter Search detectors. The results of the investigation showed that:

- Using the GPSR algorithm, it was possible to obtain SEs (mathematical equations) that can estimate the interaction locations in Super Cryogenic Matter Search detectors with high accuracy.
- Using a 5-CV process, the robust system of the five SEs has a more accurate estimation when compared to the estimation of only one SE. The RHS method proved to be very useful in finding the combination of the quality hyperparameters where the highest accuracy was achieved. From all the results, the best three cases of the SEs were selected and evaluated on the test set. The highest values of $\overline{R^2} \pm \sigma(R^2)$, $\overline{MAE} \pm \sigma(MAE)$, $\overline{RMSE} \pm \sigma(RMSE)$ were achieved in Case 1 and are equal to 0.9876 ± 0.0048, 1.0096 ± 0.1215, 1.5506 ± 0.3050.
- From the obtained results, three cases were selected based on the final mean R^2 score. From these cases, the SEs that achieved the highest estimation performance were selected as the main elements of the custom set of SEs. The results of the customized solution obtained on the entire dataset were equal to 0.991087 ± 0.001291, 0.878362 ± 0.063662, and 1.326449 ± 0.100901, respectively. The final evaluation of these equations on the entire dataset showed that this system had slightly better performance when compared to Cases 1, 2, and 3.
- Unfortunately, the custom set of SEs required all 19 input variables to compute the output. However, if only Equation (7) was used, the highest estimation accuracy could be achieved, and to compute the output, only seven input variables were required.

The proposed approach presented in this paper showed that using a simple GPSR algorithm with RHS and the 5-CV on low-end computer hardware can produce better results than the complex CNN architecture presented in [13]. The benefit of using the proposed approach is that the SEs were easily used, easy to comprehend, and require fewer computational resources than complex CNN architectures.

The main problem of the proposed approach is the initial definition of the GPSR hyperparameter ranges in the RHS method. The ranges are not unique and depend on the investigation, so each time, it has to be tuned from scratch. Depending on the dataset, the population size, number of generations, and tournament selection have to be defined. The larger the dataset size, the smaller the population, number of generations, and tournament selection size values are. Besides those hyperparameters, the Pcoef has to be defined, and this parameter is the most-sensitive one. A small increase/decrease of this value can result in a fast increase/decrease in the length of the population members.

The other important factor that influences the hyperparameter ranges is the computational resources on which the GPSR algorithm with the RHS method is executed.

Future work will be focused on synthetically enlarging the dataset size to see if the estimation accuracy could be improved. Besides that, other ML methods will be investigated, especially ensemble methods, with the idea of improving the performance metric values as much as possible.

Author Contributions: Conceptualization, N.A. and S.B.Š.; methodology, Z.C.; software, N.A. and M.G.; validation, N.A., S.B.Š. and M.G.; formal analysis, N.A., S.B.Š. and M.G.; investigation, N.A., S.B.Š. and M.G.; resources, N.A., M.G. and Z.C.; data curation, N.A., M.G. and Z.C.; writing—original draft preparation, N.A., S.B.Š. and M.G.; writing—review and editing, N.A., S.B.Š. and M.G.; visualization, N.A., S.B.Š. and M.G.; supervision, N.A. and Z.C.; project administration, Z.C.; funding acquisition, Z.C. All authors have read and agreed to the published version of the manuscript.

Funding: This research received no external funding.

Institutional Review Board Statement: Not applicable.

Informed Consent Statement: Not applicable.

Data Availability Statement: The data used in this paper was obtained from a publicly available repository located at https://www.kaggle.com/datasets/fairumn/cdms-dataset, accessed on 9 January 2023.

Acknowledgments: This research was (partly) supported by the CEEPUS network CIII-HR-0108, the European Regional Development Fund under Grant KK.01.1.1.01.0009 (DATACROSS), the project CEKOM under Grant KK.01.2.2.03.0004, the Erasmus+ project WICT under Grant 2021-1-HR01-KA220-HED-000031177, and the University of Rijeka Scientific Grants uniri-mladi-technic-22-61 and uniri-tehnic-18-275-1447.

Conflicts of Interest: The authors declare no conflict of interest.

Appendix A. Additional Description of Mathematical Functions in SEs

In GPSR, some mathematical functions such as the square root, natural logarithm, logarithm with base 2 and 10, and division are differently defined to avoid generating imaginary or inf values. To avoid imaginary values, these functions have to be applied when the SEs defined with Eqs. (3)-(7) are used. The mathematical functions are defined as follows:

- Square root:
$$y_{sqrt}(x) = \sqrt{|x|}, \tag{A1}$$

- Natural logarithm:
$$y_{\log}(x) = \begin{cases} \log(|x|) & \text{if } |x| > 0.001 \\ 0 & \text{if } |x| < 0.001 \end{cases}, \tag{A2}$$

and when natural logarithms with base 2 and 10 are used, the log in Equation (A2) is replaced with $\log_2$ or $\log_{10}$, respectively.

- Division:
$$y_{div}(x_1, x_2) = \begin{cases} \frac{x_1}{x_2} & \text{if } |x_2| > 0.001 \\ 1 & \text{if } |x_2| < 0.001 \end{cases}. \tag{A3}$$

The variables x, x_1, and x_2 in Equations (A1)–(A3) are arbitrary values and have no relation with the input dataset variables.

References

1. Roszkowski, L.; Sessolo, E.M.; Trojanowski, S. WIMP dark matter candidates and searches—current status and future prospects. *Rep. Prog. Phys.* **2018**, *81*, 066201. [CrossRef] [PubMed]
2. Akerib, D.; Alvaro-Dean, J.; Armel-Funkhouser, M.; Attisha, M.; Baudis, L.; Bauer, D.; Beaty, J.; Brink, P.; Bunker, R.; Burke, S.; et al. First results from the cryogenic dark matter search in the soudan underground laboratory. *Phys. Rev. Lett.* **2004**, *93*, 211301. [CrossRef] [PubMed]
3. Ahmed, Z.; The CDMS Collaboration. Results from the Final Exposure of the CDMS II Experiment. *arXiv* **2009**, arXiv:0912.3592.
4. Aprile, E.; Alfonsi, M.; Arisaka, K.; Arneodo, F.; Balan, C.; Baudis, L.; Bauermeister, B.; Behrens, A.; Beltrame, P.; Bokeloh, K.; et al. Dark matter results from 225 live days of XENON100 data. *Phys. Rev. Lett.* **2012**, *109*, 181301. [CrossRef] [PubMed]
5. Pyle, M.; Serfass, B.; Brink, P.; Cabrera, B.; Cherry, M.; Mirabolfathi, N.; Novak, L.; Sadoulet, B.; Seitz, D.; Sundqvist, K.; et al. Surface electron rejection from ge detector with interleaved charge and phonon channels. In Proceedings of the AIP Conference Proceedings, Stanford, CA, USA, 20–24 July 2009; Volume 1185, pp. 223–226.
6. Jahangir, O. Application of Machine Learning Techniques to Direct Detection Dark Matter Experiments. Ph.D. Thesis, UCL (University College London), London, UK, 2022.
7. Bernardini, M.; Mayer, L.; Reed, D.; Feldmann, R. Predicting dark matter halo formation in N-body simulations with deep regression networks. *Mon. Not. R. Astron. Soc.* **2020**, *496*, 5116–5125. [CrossRef]
8. Theenhausen, H.; von Krosigk, B.; Wilson, J. Neural-network-based level-1 trigger upgrade for the SuperCDMS experiment at SNOLAB. *arXiv* **2022**, arXiv:2212.07864.
9. Zhang, X.; Wang, Y.; Zhang, W.; Sun, Y.; He, S.; Contardo, G.; Villaescusa-Navarro, F.; Ho, S. From dark matter to galaxies with convolutional networks. *arXiv* **2019**, arXiv:1902.05965.
10. Lucie-Smith, L.; Peiris, H.V.; Pontzen, A. An interpretable machine-learning framework for dark matter halo formation. *Mon. Not. R. Astron. Soc.* **2019**, *490*, 331–342. [CrossRef]
11. Simola, U.; Pelssers, B.; Barge, D.; Conrad, J.; Corander, J. Machine learning accelerated likelihood-free event reconstruction in dark matter direct detection. *J. Instrum.* **2019**, *14*, P03004. [CrossRef]
12. FAIR-UMN. CDMS-Dataset. Available online: https://www.kaggle.com/datasets/fairumn/cdms-dataset (accessed on 12 January 2023).
13. FAIR-UMN, Taihui Li. FAIR Document—Identifying Interaction Location in SuperCDMS Detectors. Available online: https://github.com/FAIR-UMN/FAIR-UMN-CDMS/blob/main/doc/FAIR%20Document%20-%20Identifying%20Interaction%20Location%20in%20SuperCDMS%20Detectors.pdf (accessed on 12 January 2023).
14. Pedregosa, F.; Varoquaux, G.; Gramfort, A.; Michel, V.; Thirion, B.; Grisel, O.; Blondel, M.; Prettenhofer, P.; Weiss, R.; Dubourg, V.; et al. Scikit-learn: Machine learning in Python. *J. Mach. Learn. Res.* **2011**, *12*, 2825–2830.
15. Mendenhall, W.M.; Sincich, T.L.; Boudreau, N.S. *Statistics for Engineering and the Sciences Student Solutions Manual*; Chapman and Hall/CRC: Boca Raton, FL, USA, 2016.
16. Lakatos, R.; Bogacsovics, G.; Hajdu, A. Predicting the direction of the oil price trend using sentiment analysis. In Proceedings of the 2022 IEEE 2nd Conference on Information Technology and Data Science (CITDS), Debrecen, Hungary, 16–18 May 2022; pp. 177–182.
17. Poli, R.; Langdon, W.; Mcphee, N. *A Field Guide to Genetic Programming*; 2008. Available online: http://www.gp-field-guide.org.uk/ (accessed on 2 January 2023)
18. Chai, T.; Draxler, R.R. Root mean square error (RMSE) or mean absolute error (MAE). *Geosci. Model Dev. Discuss.* **2014**, *7*, 1525–1534.
19. Langdon, W.B. Size fair and homologous tree genetic programming crossovers. *Genet. Program. Evolvable Mach.* **2000**, *1*, 95–119. [CrossRef]
20. Crawford-Marks, R.; Spector, L. Size Control Via Size Fair Genetic Operators In The PushGP Genetic Programming System. In Proceedings of the GECCO, New York, NY, USA, 9–13 July 2002; pp. 733–739.
21. Poli, R. A simple but theoretically-motivated method to control bloat in genetic programming. In Proceedings of the European Conference on Genetic Programming, Essex, UK, 14–16 April 2003; Springer: Berlin/Heidelberg, Germany, 2003; pp. 204–217.
22. Zhang, B.T.; Mühlenbein, H. Balancing accuracy and parsimony in genetic programming. *Evol. Comput.* **1995**, *3*, 17–38. [CrossRef]
23. Anđelić, N.; Baressi Šegota, S.; Glučina, M.; Lorencin, I. Classification of Wall Following Robot Movements Using Genetic Programming Symbolic Classifier. *Machines* **2023**, *11*, 105. [CrossRef]
24. Anđelić, N.; Lorencin, I.; Baressi Šegota, S.; Car, Z. The Development of Symbolic Expressions for the Detection of Hepatitis C Patients and the Disease Progression from Blood Parameters Using Genetic Programming-Symbolic Classification Algorithm. *Appl. Sci.* **2022**, *13*, 574. [CrossRef]
25. Wang, W.; Lu, Y. Analysis of the mean absolute error (MAE) and the root mean square error (RMSE) in assessing rounding model. In Proceedings of the IOP Conference Series: Materials Science and Engineering, Kuala Lumpur, Malaysia, 13–14 August 2018; Volume 324, p. 012049.
26. Di Bucchianico, A. Coefficient of determination (R^2). *Encycl. Stat. Qual. Reliab.* **2008**, *1*, eqr173. [CrossRef]
27. McKinney, W. Data structures for statistical computing in python. In Proceedings of the Proceedings of the 9th Python in Science Conference, Austin, TX, USA, 28 June–3 July 2010; Volume 445, pp. 51–56.

28. Hunter, J.D. Matplotlib: A 2D graphics environment. *Comput. Sci. Eng.* **2007**, *9*, 90–95. [CrossRef]
29. Stephens, T. GPLearn (2015). 2019. Available online: https://gplearn.readthedocs.io/en/stable/index.html (accessed on 2 January 2023).

Disclaimer/Publisher's Note: The statements, opinions and data contained in all publications are solely those of the individual author(s) and contributor(s) and not of MDPI and/or the editor(s). MDPI and/or the editor(s) disclaim responsibility for any injury to people or property resulting from any ideas, methods, instructions or products referred to in the content.

Article

Interactive Multifactorial Evolutionary Optimization Algorithm with Multidimensional Preference Surrogate Models for Personalized Recommendation

Weidong Wu [1], Xiaoyan Sun [1,*], Guangyi Man [1], Shuai Li [1] and Lin Bao [2]

[1] School of Information and Control Engineering, China University of Mining and Technology, Xuzhou 221116, China
[2] School of Electronics and Information, Jiangsu University of Science and Technology, Zhenjiang 212000, China
* Correspondence: 3730@cumt.edu.cn

Abstract: Interactive evolutionary algorithms (IEAs) coupled with a data-driven user surrogate model (USM) have recently been proposed for enhancing personalized recommendation performance. Since the USM relies on only one model to describe the full range of user preferences, existing USM-based IEAs have not investigated how knowledge migrates between preference models to improve the diversity and novelty of recommendations. Motivated by this, an interactive multifactorial evolutionary optimization algorithm with multidimensional preference user surrogate models is proposed here to perform a multi-view optimization for personalized recommendation. Firstly, multidimensional preference user surrogate models (MPUSMs), partial-MPUSMs, and probability models of MPUSMs are constructed to approximate the different perceptions of preferences and serve for population evolution. Next, a modified multifactorial evolutionary algorithm is used for the first time in the IEAs domain to recommend diverse and novel items for multiple preferences. It includes initialization and diversification management of a population with skill factors, recommendation lists of preference grading and interactive model management of inheriting previous information. Comprehensive comparison studies in the Amazon dataset show that the proposed models and algorithm facilitate the mining of knowledge between preferences. Eventually, at the cost of losing only about 5% of the Hit Ratio and Average Precision, the Individual Diversity is improved by 54.02%, the Self-system Diversity by 3.7%, the Surprise Degree by 2.69%, and the Preference Mining Degree by 16.05%.

Keywords: evolutionary optimization; multitask optimization; interactive; surrogate model; personalized recommendation

1. Introduction

In traditional recommendation algorithms [1], content-based, collaborative filtering-based and hybrid methods, it is difficult to extract user preference features effectively. As a result, deep learning has become a hot topic in the field of recommendation due to its powerful feature extraction [2]. Although deep learning effectively fuses multi-source heterogeneous data into recommendation systems [3,4], there are few existing studies on dynamic recommendation, most of which only consider the accuracy of recommendation and lack diversity and novelty.

In the realm of personalized recommendation [5–8], the interactive evolutionary algorithms (IEAs) based on the user surrogate model (USM) are a highly effective search paradigm. The information obtained from interaction with users allows it to effectively track dynamic changes in user preferences. Due to its good performance in dynamic recommendation, it has been applied to scenarios such as movie recommendation [6], personalized pattern designs [8], preference guides [9], and energy system designs [10]. However, studies related to IEAs still consider the diversity and novelty of recommendations and the use of deep learning methods in the construction of user surrogate models less.

To increase the diversity and novelty of recommendations, items with new combinations of attributes can be obtained by fully exploiting the characteristics of the items themselves. Mining new items for a preference by transfer knowledge from other related preferences is an effective approach. The search process of IEAs is guided by a USM. Therefore, unless the USM can describe preference information from multiple perspectives, IEAs cannot effectively mine the implicit information between different preferences. Currently, it is common to build several models to solve the problem of multiple perspectives [11–13]. For example, using multiple local surrogate models, Li et al. [11] proposed an evolutionary algorithm that approaches the optimal solution of the original problem from different directions. Although surrogate models have been able to describe the problem from various perspectives, previous studies have not considered the interaction and sharing of information between models.

Although there are strategies, subspace alignment [14], and feature mapping [15] to achieve feature transfer, there is no doubt that high computation will affect the timeliness required for personalized recommendation. Transferring genetic information between individuals of population is a viable way to address knowledge sharing between USMs. A multifactorial evolutionary algorithm (MFEA) [16] is a recently emerged approach for multi-task search. In MFEA, the evolutionary search is performed on multiple optimization problems or multiple search spaces simultaneously to improve evolutionary search capabilities by exploiting the associative knowledge between the tasks. In many successful cases of multi-task optimization scenarios, MFEA has demonstrated remarkable solution quality and search efficiency over traditional evolutionary algorithms [17–19]. To the best of our knowledge, the MFEA has not been extended to the IEAs-based personalized recommendation area. As we mentioned before, IEAs with only one surrogate cannot fully explore user preference information to provide diverse and novel items. Further consideration should be given to USMs describing user preferences from different perspectives, which will be searched as multiple optimization tasks, simultaneously.

Keeping these in mind, here we propose a new IEA for enhancing diversity and novelty of personalized recommendations. Firstly, deep learning-based multidimensional preference surrogate models (MPUSMs), partial-MPUSMs, and probability models of MPUSMs are developed, all of which are composed of multiple models representing different dimensions of preferences and serving the population evolution process. Then, an interactive multifactorial evolutionary optimization algorithm with multidimensional preference user surrogate models (MPUSMs-IMFEOA) is investigated, which combines the advantages of IEA's ability to update USM and MFEA's ability to migrate the preference knowledge of the user. It is the first attempt to apply MFEA to mining user preference knowledge in IEAs for personalized recommendation.

Accordingly, the primary contributions of our research are as follows. (1) In order to properly evaluate the population individuals, partial-MPUSM and MPUSMs are created. (2) In the evolutionary search phase, the modified MFEA is applied to the field of IEAs to improve the diversity and novelty of recommendations. Moreover, the efficiency of population evolution is improved with the help of probabilistic models. In addition, the normal genetic operators are adjusted accordingly because they are not sufficient to provide sufficient item diversity. (3) In the recommendation phase, a pre-recommendation list is proposed to prevent the problem of insufficient diversity, and both Top-N list and pre-recommendation list use roulette to focus on recent preferences and distinguish the importance of different preferences. (4) Lastly, a model management method for proposed models is also investigated, which can guarantee the correctness of the models and also inherit the valid information from the previous models.

The remainder of this paper is organized as follows: Section 2 reviews related work on personalized recommendation, interactive evolutionary algorithms and multifactorial evolutionary algorithm, and finally mentions the underlying surrogate model used in this paper. Section 3 details the proposed interactive multifactorial evolutionary algorithm with the multidimensional preference surrogate models. Section 4 presents comprehensive empirical comparison experiments of the proposed algorithm. Finally, this paper is concluded and discussed in Section 5.

2. Related Work

This section will introduce four elements. Firstly, the development and shortcomings of traditional algorithms for achieving personalized recommendations are presented. Secondly, interactive evolutionary algorithms and their recent advances are described. Thirdly, multifactorial evolutionary algorithms that are most suitable for implementing knowledge migration in IEAs are introduced. Finally, the underlying model used in this paper for the construction of multidimensional surrogate models is illustrated.

2.1. Personalized Recommendation

Personalized recommendation systems typically present users with options that match their preferences as they browse short videos, listen to music or shop online without a clear goal in mind, so the recommendations should be accurate, diverse, and novel. Traditional recommendation algorithms are mainly content-based [20], collaborative filtering [21] and hybrid recommendation [22]. Although they are widely used in the recommendation field, they suffer from the problem of not being able to build user models quickly and not being able to represent item features adequately due to the difficulty of performing adequate feature extraction. In particular, collaborative filtering also often faces the problems of data sparsity (low number of user rated items) and cold start (no rating data for new items or users).

Deep learning brings new opportunities to solve these problems due to its powerful ability to extract user requirements features [2]. Yuan et al. [23] proposed a deep and wide model to successfully extract features from multiple sources of information such as user–item interaction matrix, attributes, and context. To extract dynamic information about user behavior sequences, Wang et al. [24] developed a model called an Interval- and Duration-aware Long Short-Term Memory network to capture long-term and short-term user preferences.

From the perspective of evolutionary optimization, personalized recommendation is a kind of dynamic optimization problems guided by user knowledge. The IEAs have been widely used in personalized recommendation problems to quickly track changes in user behaviors and habits to meet the individual needs [25–27]. An example is the combination of a Kano model and an interactive genetic algorithm proposed by Dou et al. [26] for enhancing the effective user-interactive process of creating customer-oriented products and responding quickly to customer's needs.

IEAs have an inherent advantage in tracking changes in user preferences, but there is relatively little research on combining deep learning to model user preferences and consider the diversity and novelty of recommendations. Hence, how to use the user preference surrogate model built based on deep learning to further guide the evolution of the population and thus uncover more diverse and novel items becomes a problem that must be solved. Since there are hidden relationships between different preferences of the user, this problem can be solved if useful knowledge can be transferred from other preferences so that a certain preference can explore new items. Thus, a new IEA is explored to search for multiple preferences while also transferring knowledge between preferences.

2.2. Interactive Evolutionary Algorithms with User Preference Surrogate Model

IEAs combine a user's evaluation information with traditional evolutionary algorithms to solve the problem of personalized recommendation involving human–computer interaction. In traditional IEAs [28], a user is involved in evaluating population individuals to guide population evolution. To alleviate user fatigue, some studies [29,30] build a user surrogate model to reduce the number of user evaluations. However, in practice, users only have a very small number of scoring behaviors. To this end, Sun et al. [31] obtained user preferences indirectly from clicks, browsing and other behaviors to build a better USM. After each population iteration, some solutions with high fitness values are selected from the USM-evaluated population and recommended to the user. Finally, the model is updated according to behavior information to ensure that the next recommended items are more in line with the user's preferences. As can be seen, the accuracy of USM is the key to ensure that the user's favorite items are searched.

From Section 2.1, it is clear that most studies using deep learning methods to model user preferences have built only one model. Nevertheless, it is difficult to use only one model to guide evolutionary algorithms to extract inter-preference information efficiently. While some studies [11–13] have constructed multi-model USM by describing the problem from different perspectives, they have not yet investigated how information can be transferred between models under the framework of IEAs. This can not only improve the efficiency of problem solving but also increase the diversity and novelty of recommendations.

In addition, to accommodate different optimization objectives, IEAs can adopt different optimization strategies to optimize population. For example, there are differential evolution algorithms [32], non-dominated sorting genetic algorithms II [33] and even the estimation of distribution algorithms [34] that directly obtain population individuals by sampling a probabilistic model describing the distribution of solutions and thus achieving population evolution.

It can be found that existing research on user preference model are not conducive to explicit expressing multiple preferences and to guiding the mining of relationships between preferences, so as to find more items with higher diversity and novelty. Accordingly, we are motivated to design a novel IEA that optimizes the population guided by multiple USMs while transferring preference knowledge when individuals are mated.

2.3. Multifactorial Evolutionary Algorithm

As mentioned in Section 2.2, we need a strategy for optimizing population that not only addresses multiple goals simultaneously, user preference surrogate models, but also transfers information between models during the evolution of the population. It is worth noting that finding satisfactory items for each dimension of preferences can provide a wider choice of perspectives, which can likewise improve the diversity of recommendations. Considering an optimization task of each preference model as a unit of work, the multifactorial evolutionary algorithm (MFEA) [16] capable of optimizing multiple related tasks simultaneously is a natural selection. Optimization of multiple tasks differs from optimization of multiple objectives [35] in that the former aims to find the optimal solution for each task, whereas the latter seeks to discover a set of Pareto solutions. MFEA is structured similarly to a classical evolutionary algorithm, making it more economical to find new items for each preference by migrating the features of different tasks in each population iteration. In the case of related problems, transferring knowledge between different problem solutions will improve performance over solving problems in isolation [36–38].

MFEA is a classic algorithm in the field of evolutionary multitasking, as described in Algorithm 1. One of the most essential innovations in MFEA is the skill factor (τ), which represents the task to which an individual is most suited to solve among all the tasks. Each task is a unique factor that affects the evolution of a single population of individuals, so the population is implicitly divided into multiple skill groups by skill factors, each of which is good at solving corresponding tasks. Two components of MFEA control the inter-tasks (different skill groups) and intra-tasks (identical skill groups) knowledge transfer for the population: assortative mating and vertical cultural transmission, as shown in Steps i and ii of Algorithm 1. Assortative mating controls two randomly selected parental candidates to perform the crossover. Two candidates with the same skill factor can mate directly. Conversely, crossovers are performed between candidates with different skill factors if the random mating probability (rmp) condition is satisfied; otherwise, mutation is performed.

Vertical cultural transmission means that individuals of the offspring randomly inherit the skill factors from their parents and evaluate only the task corresponding to the skill factor.

Algorithm 1 Basic structure of the multifactorial evolutionary algorithm
1: Generate an initial population of individuals and store it in *current-pop (P)*.
2: Evaluate every individual with respect to every optimization task in the multitasking environment.
3: Compute the skill factor (τ) of each individual.
4: **while** stopping conditions are not satisfied **do**
5:    i. Apply genetic operators on current-pop to generate an *offspring-pop (C)*. (Assortative Mating)
6:    ii. Evaluate the individuals in offspring-pop for selected optimization tasks only. (Vertical Cultural Transmission)
7:    iii. Concatenate offspring-pop and current-pop to form an intermediate-pop ($P \cup C$).
8:    iv. Update the scalar fitness (φ) and skill factor (τ) of every individual in intermediate-pop.
9:    v. Select the fittest individuals from intermediate-pop to form the next *current-pop (P)*.
10: **end while**

Since MFEA is effective at solving multiple tasks and has not been extended to IEA-based personalized recommendation, it is necessary to carry out a corresponding study to improve further the IEAs performance in optimizing personalized recommendation. Using the useful knowledge implied between preferences, knowledge migration is performed by gene selection, crossover, and mutation at the chromosome level. This is carried out to mine further desired items of the user under every preference.

2.4. DRBM-Social User Surrogate Model

Here is a brief description of the user surrogate model proposed in [13] that integrates a trained dual Restricted Boltzmann Machine model with the social knowledge of similar users (DRBM-Social). It is selected as the base component of multidimensional preference user surrogate model proposed in this paper. Restricted Boltzmann Machine (RBM) is an energy-based stochastic neural network with a two-layer network structure, as shown in part in Figure 1, that learns the probability distribution of the input data. The visible layer v contains n units and its input is the observed data, and the hidden layer h contains m units and acts as a feature extractor. Given the state (v, h) and the parameter set $\theta = \{w, a, b\}$ where w is the connection weight, a and b are biases, it is easy to determine the RBM network state based on the energy function, as shown in (1). With the input observed data, all parameters in the RBM can be obtained by the Contrastive Divergence algorithm [39]:

$$E_\theta(v, h) = -\sum_{i=1}^{n}\sum_{j=1}^{m} v_i w_{ij} h_j - \sum_{i=1}^{n} a_i v_i - \sum_{j=1}^{m} b_j h_j \tag{1}$$

where v_i and h_j are the states of the i-th visible unit and the j-th hidden one, respectively. w_{ij} represents the connection weight between them, and a_i and b_j are their respective biases.

The prediction score of DRBM-Social surrogate contains two parts: the score predicted by the dual-RBM model and the score calculated based on social knowledge. Figure 1 presents the structure of the Dual-RBM which consists of a positive RBM and a negative RBM. As a first step, the evaluated items are divided into dominant and inferior groups according to the magnitude of their scores. Then, these two groups of items are used to train the Dual-RBM separately to obtain two sets of RBM network structure parameters. The item scores predicted by DRBM-Social are derived by combining social knowledge and user preferences which contain both positive and negative preferences, as shown in (2):

$$\hat{f}(x) = \sigma(\alpha * social(x) + preference(x)) \tag{2}$$

where α is the weight used to adjust the social knowledge contribution; σ is the activation function, which can be either a sigmoid or tanh function; $social(x)$ indicates the weighted average score of M similar users on item x and will be defined in (3):

$$social(x) = \sum_{j=1}^{M} Sim(u, u_j) * R_{u_j x} \qquad (3)$$

where $Sim(u, u_j)$ is the Pearson similarity coefficient between the current user u and the neighbor u_j, and $R_{u_j x}$ is the rating of item x given by the neighbor u_j.

Then, the $preference(x)$ score of the item x is calculated based on its positive and negative preferences obtained in the positive RBM and negative RBM, respectively, as presented in (4):

$$preference(x) = \frac{\max(E_{\theta_p}) - E_{\theta_p}(x, h)}{\max(E_{\theta_p}) - \min(E_{\theta_p})} - \beta * \frac{\max(E_{\theta_n}) - E_{\theta_n}(x, h)}{\max(E_{\theta_n}) - \min(E_{\theta_n})} \qquad (4)$$

where $E_{\theta_p}(x, h)$ and $E_{\theta_n}(x, h)$ are the energy values of item x calculated by the hidden layer h in the positive and negative RBM models, respectively. In the positive RBM model, $\max(E_{\theta_p})$ and $\min(E_{\theta_p})$ are the maximum and minimum energy values of all items which are generated by the probabilistic model and then matched to the real space to obtain. β is used to adjust the proportion of the contribution of the negative preference. An item with J attributes is expressed as $x = \{x_1, x_2, \cdots, x_J\}$, where $x_{1,2,\ldots,J} \in \{0,1\}$.

Furthermore, a probability model [13] deduced by positive RBM only, as expressed in (5), is designed to efficiently reflect the degree of user preference for each attribute of the item and generate a superior population of individuals. According to (6), individuals x is generated by sampling the probability model randomly:

$$P(x) = [p(x_1 = 1), p(x_2 = 1), \cdots, p(x_J = 1)] \qquad (5)$$

$$x_j = \begin{cases} 1, & if \quad random(0,1) \geq p(x_j = 1) \\ 0, & otherwise \end{cases} \qquad (6)$$

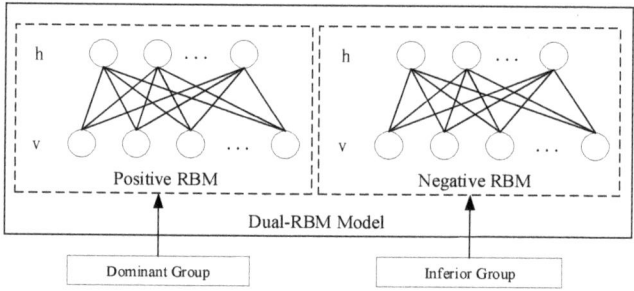

Figure 1. Structure of the Dual-RBM model.

3. Proposed MPUSMs-IMFEOA for Personalized Recommendation

In this section, the proposed Interactive multifactorial evolutionary optimization algorithm with multidimensional preference user surrogate models depicted in Figure 2 is presented in detail. The middle part of Figure 2 depicts the construction process of the multidimensional preference user surrogate model and its derived models. Firstly, K-means method and minimum Euclidean distance method are used to divide the historical evaluation items containing dominant and inferior items into K clusters representing different dimensional preference items. Secondly, K clusters of data are used to train K Dual-RBMs. Finally, based on K-trained dual-RBMs, partial-MPUSMs and MPUSMs for evaluating population individuals and items are obtained, as well as probabilistic models for generating initial population with skill factors. The proposed interactive multifactorial evolutionary optimization algorithm is illustrated in the left and right parts of Figure 2. The right side of the figure, from top to bottom, represents the improved MFEA algorithm, the two recommendation lists, and the user interaction process. The left part of the figure shows the update of the *K* clusters to further update the MPUSMs, partial-MPUSMs, and probability models using a re-clustering or fine-tuning approach based on an updated historical item set incorporating the latest evaluation results. The mathematical symbols in this paper are summarized in Table 1.

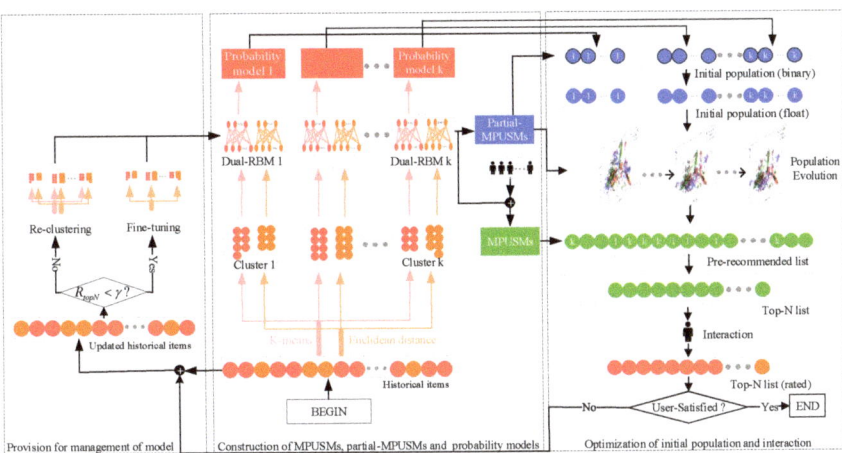

Figure 2. The framework diagram of the proposed MPUSMs-IMFEOA algorithm.

Table 1. Annotation of the symbols.

Symbol	Definition and Description
v	The visible layer of RBM
h	The hidden layer of RBM
α	Adjustment coefficient of $social(x)$
β	Adjustment coefficient of negative RBM
u	User
x	Real item
x'	Virtual individuals representing items in the population
M	Number of similar users
U_{sum}	The number of users included in each dataset used for the experiment
K	Number of preference dimensions
r	Rating of item
ε	Threshold for dividing historical items into two parts
Q	Length of the pre-recommended list
N	Length of the Top-N list
rmp	Parameter to control crossover and mutation in MFEA
R_{topN}	RMSE of the Top-N list
γ	Threshold for managing the model
pop	Number of individuals in every skill group of the population
G	Number of evolutionary generations
W	Number of latest items to calculate the roulette probability

3.1. Construction of Multidimensional Preference User Surrogate Models and Their Derivatives

Different user preference behaviors can be simulated with multidimensional preference user surrogate models (MPUSMs) to evaluate unknown items. In addition, MPUSMs will be the base for running MFEA by viewing each surrogate as an optimized objective. Here, we combine multiple Dual-RBM models and social knowledge illustrated in Section 2.4 to describe multidimensional preferences. To properly evaluate the real items and unreal individuals of population, we design the MPUSMs and partial-MPUSMs, and to improve the evolutionary efficiency of MFEA, we derive probability models of MPUSMs. The construction process of MPUSMs and their derivatives is detailed in the middle part of Figure 2 and Algorithm 2. Key Innovative steps are marked with an asterisk.

Algorithm 2 Pseudocode for the procession of creating MPUSMs and their derivatives

---

Input: Historical evaluation items of user u and other M users
Output: partial-MPUSMs, MPUSMs, and probability models of MPUSMs
    Provision of Social Knowledge
1: The $Social(x)$ for all items in the pool of items to be searched is obtained with (3).
    *Clustering
2: *Sort the historical items of the user u into dominant and inferior groups.
3: *The dominant group is clustered into K clusters by the K- means algorithm, which is repeated 10 times and the most balanced classification result is selected.
4: *Assign the inferior group to K clusters according to the distance of inferior items from the cluster centers.
    Training
5: Train K Dual-RBM models with the data from K clusters.
6: *Form a partial-MPUSMs by combining K Dual-RBM models.
7: *Combine K Dual-RBM models and the user's social knowledge to form MPUSMs.
8: Calculate the K probability models based on positive RBMs of K Dual-RBM models using (5).

3.1.1. Clustering of Historical Items

Clustering the set of historical evaluation items is a critical step for dividing the user's different preference perceptions. To conform to the characteristics of the DRBM-social model, we first divided the historical data of user u into dominant group ($r \geq \varepsilon$) and inferior group ($r < \varepsilon$) based on the magnitude of the items' rating r and the threshold ε. For the dominant group, the K-means method is used to classify the historical items into K clusters. To prevent the number of samples in a cluster from being too small to train an accurate positive RBM model, we perform K-means clustering process several times so that the most evenly distributed result is selected from it. For the inferior group, the inferior items are assigned to K clusters based on the minimum Euclidean distance between the inferior item and the clustering centers. The negative RBM model only plays a subsidiary role, so even a small number of samples within the inferior cluster does not have much impact. At this point, K clusters are obtained as the base data for training K surrogates.

3.1.2. Training and Usage of MPUSMs and Their Derivatives

The middle part of Figure 2 and the training part of Algorithm 2 clearly depict the construction process of MPUSMs, partial-MPUSMs, and K probability models. As shown on the right in Figure 2, the proposed interactive multifactorial evolutionary optimization algorithm (IMFEOA) performs the MFEA evolution process for certain generations to obtain some relatively superior candidates and then eventually provides a Top-N list for interaction with the user. As the evolving individuals of population may not be real items, social knowledge derived from existing rated items cannot be used for this process. Hence, the evolving individuals are only evaluated with the partial-MPUSMs. After the evolution loop ends, the pre-recommended list will be filled with superior individuals and matched to real items, and the corresponding social knowledge could be added to provide more adequate evaluation information. To this end, two types of surrogates are designed for separately evaluating evolving individuals (or virtual items) and pre-recommended items.

Firstly, we train K dual-RBM models with K clustered data. Then, partial-MPUSMs consist of only K dual-RBMs to approximate the user's K dimensional preferences. Their formula for assessing individuals belonging to the k-th dimension of preferences is shown in (7) and is only used to evaluate the individuals x' in the MFEA evolutionary process. For evaluating the items x from the pre-recommended list, social knowledge is used together with K Dual-RBM models to form MPUSMs, as shown in (8). Finally, the probability models of MPUSMs derived from each positive RBM are calculated with (5) and an initial population that better matches the user preferences could be generated by (6). In these two types of models, the components $\max(E_{\theta_p})$, $\min(E_{\theta_p})$, $\max(E_{\theta_n})$ and $\min(E_{\theta_n})$ of the $preference(x)$ as defined in (4) are calculated based on the samples of each cluster, which is quite different from that in our previous work [13]:

$$\hat{f}_k^p(x') = \sigma(preference(x')) \tag{7}$$

$$\hat{f}_k^c(x) = \sigma(\alpha * social(x) + preference(x)) \tag{8}$$

During the MFEA evolution process, (7) is used to directly evaluate the individuals with skill factor k. For the pre-recommended list, superior individuals are drawn from every skill group in the population, but their cluster features may have changed significantly as a result of evolution. To facilitate a reliable evaluation, these individuals should be converted to corresponding real items before finding the most appropriate preference surrogates for them to evaluate. Here, the Euclidean distance is used to find the real items for individuals in the set of items to be searched, as well as the most appropriate surrogate in the MPUSMs for the items in the pre-recommended list, for which the process is shown in (9):

$$r(x_i) = \hat{f}_k^c(x_i) \tag{9}$$

where $r(x_i)$ is the rating of the i-th item x; $k = \arg\min D_j$ denotes that the i-th item is best suited to be evaluated by the k-th surrogate model in the MPUSMs, of which $D_j = ||x_i - c_k||, (j = 1, 2, \cdots K)$ is the Euclidean distance between the i-th item and the k-th clustering center.

3.2. Interactive Multifactorial Evolutionary Optimization Algorithm with MPUSMs

In this section, we design a new IEA, Interactive multifactorial evolutionary optimization algorithm, compatible with the constructed MPUSMs to discover more diverse and novel items for recommendation. The main components of the IMFEOA include the following three parts: (1) Enhanced MFEA: For solving MPUSMs-aimed tasks and improving evolution efficiency, initial population with skill factors is obtained by sampling MPUSMs' probabilistic models that predicts the best search region. In addition, MFEA's knowledge sharing can increase the diversity of the items, but the use of common genetic operators in MFEA is still not sufficient to obtain sufficiently diverse items. MFEA with SBX crossover and Gaussian mutation is designed to optimize the personalized search to obtain more diverse and novel items. (2) MPUSM-based recommendation list: A pre-recommended list is creatively proposed in order to provide a Top-N list for user in line with diversity and current preferences. It is obtained by sampling evolved population with a roulette and resampling re-evolved population in case of insufficient diversity. (3) Interaction-assisted model management: MPUSMs and their derivatives are updated in accordance with the interaction results to learn useful information of the previous generation models while ensuring the accuracy of updated models. To the best of our knowledge, this is the first time to design an interactive MFEA for solving personalized searches. Its difference from common IEAs is that a user is not involved in the evaluation after each population iteration, which not only reduces user fatigue, but also improves the ability to uncover new items. The pseudocode of the specific process is shown in Algorithm 3. Asterisks indicate innovative key steps.

$$x'_{ij} = \begin{cases} random(0 - 0.5) & x_{ij} = 0 \\ random(0.5 - 1) & x_{ij} = 1 \end{cases} \qquad (10)$$

where x_{ij} is the value of the j-th gene locus of the i-th individual in the population.

Algorithm 3 Pseudocode of MPUSMs-IMFEOA

Input: Trained MPUSMs model, Partial-MPUSMs and probability models
Output: User-satisfied Top-N list
1: **while** Top-N list is not satisfied **do**
2:     **while** Number of items in the pre-recommended list $< Q$ **do**
        Enhanced MFEA:
3:         *By sampling pop times for each probability model, a better initial population is obtained and saved to current-pop.
4:         *Directly fill in skill factor for each individual of population.
5:         *Evaluate the initial population with partial-MPUSMs.
6:         *Encode the initial population according to (10)
7:         **while** generation $< G$ **do**
            Optimization of diversity population
8:             *Apply SBX and Gaussian mutation to current-pop to generate an offspring-pop.
9:             *Decode offspring-pop according to the inverse (10).
10:           *offspring-pop individuals are evaluated with partial-MPUSMs.
11:           Update the scalar fitness and skill factor of each individual in the merged current-pop and offspring-pop.
12:           Select the fittest individuals to form next current-pop.
13:         **end while**
14:         *Select the superior individuals from the evolved population with multiple skill groups by roulette to fill the pre-recommended list.
15:     **end while**
        Pre-recommended list:
16:     Convert Q individuals to Q real items by matching them with the closest item in the item pool.
17:     The Q items are evaluated using the MPUSM.
18:     ***Top-N list:*** Select the highest-rated N items in each category from the pre-recommended list using roulette.
19:     **Interactive Evaluations:** The Top-N list is submitted to the user for evaluation.
20:     ***Model Management:*** Update the partial-MPUSMs, MPUSMs and probability models based on interaction results. In case $R_{topN} < \gamma$, then the models are fine-tuned. In case $R_{topN} > \gamma$, re-clustering and retraining the model are necessary.
21: **end while**

3.2.1. Enhanced MFEA

To further improve evolutionary efficiency, initial population with skill factors is generated based on multiple probability models that make individuals inherently suitable for solving the corresponding tasks. Thus, by sampling the corresponding probabilistic models with (6), this paper establishes K skill groups in the population, each with its own skill factor. In other words, the individual is created by the k-th probability model, and thereby its skill factor is k. Individuals are then evaluated with the partial-MPUSMs to prepare for the next step of population evolution.

In particular, the real items are represented as discrete binary vectors, and an item with J attributes is expressed as $x = \{x_1, x_2, \cdots, x_J\}$, where $x_{1,2,\ldots,J} \in \{0, 1\}$. They are encoded continuously with (10) so that continuous crossover and mutational operators can be accommodated to ensure population recommendation diversity. Consequently, individuals have two encoding representations: the discrete form for evaluation and the continuous form for evolution.

Given that personalized recommendations require both satisfying user's preferences and conforming to the diversity of recommendation, it is important to ensure population diversity in order to select a sufficient number of items for each preference dimension. As a result, a targeted adjustment must be made to the MFEA. After specific experiments, simulated binary crossover (SBX) and Gaussian mutation are chosen to accomplish the migration of preference knowledge, as described in Section 4.2. It is noticeable that individuals in evolution are unrealistic, so we use partial-MPUSMs to evaluate them.

As crossover and mutation occurs between skill groups, knowledge transfer will occur between different dimensional preferences. Eventually, with the continuous iterative evolution of population, the implicit preferences of user will be uncovered and a population with excellent individuals in multiple preference dimensions will be formed.

3.2.2. MPUSM-Based Recommendation List

Two selection processes for Top-N items are designed here: coarse and fine selection to obtain a pre-recommended list and a Top-N list, respectively. During the coarse selection stage, although population diversity has already been enhanced by changing the coding form and adjusting the crossover and mutation operators, sometimes, an evolved population may not be sufficient to provide enough unduplicated individuals. For this reason, a new initial population generated by probability models is again evolved to provide more non-repeating individuals. In addition, MPUSM separates user's preferences, so that user's attention to different dimensions of preferences can be determined based on recently assessed items. Therefore, the roulette wheel selection is applied, which not only takes into account each preference's importance but also provides more randomness. According to (11), the probability of roulette is calculated. Moreover, the preference dimension to which the item belongs is determined by calculating the magnitude of the Euclidean distance of the item from the K clustering centers:

$$F = [f_1, f_2, \cdots, f_K] \tag{11}$$

where f_k denotes the probability of attention to the k-th dimensional preference and $f_k = w_k/W$. w_k represents the number of items belonging to the k-th cluster in the latest W historical evaluations.

The population is divided into K parts by skill factors, so it is easy to extract diverse and non-repeating individuals for the pre-recommended list using the roulette. At this time, individuals with higher rating are given priority for extraction. Once the population has no more non-replicated individuals, a newly created population is re-evolved for filling the pre-recommended list. After the pre-recommended list is filled, the individuals are transformed into nearest real items in an item pool using the minimum Euclidean distance. Then, they need to be evaluated with MPUSMs, at which point the list is still divided into K groups.

During the fine selection stage, a Top-N list containing the N most desired real items is generated. Currently, a pre-recommended list contains items on multiple dimensional preferences, and each item is evaluated with utmost precision MPUSMs. Thus, the roulette is used again to draw N items from the pre-recommended list to form the Top-N list.

3.2.3. Interaction-Assisted Model Management

Model updating is an essential part of tracking user preference changes. The role of this part is to recommend the Top-N list to the user and to update the MPUSMs, partial-MPUSMs and probability models according to the interaction results. In order to retain the helpful information from the previous models while updating the model, we keep information that correctly classifies preferences in case the MPUSMs are accurate. In this way, the user's preferences are retained, and the latest information can be added. That is, if the models are accurate, they will be fine-tuned, while if they are not accurate, they must be completely retrained. The specific work is as follows.

In this work, the root mean square error (RMSE) of the true and predicted ratings of the items for training MPUSMs is used as the threshold γ, and the RMSE of the Top-N list is noted as R_{topN}. If $R_{topN} < \gamma$; this means that the model preference classification and prediction is accurate, so the MPUSMs and their derivatives can be fine-tuned directly to extend the users preferences. In addition, if $R_{topN} < \gamma$, this indicates that the prediction results of the models are inaccurate. That is, the classification of user preferences does not match user habits and needs to be re-clustered.

Following is a detailed explanation of the fine-tuning and re-clustering process. Firstly, the obtained evaluation results of the Top-N list are combined with the historical evaluation items. If fine-tuning is needed, then the updated historical evaluation items are grouped into K classes by following the previous clustering centers. If re-clustering is necessary, then the clustering process of Algorithm 2 is executed for the updated historical evaluation items. Finally, the models are updated by running the training process in Algorithm 2, either after fine-tuning or re-clustering.

3.3. Evaluation Indicators

Three types of evaluation metrics, namely accuracy, diversity, and novelty, are used to assess the Top-N list for comprehensively evaluating the performance of MPUSMs-IMFEOA. Among them, the accuracy [13] is evaluated by three indicators: Root Mean Square Error ($RMSE$), Hit Ratio (HR), and Average Precision (AP). The Diversity [40] metrics are set as individual diversity (Div) and Self-System Diversity (SSD). The novelty includes surprise degree (Sur) [40] and preference mining degree (PMD), where the PMD is firstly and specifically proposed to evaluate the effectiveness of our algorithm in mining inter-preference knowledge. Below are details of the evaluation metrics.

3.3.1. Root Mean Square Error (RMSE)

$RMSE$ calculates the error between the predicted rating and the user's real rating:

$$RMSE = \sqrt{\frac{\sum_{i \in S}(r_i - \hat{r}_i)^2}{|S|}} \qquad (12)$$

where r_i denotes the user's true rating of item i, and $\hat{r}_i$ is the predicted one. S represents the set of items to be evaluated, and $|S|$ is the number of items in the set S.

3.3.2. Hit Ratio (HR)

HR is the ratio of the number of satisfying items in the Top-N list to all the items that the user likes in the item pool of items to be searched. The larger the HR, the more satisfying items in the Top-N list:

$$HR = \frac{Hit}{|GT|} \qquad (13)$$

where Hit is the number of satisfying items in the Top-N list, and $|GT|$ is the number of items in the item pool that the user is satisfied with.

3.3.3. Average Precision (AP)

AP measures the impact of item ranking on user experience perception, and a higher AP value indicates a higher ranking of items for user satisfied items:

$$AP = \frac{1}{|L|} \sum_{i=1}^{|L|} \frac{i}{position(i)} \qquad (14)$$

where $position(i)$ is the ranking position of the user preferred i-th item in the Top-N list. $|L|$ is the number of items in the Top-N list.

3.3.4. Individual Diversity (Div)

Div measures the diversity of items in Top-N list items, focusing on whether items with different dimensional preferences can be recommended to the user to avoid homogenization:

$$Div(L) = 1 - \frac{\sum\limits_{i,j \in L, i \neq j} s(i,j)}{|L|(|L|-1)} \tag{15}$$

where $s(i,j) \in [0,1]$ defines the similarity of items in the Top-N list, and $s(i,j) = 1 - d(i,j)/\sqrt{D}$. Among them, $d(i,j)$ denotes the Euclidean distance between the i-th item and the j-th item, and D is the dimension of the item vector.

3.3.5. Self-System Diversity (SSD)

SSD represents the ratio of items recommended this time that are not similar to those recommended last time. The user's interest in the item changes over time, and they prefer new recommendations that differ from those in the past. The larger the SSD, the better the temporal diversity of the recommended list:

$$SSD = \frac{|L_t / L_{t-1}|}{|L_t|} \tag{16}$$

where L_{t-1} is the last recommendation list of L_t, and $L_t / L_{t-1} = \{i \in L_t | i \notin L_{t-1}\}$.

3.3.6. Surprise Degree (Sur)

Sur means that, if the recommended items are not similar to the user's historical preferences, but the user is satisfied with that, the recommendation is said to have a high degree of surprise:

$$Sur(x) = \frac{1}{|L||H|} \sum_{i=1}^{|L|} \sum_{h=1}^{|H|} \frac{r_i}{\frac{r_h}{r_{\max}} s(i,h)} \tag{17}$$

where r_i represents the user rating of the i-th recommended item. r_h represents the rating of the h-th historical item, and $r_{\max}$ is the maximum rating in the set of historical items. $|H|$ is the number of items projects evaluated historically.

3.3.7. Preference Mining Degree (PMD)

PMD measures the algorithm's ability to mine user preference information. By calculating the degree of dissimilarity between the Top-N list and the set of items evaluated historically, it evaluates whether the algorithm can predict the user's unexpressed preferences by learning the user's historical preferences. A larger PMD value indicates that the algorithm is more capable of mining user preferences:

$$PMD = 1 - \frac{1}{|L||H|} \sum_{i=1}^{|L|} \sum_{h=1}^{|H|} s(i,h) \tag{18}$$

where $|H|$ is the number of items that have been rated by the user. $s(i,h)$ represents the similarity between the item i in the Top-N list and the historically evaluated item h.

4. Experiments and Results

To verify the effectiveness of the proposed algorithm in the field of personalized recommendation, comprehensive empirical comparison studies are conducted in this section. The dataset sources and processing, as well as the experiment-related parameter settings and compared algorithms and surrogate models, are first described in Section 4.1. Then, four sets of experiments are designed. (1) Effect of genetic operators on population diversity; (2) Validate the accuracy of MPUSM evaluation items compared to DRBM-Social; (3) The recommended performance of the proposed MPUSMs-IMFEOA is verified by comparing it with SC-DRBMIEDA [13], which is also based on the DRBM-social model; (4) Visually demonstrate the search capability of MPUSMs-IMFEOA by comparing it to other IEAs.

4.1. Experimental Setup

Three datasets from Amazon [41] are selected for the experiment, and the statistical information of the datasets is described in Table 2. Based on the dataset, the experiments mainly use item IDs, user IDs, review time, categories, and rating information. Each experimental dataset is composed of evaluation data from U_{sum} randomly selected users in the dataset. In each experimental dataset, one user is selected as the experimental subject, and the remaining users who are similar will provide social knowledge for that user. If all items evaluated by a user have a total of J categories, the items are expressed as $x = \{x_1, x_2, \cdots, x_J\}$, where $x_{1,2,\ldots,J} \in \{0,1\}$. The evaluation data of the experimental subjects are processed as follows.

Table 2. # of users, items and ratings of the datasets.

Dataset	# of Users	# of Items	# of Ratings
Digital_Music (Music)	478,235	266,414	836,006
CDs_and_Vinyl (CDs)	1,578,597	486,360	3,749,004
Kindle_Store (Kindle)	1,406,890	430,530	3,205,467

Considering the actual situation of interactive behavior, the set C of historical evaluation items of a user is sorted by review time. Then, it is divided to obtain multiple training sets (sliding window) and item pools containing the items to be searched at different interactions, as shown in Figure 3. In these experiments, the set C is divided into two parts, one for the first 70% of items as the set C_{HI} regarded as the historical interactive items, and the other as the fixed set C_F of items to be searched. The last 30% of the set C_{HI}, named C_I, is considered as the result of 10 actual interactions. The training items are obtained from the set C_{HI} through a sliding window with a fixed length of $0.4 * |C_{HI}|$. $0.1 * |C_I|$ items following the sliding window is seen as the actual result of the interaction, and all the data after the sliding window in C are regarded as an item pool to be searched, so that the item pool decreases gradually as the interaction proceeds. To be relevant, the updated history items in experiments include sliding window data and the Top-N list, except for the first interaction. However, the R_{topN} used for model management is still calculated from the Top-N list rather than real interaction results.

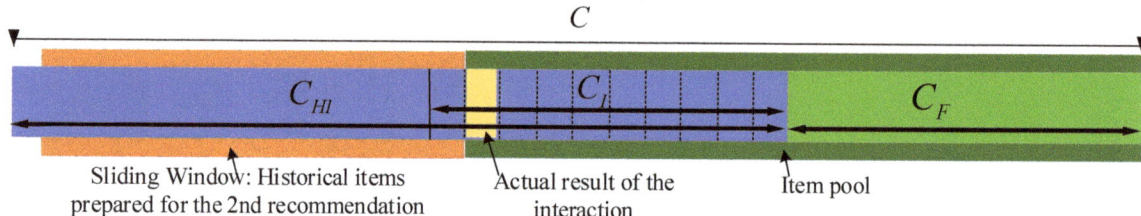

Figure 3. The data set division of the user.

In these experiments, MPUSMs and their derived models are created based on Dual-RBM [13], so the parameters (number of units in the hidden and visible layers) of the Dual-RBM and the training method are the same as in [13]. In addition, the parameters (α, β) related to the calculation of scores for MPUSMs and partial-MPUSMs are referred to [13]. Moreover, IMFEOA is built based on MFEA [16], so the parameter rmp is referenced to [16]. In addition, the population to be optimized by IMFEOA will evolve until no new individuals can be searched for, so that unknown items can be fully explored ($G = 100$). The remaining parameters are deliberately set to be identical to SC-DRBMIEDA [13] in order to be able to compare with it in the same context. Detailed parameter configurations are given in Table 3. The user surrogate models and algorithm descriptions used for comparison in these experiments are shown in Table 4.

Table 3. Experimental parameters of the proposed algorithm.

Parameter	Value
α	0.3
β	0.2
K	3
ε	3
U_{sum}	200
rmp	0.3
pop	50
G	100
Q	100
N	10
W	50
Number of visible units of RBM	# of attributes
Number of hidden units of RBM	1.2 * # of attributes

1 The '# of attributes' represents the number of attributes of the items. 2 The '*' is the multiplication sign.

Table 4. The comparison models and algorithms used in this paper.

DRBM-Social [13]	It uses a model based dual-RBM and social knowledge to describe the user's preference information
RBM [42]	It uses an RBM-based model to describe the user's preference information
MPUSMs based on RBM	They use RBM as the underlying model for building MPUSMs
SC-DRBMIEDA [13]	It is an IEA that uses the same DRBM-Social as the proposed algorithm, but it does not implement knowledge transfer between preferences.
DRBM-Social-IGA	It is an IEA using a general genetic algorithm to optimize DRBM-Social.
RBM-IGA	It is an IEA using a general genetic algorithm to optimize RBM.
MPUSMs-IMFEOA based on RBM	It differs from the algorithm proposed in this paper in that it replaces the underlying surrogate model for constructing MPUSMs

4.2. Experiment 1—Effect of Genetic Operators on the Variety of Individuals in the Population

The genetic operators have a significant impact on the performance of the proposed IMFEOA algorithm. It directly determines the diversity of individuals in the population. As described in Section 3.2.1, individuals have two forms of coding. In this experiment, Two-Point Crossover, Uniform Crossover, and the method of sampling RBM probability model (Prob) applicable to binary vectors are chosen, as well as the Simulated Binary Crossover (SBX) and Gaussian Mutation (Gauss) methods suitable to float vector.

During this experiment, we randomly select the user with ID A9Q28YTLYRE07 in the music dataset, and the first sliding window data are selected after processing according to the method in Figure 3. Common recommended settings used for all genetic operators. Combining the three crossover operators and two mutation methods, the effect of the experimental genetic operators on the variety of individuals in the population is shown in Figure 4. It is clear that only the SBX and Gaussian Mutation operators are capable of maintaining the diversity of the population.

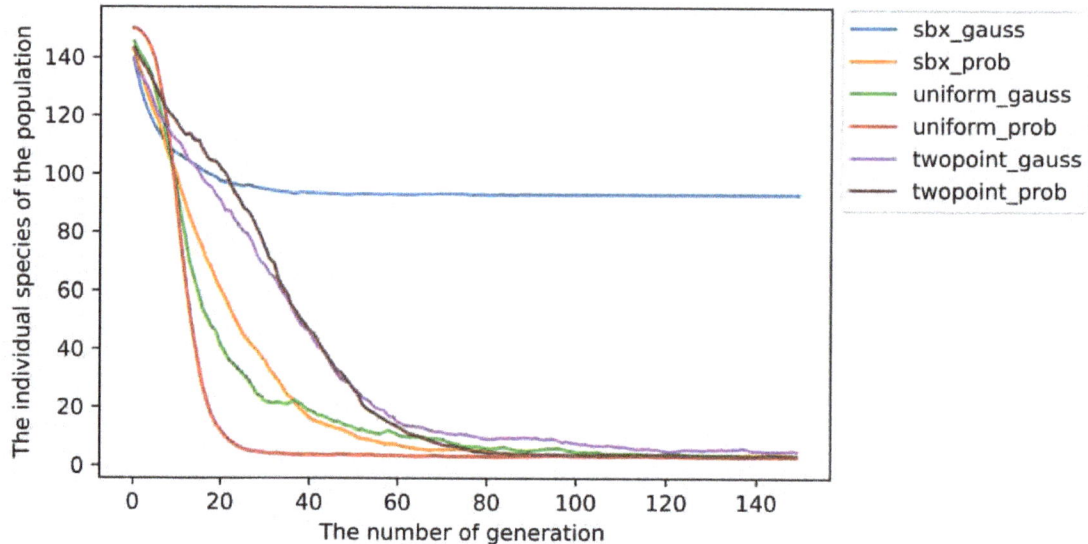

Figure 4. Effect of six gene operator combinations on the number of non-repeating individuals in the population.

4.3. Experiment 2—Evaluation of the Rating Accuracy of MPUSMs

Test the accuracy of MPUSMs with the DRBM-Social surrogate on different vector dimensions, and they represent the accuracy of multiple local models and a single overall model, respectively. In each dataset, a user with different numbers of categories is selected. The IDs of the three users are 'A9Q28YTLYRE07' (Music), 'A2582KMXLK2P06' (CDs), and 'A320TMDV6KCFU' (Kindle), and their corresponding number of categories are 187, 233, and 149, respectively. To enable a reasonable comparison, the local model of MPUSMs has the same parameters as DRBM-Social and the same parameters in different vector dimensions. As the user has been configured to perform ten interactions, there are ten historical item sets and item pools. Using historical item sets as a basis for rating items in the item pools. Figure 5 shows the situation for three users. Multi and Solo in the figure denote MPUSMs and DRBM-Social.

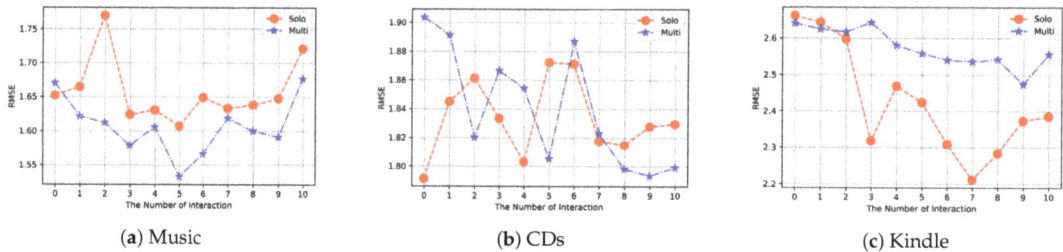

(a) Music (b) CDs (c) Kindle

Figure 5. RMSE performance of MPUSMs and DRBM-Social on three users with different vector dimensions during 10 interactions.

In Figure 5, three scenarios emerge in terms of predicting rating results: Better performance from MPUSMs (Figure 5a), Similar performance of MPUSMs and DRBM-Social (Figure 5b), and Better performance from DRBM-Social (Figure 5c). It can be seen that the accuracy of MPUSMs rating is not directly related to the vector dimension size. The better performance of MPUSMs is due to the fact that it is constructed by dividing user preferences in a manner similar to an attention mechanism. In this way, we can simulate the behavior of users who evaluate items based on only a few of their preferences. Conversely, MPUSMs do not always perform better. Due to the fact that the K-means clustering may not correctly classify user's preferences or the limitations of the local surrogate model, this may sometimes lead to poor prediction. Nevertheless, in general, MPUSMs still have an accuracy comparable to the global surrogate model.

4.4. Experiment 3—Performance of Personalized Recommendation

To evaluate the performance of the proposed algorithm for personalized recommendation results, the TOP-N lists generated from 10 interactions are comprehensively evaluated using three types of evaluation metrics. This experiment uses the same dataset as Experiment 2 and compares it with the SC-DRBMIEDA algorithm, and the two algorithms are run 30 times independently. The biggest difference between the two algorithms is that the excellent initial population of the proposed algorithm evolves further to mine preference knowledge, while the population of SC-DRBMIEDA does not have this process.

In comparison with it, the advantage of migrating inter-preference knowledge is easiest to manifest. In particular, to ensure the comparability of the two algorithms, the population size of SC-DRBMIEDA is the same as the size of the pre-recommended list, except that the model parameters are the same as in Experiment 2. That is, the source size of the TOP-N list is the same. Figures 6–8 depict the performance of the two personalized recommendation algorithms on Music, CDs, and Kindle datasets, respectively. A summary of the combined performance of each metric on the ten interaction results can be found in Table 5, where IMFEOA and IEDA stand for MPUSMs-IMFEOA and SC-DRBMIEDA, respectively. Best and Ave are the maximum and average of the ten interaction results, Growth represents the average percentage improvement of the proposed algorithm over SC-DRBMIEDA, and Total denotes the average growth rates of the combined 3 datasets. Bolded values in the table indicate higher performance. In this experiment, the Mann–Whitney U nonparametric test with a confidence level of 0.95 is used to demonstrate that the proposed algorithms are significantly different from each other, and the mark '*' indicates that the algorithm is significantly different from other algorithms.

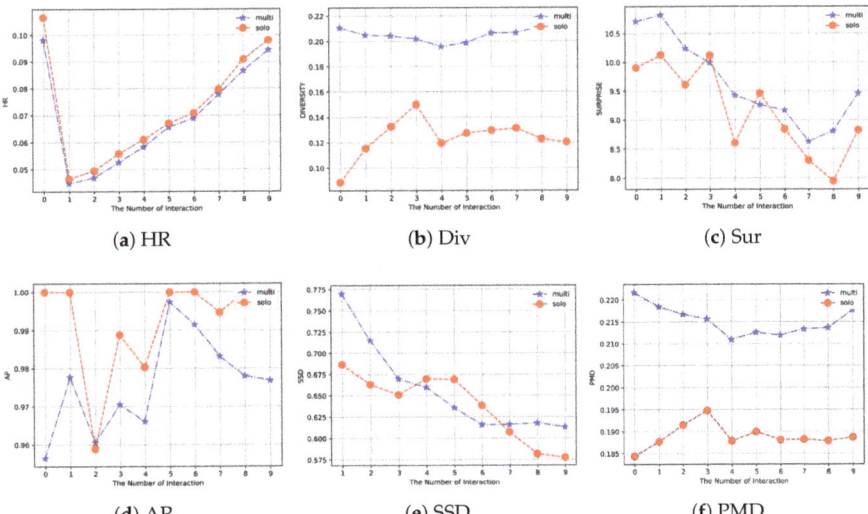

Figure 6. The recommendation performance of the two algorithms on six metrics when performing interactions in the Music dataset.

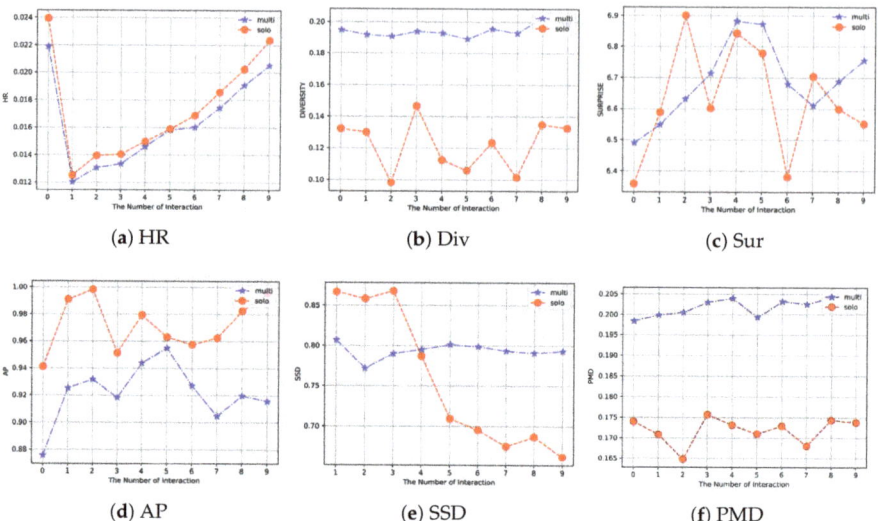

Figure 7. The recommendation performance of the two algorithms on six metrics when performing interactions in the CD dataset.

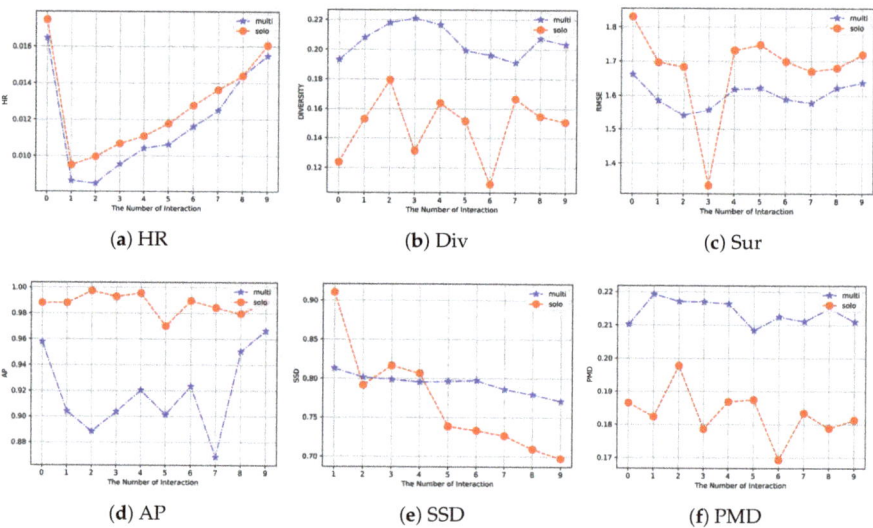

Figure 8. The recommendation performance of the two algorithms on six metrics when performing interactions in the Kindle dataset.

Combining Figures 6–8, the following conclusions could be drawn. For one, the MPUSMs-IMFEOA algorithm consistently excels in Div metric, and the curve is smoother. This indicates the effective improvement of the individual diversity and its stability of recommendation results by dividing preferences through MPUSMs and selecting appropriate genetic operators. It also suggests that the algorithm improves the search ability for each preference. Second, as shown by the performance of SSD, the temporal diversity of SC-DRBMIEDA decreases significantly with the reduction of the project pool size, while the proposed MPUSMs-IMFEOA is more likely to maintain temporal diversity and is less affected by the size of the project pool. Third, the proposed algorithm consistently performs better on the *PMD* metric. The significant improvement in *PMD* implies that the algorithm is significantly more capable of mining unknown user preferences. It indicates that knowledge transfer between preferences can effectively facilitate the discovery of entirely new items.

Table 5. Combined evaluation results of six metrics for both algorithms.

Metric \ Daraset	Music IMFEOA Best	Music IMFEOA Ave	Music IEDA Best	Music IEDA Ave	Music Growth	CDs IMFEOA Best	CDs IMFEOA Ave	CDs IEDA Best	CDs IEDA Ave	CDs Growth	Kindle IMFEOA Best	Kindle IMFEOA Ave	Kindle IEDA Best	Kindle IEDA Ave	Kindle Growth	Total
HR	0.098	0.069	0.106	0.073 *	−4.33%	0.022	0.016	0.024	0.017 *	−5.47%	0.016	0.012	0.017	0.013 *	−7.14%	−5.65%
AP	0.997	0.976	1	0.992 *	−1.66%	0.955	0.922	0.998	0.972 *	−5.19%	0.967	0.919	0.997	0.987 *	−6.98%	−4.61%
Div	0.215	0.203 *	0.15	0.124	63.98%	0.203	0.194 *	0.147	0.122	59.65%	0.221	0.205 *	0.179	0.148	38.42%	54.02%
SSD	0.77	0.591	0.687	0.575	2.93%	0.807	0.714 *	0.868	0.681	4.88%	0.813	0.714 *	0.91	0.693	3.30%	3.70%
Sur	10.938	9.686 *	10.128	9.175	5.57%	6.883	6.687	6.9	6.631	0.86%	6.659	6.512 *	6.55	6.406	1.65%	2.69%
PMD	0.222	0.215 *	0.195	0.183	13.96%	0.205	0.202 *	0.176	0.172	17.51%	0.219	0.214 *	0.198	0.183	16.68%	16.05%

1 The mark '*' indicates that the algorithm is significantly different from other algorithms. 2 Bolded values in the table indicate higher performance.

As shown in Table 5, the MPUSMs-IMFEOA has a lower accuracy according to the Ave of *HR* and *AP* metrics. Even so, the MPUSM-IMFEOA algorithm is still sufficient to recommend desired items. This is directly reflected in the *AP* values above 0.9 for all three datasets, as the *AP* values can laterally reflect the number of satisfying items of users in the recommendation list. Furthermore, the proposed algorithm performs poorly in *HR* and *AP* metrics for two reasons. One may be due to the reduced scoring accuracy of the MPUSMs. As seen in Figure 5, MPUSMs have the best accuracy in the Music dataset, and MPUSMs-IMFEOA also has the least degradation in the performance of *HR* and *AP* metrics in the Music dataset, as reflected in the Growth of the three datasets in Table 5. The other may be caused by the introduction of low-rated individuals into the pre-recommended list resulting in an insufficient number of competent individuals. In other words, the distribution of evolved population will be somewhat different from the distribution of actual excellent items. For example, Figure 5a indicates that MPUSMs perform better than DRBM-Social on *RMSE*, but proposed algorithm performance on *HR* and *AP* metrics is worse.

From the Growth rate of *SSD* in Table 5, it can be seen that the proposed algorithm shows a slight improvement in each dataset, which further demonstrates that MPUSMs' interactive dynamic update improves its ability to avoid homogeneous recommendations. Similarly, the improvement in *Sur* indicates that MPUSMs-IMFEOA is more likely to uncover items that are not similar to users' historical preferences but that they are satisfied with. It not only confirms the accuracy of MPUSMs, but also reveals the algorithm's ability to explore unknown items at the same time.

Overall, it can be seen from the Total values in Table 5 that, at the expense of 5% accuracy, MPUSMs-IMFEOA simultaneously improves the diversity and novelty of the recommended results. In addition, the reduction of *HR* and *AP* is limited and does not actually have a negative impact on the user experience. Hence, MPUSMs-IMFEOA is capable of guaranteeing the accuracy of recommendations, exploiting knowledge among preferences to improve diversity and novelty and ultimately bringing the user a more satisfying personalized recommendation experience.

4.5. Experiment 4—Search Performance of MPUSMs-IMFEOA

To demonstrate the search performance produced by the proposed IMFEOA in conjunction with the MPUSMs, the distribution of individuals in the population evolution process is observed (Figures 9–12). This experiment is validated by changing the underlying surrogate models of MPUSMs and the evolutionary algorithm. This experiment is validated by changing the underlying surrogate models of MPUSMs and the evolutionary algorithm. That is, single and multi-model guided population evolution under different underlying surrogate models exhibits performance suitable for personalized recommendations. Therefore, introduce Restricted Boltzmann Machine (RBM) [42] to build surrogates and Genetic Algorithm (GA) [30] to optimize the population. This is because a surrogate model built with an RBM is equally easy to calculate a probabilistic model such that the starting point of the initial population being evolved is the same. In addition, a genetic algorithm can use the same crossover and mutation operators to optimize populations, suitable for comparison with MFEA.

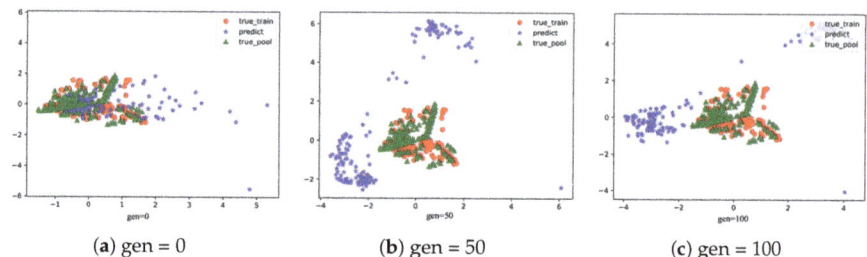

Figure 9. Population evolution of DRBM-Social-IGA based on the DRBM-Social surrogate model.

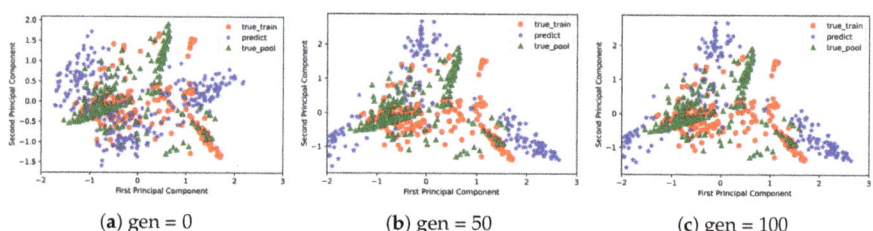

Figure 10. Population evolution of MPUSMs-IMFEOA based on the DRBM-Social surrogate model.

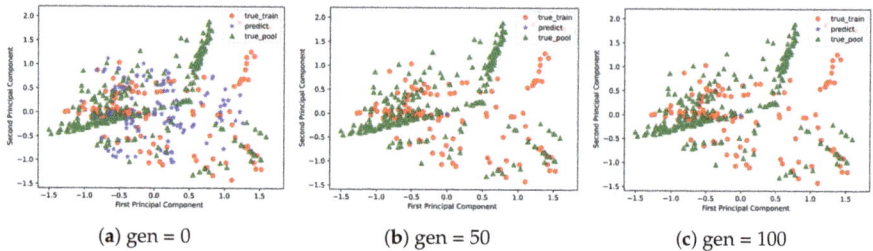

Figure 11. Population evolution of RBM-IGA based on the RBM surrogate model.

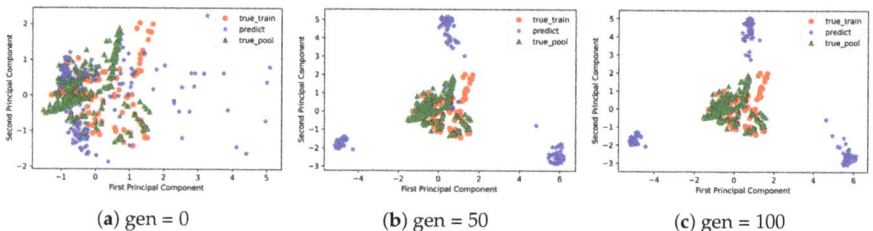

Figure 12. Population evolution of MPUSMs-IMFEOA based on the RBM surrogate model.

Specifically, there are MPUSMs-IMFEOA and DRBM-Social-IGA based on the DRBM-Social surrogate model, and MPUSMs-IMFEOA and RBM-IGA based on the RBM surrogate model. To make the IMFEOA and IGA algorithms comparable, they both use the same crossover operator and mutation operator and set the same crossover probability ($p = 1/3 + 2/3 * rmp$) and mutation probability ($1 - p$) for IGA as for IMFEOA. In addition, the visible/hidden layer units of the RBM are the same as those of the Dual-RBM, and probability models are set for them. The dataset is the same as Experiment 1. The true-train in the legend is the real historical items, and the true-pool denotes the items in the item pool, and the predict indicates the evolving individuals.

As can be seen from Figures 9 and 10, the MPUSMs-IMFEOA based on the DRBM-Social surrogate model is more likely to keep the population distribution similar to the actual item distribution, ensuring that the produced individuals are valid. From Figures 11 and 12, the evolved population in Figure 12 has more unduplicated individuals, indicating that the IMFEOA is more likely to maintain population diversity. Comparing Figures 10 and 12 shows that they differ only in the surrogate model, but both have a richly diverse population. It further indicates that IMFEOA can effectively enhance population diversity. However, MPUSMs-IMFEOA based on the RBM surrogate model is insufficient to maintain the correct true distribution of populations, suggesting that the performance of MPUSMs-IMFEOA is also highly dependent on the surrogate model.

5. Discussion and Conclusions

In this article, we introduce the knowledge transfer method into the interactive evolutionary algorithm for the first time to improve the diversity and novelty of personalized recommendations. With the aim of this, a new IEA (MPUSMs-IMFEOA) is proposed based on a modified MFEA and deep learning-driven user preference surrogate models.

Firstly, to describe preference information from different perspectives, MPUSMs and partial-MPUSMs are first developed. They are used to evaluate unknown items at different population evolutionary stages, and partial-MPUSMs are the objectives to be simultaneously optimized by MFEA for achieving knowledge transfer between preferences. Then, probability models of MPUSMs are constructed to generate initial populations with skill factors to improve the evolutionary efficiency of MFEA.

Secondly, a new IEA method based on MPUSMs and its derived models and joint multitasking learning is proposed. Accordingly, improved MFEA, recommendation lists generation method, and interaction-assisted model management method are designed for the characteristics of the models mentioned above.

Finally, comprehensive empirical studies are conducted in the Amazon dataset. Experimental results show that the proposed algorithm can increase diversity and novelty while maintaining accuracy, which demonstrates the effectiveness of building multiple surrogate models to facilitate the mining of inter-preference knowledge. Specifically, the MPUSMs ensure the accuracy of its evaluation results by conforming to the characteristic that users rely only on partial preferences to evaluate items. In addition, with a loss of only about 5% recommendation accuracy (HR and AP), individual diversity (Div) improved by 54.02%, temporal diversity (SSD) by 3.7%, surprise degree (Sur) by 2.69%, and preference mining degree (PMD) by 16.05%. Comparison with other evolutionary algorithms also shows that the combination of MPUSMs and its derived models with IMFEOA makes it easier to mine user's unknown preferences, and the search performance of IMFEOA is highly dependent on the underlying surrogate model.

MPUSMs-IMFEOA also has certain limitations. In terms of surrogate models, their model parameters cannot be automatically adjusted. As shown in Experiment 1, the accuracy of the models shows a difference in performance on different datasets with the same parameters. Although they can provide sufficient recommendation accuracy, it still has impact on the final recommendation results. In addition, the classification method of the data of the training models may not fit the user' habits. At the algorithmic level, IMFEOA may unearth items that are not actually possible to exist, and thus the corresponding items found in the item pool using the shortest distance method are equally inaccurate. Therefore, the search area of IMFEOA needs to be adjusted.

In the future, we hope to optimize three aspects under the framework of the MPUSMs-IMFEOA algorithm. The first is to apply other user surrogate models to the MPUSMs to address issues such as model accuracy, cold start, and inappropriate preference classification. Secondly, further improving the efficiency and effectiveness of knowledge transfer is a research direction. Finally, it is possible to study how to update the model more effectively during the interaction.

Author Contributions: Conceptualization, W.W., X.S. and L.B.; Data curation, G.M.; Formal analysis, W.W.; Investigation, W.W.; Methodology, W.W.; Resources, L.B.; Software, W.W.; Supervision, X.S.; Validation, G.M. and S.L.; Visualization, S.L.; Writing—original draft, W.W.; Writing—review and editing, X.S. All authors have read and agreed to the published version of the manuscript.

Funding: This research was funded by the National Natural Science Foundation of China, Grant No. 61876184.

Data Availability Statement: The data presented in this study are openly available in Amazon Review Data at https://nijianmo.github.io/amazon/index.html, accessed on 28 December 2021, reference number [41].

Conflicts of Interest: The authors declare no conflict of interest. The funders had no role in the design of the study; in the collection, analyses, or interpretation of data; in the writing of the manuscript; or in the decision to publish the results.

References

1. Yu, M.; He, W.; Zhou, X.; Cui, M.; Wu, K.; Zhou, W. Review of recommendation systems. *Comput. Appl.* **2021**, 1–16
2. Batmaz, Z.; Yurekli, A.; Bilge, A.; Kaleli, C. A review on deep learning for recommender systems: Challenges and remedies. *Artif. Intell. Rev.* **2019**, *52*, 1–37. [CrossRef]
3. Huang, W.; Wu, J.; Song, W.; Wang, Z. Cross attention fusion for knowledge graph optimized recommendation. *Appl. Intell.* **2022**, *52*, 10297–10306. [CrossRef]
4. Khanal, S.S.; Prasad, P.; Alsadoon, A.; Maag, A. A systematic review: Machine learning based recommendation systems for e-learning. *Educ. Inf. Technol.* **2020**, *25*, 2635–2664. [CrossRef]
5. Bao, L.; Sun, X.; Gong, D.; Zhang, Y. Multi-Source Heterogeneous User Generated Contents-Driven Interactive Estimation of Distribution Algorithms for Personalized Search. *IEEE Trans. Evol. Comput.* **2021**, *26*, 844–858. . [CrossRef]
6. Chen, Y.; Sun, X.; Gong, D.; Yao, X. DPM-IEDA: Dual Probabilistic Model Assisted Interactive Estimation of Distribution Algorithm for Personalized Search. *IEEE Access* **2019**, *7*, 41006–41016. [CrossRef]
7. Gabor, T.; Altmann, P. Benchmarking Surrogate-Assisted Genetic Recommender Systems. In *Proceedings of the 2019 Genetic and Evolutionary Computation Conference Companion (Geccco'19 Companion)*; Assoc Computing Machinery: New York, NY, USA, 2019; pp. 1568–1575. [CrossRef]
8. Cai, H. User Preference Adaptive Fitness of Interactive Genetic Algorithm Based Ceramic Disk Pattern Generation Method. *IEEE Access* **2020**, *8*, 95978–95986. [CrossRef]
9. Trachanatzi, D.; Rigakis, M.; Marinaki, M.; Marinakis, Y. An Interactive Preference-Guided Firefly Algorithm for Personalized Tourist Itineraries. *Expert Syst. Appl.* **2020**, *159*, 113563. [CrossRef]
10. Aghaei Pour, P.; Rodemann, T.; Hakanen, J.; Miettinen, K. Surrogate Assisted Interactive Multiobjective Optimization in Energy System Design of Buildings. *Optim. Eng.* **2022**, *23*, 303–327. [CrossRef]
11. Li, G.; Zhang, Q. Multiple Penalties and Multiple Local Surrogates for Expensive Constrained Optimization. *IEEE Trans. Evol. Comput.* **2021**, *25*, 769–778. [CrossRef]
12. Ji, X.; Zhang, Y.; Gong, D.; Sun, X. Dual-Surrogate-Assisted Cooperative Particle Swarm Optimization for Expensive Multimodal Problems. *IEEE Trans. Evol. Comput.* **2021**, *25*, 794–808. [CrossRef]
13. Bao, L.; Sun, X.; Chen, Y.; Gong, D.; Zhang, Y. Restricted Boltzmann Machine-Driven Interactive Estimation of Distribution Algorithm for Personalized Search. *Knowl.-Based Syst.* **2020**, *200*, 106030. [CrossRef]
14. Liang, Z.; Zhu, Y.; Wang, X.; Li, Z.; Zhu, Z. Evolutionary Multitasking for Multi-objective Optimization Based on Generative Strategies. *IEEE Trans. Evol. Comput.* **2022**. [CrossRef]
15. Chen, Z.; Zhou, Y.; He, X.; Zhang, J. Learning task relationships in evolutionary multitasking for multiobjective continuous optimization. *IEEE Trans. Cybern.* **2020**, *52*, 5278–5289. [CrossRef]
16. Gupta, A.; Ong, Y.S.; Feng, L. Multifactorial Evolution: Toward Evolutionary Multitasking. *IEEE Trans. Evol. Comput.* **2016**, *20*, 343–357. [CrossRef]
17. Lin, J.; Liu, H.L.; Tan, K.C.; Gu, F. An Effective Knowledge Transfer Approach for Multiobjective Multitasking Optimization. *IEEE Trans. Cybern.* **2021**, *51*, 3238–3248. [CrossRef]
18. Zhou, L.; Feng, L.; Tan, K.C.; Zhong, J.; Zhu, Z.; Liu, K.; Chen, C. Toward Adaptive Knowledge Transfer in Multifactorial Evolutionary Computation. *IEEE Trans. Cybern.* **2021**, *51*, 2563–2576. [CrossRef]
19. Xu, Q.; Wang, N.; Wang, L.; Li, W.; Sun, Q. Multi-Task Optimization and Multi-Task Evolutionary Computation in the Past Five Years: A Brief Review. *Mathematics* **2021**, *9*, 864. [CrossRef]
20. Van Dat, N.; Van Toan, P.; Thanh, T.M. Solving distribution problems in content-based recommendation system with gaussian mixture model. *Appl. Intell.* **2022**, *52*, 1602–1614. [CrossRef]

21. Khojamli, H.; Razmara, J. Survey of similarity functions on neighborhood-based collaborative filtering. *Expert Syst. Appl.* **2021**, *185*, 115482. [CrossRef]
22. Song, C.; Yu, Q.; Jose, E.; Zhuang, J.; Geng, H. A Hybrid Recommendation Approach for Viral Food Based on Online Reviews. *Foods* **2021**, *10*, 1801. [CrossRef] [PubMed]
23. Yuan, W.; Wang, H.; Hu, B.; Wang, L.; Wang, Q. Wide and deep model of multi-source information-aware recommender system. *IEEE Access* **2018**, *6*, 49385–49398. [CrossRef]
24. Wang, D.; Xu, D.; Yu, D.; Xu, G. Time-aware sequence model for next-item recommendation. *Appl. Intell.* **2021**, *51*, 906–920. [CrossRef]
25. Zeng, D.; He, M.; Zhou, Z.; Tang, C. An Interactive Genetic Algorithm with an Alternation Ranking Method and Its Application to Product Customization. *Hum.-Centric Comput. Inf. Sci.* **2021**, *11*. [CrossRef]
26. Dou, R.; Zhang, Y.; Nan, G. Application of Combined Kano Model and Interactive Genetic Algorithm for Product Customization. *J. Intell. Manuf.* **2019**, *30*, 2587–2602. [CrossRef]
27. Zhu, X.; Li, X.; Chen, Y.; Liu, J.; Zhao, X.; Wu, X. Interactive Genetic Algorithm Based on Typical Style for Clothing Customization. *J. Eng. Fibers Fabr.* **2020**, *15*, 1558925020920035. [CrossRef]
28. Takagi, H. Interactive evolutionary computation: Fusion of the capabilities of EC optimization and human evaluation. *Proc. IEEE* **2001**, *89*, 1275–1296. [CrossRef]
29. Tomczyk, M.K.; Kadziński, M. Decomposition-based interactive evolutionary algorithm for multiple objective optimization. *IEEE Trans. Evol. Comput.* **2019**, *24*, 320–334. [CrossRef]
30. Gong, D.; Yuan, J.; Sun, X. Interactive Genetic Algorithms with Individual's Fuzzy Fitness. *Comput. Hum. Behav.* **2011**, *27*, 1482–1492. [CrossRef]
31. Sun, X.; Lu, Y.; Gong, D.; Zhang, K. Interactive genetic algorithm with CP-nets preference surrogate and application in personalized search. *Control Decis.* **2015**, *30*, 1153–1161.
32. Funaki, R.; Sugimoto, K.; Murata, J. Estimation of Influence of Each Variable on User's Evaluation in Interactive Evolutionary Computation. In Proceedings of the 2018 9th International Conference on Awareness Science and Technology (iCAST), Fukuoka, Japan, 19–21 September 2018; pp. 167–174. [CrossRef]
33. Alizadeh, V.; Kessentini, M.; Mkaouer, M.W.; Ocinneide, M.; Ouni, A.; Cai, Y. An Interactive and Dynamic Search-Based Approach to Software Refactoring Recommendations. *IEEE Trans. Softw. Eng.* **2020**, *46*, 932–961. [CrossRef]
34. Chen, Y.; Sun, X.; Gong, D.; Zhang, Y.; Choi, J.; Klasky, S. Personalized Search Inspired Fast Interactive Estimation of Distribution Algorithm and Its Application. *IEEE Trans. Evol. Comput.* **2017**, *21*, 588–600. [CrossRef]
35. Deb, K.; Pratap, A.; Agarwal, S.; Meyarivan, T. A Fast and Elitist Multiobjective Genetic Algorithm: NSGA-II. *IEEE Trans. Evol. Comput.* **2002**, *6*, 182–197. [CrossRef]
36. Gupta, A.; Ong, Y.S. Back to the Roots: Multi-X Evolutionary Computation. *Cogn. Comput.* **2019**, *11*, 1–17. [CrossRef]
37. Bali, K.K.; Ong, Y.S.; Gupta, A.; Tan, P.S. Multifactorial Evolutionary Algorithm With Online Transfer Parameter Estimation: MFEA-II. *IEEE Trans. Evol. Comput.* **2020**, *24*, 69–83. [CrossRef]
38. Gupta, A.; Zhou, L.; Ong, Y.S.; Chen, Z.; Hou, Y. Half a Dozen Real-World Applications of Evolutionary Multitasking, and More. *IEEE Comput. Intell. Mag.* **2022**, *17*, 49–66. [CrossRef]
39. Hinton, G.E. Training products of experts by minimizing contrastive divergence. *Neural Comput.* **2002**, *14*, 1771–1800. [CrossRef]
40. Jiang, S.; Song, J. Evaluation Metrics for Personalized Recommendation Systems. *J. Phys. Conf. Ser.* **2021**, *1920*, 012109. [CrossRef]
41. Ni, J.; Li, J.; McAuley, J. Justifying recommendations using distantly-labeled reviews and fine-grained aspects. In *Proceedings of the 2019 Conference on Empirical Methods in Natural Language Processing and the 9th International Joint Conference on Natural Language Processing (EMNLP-IJCNLP)*; Association for Computational Linguistics: Hong Kong, China, 2019; pp. 188–197.
42. Bao, L.; Sun, X.; Chen, Y.; Man, G.; Shao, H. Restricted Boltzmann Machine-Assisted Estimation of Distribution Algorithm for Complex Problems. *Complexity* **2018**, *2018*, 2609014. [CrossRef]

Disclaimer/Publisher's Note: The statements, opinions and data contained in all publications are solely those of the individual author(s) and contributor(s) and not of MDPI and/or the editor(s). MDPI and/or the editor(s) disclaim responsibility for any injury to people or property resulting from any ideas, methods, instructions or products referred to in the content.

Article

Genetic Algorithms Optimized Adaptive Wireless Network Deployment

Rahul Dubey * and Sushil J. Louis

Department of Computer Science and Engineering, University of Nevada Reno, NV 89512, USA; sushil@unr.edu
* Correspondence: rdubey018@nevada.unr.edu

Abstract: Advancements in UAVs have enabled them to act as flying access points that can be positioned to create an interconnected wireless network in complex environments. The primary aim of such networks is to provide bandwidth coverage to users on the ground in case of an emergency or natural disaster when existing network infrastructure is unavailable. However, optimal UAV placement for creating an ad hoc wireless network is an NP-hard and challenging problem because of the UAV's communication range, unknown users' distribution, and differing user bandwidth requirements. Many techniques have been presented in the literature for wireless mesh network deployment, but they lack either generalizability (with different users' distributions) or real-time adaptability as per users' requirements. This paper addresses the UAV placement and control problem, where a set of genetic-algorithm-optimized potential fields guide UAVs for creating long-lived ad hoc wireless networks that find all users in a given area of interest (AOI) and serve their bandwidth requirements. The performance of networks deployed using the proposed algorithm was compared with the current state of the art on several experimental simulation scenarios with different levels of communication among UAVs, and the results show that, on average, the proposed algorithm outperforms the state of the art by 5.62% to 121.73%.

Keywords: genetic algorithms; optimization; UAVs; wireless networks; potential fields

Citation: Dubey, R.; Louis, S.J. Genetic Algorithms Optimized Adaptive Wireless Network Deployment. *Appl. Sci.* **2023**, *13*, 4858. https://doi.org/10.3390/app13084858

Academic Editor: Vincent A. Cicirello

Received: 15 March 2023
Revised: 5 April 2023
Accepted: 7 April 2023
Published: 12 April 2023

Copyright: © 2023 by the authors. Licensee MDPI, Basel, Switzerland. This article is an open access article distributed under the terms and conditions of the Creative Commons Attribution (CC BY) license (https://creativecommons.org/licenses/by/4.0/).

1. Introduction

UAVs are gaining traction as a useful tool in many different domains, including search and rescue [1,2], surveillance [3], cargo transport [4], surveying and mapping [5,6], disaster relief [7], and Internet of Things (IoTs) [8,9]. This paper specifically focuses on the search and rescue application domain, where the challenge is to establish an on-demand UAV-based wireless network in remote areas or when existing communication infrastructure fails to operate due to natural calamity or stress on the existing network.

In network deployment, a set of UAVs is deployed within a given AOI to provide wireless connectivity to ground users for an extended period. For known user distributions, centralized network deployment algorithms are used to generate wireless mesh networks. However, deploying networks to serve users with unknown distributions is challenging. While many techniques for network deployment in unknown environments have been proposed in the literature, generalizability and real-time adaptation to uniformly and non-uniformly distributed users remain challenges. This paper aims to present an approach that is generalizable to different types of user distributions and adaptable to users' requirements.

The deployment of a UAVs-based ad hoc network in unknown environments is usually divided into two sub-challenges. The first challenge is to find the users' distribution. This is a relatively easier task and several geometrical-based algorithms [10,11] have been proposed in the literature to deploy static networks. The second challenge is to adaptively re-deploy UAVs in real-time to connect users with the command center [12], which is an NP-hard problem. Compared to static networks, less work exists on the part of the re-configuration of networks that adapts to users' requirements in real-time. The reconfiguration of networks is difficult due to the limited communication range between UAVs, limited battery life of UAVs, and various environmental conditions. This problem needs more attention and has many real-world applications.

This paper presents a two-phase approach to network deployment, where potential fields (PFs) govern the movement of UAVs. In the first phase, a set of potential field parameters is derived for static network deployment. For the second phase, a genetic optimization approach is presented, genetic adaptive network deployment (GANet), that optimally deploys UAVs to create an ad hoc wireless network. The problem has been formulated as an optimization problem with the aim of maximizing the sum of bandwidth coverage provided to users and the longevity of the deployed network. To provide more bandwidth coverage, UAVs move under the influence of a set of potential fields toward users in the AOI, and to increase the longevity of the deployed network, UAVs position themselves to share the bandwidth demanded by users while maintaining connectivity to the command center.

In the simulation, each UAV's movement is controlled by a set of potential fields, where each potential field has two parameters [13,14], and these parameters can be tuned to achieve the desired behavior from UAVs. However, searching for parameters that optimally control the movement of UAVs to create optimal wireless networks is difficult for two reasons: (1) due to the non-linearity of potential fields and (2) because of the large search space. In the literature, it has been shown that genetic algorithms (GAs) work well with non-linear problems and problems with a large search space. Thus, a genetic algorithm has been used to evolve solutions. The potential field parameters are encoded in a real-valued chromosome and the GA searches through the space of potential field parameters. Note that other search algorithms can also be applied for searching through this space, and, in the results section, the GAs performance is compared with two other search algorithms. The network deployment problem is divided into two phases to deal with two different problems: (1) the search phase, for searching users on the ground by deploying a static network, and (2) the service phase, to reconfigure a deployed network adaptively to better serve users.

The proposed two-phase approach deploys networks in real-time and adapts to users' requirements. However, this does not guarantee generalizability and robustness. To obtain generalizable solutions (a set of potential field parameters) that work for unknown user distributions, four different training scenarios with varying user distributions (uniform and non-uniform) were generated. The aim is to tune the potential field parameters to be distribution-agnostic and enable UAVs to better serve users. To measure the generalizability and robustness of the proposed approach, the performance of the best evolved solution was measured on 100 different test scenarios with varying user distributions. These test scenarios were designed to simulate real-world scenarios and provide a comprehensive evaluation of the approach's performance.

The experimental results show that the networks deployed using the best solution evolved using GANet on four training scenarios outperformed the networks deployed using the current best network deployment algorithm, adaptive triangular network deployment (ATRI) [15], on training scenarios. More importantly, it was found that this performance advantage also holds over the never before seen 100 test scenarios. Furthermore, different levels of information exchange between UAVs have been considered to measure the differences in the performance of deployed networks. Specifically, we compared the performance of networks deployed considering one-hop (neighbor) UAV-to-UAV communication with two-hop (neighbor's neighbors) communication. The results show that the performance of two-hop GANet is statistically significantly better than one-hop GANet and ATRI.

This work makes two significant contributions. First, we propose a unified approach that can be used for both static and dynamic network deployment. Second, our proposed approach is adaptive, generalizable, and demonstrates a superior performance compared to the current state-of-the-art network deployment algorithms when deploying dynamic networks. The remainder of this paper is organized as follows. Section 2 describes prior work in mesh networks and potential fields-based UAVs movement. Section 3 sets up the problem and Section 4 describes the UAV's movement model under the influence of potential fields. Section 5 explains the two-phase proposed algorithm. The experimental setup and results constitute Section 6 and the last section provides conclusions.

2. Related Work

Many challenges arise when deploying a wireless mesh network in unknown environments. Optimal UAV placement, fast deployment, maintaining a mesh network connected to a command center, minimizing the number of deployed UAVs, maximizing the lifetime of deployed networks, routing [16], and channel allocation all present significant challenges [12,17]. In the literature, several techniques have been introduced for positioning UAVs based on Delaunay triangulation (DT) [15], the circle packing theorem (CPT) [10], and Voronoi diagrams (VDs) [11] to create static networks. A network deployed using DT has no coverage gap while having minimum overlap between UAV's sensing areas. CPT computes the locations of UAVs while ensuring no overlap and a minimum coverage gap. A VD divides the entire AOI into different segments and places a UAV in each segment to provide bandwidth coverage to users. Such static network deployment is primarily used for area coverage. Lam [18] deployed a heterogeneous sensor network using circle packing by filling the given AOI with circles of different radii corresponding to different UAV types. Lyu [19] proposed a placement optimization technique that first deploys a network using CPT and then adjusts the altitude of deployed UAVs to cover the entire AOI with fewer UAVs while providing wireless coverage to ground terminals. These are computational geometry model-based techniques used to compute the position of UAVs and they do not adapt to users' locations or bandwidth requirements. Our proposed GANet, on the other hand, adapts to both while maximizing network longevity.

Researchers have presented modifications to make static networks more adaptive to users' requirements. Ming [15] introduced the state-of-the-art adaptive triangular network deployment algorithm, ATRI, by taking inspiration from Delaunay triangulation. Using one-hop neighbor's state information, ATRI generates a mesh network by adjusting the distance between neighboring UAVs in order to better share bandwidth within a given AOI. Unlike our approach, ATRI, however, does not work well when users are distributed non-uniformly or clustered in groups. In [20], the authors initially deployed UAVs using CPT and presented a mathematical model to adjust the altitude of deployed UAVs for near-optimal coverage. Increasing the altitude of UAVs increases the coverage range but decreases the signal strength, resulting in a poorer coverage quality. In this work, UAVs were deployed at a fixed altitude and only moved in two-dimensional space. Bartolini [21] proposed a Voronoi polygon-based adaptive network deployment algorithm to deploy heterogeneous mobile sensors. The algorithm computes different polygons for different sensors and moves a sensor only when the sensor does not detect any users on the ground. Amar [22] presented a dynamic algorithm to serve a sub-region within the AOI that requires more bandwidth. Both of these papers assume a uniform users' distribution whereas this paper deals with non-uniform users' distribution as well. Section 5 of this paper describes how to use potential fields to deploy networks in the first phase to mimic DT and then re-deploy UAVs using a set of potential fields optimized by a genetic algorithm to maximize the bandwidth coverage and longevity during the second phase.

The potential-field-based real-time control of mobile agents was first introduced by Khatib [23]. Owing to their simplicity, many researchers used potential fields to control UAVs and other autonomous agents in different domain-specific tasks [14,24,25]. Howard [26] used potential fields to deploy mobile sensor networks to cover an area. In their approach, an agent experienced repelling potential fields based on the distance from other agents and obstacles and moved toward unexplored areas. Poduri [27] introduced a mobile network deployment algorithm with the constraint that each agent has at least K neighbors, where K is a user-defined number. Poduri's paper provides evidence that potential fields can be used to generate a static mesh network similar to that generated by Delaunay triangulation or the circle packing theorem. Zhao [28] presented a centralized algorithm and a potential-field-based distributed algorithm for UAV deployment while maintaining connectivity among UAVs. However, the authors considered only two different types of user distributions: uniformly random and in three clusters spread around the AOI with a command center in the middle. All of these approaches work well in a relatively uniform distribution of users but not as well when users are distributed non-uniformly.

Machine-learning-based approaches have been used for optimal UAV deployment [29]. Genetic algorithms have been particularly popular for deploying static networks in the AOI. For example, Reina [30] presented a multi-layout multi-subpopulation genetic algorithm that optimizes coverage, fault-tolerance, and redundancy. Dina [31] used a variable-length genetic algorithm to optimize the area coverage and deployment cost using non-homogeneous sensors. Subash [32] proposed a GA-based deployment algorithm that maximizes longevity by activating and deactivating sensor nodes. Ruetten [33] developed a GA-based area optimizer that maximizes the covered area. Aziz [34] used a hybrid of particle swarm optimization (PSO) and Voronoi diagrams to find the optimal deployment of nodes/UAVs for best coverage, and Li [35] used the artificial bee colony (ABC) algorithm to maximize throughput. Abdulrab [36] used the Harris hawk's optimization (HHO) algorithm to find the optimal sensor placement for maximum coverage and connectivity. These approaches typically encode locations in chromosomes and are not generalizable. Table 1 compares different network deployment methods based on their control mechanism, applicability toward uniform and non-uniform user distributions, and generalizability, providing readers with a clear overview of the current state of research in this field.

Unlike previous works that solely focus on UAV position optimization, this paper introduces a genetic algorithm to evolve potential field parameters that guide real-time collision-free UAV movement toward optimal positions. By fine-tuning the potential parameters, the proposed algorithm can optimally place UAVs regardless of the distribution of users and other UAVs in the area of interest. The resulting UAV network is deployed in real-time and is highly generalizable. This approach not only provides UAV positions but also ensures safe and efficient movement to these positions.

Table 1. Comparison of wireless network deployment methods.

Papers	Methods	Network Static (S) Dynamic (D)	Control Centralized (CC) Distributed (DC)	User Distribution Uniform (U) Non-Uniform (NU)	Generalizable
[10]	CPT	S	DC	U, NU	No
[11]	VD	S	DC	U, NU	No
[15]	ATRI	D	DC	U, NU	No
[18]	CPT	S	CC	U, NU	No
[19]	CPT	D	CC	U	No
[20]	CPT	D	CC	U	No
[21]	VD	D	DC	U	No
[22]	DT	D	DC	U	No
[26]	PF	S	DC	U, NU	Yes
[27]	PF	S	DC	U, NU	Yes
[28]	PF	D	DC	U, NU	No
[30]	GA	S	CC	U, NU	No
[31]	GA	S	CC	U, NU	No
[32]	GA	S	CC	U, NU	No
[33]	GA	S	CC	U	No
[34]	PSO	S	CC	U, NU	No
[35]	ABC	S	CC	U, NU	No
[36]	HHO	S	CC	U	No
proposed	GA and PF	D	DC	U, NU	**Yes**

The next section presents the problem formulation, fitness computation, and an elitist genetic algorithm.

3. Problem Formulation

3.1. System Model

The network deployment problem was formulated as an optimization problem to maximize the performance of the network in terms of bandwidth coverage and longevity. Assume that a set of N identical UAVs $U = \{u_1, u_2, \ldots, u_n\}$ need to be deployed in a given AOI of 2000×2000 m^2 with a command center located in the middle at $(1000, 1000)$ [37]. All UAVs are initially located at the center within a 10×10 m^2 area and fly at a constant altitude of 100 m. Each UAV is equipped with sensors that can detect users on the ground within a ground-sensing range (U_{gr}) of 100 m and can communicate with its neighbors within an air-to-air communication range (U_{ar}) of 300 m [22]. A total of m users, each with a position and a bandwidth requirement (p_i, b_i), are distributed over the AOI. Therefore, the users can be represented by a list of pairs: $\{(p_1, b_1), (p_2, b_2), \ldots, (p_m, b_m)\}$. These simulation parameters can be easily changed. In the results section, we demonstrate the performance of our proposed algorithm with various AOIs, numbers of UAVs and users, and different user distributions. Additionally, other parameters, such as the range of UAV–UAV communication, range of UAV–user communication, and initial energy, can also be adjusted as needed. This paper treats the network deployment problem as a search problem and thus channel modeling for communication among UAVs and between UAVs and users is a subject for future work.

3.2. Fitness Computation

Assume that each UAV has a limited initial energy of 10^6 joules and consumes energy while hovering/moving and providing data to users [22]. Intuitively, if more UAVs share the bandwidth demanded by users, the per UAV bandwidth service can be reduced, leading to less energy consumption and longer UAV flight times. Thus, the aim is to deploy UAVs to increase the bandwidth coverage while sharing bandwidth demands. Each UAV is classified as either an active UAV (AU) or an inactive UAV (IU) based on whether or not the UAV is serving a user. If a UAV finds users within its ground-sensing range, U_{gr}, the UAV is an AU; otherwise, it is an IU. The number of AUs is proportional to the network lifetime. The more AUs, the more bandwidth that can be distributed among AUs, leading to less power consumption per AU and thus a longer battery life to stay aloft and keep the network alive. A variable $\alpha_i \in \{0, 1\}$ represents the status of UAVs, where α_i is 1 for active UAVs and 0 for inactive UAVs.

$$f_{bw} = \sum_{i=1}^{N} u_{bi} \tag{1}$$

$$f_\alpha = \sum_{i=1}^{N} \alpha_i \tag{2}$$

$$\text{Maximize} \quad f = \left[\frac{f_{bw}}{\sum_{j=1}^{m} b_j} + \frac{f_\alpha}{N} \right] \tag{3}$$

Equation (1) calculates the aggregate bandwidth that is provided to users, where i sums over N UAVs and u_{bi} denotes the bandwidth served by the ith UAV. The objective function value f_{bw} increases as the bandwidth coverage expands. In the same vein, Equation (2) tallies the total number of active UAVs, f_α, in the deployed network. As the number of f_α increases, the per AU data rate decreases and the energy consumption decreases as well. To ensure that the bandwidth coverage lasts for an extended period, f_{bw} and f_α are combined in Equation (3) to form the fitness function, where b_j is the bandwidth requirement of the jth user and m is the number of users.

It is worth mentioning that the first and second terms of the fitness function were normalized due to the uncertainty surrounding the maximum bandwidth demand from users and the number of available UAVs for deployment. By normalizing the components of the fitness function, the maximum and minimum values are always confined within a fixed range. The maximum value of each term in the Equation (3) is 1, which limits the fitness value to a range of [0–2]. When computing the objective function using Equation (3), several constraints must be satisfied. Each UAV must be positioned within the specified AOI; thus, the UAV's x and y coordinates must be $\in (0, 2000)$. Our simulation used the xy plane for the horizontal plane, $u_{bi} \leq u_{bmax}$, where $u_{bmax} = 5$, which is the maximum bandwidth a UAV can provide in Mbps [22].

3.3. Genetic Algorithms

A $(\mu + \lambda)$ elitist genetic algorithm shown in Algorithm 1 searches through the space of potential field parameter values, which is encoded in the real-value chromosome. The $(\mu + \lambda)$ elitist genetic algorithm is a variant of the genetic algorithm that combines the best individuals from the parent population and offspring population to form a new population. It uses elitism to ensure that the best individuals are preserved from one generation to the next [38]. In Algorithm 1 P_x, P_m are probabilities of crossover and mutation, g is the number of generations, h is the number of hops for communication between UAVs, μ is the population size, and λ is the number of children generated through recombination. A number of prototyping experiments were conducted to explore different μ and g values, types of selection strategies, crossover types, and mutation algorithms and associated probabilities. Based on these preliminary experiments, simulated binary crossover, polynomial mutation, and binary tournament selection were chosen.

Algorithm 1 An elitist genetic algorithm

Input : $P_x, P_m, g, \mu, \lambda, h$
$P_0 \leftarrow Initialize(\mu)$
$Evaluate(P_0, H)$
$P_c = []$
for gen in g **do**
    for i in λ **do**
        $p_1, p_2 \leftarrow Tournament\ Selection(P_t)$
        $c_1, c_2 \leftarrow SBX\ Crossover(p_1, p_2, P_x)$
        $c_1, c_2 \leftarrow Polynomial\ Mutation(c_1, c_2, P_m)$
        $P_c.add(c_1, c_2)$
        $i = i + 2$
    end
    $Evaluate(P_c, h)$
    $P_{t+1} \leftarrow BestHalf(P_t, P_c)$
    $P_c = []$
end

The aim is to show that the genetic algorithm evolved potential field parameters that work across a wide range of user distributions in the AOI, and the results (Section 6) show that using the average performance over the four training scenarios translates robustly to testing scenarios. The next section explains the UAV's movement under the influence of potential fields. Subsequently, Section 5 explains how to derive these potential field parameters to deploy a static network in the search phase and how to use GA-tuned parameters to reconfigure the deployed network.

The aim is to evolve high-quality and *robust* solutions for wireless network deployment. However, earlier work has shown that solutions evolved on one scenario (user distribution) may not be robust [13,39]; that is, potential field parameters evolved on one scenario may over-fit and thus not work well in other scenarios. Therefore, four *training* scenarios with different user distributions were created to evolve robust solutions, as shown in Figure 1. The genetic algorithm seeks to maximize the average objective function value, computed using Equation (3), over all four scenarios. The robustness of the best-evolved solution was tested on 100 different unseen *test* scenarios, and three of those are shown in Figure 2.

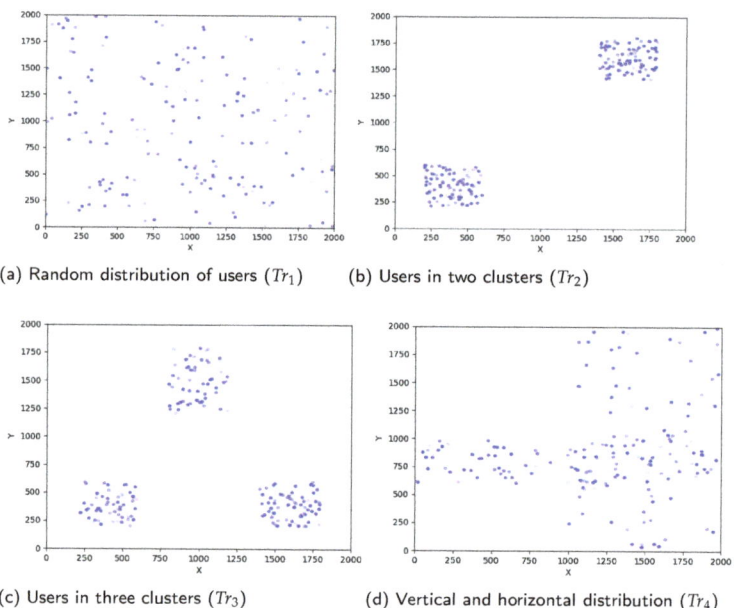

Figure 1. Four training scenarios, where blue dots show the location of users. Lighter color represents low-bandwidth-demanding users and darker color shows higher bandwidth demand.

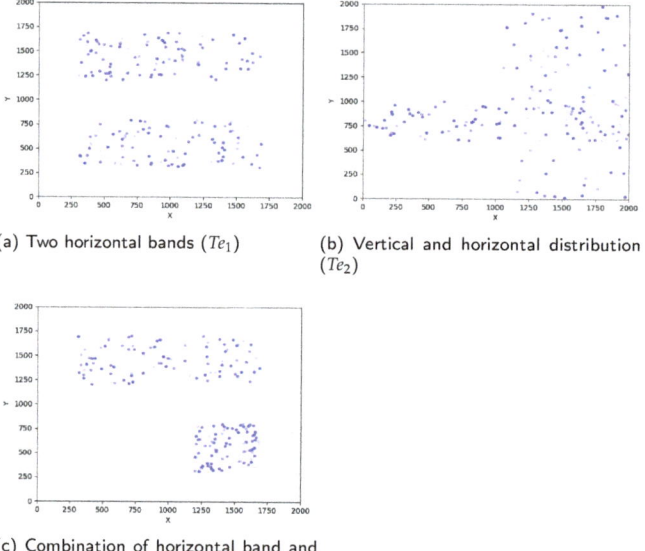

Figure 2. The figures show three testing scenarios (T_{e1}, T_{e2}, and T_{e3}). The robustness of the best-evolved solution will be tested in these scenarios.

4. Potential-Field-Based UAV Movement Modeling

Potential fields have been used to guide autonomous agents [14,25] in complex environments and in real-time. This section briefly explains how potential fields guide UAVs. Assume that, initially, N UAVs are placed near the center of the AOI and are hovering at 100 m altitude with zero horizontal speed (sp). The maximum speed of a UAV is fixed to 15 m/s, with an acceleration of 0.1 m/s^2. Assume that a UAV (kth) can communicate with n neighbors directly and experience an attractive (P_a) and repulsive (P_r) potential field based on the distance from each of its neighbors. The vector sum of these potential fields (P_k) is given by Equation (4), which provides a heading or direction for the kth UAV ($\in (0, 2\pi)$) to move along.

$$\begin{aligned} \vec{P_k} &= \vec{P_a} + \vec{P_r} \\ \vec{P_k} &= \sum_{i=0, i \neq k}^{n} \vec{P_{aki}} + \sum_{i=0, i \neq k}^{n} \vec{P_{rki}} \end{aligned} \quad (4)$$

In the literature, potential fields of the form $c * d^e$ have been used to control the movement of autonomous agents [40], where coefficient (c) and exponent (e) are optimizable parameters that determine the field effect. In Equation (4), both P_a and P_r can be substituted by the form of cd^e as shown by Equation (5).

$$\vec{P_k} = \sum_{i=0, i \neq k}^{n} \vec{a_{ki}} c_a d_{ki}^{e_a} + \sum_{i=0, i \neq k}^{n} \vec{a_{ik}} c_r d_{ki}^{e_r} \quad (5)$$

In the above equation, $\vec{a_{ki}}$ is a unit vector pointing from the kth UAV to the ith neighbor and vice-versa for $\vec{a_{ik}}$, $d_{ki} = d_{ik}$ is the distance between the kth and ith UAVs, c_a, c_r are potential field coefficients, and e_a, e_r are potential field exponents. The genetic algorithm optimizes these coefficients and exponents to control each UAV's movement, and $\vec{P_k}$'s direction specifies the kth UAV's heading to move along. In our model, UAVs change their current heading to this new heading (computed using Equation (5)) instantaneously with velocity ($\vec{v}$) and position ($\vec{p}$) computed using Equation (6).

$$\begin{aligned} \vec{v} &= (sp \times cos(heading), 100, sp \times sin(heading)) \\ \vec{p} &= \vec{p} + \vec{v} \times \Delta t \end{aligned} \quad (6)$$

Since UAVs move under the influence of potential fields, their movement comes to a halt when the resultant field value (P_k) reaches zero or falls below a threshold value. So far, the background of the problem, the fitness criteria, training, and testing scenarios, and the UAV's movement model have been discussed. The next section presents the proposed network deployment algorithm in two phases to deploy on-demand wireless networks.

5. Proposed Network Deployment

The proposed network deployment works in two phases. The first phase deals with static network deployment to find the user distribution in an unknown environment. In the second phase, we present the genetic adaptive network deployment algorithm, GANet (shown in Algorithm 2), which adaptively reconfigures the deployed static network (in the first phase) as per user demand.

Algorithm 2 Genetic Adaptive Network Deployment (GANet)

Input : $UAVs, Users, CS, MS, MT, h$
Output: fit/MS
fit = 0
 for *scenario in MS* **do**
  t, bwc, AUs = 0
  while $t \leq MT$ **do**
   $AssociateUsers(UAVs, Users)$
   $FindNeighbors(UAVs)$
   $ActivePotentials(h)$
   $Dir = ComputeDir(CS)$
   $MoveAll(Dir)$
   t=t+1
  end
  for *i in UAV* **do**
   bwc += u_{bi}
   AUs += α_i
  end
  fit += (bwc/MaxBW) + (AUs/UAVs)
 end

5.1. First Phase: Search

In this phase, UAVs move to cover the entire AOI, also called static network deployment. Potential fields are used to deploy UAVs for maximal area coverage with minimal overlap. This is performed to find the distribution of users within the area.

5.1.1. Derivation of Parameters

To generate a static network using potential fields, the sum of potential fields must be zero; that is, $\vec{P_a} + \vec{P_r} = 0$ (from Equation (4)) for each UAV. We thus need to find values of c_a, c_r, e_a, e_r such that, when $d_{ki} = \sqrt{3}\, U_{gr}$, $\|\vec{P_a}\| = \|\vec{P_r}\|$ as shown by Equation (7). It has been shown that, for $d_{ki} = \sqrt{3}\, U_{gr}$, the coverage is optimal [15]. Equation (7) is derived from Equation (4) by substituting $P_k = 0$.

$$\begin{aligned}
\vec{a}_{ki} c_a d_{ik}^{e_a} + \vec{a}_{ik} c_r d_{ik}^{e_r} &= 0 \\
\vec{a}_{ki} c_a d_{ik}^{e_a} &= \vec{a}_{ik} c_r d_{ik}^{e_r} \\
c_a (\sqrt{3} U_{gr})^1 &= c_r (\sqrt{3} U_{gr})^{-2} \\
c_r &= c_a (\sqrt{3} U_{gr})^3
\end{aligned} \quad (7)$$

In the equation, the directions of unit vectors $\vec{a}_{ki}$ and $\vec{a}_{ik}$ are opposite to each other ($\vec{a}_{ik} = -\vec{a}_{ki}$). Earlier work in potential-fields-based movement control has shown that the attractive potential should be proportional to the distance and the repulsive potential field should be inversely proportional to the distance squared to provide collision-free cohesive movement [41,42]; thus, e_a and e_r are set to 1 and -2, respectively.

In the simulation experiment, $U_{gr} = 100$ meters, and Equation (7) shows the relationship between c_r and c_a. Assuming that $c_a = 1$, we obtain $c_r = 3 \times \sqrt{3} \times 100^3$.

5.1.2. Deployment of Static Networks of Different Sizes

These derived parameters can be used to guide UAVs to create a static network of different sizes. For example, Figure 3a shows 10 randomly distributed UAVs, where each UAV starts at the locations indicated by the blue dots. The UAVs then move under the influence of attractive and repulsive potential fields from neighbors using the above parameters, and Figure 3b shows the resulting static network when the UAVs stop moving. This positioning is identical to Delauney triangulation [15], and this shows that potential-field-based static network deployment can subsume Delaney-triangulation-based network deployment. Furthermore, potential field parameters can also be derived to deploy networks similar to CPT [10]. Note that there exist multiple sets of values of the coefficients and exponents that will result in similar static network deployments.

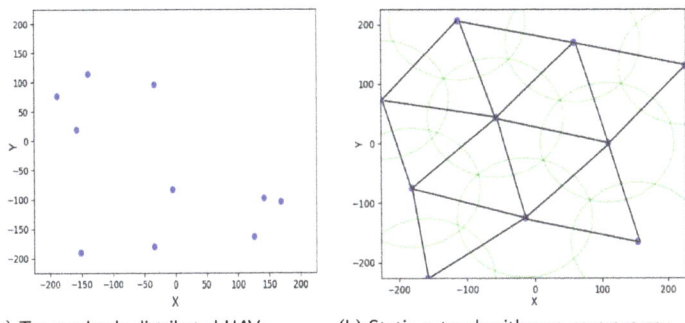

(a) Ten randomly distributed UAVs (b) Static network with zero coverage gap

Figure 3. UAVs moving under the influence of potential fields with parameters (c_a, e_a, c_r, e_r) as $(1, 1, 3 \times \sqrt{3} \times 100^3, -2)$ to create a network with zero coverage gap.

The same parameters can also be used to create static networks with different numbers of UAVs. For example, 156 UAVs are deployed in an AOI of 2000×2000 m^2, where UAVs start within an area of 10×10 m^2 near the center of the AOI. The search phase runs for 1500 time steps and Figure 4 shows the UAVs' positioning (green circles) at the end, where users (blue dots) are distributed in three clusters. Green circles show UAV coverage areas with the UAV in the center of each circle, and the red circle represents the command center. A black circle shows a UAV serving no users within U_{gr}, thus acting as an inactive UAV. Similarly, the magenta circle shows a UAV serving multiple users and is an example of an active UAV. Increasing the number of UAVs that serve the same users leads to bandwidth sharing that decreases the active UAV energy consumption and thus increases the network longevity. The second phase of the network deployment seeks to increase the number of active UAVs by moving UAVs from areas with lower bandwidth requirements (fewer users) to areas with higher bandwidth requirements (more users) while maintaining network connectivity with the command center.

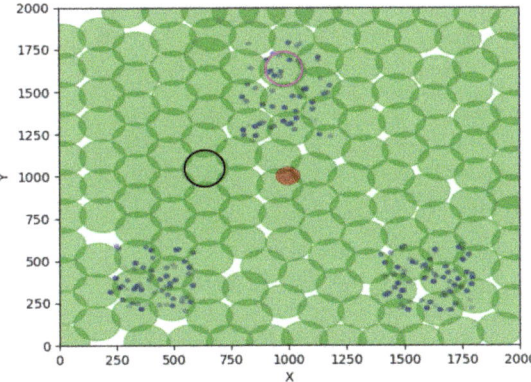

Figure 4. Deployment of UAVs at the end of the first phase on the third training scenario. Blue dots represent users, green circles represent UAVs' coverage areas, and the red circle represents the location of the command center. A black circle shows an inactive UAV and the magenta circle shows an inactive UAV.

5.2. Second Phase: Service

In this phase, a different set of potential fields optimized by a genetic algorithm redeploys UAVs to better serve users found in the first phase. The movement of UAVs in the service phase is specified by Algorithm 2, GANet. The algorithm first re-configures the static networks by controlling the movement of UAVs using a set of potential field parameters encoded in a candidate solution and then computes the fitness of the candidate solution.

5.2.1. Re-Configuration of Static Network

In Algorithm 2, MS is the number of training scenarios (4), MT is the maximum simulation time steps (1500), CS is a candidate solution provided by the GA that encodes potential field coefficients and exponents, and h is the number of communication hops allowed between UAVs. At each simulation step (t), every UAV scans for users (AssociateUsers) within U_{gr} to find their positions and bandwidth requirements. The time complexity of this operation is $\mathcal{O}(m)$. Similarly, FindNeighbors finds other UAVs within U_{ar} distance and records them as neighbors with the time complexity of $\mathcal{O}(n)$. ActivePotentials finds a set of potential fields depending on the hop count (h), and ComputeDir computes the resultant potential field for each UAV by summing all potential fields acting on a UAV using Equation (8).

$$\vec{P}_k = \vec{P}_b + \sum_{i=1}^{h}(\vec{P}_{u_i} + \delta_r \vec{P}_{r_i}) + \delta_c \vec{P}_c \qquad (8)$$

Next, MoveAll(Dir) moves the kth UAV in the direction of $\vec{P}_k$. These three previous functions take a constant time for execution. After MT time steps, the algorithm computes the fitness for each training scenario by summing the normalized bandwidth coverage and the normalized number of active AUs. Finally, once the evaluation of all four scenarios is over, the algorithm returns the average of these fitness values as the fitness of this candidate solution to the genetic algorithm.

Note that, in Equation (8), h refers to the number of hops for communication. The vector sum of the potential fields for UAV k given by $\vec{P}_k$ provides a desired heading for the kth UAV to turn toward. Each term in the equation describes a potential field whose parameters need to be optimized. To move the kth UAV toward users, an attractive potential field is generated based on users' bandwidth requirements ($\vec{P}_b$), and to reduce the per UAV bandwidth coverage and save energy, the kth UAV uses an attractive potential field toward neighboring UAVs whose magnitude depends on the neighbor's bandwidth load ($\vec{P}_u$), as shown by Figure 5. At the same time, to avoid collisions, a repulsive potential field, $\vec{P}_r$, ensures that UAVs do not collide with each other. This potential field only applies to UAVs that are less than d^* distance away, and we used the Dirac delta function, δ_r, to model this, where $\delta_r (\in \{0,1\})$. When UAV j is less than d^* distance away, then $\delta_r = 1$; otherwise, it is 0. Lastly, when a UAV has no users within U_{gr} and no UAVs within U_{ar}—that is, the UAV is inactive—it moves toward the command center as a default behavior. This is achieved with an attractive potential field ($\vec{P}_c$) based on the distance to the command center. Here, δ_c is another Dirac delta function used to ensure that this potential field only acts on inactive UAVs.

5.2.2. h-hop Network Deployment

To account for the limited communication range of UAVs, Equation (8) is modified for one-hop and two-hop neighbors, resulting in Equations (9) and (10), respectively. These equations represent potential fields in the form of $c \times d^e$, where the values of the coefficients and exponents are optimized to generate the desired UAV movements.

$$\vec{P}_k = \sum_{i=0}^{m_k} c_1 \vec{b}_i^{e_1} + \sum_{j=0}^{n_k}(c_2 \vec{u}_{kj}^{e_2} + \delta_r c_3 \vec{r}_{kj}^{e_3}) + \delta_c c_4 \vec{s}_k^{e_4} \qquad (9)$$

$$\vec{P}_k = \sum_{i=0}^{m_k} c_1 \vec{b}_i^{e_1} + \sum_{j=0}^{n_k}(c_2 \vec{u}_{kj}^{e_2} + \delta_r c_3 \vec{r}_{kj}^{e_3}) + \sum_{j'=0}^{n_{k'}}(c_4 \vec{u}_{kj'}^{e_4} + \delta_r c_5 \vec{r}_{kj'}^{e_5}) + \delta_c c_6 \vec{s}_k^{e_6} \qquad (10)$$

Here, $c_1, c_2, c_3, c_4, c_5, c_6$ are the coefficients, $e_1, e_2, e_3, e_4, e_5, e_6$ are the exponents of the potential fields, m_k is the number of users within the kth UAV sensing range, n_k is the number of one-hop neighbors, and n'_k is the number of two-hop neighbors. $\vec{b}$ is the bandwidth required by the ith user and points towards the user. $\vec{u}$ is the bandwidth being served by the jth one-hop neighbor of kth UAV and points toward this neighbor. $\vec{r}$ is the vector difference between the positions of UAV k and UAV j and points from j to k. In the last term, $\vec{s}$ is the vector difference between the command center position and UAV k pointing from k to the command center.

Note that, for $h = 1$ and $h = 2$, the kth UAV experiences four and six different potential fields, respectively. In general, for h-hop communication, the kth UAV will experience $2(h+1)$ different potential fields. Since each potential field has two tunable parameters (c, e), a total of $2 \times 2(h+1)$ potential field parameters will need to be optimized. Due to the non-linearity of potential fields, optimizing their parameters is difficult and thus a genetic algorithm is used. Coefficients and exponents are encoded in chromosomes, where $c's \in (-8192, 8192)$, $e's \in (-5.12, 5.12)$, and d^* ranges between $0 \leq d^* \leq 2^8$. Once GA evolved solutions, we compared the deployed network performance using GANet against the performance of the current state-of-the-art ATRI algorithm.

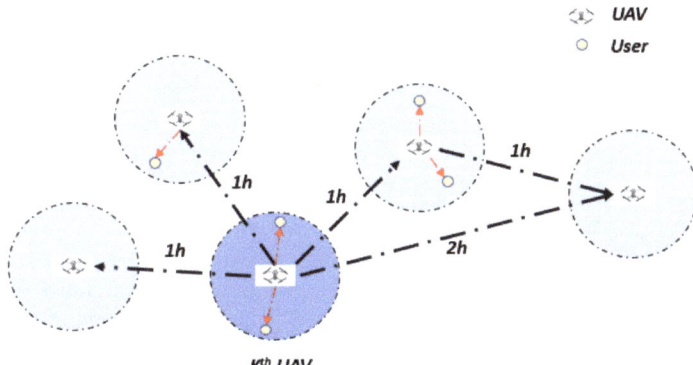

Figure 5. The figure shows five UAVs with blue-colored regions representing UAV's ground coverage and five users represented by light yellow-colored dots. Red and black arrows show directions of P_b and P_u. Kth UAV has two users within its U_{gr} ($m_k = 2$): three one-hop ($n_k = 3$) and one two-hop ($n_{k'} = 1$) neighbors. In this example, $\delta_r = 0$ and $\delta_c = 0$.

6. Experimental Results and Discussion

This section first evaluates the performance of genetic algorithm parameter tuning on a simple tractable problem and compares it against a random exhaustive search and a hill climber (gradient ascent). Later, we conducted more extensive experiments on 4 training and 100 testing scenarios. In order to carry this out, we created various training and testing scenarios with diverse user distributions using the Unity 3D game engine [43]. The entire simulation was implemented in the C# language. Since network deployment simulation is computationally expensive, we parallelized the simulation using Open MPI [44] to increase the evaluation speed.

6.1. A Simple Test Problem

Starting with a test problem of deploying 10 UAVs to provide bandwidth coverage to 10 users in a small area of 450×450 m^2, where the command center is located at $(0,0)$, a static network was deployed using the derived potential field parameters. In Figure 6a, blue dots represent users, the red circle is the command center, and green circles show the coverage of UAVs. The objective is to find another set of potential field parameters that can re-configure the static network using one-hop information to maximize the fitness.

Three different search methods were employed to search for optimal potential field parameters that result in the optimum fitness of 2. Note that the maximum fitness according to Equation (3) is 2. Simulation experiments were conducted ten times with different initializations, and the average performance was recorded. All three techniques found optimal solutions and a network deployed using an optimal solution is shown by Figure 6b, where each UAV is serving users and sharing the total bandwidth demanded by users to reduce the per UAV data rate. Figure 6c compares the performance of the three algorithms and shows that the genetic algorithm finds the optimum in significantly fewer evaluations than the others. The genetic algorithm (39) is more than twice as fast as its nearest competitor (92). These results demonstrate the effectiveness of GA for searching potential field parameters. The next subsection discusses the results obtained from the four training scenarios.

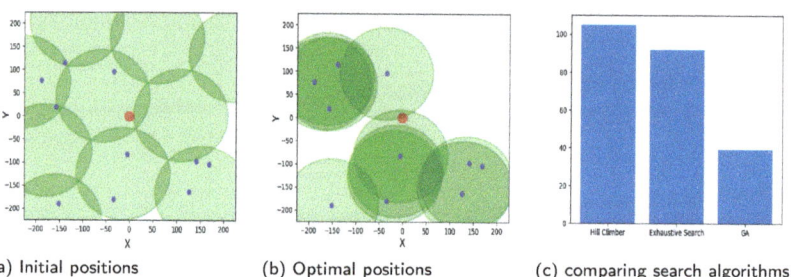

(a) Initial positions   (b) Optimal positions   (c) comparing search algorithms

Figure 6. UAV deployment within an AOI with 10 users (blue dots). (**a**) shows initial UAV positions covering the area, (**b**) shows the final UAV positions focusing on user locations and bandwidth sharing, and (**c**) compares the three search algorithms.

6.2. Experiments on Training Scenarios

Experiments were conducted using 156 UAVs and 200 users on four training scenarios. A total of 156 UAVs were selected as they provide sufficient coverage to completely cover the AOI without any gaps in coverage. In each scenario, the simulation ran for 1500 steps, and, in each time step, UAVs moved toward users under the influence of a GA-optimized set of potential fields as given in Algorithm 2. To find GA parameter settings, the population size varied between 10 to 60, the number of generations between 10 to 90, and, through a grid search, a population size of 20 and 20 generations was chosen. Offspring were produced using simulated binary crossover, with a crossover probability of 0.90, and polynomial mutation, with a probability of 0.05 [38].

6.2.1. Comparing the Performance of Evolved Solutions

The simulation experiments ran 10 times with different random seeds to evolve solutions considering one-hop and two-hop communication between UAVs. Figure 7 and Table 2 show the fitness obtained using $Best_{1h}$, $Best_{2h}$, and ATRI [15] on the four training scenarios with 156 UAVs and 200 users, where $Best_{1h}$ and $Best_{2h}$ are the best solutions evolved using GA considering one-hop and two-hop communication respectively. Table 2 shows that, on T_{r1}, the $Best_{2h}$ performance is 14.2% higher than $Best_{1h}$, and 20.62% higher than ATRI.

In addition, the performance of $Best_{1h}$ is 5.62% higher than ATRI. Similar performance improvements are observed on T_{r2}, T_{r3}, and T_{r4}. On average, the $Best_{2h}$ performance is 18.90% higher than $Best_{1h}$ and 41.30% higher than ATRI. Similarly, the $Best_{1h}$ fitness is 18.88% more than ATRI.

Table 2. Comparing fitness obtained with 156 UAVs and 200 users on training scenarios.

Users	Methods	Tr_1	Tr_2	Tr_3	Tr_4
	ATRI	1.60	1.12	1.34	1.47
200	GANet-1 hop	1.69	1.56	1.59	1.71
	GANet-2 hop	**1.93**	**1.95**	**1.97**	**1.94**

Close inspection indicates that the fitness of $Best_{2h}$ is higher because networks deployed using $Best_{2h}$ provide more bandwidth coverage to users compared to others and have larger numbers of AUs as shown in Figure 8. Note that fitness is the combination of the percentage bandwidth coverage and percentage of active UAVs. Figure 8 compares (a) the bandwidth served and (b) the number of active UAVs in networks deployed using the $Best_{2h}$, $Best_{1h}$, and ATRI in four scenarios, respectively.

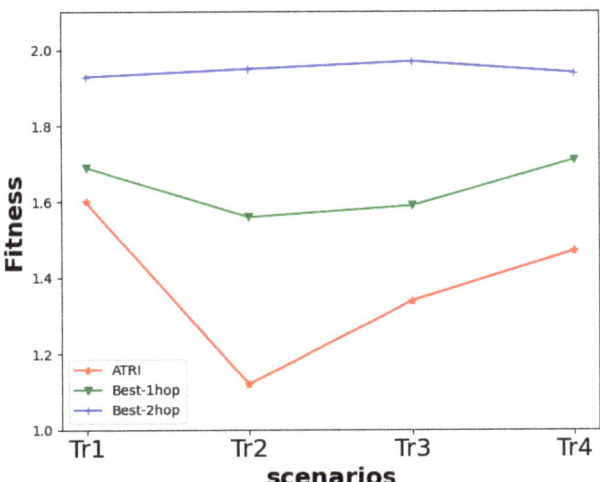

Figure 7. Comparing the best fitness obtained using GANet 1-hop, GANet 2-hop, and ATRI.

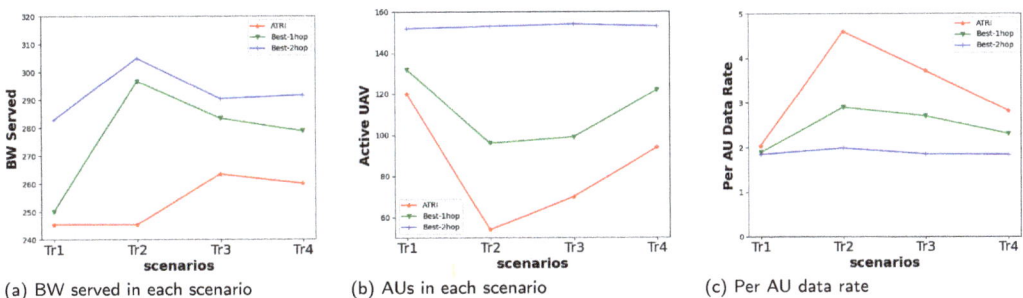

(a) BW served in each scenario (b) AUs in each scenario (c) Per AU data rate

Figure 8. Comparing results on four training scenarios: (**a**) bandwidth served by deployed networks, (**b**) the number of AUs in the networks, and (**c**) per AU data rate.

The larger the number of AUs, the smaller the per AU data rate provided to users, leading to less energy consumption and increasing the longevity of the network. Figure 8c shows that the per AU data rate for $Best_{2h}$ is smaller than the other two, indicating a higher number of AUs and thus a higher fitness. The figure also shows that the per AU data rate for $Best_{1h}$ is smaller than ATRI, again indicating higher fitness than ATRI. These results provide evidence that the proposed genetic-algorithm-optimized network deployment approach is better than ATRI, the current best dynamic network deployment technique.

Additional experiments were conducted by considering three-hop communication between UAVs, but no significant performance improvement was observed.

To determine why ATRI's performance is inferior to GANet, UAVs distributions in the networks deployed using ATRI, $Best_{1h}$, and $Best_{2h}$ on the first training scenario (T_{r1}) are shown by Figure 9. The figure illustrates that solutions ($Best_{1h}$, $Best_{2h}$) evolved using GA allow UAVs to move close to users that are distributed throughout the AOI (see Figure 9b,c) while maintaining connectivity with the command center. The evolved potential field parameters guide UAVs toward the area with more user concentration to serve them better. This leads to a higher number of AUs in the deployed network, better network capacity utilization, and higher fitness. However, in the case of ATRI as shown by Figure 9a, the distribution of UAVs does not cover corners, has a circular deployment pattern, and also contains a larger number of IUs (inactive UAVs), resulting in less fitness compared to GANet. It is important to note that ATRI uses one-hop communication for the network deployment, and thus a comparison of ATRI with GANet one-hop is best suited. The next subsection explicitly compares ATRI and GANet one-hop.

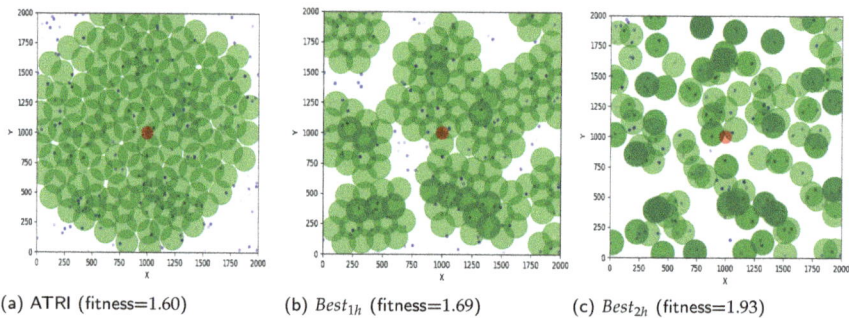

(a) ATRI (fitness=1.60) (b) $Best_{1h}$ (fitness=1.69) (c) $Best_{2h}$ (fitness=1.93)

Figure 9. (**a**–**c**) show networks on T_{r1} using ATRI, GANet 1-hop, GANet 2-hop. GANet obtains better performance by focusing more on areas with users and avoiding areas with no users.

6.2.2. GANet-1 hop vs. ATRI

Figure 10 shows networks deployed using ATRI and $Best_{1h}$ on the second (T_{r2}) and third (T_{r3}) training scenarios. In T_{r2}, which has two user clusters, ATRI's performance is 28.2% lower than $Best_{1h}$'s. A similar pattern can be seen for T_{r3} as well. ATRI-deployed UAVs are more uniformly distributed across the AOI, whereas, with GANet, UAVs have learned parameter values that enable them to cluster around users while maintaining connectivity to the command center. This increases the number of AUs (active UAVs) in the network and thus increases the fitness and reduces the per active UAV data rate to users. Note that, on T_{r1}, where users are more uniformly distributed, ATRI's performance is comparable to GANet. However, in the remaining scenarios, where the users' distribution is non-uniform, GANet performed better, pointing out the weakness of the ATRI technique over non-uniform users' distributions.

6.3. Generalizability of Evolved Solutions

Previous subsections have demonstrated that $Best_{2h}$ outperforms other solutions when there are 200 users and 156 UAVs. However, the number of users and UAVs can vary depending on situations; thus, in order to find the robustness and generalizability of evolved solutions in different environments, the same set of experiments was conducted using the same evolved solutions ($Best_{1h}$, $Best_{2h}$) but with varied numbers of users, UAVs, and distributions of users. We changed the number of users and UAVs by 25% at a time (200, 150, 100, and 50 users and 156, 117, and 78 UAVs) and recorded the fitness of each unique pair of UAVs and users. Tables 3 and 4 compare the fitness obtained with varied numbers of UAVs and users. Based on Table 3, on average, over four scenarios, the performance of $Best_{2h}$ is between 17.11–45.94% larger than $Best_{1h}$ and ATRI. Similarly, according to Table 4, the $Best_{2h}$ fitness is 32.75% to 121.73% higher than ATRI, and between 14.46–27.27% more than $Best_{1h}$. We also observed a similar performance improvement pattern between $Best_{1h}$ and ATRI, where $Best_{1h}$'s fitness is higher by 14.06% to 73.91%.

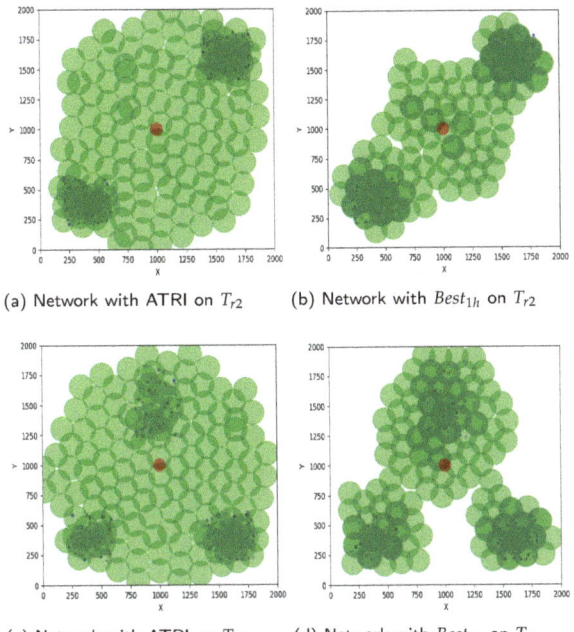

(a) Network with ATRI on T_{r2} (b) Network with $Best_{1h}$ on T_{r2}

(c) Network with ATRI on T_{r3} (d) Network with $Best_{1h}$ on T_{r3}

Figure 10. Network deployments of $Best_{1h}$ and ATRI on T_{r2} and T_{r3}. In both scenarios, $Best_{1h}$ performed better than ATRI because GANet allows UAVs to move toward users with bandwidth demand.

Table 3. Comparing fitness with different numbers of UAVs and 200 users on training scenarios.

Users	Methods	Tr_1	Tr_2	Tr_3	Tr_4	Tr_1	Tr_2	Tr_3	Tr_4
		\multicolumn{4}{c}{117 UAVs}			78 UAVs		
200	ATRI	1.50	1.02	1.29	1.40	1.32	0.76	1.13	1.26
	GANet-1 hop	1.52	1.53	1.61	1.62	1.47	1.25	0.98	1.57
	GANet-2 hop	**1.64**	**1.88**	**1.90**	**1.84**	**1.65**	**1.53**	**1.59**	**1.69**

Table 4. Comparing fitness with different numbers of UAVs and users on training scenarios.

Users	Methods	Tr_1	Tr_2	Tr_3	Tr_4	Tr_1	Tr_2	Tr_3	Tr_4	Tr_1	Tr_2	Tr_3	Tr_4
		156 UAVs				117 UAVs				78 UAVs			
150	ATRI	1.52	1.11	1.29	1.43	1.38	1.09	1.28	1.33	1.20	0.71	1.14	1.28
	GANet-1 hop	1.58	1.44	1.53	1.62	1.52	1.52	1.62	1.71	1.36	1.38	0.93	1.60
	GANet-2 hop	**1.97**	**1.96**	**1.95**	**1.93**	**1.69**	**1.88**	**1.80**	**1.91**	**1.63**	**1.66**	**1.64**	**1.66**
100	ATRI	1.34	1.13	1.24	1.42	1.26	1.40	1.28	1.34	1.10	0.81	0.93	1.23
	GANet-1 hop	1.58	1.29	1.39	1.55	1.42	1.33	1.48	1.57	1.24	1.31	1.50	1.58
	GANet-2 hop	**1.84**	**1.98**	**1.93**	**1.98**	**1.78**	**1.86**	**1.78**	**1.91**	**1.61**	**1.49**	**1.66**	**1.78**
50	ATRI	1.18	1.13	1.16	1.20	1.05	1.06	1.12	1.06	0.81	0.37	0.71	0.84
	GANet-1 hop	1.23	1.20	1.29	1.38	1.22	1.27	1.36	1.25	1.08	1.34	1.27	1.15
	GANet-2 hop	**1.32**	**1.92**	**1.29**	**1.64**	**1.51**	**1.77**	**1.85**	**1.60**	**1.21**	**1.70**	**1.67**	**1.55**

These tables show that with changes in either the number of users or both the number of UAVs and users, the $Best_{2h}$ performance is still the best, indicating that evolved solutions are robust and generalizable. Figure 11a shows the distribution of fitness for each unique pair of users and UAVs using ATRI, $Best_{1h}$, and $Best_{2h}$ on the training scenarios. Here, $Best_{2h}$ performed statistically significantly better than both $Best_{1h}$ and ATRI with p-value < 0.05. This provides evidence that, even with a different numbers of users and UAVs, GANet works well in training scenarios.

Similar experiments were conducted on 100 testing scenarios, and the fitness of each unique pair of UAVs and users was recorded on every testing scenario. Figure 11b shows the fitness distribution on 100 testing scenarios, where again the performance of $Best_{2h}$ is statistically significant, with a p-value < 0.05 against the other two. These results provide evidence that once GANet optimizes potential field values for a set of UAVs on a set of training scenarios, the performance remains the same on new unseen scenarios with different numbers of users, UAVs, and user distributions.

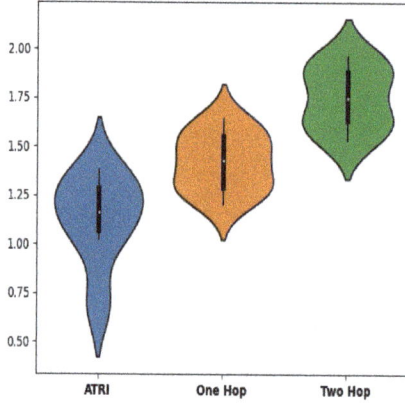
(a) Distribution of fitness on T_r scenarios

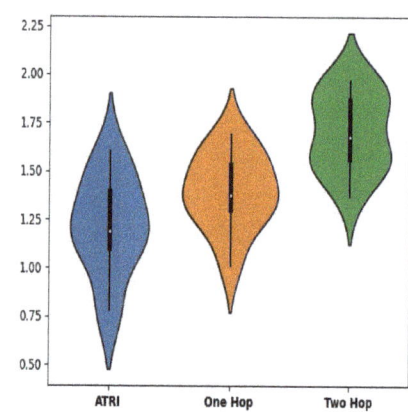
(b) Distribution of fitness on T_e scenarios

Figure 11. (**a**,**b**) show the distribution of fitness for each unique pair of UAVs and users on 4 training and 100 testing scenarios, respectively. In each case, $Best_{2h}$ better indicates better transferability of evolved solutions.

6.4. Maximizing the Minimum Objective

In all previous experiments, the goal was to optimize the fitness score computed as the *average* performance (Equation (3)) over four training scenarios. Another approach is to maximize the *minimum* performance [45,46]. However, when comparing the GANet one-hop performance between maximizing the minimum (maximin) and maximizing the average, we found that maximizing the average performs significantly better. Figure 12 shows that the best solution evolved using the averaged fitness, green bar, performed better than the best solution evolved with the maximin fitness, blue bar, on all four scenarios. On average, the averaged fitness performed 45.13% better than the maximin fitness. We believe that the averaging performance better preserves diversity in the population, leading to more exploration by the genetic algorithm and ultimately resulting in a better performance.

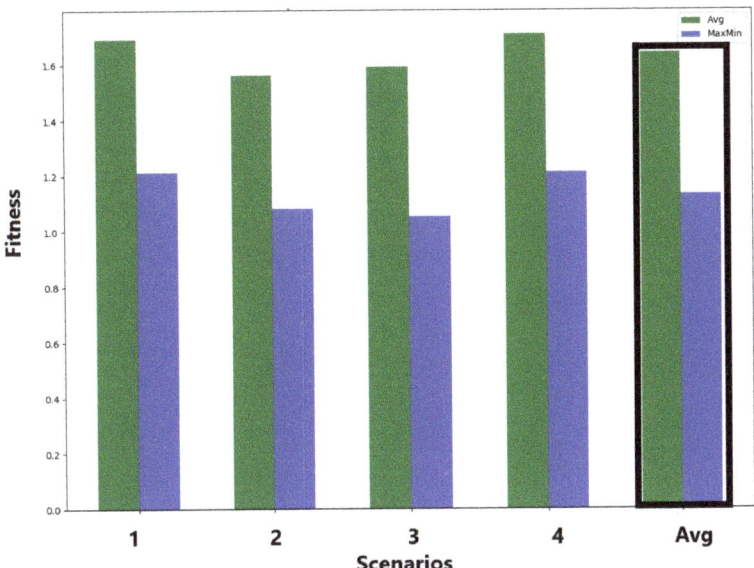

Figure 12. GANet performance when maximizing the average performance across four training scenarios (green) compared to maximizing the minimum performance over four (blue).

7. Conclusions and Future Work

This paper introduces a unified approach for deploying networks using genetic algorithms optimization. Unlike prior works, the proposed approach unifies the representation across both the search and service phases of network deployment using potential fields. The proposed GANet optimizes potential field parameters that work robustly for different user distributions. In the first phase, the potential field parameters are derived, and, in the second phase, an elitist genetic algorithm optimizes another set of non-linear potential field parameters to maximize the bandwidth while increasing the network longevity.

The results indicate that GANet deployment is significantly better than the current state-of-the-art network deployment algorithm, ATRI. Additionally, the study demonstrates how different levels of information exchange between UAVs affect performance. GANet with two-hop UAV communications outperformed one-hop UAV communication. The experiments show that the offline evolved solutions are robust and work well on 100 never-before-seen test scenarios with different numbers of UAVs and users, providing evidence for the potential-fields-based approach's robustness and generalizability. GANet adapts better when users are non-uniformly distributed and matches or exceeds ATRI's performance with more uniform distributions.

In the future, we plan to extend this work by considering dynamic bandwidth requirements for users and maintaining coverage and bandwidth as users move towards the rescue.

Author Contributions: Conceptualization R.D., S.J.L.; methodology, R.D., S.J.L.; formal analysis, R.D.; investigation, R.D.; resources, R.D., S.J.L.; data curation, R.D.; writing—original draft preparation, R.D.; writing—review and editing, R.D., S.J.L.; visualization, R.D.; supervision, S.J.L.; project administration, S.J.L.; funding acquisition, S.J.L. All authors have read and agreed to the published version of the manuscript.

Funding: This material is based upon research supported by, or in part by, the U. S. Office of Naval Research under award number N00014-22-1-2122.

Institutional Review Board Statement: Not applicable.

Informed Consent Statement: Not applicable.

Data Availability Statement: All data used to support the findings of the study is included in this paper.

Conflicts of Interest: The authors declare no conflict of interest.

Abbreviations

The following abbreviations are used in this manuscript:

GA	Genetic Algorithms
GANet	Genetic Adaptive Network Deployment
AOI	Area of Interest
ATRI	Adaptive Triangulation
PF	Potential Field
UAV	Unmanned Aerial Vehicle
DT	Delaunay Triangulation
CPT	Circle Packing Theorem

References

1. Silvagni, M.; Tonoli, A.; Zenerino, E.; Chiaberge, M. Multipurpose UAV for search and rescue operations in mountain avalanche events. *Geomat. Nat. Hazards Risk* **2017**, *8*, 18–33. [CrossRef]
2. Tomic, T.; Schmid, K.; Lutz, P.; Domel, A.; Kassecker, M.; Mair, E.; Grixa, I.L.; Ruess, F.; Suppa, M.; Burschka, D. Toward a Fully Autonomous UAV: Research Platform for Indoor and Outdoor Urban Search and Rescue. *IEEE Robot. Autom. Mag.* **2012**, *19*, 46–56. [CrossRef]
3. Li, X.; Savkin, A.V. Networked Unmanned Aerial Vehicles for Surveillance and Monitoring: A Survey. *Future Int.* **2021**, *13*, 174. [CrossRef]
4. Lee, S.J.; Lee, D.; Kim, H.J. Cargo Transportation Strategy using T3-Multirotor UAV. In Proceedings of the 2019 International Conference on Robotics and Automation (ICRA), Montreal, QC, Canada, 20–24 May 2019; pp. 4168–4173. [CrossRef]
5. Carrera-Hernández, J.; Levresse, G.; Lacan, P. Is UAV-SfM surveying ready to replace traditional surveying techniques? *Int. J. Remote Sens.* **2020**, *41*, 4820–4837. [CrossRef]
6. Nex, F.; Remondino, F. UAV for 3D mapping applications: A review. *Appl. Geomat.* **2014**, *6*, 1–15. [CrossRef]
7. Erdelj, M.; Król, M.; Natalizio, E. Wireless sensor networks and multi-UAV systems for natural disaster management. *Comput. Netw.* **2017**, *124*, 72–86. [CrossRef]
8. Boursianis, A.D.; Papadopoulou, M.S.; Diamantoulakis, P.; Liopa-Tsakalidi, A.; Barouchas, P.; Salahas, G.; Karagiannidis, G.; Wan, S.; Goudos, S.K. Internet of things (IoT) and agricultural unmanned aerial vehicles (UAVs) in smart farming: A comprehensive review. *Int. Things* **2020**, *18*, 100187. [CrossRef]
9. Motlagh, N.H.; Bagaa, M.; Taleb, T. UAV-Based IoT Platform: A Crowd Surveillance Use Case. *IEEE Commun. Mag.* **2017**, *55*, 128–134. [CrossRef]
10. Gáspár, Z.; Tarnai, T. Upper bound of density for packing of equal circles in special domains in the plane. *Period. Polytech. Civ. Eng.* **2000**, *44*, 13–32.
11. Chevet, T.; Maniu, C.S.; Vlad, C.; Zhang, Y. Voronoi-based UAVs formation deployment and reconfiguration using MPC techniques. In Proceedings of the IEEE International Conference on Unmanned Aircraft Systems (ICUAS), Dallas, TX, USA, 12–15 June 2018; pp. 9–14.
12. Mozaffari, M.; Saad, W.; Bennis, M.; Nam, Y.H.; Debbah, M. A tutorial on UAVs for wireless networks: Applications, challenges, and open problems. *IEEE Commun. Surv. Tutor.* **2019**, *21*, 2334–2360 [CrossRef]
13. Dubey, R.; Louis, S.J.; Sengupta, S. Evolving Dynamically Reconfiguring UAV-hosted Mesh Networks. In Proceedings of the 2020 IEEE Congress on Evolutionary Computation (CEC), Glasgow, UK, 19–24 July 2020; pp. 1–8. [CrossRef]
14. Dubey, R.; Ghantous, J.; Louis, S.; Liu, S. Evolutionary Multi-objective Optimization of Real-Time Strategy Micro. In Proceedings of the 2018 IEEE Conference on Computational Intelligence and Games (CIG), Maastricht, The Netherlands, 14–17 August 2018; pp. 1–8.
15. Ma, M.; Yang, Y. Adaptive triangular deployment algorithm for unattended mobile sensor networks. *IEEE Trans. Comput.* **2007**, *56*, 847–946. [CrossRef]
16. Binh, L.H.; Truong, T.K. An efficient method for solving router placement problem in wireless mesh networks using multi-verse optimizer algorithm. *Sensors* **2022**, *22*, 5494. [CrossRef]
17. Wang, J.; Jiang, C.; Han, Z.; Ren, Y.; Maunder, R.G.; Hanzo, L. Taking drones to the next level: Cooperative distributed unmanned-aerial-vehicular networks for small and mini drones. *IEEE Veh. Technol. Mag.* **2017**, *12*, 73–82. [CrossRef]

18. Lam, M.L.; Liu, Y.H. Heterogeneous Sensor Network Deployment Using Circle Packings. In Proceedings of the 2007 IEEE International Conference on Robotics and Automation, Rome, Italy, 10–14 April 2007; pp. 4442–4447.
19. Lyu, J.; Zeng, Y.; Zhang, R.; Lim, T.J. Placement optimization of UAV-mounted mobile base stations. *IEEE Commun. Lett.* **2016**, *21*, 604–607. [CrossRef]
20. Mozaffari, M.; Saad, W.; Bennis, M.; Debbah, M. Efficient deployment of multiple unmanned aerial vehicles for optimal wireless coverage. *IEEE Commun. Lett.* **2016**, *20*, 1647–1650. [CrossRef]
21. Bartolini, N.; Calamoneri, T.; La Porta, T.F.; Silvestri, S. Autonomous deployment of heterogeneous mobile sensors. *IEEE Trans. Mob. Comput.* **2010**, *10*, 753–766. [CrossRef]
22. Patra, A.N.; Regis, P.A.; Sengupta, S. Distributed allocation and dynamic reassignment of channels in UAV networks for wireless coverage. *Pervasive Mob. Comput.* **2019**, *54*, 58–70. [CrossRef]
23. Khatib, O. Real-Time Obstacle Avoidance for Manipulators and Mobile Robots. *Int. J. Robot. Res.* **1986**, *5*, 90–98. [CrossRef]
24. Woods, A.C.; La, H.M. A novel potential field controller for use on aerial robots. *IEEE Trans. Syst. Man, Cybern. Syst.* **2017**, *49*, 665–676. [CrossRef]
25. Dubey, R.; Louis, S.; Gajurel, A.; Liu, S. Comparing Three Approaches to Micro in RTS Games. In Proceedings of the 2019 IEEE Congress on Evolutionary Computation (CEC), Wellington, New Zealand, 10–13 June 2019; pp. 777–784.
26. Howard, A.; Matarić, M.J.; Sukhatme, G.S. *Mobile Sensor Network Deployment Using Potential Fields: A Distributed, Scalable Solution to the Area Coverage Problem*; Springer: Berlin/Heidelberg, Germany, 2002; pp. 299–308.
27. Poduri, S.; Sukhatme, G.S. Constrained coverage for mobile sensor networks. In Proceedings of the IEEE International Conference on Robotics and Automation, ICRA'04, New Orleans, LA, USA, 26 April–1 May 2004; Volume 1, pp. 165–171.
28. Zhao, H.; Wang, H.; Wu, W.; Wei, J. Deployment algorithms for uav airborne networks toward on-demand coverage. *IEEE J. Sel. Areas Commun.* **2018**, *36*, 2015–2031. [CrossRef]
29. Anbalagan, S.; Bashir, A.K.; Raja, G.; Dhanasekaran, P.; Vijayaraghavan, G.; Tariq, U.; Guizani, M. Machine-learning-based efficient and secure RSU placement mechanism for software-defined-IoV. *IEEE Internet Things J.* **2021**, *8*, 13950–13957. [CrossRef]
30. Reina, D.; Tawfik, H.; Toral, S. Multi-subpopulation evolutionary algorithms for coverage deployment of UAV-networks. *Ad Hoc Netw.* **2018**, *68*, 16–32. [CrossRef]
31. Deif, D.S.; Gadallah, Y. Wireless Sensor Network Deployment Using A Variable-Length Genetic Algorithm. In Proceedings of the 2014 IEEE Wireless Communications and Networking Conference (WCNC), Istanbul, Turkey, 6–9 April 2014; pp. 2450–2455. [CrossRef]
32. Harizan, S.; Kuila, P. Coverage and connectivity aware energy efficient scheduling in target based wireless sensor networks: An improved genetic algorithm based approach. *Wirel. Netw.* **2019**, *25*, 1995–2011. [CrossRef]
33. Ruetten, L.; Regis, P.A.; Feil-Seifer, D.; Sengupta, S. Area-optimized UAV swarm network for search and rescue operations. In Proceedings of the 2020 10th Annual Computing and Communication Workshop and Conference (CCWC), Las Vegas, NV, USA, 6–8 January 2020; pp. 613–618.
34. Ab Aziz, N.A.B.; Mohemmed, A.W.; Alias, M.Y. A wireless sensor network coverage optimization algorithm based on particle swarm optimization and Voronoi diagram. In Proceedings of the 2009 International Conference on Networking, Sensing and Control, Okayama, Japan, 26–29 March 2009; pp. 602–607.
35. Li, J.; Lu, D.; Zhang, G.; Tian, J.; Pang, Y. Post-disaster unmanned aerial vehicle base station deployment method based on artificial bee colony algorithm. *IEEE Access* **2019**, *7*, 168327–168336. [CrossRef]
36. Abdulrab, H.Q.; Hussin, F.A.; Abd Aziz, A.; Awang, A.; Ismail, I.; Saat, M.S.M.; Shutari, H. Optimal coverage and connectivity in industrial wireless mesh networks based on Harris' hawk optimization algorithm. *IEEE Access* **2022**, *10*, 51048–51061. [CrossRef]
37. Patra, A.N.; Regis, P.A.; Sengupta, S. Dynamic self-reconfiguration of unmanned aerial vehicles to serve overloaded hotspot cells. *Comput. Electr. Eng.* **2019**, *75*, 77–89. [CrossRef]
38. Deb, K.; Pratap, A.; Agarwal, S.; Meyarivan, T. A fast and elitist multiobjective genetic algorithm: NSGA-II. *IEEE Trans. Evol. Comput.* **2002**, *6*, 182–197. [CrossRef]
39. Tian, X.; Guo, Y.; Negenborn, R.R.; Wei, L.; Lin, N.M.; Maestre, J.M. Multi-scenario model predictive control based on genetic algorithms for level regulation of open water systems under ensemble forecasts. *Water Resour. Manag.* **2019**, *33*, 3025–3040. [CrossRef]
40. Louis, S.J.; Liu, S. Multi-objective evolution for 3D RTS micro. In Proceedings of the 2018 IEEE Congress on Evolutionary Computation (CEC), Rio de Janeiro, Brazil, 8–13 July 2018; pp. 1–8.
41. Liu, S.; Louis, S.; Ballinger, C. Evolving Effective Micro Behaviors in Real-Time Strategy Games. *IEEE Trans. Comput. Intell. Games* **2016**, *8*, 351–362. [CrossRef]
42. Dubey, R.; Louis, S.J. Evolving potential field parameters for deploying UAV-based two-hop wireless mesh networks. In Proceedings of the Genetic and Evolutionary Computation Conference Companion (GECCO 21), Lille, France, 10–14 July 2021; pp. 311–312.
43. Unity3D. Available online: https://unity.com/ (accessed on 25 March 2023).
44. OpenMPI. Available online: ttps://www.open-mpi.org/ (accessed on 25 March 2023).

45. Wang, H.; Feng, L.; Jin, Y.; Doherty, J. Surrogate-assisted evolutionary multitasking for expensive minimax optimization in multiple scenarios. *IEEE Comput. Intell. Mag.* **2021**, *16*, 34–48. [CrossRef]
46. Pióro, M.; Żotkiewicz, M.; Staehle, B.; Staehle, D.; Yuan, D. On max–min fair flow optimization in wireless mesh networks. *Ad Hoc Netw.* **2014**, *13*, 134–152. [CrossRef]

Disclaimer/Publisher's Note: The statements, opinions and data contained in all publications are solely those of the individual author(s) and contributor(s) and not of MDPI and/or the editor(s). MDPI and/or the editor(s) disclaim responsibility for any injury to people or property resulting from any ideas, methods, instructions or products referred to in the content.

Article

Multi-Objective Optimization of Electric Vehicle Charging Station Deployment Using Genetic Algorithms

Vasiliki Lazari and Athanasios Chassiakos *

Department of Civil Engineering, University of Patras, 265 00 Patras, Greece
* Correspondence: a.chassiakos@upatras.gr

Abstract: The incorporation of electric vehicles into the transportation system is imperative in order to mitigate the environmental impact of fossil fuel use. This requires establishing methods for deploying the charging infrastructure in an optimal way. In this paper, an optimization model is developed to identify both the number of stations to be deployed and their respective locations that minimize the total cost by utilizing Genetic Algorithms. This is implemented by combining these components into a linear objective function aiming to minimize the overall cost of deploying the charging network and maximize service quality to users by minimizing the average travel distance between demand spots and stations. Several numerical and practical considerations have been analyzed to provide an in-depth study and a deeper understanding of the model's capabilities. The optimization is done through commercial software that is appropriately parametrized to adjust to the specific problem. The model is simple yet effective in solving a variety of problem structures, optimization goals and constraints. Further, the quality of the solution seems to be marginally affected by the shape and size of the problem area, as well as the number of demand spots, and this may be considered one of the strengths of the algorithm. The model responds expectedly to variations in the charging demand levels and can effectively run at different levels of grid discretization.

Keywords: electric vehicles; charging station allocation; facility-to-site allocation; optimization; genetic algorithms; evolutionary computing; grid discretization

Citation: Lazari, V.; Chassiakos, A. Multi-Objective Optimization of Electric Vehicle Charging Station Deployment Using Genetic Algorithms. *Appl. Sci.* **2023**, *13*, 4867. https://doi.org/10.3390/app13084867

Academic Editors: Vincent A. Cicirello and Adel Razek

Received: 1 February 2023
Revised: 22 March 2023
Accepted: 5 April 2023
Published: 13 April 2023

Copyright: © 2023 by the authors. Licensee MDPI, Basel, Switzerland. This article is an open access article distributed under the terms and conditions of the Creative Commons Attribution (CC BY) license (https://creativecommons.org/licenses/by/4.0/).

1. Introduction

The use of electric vehicles is rising in many cities around the world, underpinned by the urgent need to reduce the levels of air and noise pollution and tackle the ever-growing energy-related greenhouse gas (GHG) emissions from conventionally fueled vehicles. Research conducted on large city air pollution metrics indicates that the highest contribution comes from the transportation system, where multiple internal combustion engines work with diesel fuel and spark-ignition engines mainly work with petrol [1]. Electric vehicles (EVs) promise high efficiency, energy savings, low noise, and zero emissions; however, the lack of supporting charging infrastructure is holding back their prompt and widespread adoption. Although EV charging infrastructure is being installed in several countries, most of them have not been able to install the required number of EV charging stations, except in some states. The world's highest charging station density (19–20 charging stations per 100 km) appears in the Netherlands, while China is the second-best performing country with 3–4 charging stations per 100 km, and the UK will have approximately 3 charging points per 100 km by 2030 [2]. Low or almost zero station deployment levels reduce EV adoption. For that reason, it is imperative to support the deployment of an extended network of charging stations to attract private or public vehicle drivers to use EVs. This is a prerequisite to solving efficiently the facility allocation problem, meaning that the number and location of the charging stations composing the respective network should be optimized first while considering certain constraints (budget limitation, charging station capacity, dispersion of the charging demand, etc.). Finding the optimal location for a charging station is a decision that may require considering various and potentially contradictory factors such as driver satisfaction with charging, operator concerns about economics, fleet losses from power outages, grid safety issues, and traffic problems in the transportation system [3]. Also, electric vehicles' battery status is a factor that should be taken into account when considering such problems, as the competitive relationship between charging time and performance degradation may render the battery's optimal fast charging challenging [4], thus preventing the high adoption of EVs even if the respective infrastructure has been enhanced. In [5], the researchers have proposed a model for deciding the location and capacity of the EV fast charging stations, along with the optimal multistage expansion of the distribution network to cope with future load growth. This is formulated in three interconnected layers (Layer 1: optimal locations of FCS, Layer 2: optimal numbers of FCSs and CPs, and Layer 3: optimal planning of DN expansion), with a sub-layer of optimal assignment of EVs at FCSs. The solution is carried out along with validation using different metaheuristic methods (Differential Evolution, Symbiotic Organism Search, Arithmetic Optimization Algorithm) and integer linear programming. As emphasized in [6], the EV charging station placement problem requires the charging network to be pervasive enough such that an EV can easily access a charging station within its driving range and also widely spread so that EVs can compete to displace internal combustion engine (ICE) vehicles.

2. Background

To improve EV charging efficiency, researchers have been working on the Electric Vehicle Charging Station (EVCS) distribution problem, investigating different perspectives in order to reach an optimal allocation scheme. In terms of planning scenarios, several factors and targets have been considered, including the optimal location, number, service type, and capacity of each EVCS to satisfy the user's needs and ultimately the growth of EV penetration. In terms of solution algorithms, the global research effort has developed a variety of methods and algorithms in order to solve the charging station sizing and placement problem [7]. In particular, the formulated optimization problems for the placement of EVCS can form a single or multi-objective, linear or nonlinear, convex or concave assembly. According to the used variables, the formulated problem can be continuous, integer, discrete, or combinatory [8]. The most frequently used algorithms in existing studies can be divided into two types: heuristic/meta-heuristic and mathematical algorithms. Heuristic algorithms are relatively more popular than mathematical ones due to their ability to find a global or near-global optimum solution even in complex problems.

Several studies have attempted to estimate the optimal location of the EVCS by examining alternative distribution scenarios based on a predefined set of candidate EVCS sites. The work in [9] develops a bi-level mathematical model to optimize the location of charging stations for EVs with consideration of the driving range. The upper level is to maximize the flows served by charging stations, while the lower level depicts route choice behavior given the location of the charging station. Genetic Algorithms have also been employed in [10] and [11] to calculate the necessary number of charging stations and best placement positions to satisfy the clients' demand using origin–destination (OD) data of conventional vehicles and real-world driving data of 196 battery EVs. The research in [12] uses an intelligent multi-objective optimization method to handle the problem by integrating a multi-objective particle swarm optimization (MOPSO) process to obtain a set of Pareto optimal solutions and an entropy weight method-based evaluation process to select the final solution from the Pareto optimal solutions. The effort in [13] develops an optimization model of charging station location that considers the waiting time for EVs in the queue for charging as an influencing factor. The minimization of the time cost to EV drivers is performed with the utilization of the hybrid evolutionary algorithm SCE-UA. Another work in [14] presents a case study for planning the locations of public electric vehicle charging stations with the employment of three different classic facility location models (set covering model, maximal covering location model, and p-median model) to test the model's effectiveness. The optimal positioning of EVCS in an urban area is analyzed in [15] by introducing weighting maps (cost values, distance) for managing different social requirements into the optimization process while utilizing evolutionary algorithms (Particle Swarm Optimization (PSO), Genetic Algorithms (GA), Biogeography-based Optimization (BBO), and Social Network Optimization (SNO)). The proposed solution to the deployment problem applies a greedy approach that consists of placing all the CS, one at a time, such that each one is locally optimal and is tested through various scenarios with a different predefined number of EVCS to appoint. A multi-period optimization problem is proposed in [16] for EVCS placement, in which the distribution of charging demand is modeled with a combination of node-based and flow-based approaches, so as to model the needs of EVs to recharge on intermediary stops on long-haul travels. For this problem, a mixed-integer linear programming (MILP) formulation is considered. The research in [17] applies a Geographic Information System (GIS)-based Multi-Criteria Decision Analysis process using the analytical hierarchy process (AHP) to address the electric vehicle charging station site selection in light of 15 environmental, economic, and urbanity criteria. A hierarchical optimization model, integrating three levels of analysis, is also developed in [18] to assist city planners with charging station location selection and system design. In this work, 10 different locations were considered for EVCS deployment, consisting of shopping malls and cultural centers. A study on expanding existing EVCS networks is realized in [19], following two different expansion strategies to choose additional sites for charging station placement, with the candidate locations being based on real-world data on charging station utilization and correlation with places of common interest.

Other research studies integrate into the EVCS allocation problem the optimization of its sizing by assigning the optimal charging piles per infrastructure deployed. The work in [20] examines the optimal EV charging infrastructure location and capacity determination problem, assuming charging queuing behavior with finite queue length constraints. The GA is used to minimize the total cost, which considers charging queuing behavior with finite queue length and various siting constraints in a small-scale case study (20 EVCS candidate locations). The effort in [21] utilizes Genetic Algorithms to solve the fast-charging station location-and-sizing problem to maximize EV charging station owner profits across a region for BEV owners who wish to charge *en route*, taking into consideration elastic demand, station congestion, and network equilibrium. The research in [22] proposes a multi-criterion-oriented optimization approach to determine the optimal charging station placement and charging piles assignment under multiple constraints, such as recharging demand and cruising range, by employing two different algorithms, the Lazy Greedy with Direct Gain (LGDG) and Lazy Greedy with Effective Gain (LGEG), both based on the greedy method. Public trajectory data collected from the taxi cabs have been utilized for building up and verifying the research work based on the hypothesis that the traveler/driver behavior remains unchanged when switching to driving/taking electric vehicles. The work in [23] develops a model based on an artificial immune algorithm to identify the optimal solution considering the overall user satisfaction: charging convenience, charging cost, and charging time. In [24], the planning and sizing model is established to minimize the annual cost of the charging station under multiple constraints that consider the actual load, charging power, and charging distance through the utilization of a neighborhood mutation immune clone selection algorithm. The Analytic Hierarchy Process (AHP) is adopted in [25] to rank 10 EVCS alternative locations in a large city district by assigning relative weights as input for the mathematical model to determine the number of charging stations to install and their relevant capacities.

In the context of optimal EVCS placement planning, the charging facility type is used to build up the optimization objectives in several research studies. In the study [26], three typical kinds of charging facilities have been considered: slow-charging facility (SCF), normal-charging facility (NCF), and fast-charging facility (FCF). The resulting optimization model employs mixed-integer second-order cone programming (MISOCP) and aims to minimize the annualized social cost of the whole EV charging system (investment cost of EVCSs, grid reinforcement cost, O&M cost of EVCSs, network losses cost). The research in [27] establishes a GIS-based multi-objective Particle Swarm Optimization model aiming at minimizing the total cost of charging station investment and maximizing the service coverage. Four types of charging stations are considered, each with different specifications, characteristics, and costs (AC, DC, rated voltage, etc.). The research in [28] searches for locations, capacity options, and service types for EV charging stations, after estimating charging demand from GPS trajectory data. To solve the corresponding mixed-integer linear programming (MILP) model, a hybrid evolutionary algorithm that combines the non-dominated sorting genetic algorithm-II (NSGA-II) with linear programming (LP) and neighborhood search is proposed. The method adopted in [29] addresses the EVCS distribution problem through an approach in which, apart from the optimal location, the number of the required charging stations is calculated to cover charging needs. The EVCS placement is estimated through the employment of Genetic Algorithms that estimate the EVCS longitudes and latitudes.

In summary, existing research generally considers optimization models that develop the optimal placement of charging stations upon a predefined set of candidate sites (facility-to-location configuration). Also, the number of stations composing the network is generally considered fixed, with the exception of the research studies in [20] and [29], in which the optimization variables include the number of charging stations, with [20] formulating the problem as a facility-to-location one. This paper aims to fill these gaps by developing an optimization model that determines the optimal number and placement of charging stations freely within the application area (facility-to-site formulation) to cover the distributed EV charging demand. The problem is developed in a bi-objective model with two objectives: minimizing the total cost for the charging station deployment (construction, operational, and maintenance costs) and maximizing service quality by minimizing the average traveling distance (and thus, cost) between demand spots and station locations, while considering any station capacity constraints. This analysis intends to enrich previous research efforts in several directions, in terms of both numerical and practical considerations. The work includes a number of case studies and numerical applications to assess the capability, applicability, and scalability aspects of the model. Overall, the present research aims at an in-depth study and better understanding of the problem characteristics and the quality and practicality of the solutions in a real-life environment.

The rest of the paper is organized as follows. Section 3 presents the problem description and the proposed model formulation. In Section 4, 10 case studies are developed and run with the corresponding results presented and commented on. Section 5 provides a brief discussion of the obtained results and the main conclusions of the study.

3. Proposed Model

Research findings show that EV adoption requires deploying charging infrastructure (in terms of number and location) in a manner that is financially feasible as an investment and also manages to satisfy demands and overcome EV users' range anxiety. To support this goal, a bi-objective model is introduced for deploying EV charging stations for minimizing the total charging station network cost and maximizing the service quality/user satisfaction by reducing travel distance.

The problem of charging station allocation in space falls within the general class of the Quadratic Assignment Problem (QAP). There are two general methods for dealing with this problem, the facility-to-location and the facility-to-site approach. In the first case, a number of potential sites for facility placement are predefined and the optimization algorithm seeks among them the best placement of the required facilities (stations). In the latter case, the algorithm freely assigns the facility to the solution space aiming at developing the best outcome based on the objective parameters and function. In the present formulation, the facility-to-site approach is primarily employed as it is difficult and rather impractical to predefine a reasonable number of potential station locations, except perhaps a few spots with known high charging demand (e.g., city downtown, shopping malls, etc.) or to exclude certain spots from the solution space if station deployment is infeasible (e.g., area topology, grid restrictions) or experience very low EV demand around it. The main shortcoming (and risk) of the facility-to-location approach is that, by selecting certain points for station deployment, one can omit other points that can potentially lead to better outcomes. In contrast, the facility-to-site approach can search all possible solutions but generally at the expense of higher computation effort. For the purpose of comparison, the problem has also been formulated as a facility-to-location one, based on grid discretization size, and the merits and demerits of both approaches are explored in Case Study 8 below.

In this model development, an area of interest (A) is considered to represent a city or a neighborhood. Within this area, a number of charging stations (either predefined or open to the optimization process) are to be spatially allocated. In its general form, the model seeks two outcomes, the optimal number of stations and their spatial location to best serve the existing EV demand. Two cost (fitness) functions are considered, one for station deployment (f1) and the other for EV users traveling from each origin to its nearest station (f2), which is proportional to the traveling distance. As the decision for station placement is a one-time decision in advance of the operational phase, Euclidean distances have been considered in this formulation. The actual distances as well as the traffic conditions are rather dynamic and may be more effectively considered at the later stage of EV distribution to stations analysis.

The mathematical form of the module is as follows. The demand points are set at predetermined coordinates (DPx, DPy). The decision parameters are the coordinates (CSx, CSy) of the charging stations. Two cost functions f1 and f2 are calculated as indicated in Equations (1) and (2).

$$f1 = C_{station} \cdot Si = (C1 + C2 + C3) \cdot Si \tag{1}$$

where $C_{station}$ is the total cost for station deployment, comprising the construction (C1), operational (C2), and maintenance (C3) costs, and Si is the number of stations. Although the above parameters have been taken here with fixed values in most of the analyses, it is easy to extend the problem structure to differentiate the costs depending on the size and deployment area of each station, as, for instance, in Case Study 4.

$$f2 = \sum_1^N d_{ij} = \sum_1^N \sqrt{(CSxi - DPxj)^2 + (CSyi - DPyj)^2} \tag{2}$$

where d_{ij} is the distance between the demand point j and its closest charging station i for each of the N demand points.

The objective function has the form:

$$\min F = w1 \cdot f1 + w2 \cdot f2, \tag{3}$$

where f1 and f2 are the station deployment and user traveling costs, respectively, and wi's are the weights of the sub-objectives.

There are no major constraints in the model other than constraining the search area within the city bounds. If there are infeasible zones within the search area, these zones are excluded by considering an extremely high station deployment cost (Case Study 10). Some additional constraints are introduced as part of specific case studies that are described in the next section. For instance, constraints on station capacity may be introduced to avoid large deviations in station utilization.

The optimal solution is sought via the application of Genetic Algorithms. These algorithms have been proven consistent and efficient in solving a wide range of NP problems, such as the one of electric vehicle charging station deployment. They rely on bio-inspired operators, including selection, crossover, and mutation across decision variable values (Figure 1). The chromosome consists of two parts: a string of the (x, y) coordinates of the charging stations and a string of binary values indicating whether a specific station is going to be deployed or not. For the latter part of the chromosome, if a station gets a value of 1, it is ordinarily evaluated based on its proximity to the demand points; otherwise, it is artificially excluded from evaluation either by using the "if" command or by assigning an additional high transport cost (penalty) to any demand assigned to this station. In each algorithm iteration, the distances from each station to all demand spots are calculated, and each spot is assigned to the nearest station. In this model, parents are chosen using a linear ranking-based mechanism. Then, 10% of the fittest members of the population are carried over to the next generation without any changes (crossover or mutation) to them (elitism). This offers a smoother selection probability curve and prevents good organisms from completely dominating evolution at an early point. The crossover process is implemented using a uniform crossover routine, meaning that instead of chopping the list of variables in a given scenario at some point and dealing with each of the two blocks ("single-point" or "double-point" crossover), two groups are formed by randomly selecting items to be in one group or another. The uniform crossover method is considered better at preserving and generating any schema from the two parents when compared to traditional x-point crossovers that could bias the search with the irrelevant position of the variables. The mutation process is performed by looking at each variable individually. A random number between 0 and 1 is generated for each of the variables in the organism, and if a variable gets a number that is less than or equal to the mutation rate (for example, 0.1), that variable is mutated. Mutating a variable involves replacing it with a randomly generated value (within its valid min-max range).

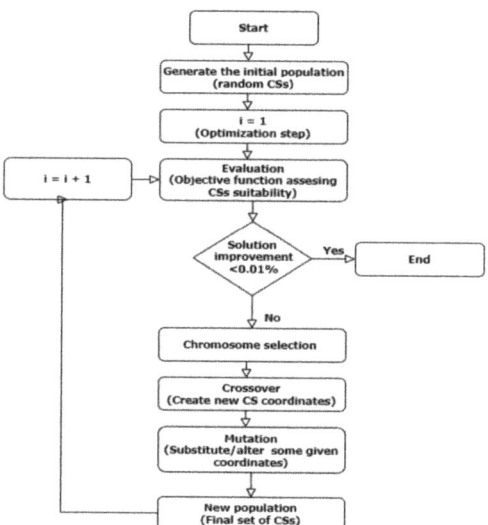

Figure 1. Genetic Algorithm flowchart.

Part of the analysis is the fine-tuning of the algorithm parameters for the specific problem to achieve better convergence within a reasonable computation time. In particular, the genetic algorithm is parameterized to use 50 chromosomes to form the initial population with crossover and mutation rates of 0.5 and 0.1, respectively. An iterative procedure of 200,000 trials, or 60 min of runtime, is used for all the scenarios that have been tested. Due to the stochastic nature of the GA, the algorithm was run a number of times (typically three or five) in each specific scenario, and the solution with the best fitness value was recorded. The result deviations were also recorded to provide an indication of the solution variability. The implementation has been done with commercial GA software (Palisade Evolver Version 8.3.2), which runs as an add-in to the Excel program. The employment of commercial software is considered adequate at this development stage, which mainly aims to set up the problem while providing more accessibility for practical use. Further, the current evaluation indicates that the solution surface is quite smooth without notable local minima; therefore, the software may be capable of reaching near-optimal solutions.

Finally, the employment of the Excel software facilitates the problem setting and test-case configuration.

4. Case Study Development and Results

To illustrate the algorithm application and demonstrate the contribution of the proposed model to the electric vehicle charging station (EVCS) deployment, a case study consisting of a 200-EV fleet is considered. The area of interest in which the EVCS network is deployed is a square one, extending from −50 to 50 in a coordinate system in which 1 length-unit may indicatively represent 100 m to simulate a medium-sized city of 100 km^2. The charging demand has been pseudo-randomly distributed all over the area of interest, while some specific sites concentrate a higher charging demand, representing places of different levels of interest. This type of allocation develops a balance between the necessary scattered distribution and some variation in demand concentration from point to point. In particular, there are 100 demand spots where 1, 2, or 3 EVs have been randomly allocated (Table S1). The number of charging stations varies from 1 to 25 in the core analysis, while larger numbers of stations are examined as part of a case study. The initial station allocation for the optimization deployment can be anyone, starting from the case of all stations coinciding at the center coordinates (this alternative is mostly used), progressing to a symmetrically scattered positioning within the application space, up to fully random placing in the area (Figure 2).

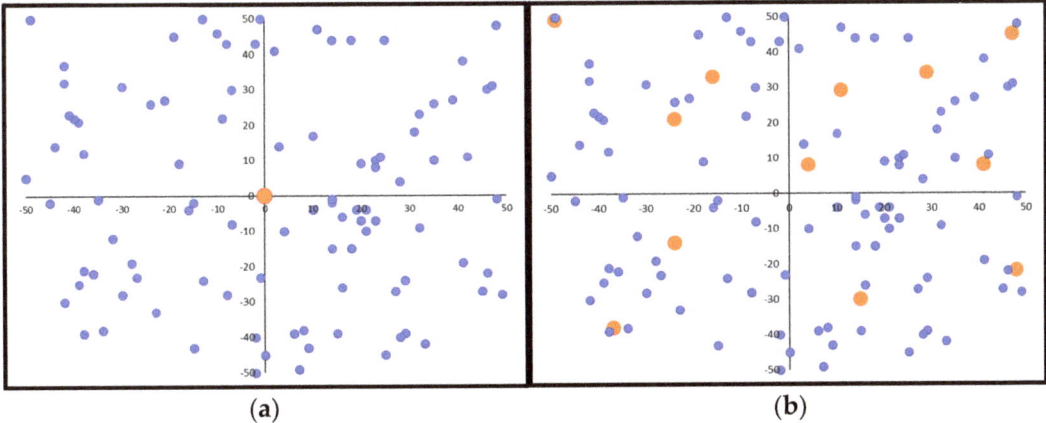

Figure 2. Demand scatter diagram (blue color) and charging station deployment (orange color) for the initial solution of the optimization: (**a**) center coordinates; (**b**) random distribution.

The efficiency of the proposed model has been tested through a number of case studies that are summarized in Table 1 and described in the following subsections. The aim of the analysis is to provide insight and evidence of the algorithm capability in terms of both the mathematical solution vigor and its response potency to practical considerations. In each case, multiple runs have been made (usually three) to account for the result variability due to the probabilistic nature of the genetic algorithm. Different values of the GA parameters (initial population size, crossover, mutation rates, finishing criterion) have been analyzed and appropriately set.

Table 1. Case study objectives and considerations.

Case Study No.	Objectives	Considerations
1	Optimal station allocation	Predefined number of stations, different initial stations positioning
2	Optimal station allocation	Different station capacity constraints
3	Optimal station number and allocation	Different weighting coefficients w1 and w2
4	Optimal station allocation	Different station deployment cost zones
5	Optimal station allocation	Varying EV charging demand levels
6	Optimal station allocation	Varying service area shape
7	Optimal station allocation	Large service area and number of stations
8	Optimal station allocation	Different grid discretization levels
9	Optimal station allocation	Two-phase station network deployment.
10	Optimal station allocation	Varying EV range capacity, infeasible zones

4.1. Case Study 1: Station Allocation for EV Traveling Distance (Cost) Minimization at Different Numbers of Stations

Initially, an exhaustive investigation of all scenarios representing the number of charging stations, from 1 to 25, was performed to develop the best station allocations (i.e., the lowest cumulative traveling distance of all EVs) at every number of stations. There are alternative configurations of station allocation that lead to approximately the same travel cost. Figure 3 presents three of them being developed by an initial station placement at the center of the area, a symmetrical initial station positioning, and a fully random initial station distribution within the analysis area. It appears that the algorithm converges to virtually the same station settlement regardless of the initial station distribution.

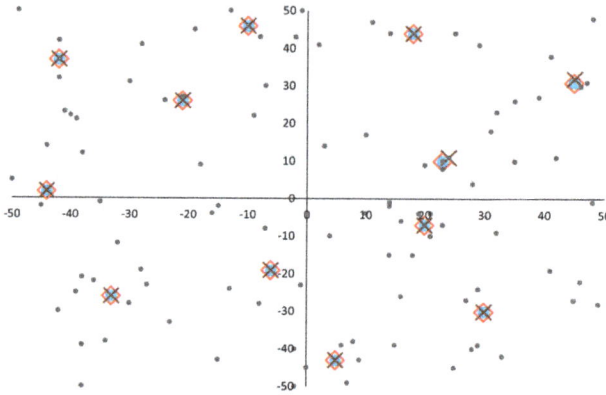

Figure 3. Optimal charging station deployment for the case of 12 stations and different initial station distributions: center placement (brown color), symmetrical placement (red color), random placement (blue color), demand points (grey color).

Obviously, as the number of stations increases and the network becomes denser, the EV user can find a nearby station to charge and therefore reduce the travel distance. Figure 4 presents the average travel distance per EV, depending on the number of stations. The distance ranges from 4.4 to 20.6 for 25 and 3 stations, respectively (these numbers increase radically for 1 or 2 stations, not shown in the diagram). Figure 5 depicts 2 indicative networks deployed through Case Study 1 applications, which consist of 12 and 20 stations, respectively. To provide a rough indication of the required computational time, the algorithm requires on a typical laptop computer 90, 150, 350, and 1100 s for 10, 15, 20, and 25 station networks, respectively.

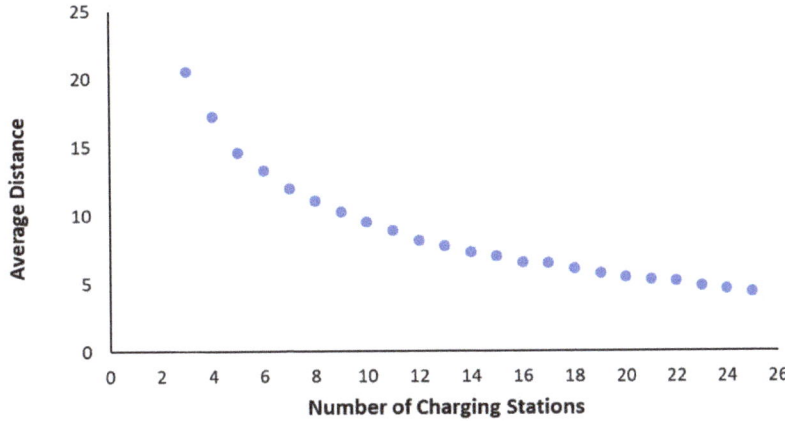

Figure 4. Trade-off diagram between the number of charging stations and the average EV travel distance.

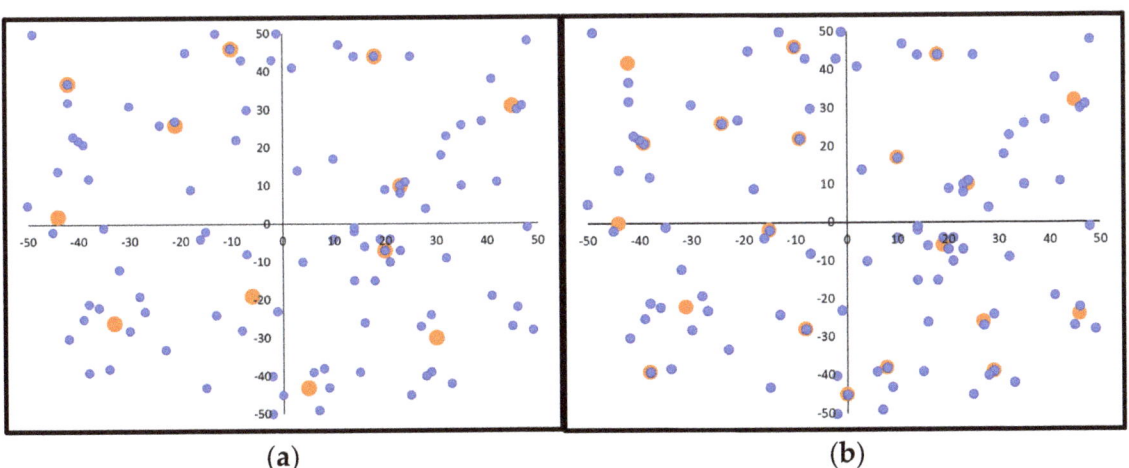

Figure 5. Indicative charging station deployment optimization results: (**a**) 12-station network; (**b**) 20-station network (demand points in blue color, charging stations in orange color).

4.2. Case Study 2: Station Allocation for EV Travel Distance (Cost) Minimization at Different Station Capacity Levels

The proposed methodology focuses on charging station allocation based on the minimum travel distance of the participating EVs. In every case, the assignment of each EV is given to its closest station. In this way, however, there may be cases where some stations can be loaded with high charging demand while others may serve very few EVs. Both cases may present adverse financial consequences. For this reason, in the present case study, the algorithm is expanded with constraints about station capacity, i.e., the maximum number of EVs that can be served (a similar analysis can be done for the minimum number of EVs). The result of such a constraint is that some EVs may need to travel to another station but the closest one, if the capacity of the latter is exceeded. The objective of the optimization, in this case, is to assign the extra EV to the second-closest station, if the capacity level permits.

In this analysis, the average number of EVs per charging station is considered (depending on the number of stations). Next, a station capacity tolerance (float) above the average value is assumed as a percentage increase. The results of the investigation are portrayed in Figure 6 for four indicative numbers of charging stations. As expected, the stricter the station capacity level, the longer the total (and average) distance traveled. The distance increases from the fully free EV allocation to a very restrictive station capacity of 10% above average, ranging between 7% and 21% without a specific pattern along the number of stations.

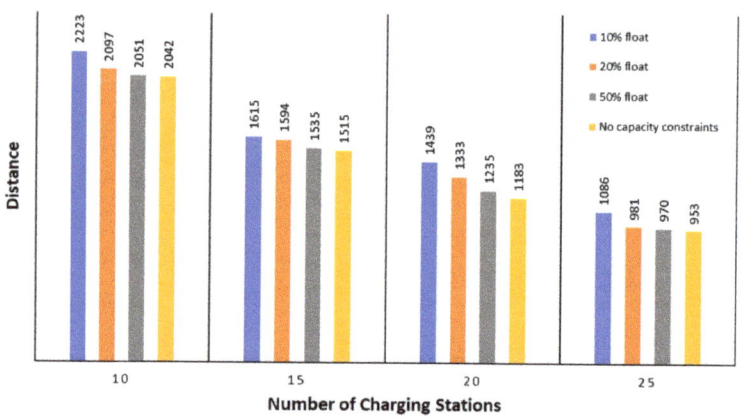

Figure 6. Total distance traveled by EV users as a function of the number and capacity of charging stations.

4.3. Case Study 3: Optimization of Station Number and Allocation Based on Station Deployment Cost and EV Travel Cost

Unlike most previous studies that solely focused on charging station allocation, the present algorithm can simultaneously handle the multi-objective problem of obtaining the optimal number of stations along with their optimal placement. This problem follows the classical trade-off between the cost of the charging station development, which relates to the number of stations and their construction cost, and the cost of EV total travel distance, which decreases with the number of charging stations. This means that as the EVCS network becomes denser, the EV user can more easily find a nearby available station to charge at and, therefore, reduce the daily travel distance. However, the number of charging stations should be determined and kept at a certain level, based on the charging demand, the size of the area of interest/service coverage, as well as any budget availability constraints. A number of runs have been made for different values of the w1/w2 ratio of Equation (3) (i.e., the relative weights of the model objective function). For instance, with w1/w2 = 100, the optimal number of stations is 12 (based on repetitive runs). The full picture of this investigation is depicted in Figure 7. At a low value of w1/w2 = 25, the optimal number of stations is 25, while at w1/w2 = 300, a network of 5 stations provides the best solution. The wi's are the input values representing each specific charging station deployment area and provide the appropriate number of stations that better serves the tradeoff between deployment cost and EV travel cost. In terms of computational effort, the multi-objective configuration requires more time (as it solves a more complex problem), which, compared to the single-objective travel distance problem, presents an increased time by an order of 1.3 to 1.5 times in average values.

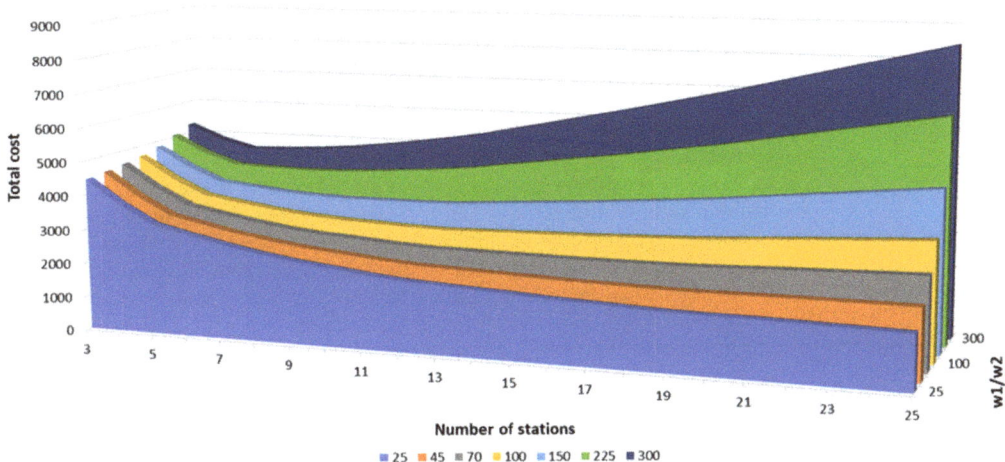

Figure 7. Optimal number of charging stations for different w1/w2 ratios.

4.4. Case Study 4: Station Allocation for EV Travel Distance (Cost) Minimization for Different Station Deployment Cost Zones

The previous analyses assume the same cost for deploying all charging stations within a scenario. However, it is known that different land uses and human activity concentrations create different cost zones within a city. To account for such cost variances, the present analysis considers four different cost zones in the form of inner to outer rectangles, with the cost being the highest in the inner zone and decreasing outward. Another parameter that plays a role in the charging station allocation is the cost rate among zones. As part of this study, different zone cost sets have been analyzed, with the expected result of station construction toward the cheaper zones, especially if the cost proportions are high. A representative allocation solution for this case is presented in Figure 8.

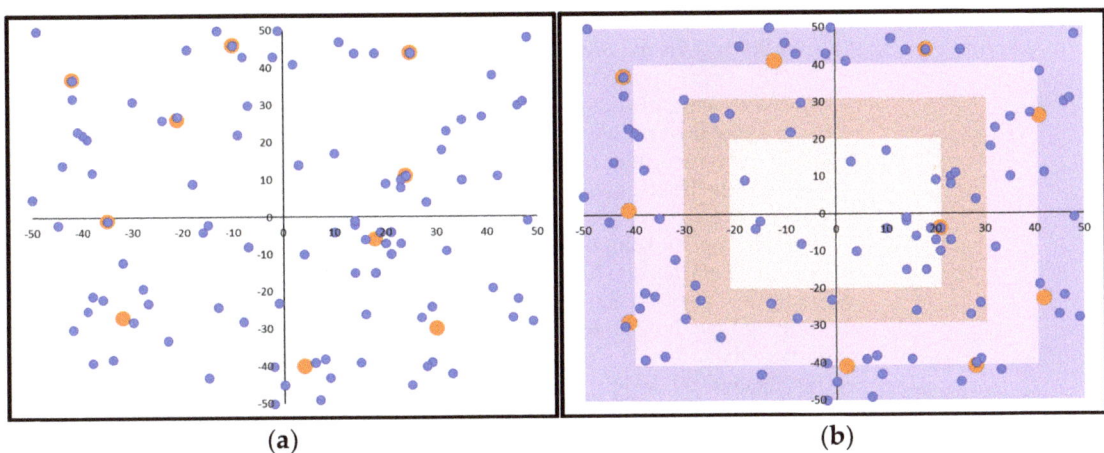

Figure 8. Optimal charging station deployment for 10 stations: (**a**) fixed station cost; (**b**) variable station cost (demand points in blue color, charging stations in orange color, highest cost zone in beige color, lowest cost zone in purple color).

4.5. Case Study 5: Station Allocation for EV Travel Distance (Cost) Minimization for Highly Unequal Demand Dispersion

The demand distribution is quite unequal in certain cases. For example, places of high interest (e.g., downtown premises, shopping malls, sport arenas, etc.) may attract a high charging demand. This may be modeled by a disproportionally increased demand at specific sites. To deal with this case, an example is formulated based on the original input data but considering 10 random locations having 5 or 10 times higher demand than all other spots. Here, 15 charging stations are considered for serving the EV demand. The station allocation results are indicated in Figure 9. The dark blue dots indicate low-demand spots, the light blue ones are high-demand spots, and the orange circles indicate station places. It can be seen that 10 out of 15 stations are placed exactly on or close to high-demand spots, while the other 5 stations are optimally placed to cover the sparse demand.

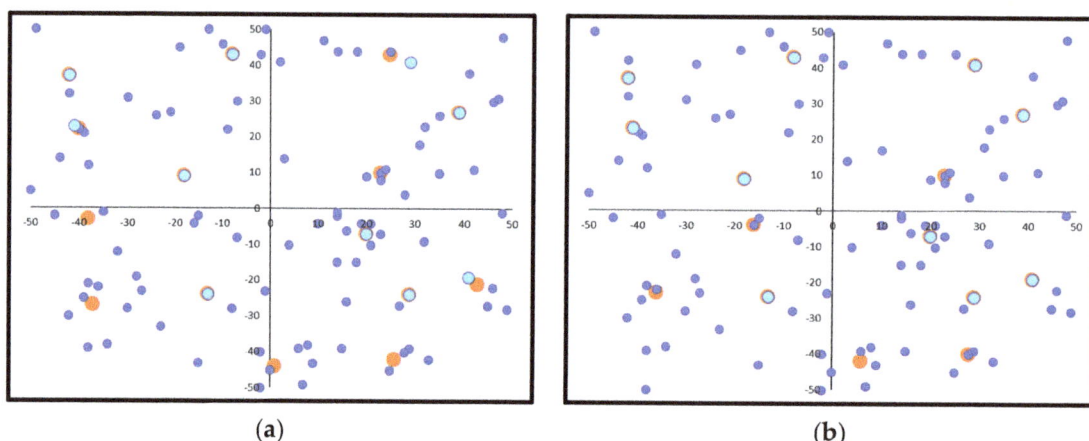

Figure 9. Indicative charging station deployment results for Case Study 5: (**a**) 5 times higher demand in specific points; (**b**) 10 times higher demand in specific points (normal demand points in dark blue color, increased demand points in light blue color, charging stations in orange color).

4.6. Case Study 6: Station Allocation for EV Travel Distance (Cost) Minimization for Irregular Service Area Shape

An interesting research question with practical importance is the adjustment of the solutions to the uneven shape of a typical city or neighborhood area. There are two ways to deal with this issue, either consider the city shape as it is or develop a square (or rectangle) that surrounds the city shape. In the first case, the solution space is reduced to the city shape; however, a large number of constraints should be added to model the city limits. In the latter case, the solution space increases, but the simplicity of the algorithm in terms of constraints is retained. As part of this analysis, the solution space of the basic scenario was extended, considering solution space dimensions ranging from -100 to 100 in both directions (4 times larger than the original area) and -200 to 200 in both directions (16 times larger than the original). The demand points and size remained as originally (within the -50 to 50 range) while different initial station allocations were considered, from all stations being at point (0, 0) to randomly assigned within the large box. Figure 10 provides an indicative solution for the case of an area ranging from -100 to 100, in which 10 charging stations are to be deployed, while Table 2 summarizes some results of the process regarding the fitness value of the total distance traveled. Not surprisingly, the algorithm, after some exploration of the whole space, has always converged to a solution very similar to the initial one and, interestingly, within the same computation time. This can highlight the GA converging capability regardless of the space size and the initial station placement.

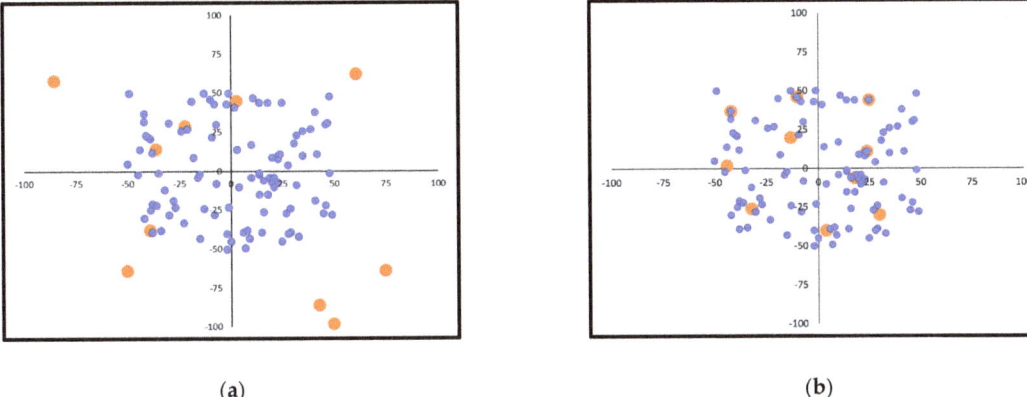

(a)　　　　　　　　　　　　　　　　　(b)

Figure 10. Indicative charging station deployment example for a −100 to 100 field range: (**a**) initial random solution; (**b**) final optimal solution (demand points in blue color, charging stations in orange color).

Table 2. Optimization distance results for Case Study 6.

Number of Stations	Field Range	Total Distance
10	−50 to 50	2042
	−100 to 100	2054
	−200 to 200	2053
20	−50 to 50	1170
	−100 to 100	1161
	−200 to 200	1165

4.7. Case Study 7: Station Allocation for EV Travel Distance (Cost) Minimization for Large Service Area Size and Number of Stations

Regarding the size of the area for which the analysis is performed, the following aspects need to be considered: depending on the city (or neighborhood) size, the number of stations can vary from a few to many. In fact, this decision has also to do with the demand level, in the sense that if the demand is low and dispersed, it is not cost-effective to build a large number of stations. Instead, the users should travel a longer distance to reach a station. Within the present analysis, a relatively small number of stations have been considered (up to 25). This is very realistic for neighborhoods and small-to-medium-size cities, e.g., 10 by 10 km (thus every distance unit in the above analysis corresponds to 100 m). To evaluate the scalability performance of the algorithm, larger station numbers have been further considered (within the same plan structure and demand size and distribution). In particular, 50, 100, and 200 stations were examined. The total traveling distances for 50, 100, and 200 stations are 411, 197, and 105, respectively. Nevertheless, since the number of demand generated spots is 100, there is no need to go over that number in terms of stations built. The indicative results of this analysis are shown in Figure 11.

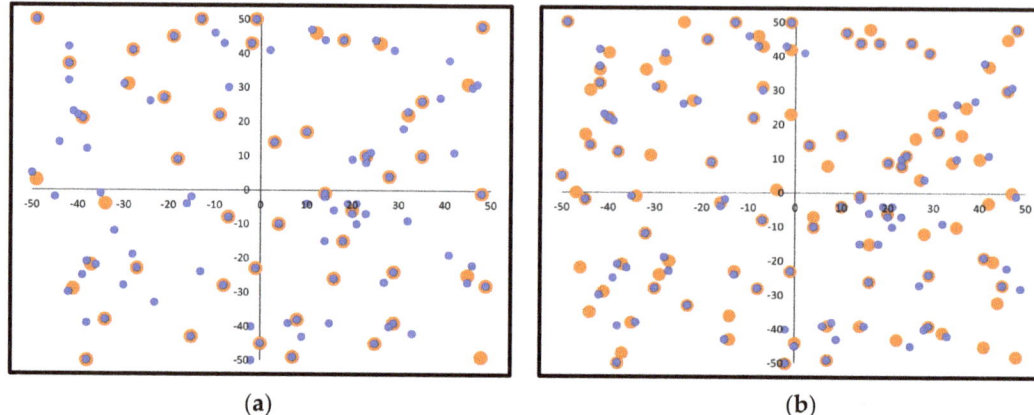

Figure 11. Indicative charging station deployment optimization results for Case Study 7: (**a**) 50-station charging network; (**b**) 100-station charging network (demand points in blue color, charging stations in orange color).

A similar analysis assumes that there are 50 (100) demand spots of 4 (2) EVs each. The same number of stations is considered, with the expectation being that one station is placed exactly at each demand spot. Although no full convergence on this goal is obtained, the solution indicates that most stations coincide with demand spots, while a few others attract EVs from more than one demand point, making the use of some other stations unnecessary. In Figure 12, the station allocations for the cases of 50 or 100 uniform demand points (with a total demand of 200 EVs) are presented to indicate the outcome. In the case of 50 stations, 4 of them are redundant (they do not attract any demand) and can be omitted. In the case of 100 stations, 23 of them are redundant.

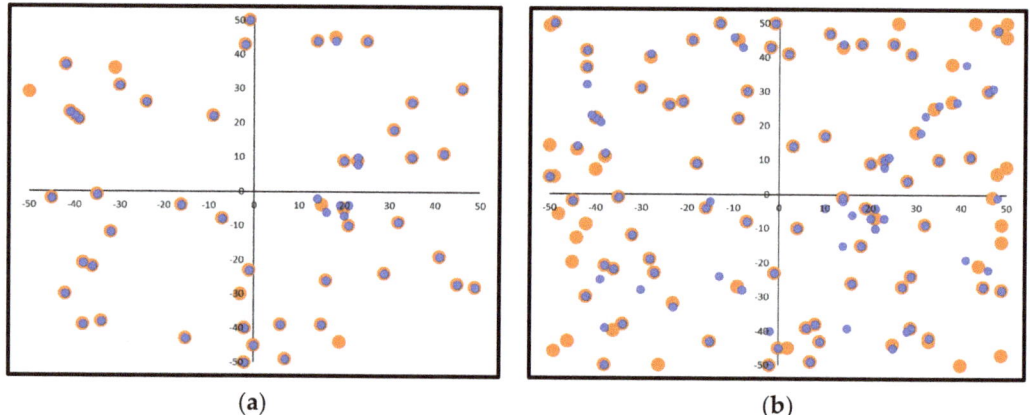

Figure 12. Indicative charging station deployment optimization results for Case Study 7: (**a**) 50 demand points; (**b**) 100 demand points (demand points in blue color, charging stations in orange color).

4.8. Case Study 8: Station Allocation for EV Travel Distance (Cost) Minimization at Grid Discretization Levels

The issue of grid discretization is important for the following reasons. The area size under analysis may vary considerably along city sizes and the achievement of a certain level of detail is a desirable aspect. On the other hand, fine discretization may lead to difficulties in achieving an optimal solution in terms of both, the fitness value and the computational time. In this regard, a number of grid discretization alternatives have been examined as part of this investigation. All the solutions presented above have been obtained by considering station coordinates with integer values. In the present analysis, different discretization levels are considered in the basic scenario. As the discretization level becomes coarser, the problem progressively moves from a facility-to-site to a facility-to-location formulation. For instance, if the grid unit is 10, there are 11 by 11 possible locations for allocating the charging stations in the −50 to 50 range of the rectangle area, and the objective is to select which of these points will host a station. Such rough discretization may be preferable if the application space is widespread, e.g., in large cities. As the number of possible locations decreases, the number of alternative solutions also decreases, as does the optimization quality in terms of the fitness value. On the other hand, a possible benefit of such rough discretization may be any reduction in computation time. Not surprisingly, the computational time is not improved because the required computations are virtually the same, as the only change is the set of values that the decision parameters (station coordinates) can choose from. In fact, even with a fine decimal level of discretization, the computation burden is not increased, and the resulting fitness value is comparable to that of the integer discretization. Figure 13 illustrates the results for 10 stations and discretization levels of 1 and 10 units. The total traveling distance for discretization levels of 1, 2, 5, and 10 units is 2042, 2045, 2078, and 2193, respectively. The lowest of the above values is also achieved with any decimal discretization of the application area.

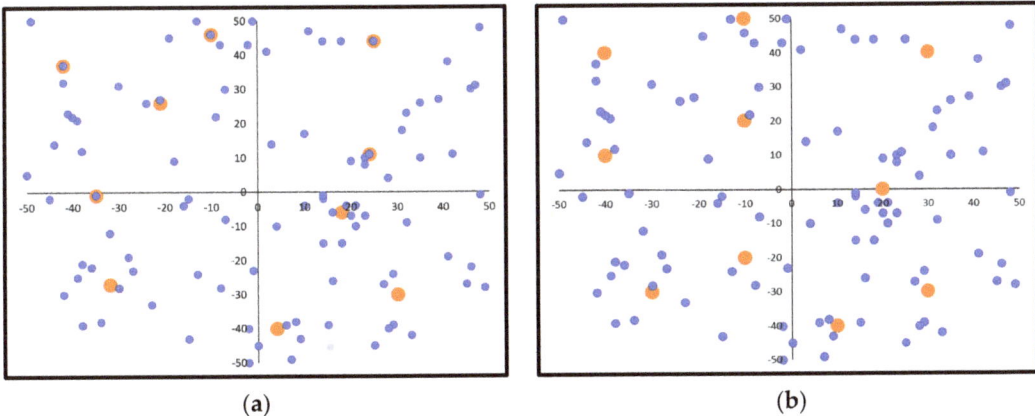

Figure 13. Indicative deployment optimization results of a 10-station network for different grid discretization: (**a**) 1 unit discretization; (**b**) 10 units discretization (demand points in blue color, charging stations in orange color).

4.9. Case Study 9: Station Allocation for EV Travel Distance (Cost) Minimization at Progressive Station Network Deployment

This analysis refers to the practical consideration of the dynamic demand evolution over time. The EV market and charging demand are still in their infancy in many places around the world but they grow fast. Although a full-scale station development could be done now with the expectation of full market growth in a few years, an alternative cost-effective solution is the ongoing (phased) station development. This approach balances the demand growth and the service provision without binding unnecessary resources for building several stations from the beginning to serve initially sparse EV charging demand. This progressive development may also be of benefit if the EV demand, besides increasing, presents a different space allocation in the future, as a result of city development. For the purposes of the present study, it is assumed that the EV demand retains the same space allocation which is also a realistic scenario. The analysis is performed in two stages. Initially, 10 charging stations are optimally allocated (with reference to the basic scenario). In the second phase, these 10 stations are considered fixed, and another 10 stations are further allocated in an optimal way to improve the total EV traveling distance. The final solution is compared to the one with a single-phase allocation of 20 charging stations. Figure 14 comparatively illustrates the two scenarios, one deploying 20 stations at once and a second deploying 10 stations nowadays and another 10 stations in a future stage. It can be seen that the total traveling distance in these two cases is 1170 and 1236, respectively, indicating an increase in the second case of about 6%.

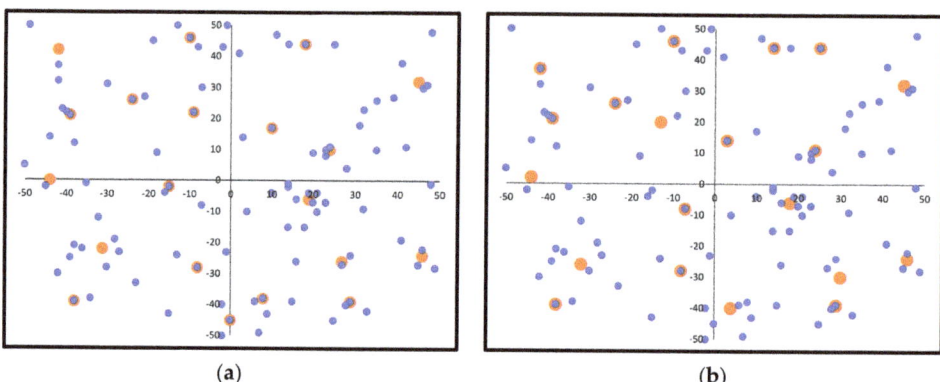

Figure 14. Indicative deployment optimization results of a 20-station network: (**a**) single-phase allocation; (**b**) two-phase allocation (demand points in blue color, charging stations in orange color).

4.10. Case Study 10: Station Allocation for EV Travel Distance (Cost) Minimization Considering EV Range Capacity and Infeasible Zones for Station Deployment

In this part of the analysis, two further cases are considered, the implication of EV range limitations to reach a station and the potential existence of infeasible zones for station deployment. In regard to the traveling range limitation, the 12-station network that was examined in Case Study 1 is revisited. Different range thresholds are considered to indicate the maximum distance the EVs can travel. Table 3 illustrates some indicative results. If the permittable range is higher than 27.31 units, which is the maximum EV traveling distance in the case of free station allocation, the optimization results coincide with those of Case Study 1 in terms of station spatial distribution and the objective function value. Below the threshold of 27.31 units, station reallocation is needed to satisfy the maximum range constraint at the expense, however, of the total (and average) traveling distance. The lowest limit of the maximum traveling distance that can lead to a feasible solution is 17.03 and can be found by minimizing the max trip distance of all EVs.

Table 3. Optimization results for different EV traveling range constraints.

EV Range Constraint (Units)	Total Distance (Units)	Average Distance (Units)	Maximum Distance (Units)
∞	1762	8.81	27.31
27.31	1762	8.81	27. 31
26	1769	8.85	25.50
24	1779	8.90	23.71
22	1797	8.98	21.93
20	1871	9.35	19.92
18	2068	10,34	18
17.03	2208	11.04	17.03

The second part of this Case study examines the practical implication of having infeasible zones for station deployment within the search area. This case can be considered a special instance of Case Study 4 (different cost zones) by setting an extremely high-cost value within the infeasible zones. An indicative example is presented in Figure 15 where the inner zone is considered to be infeasible for station deployment. The results indicate that five stations initially allocated in this zone (green color) are moved outside it with some of them placed just outside the infeasible region to serve EVs from this area. The corresponding fitness values for the two cases in terms of total distance are 948 and 1019, respectively.

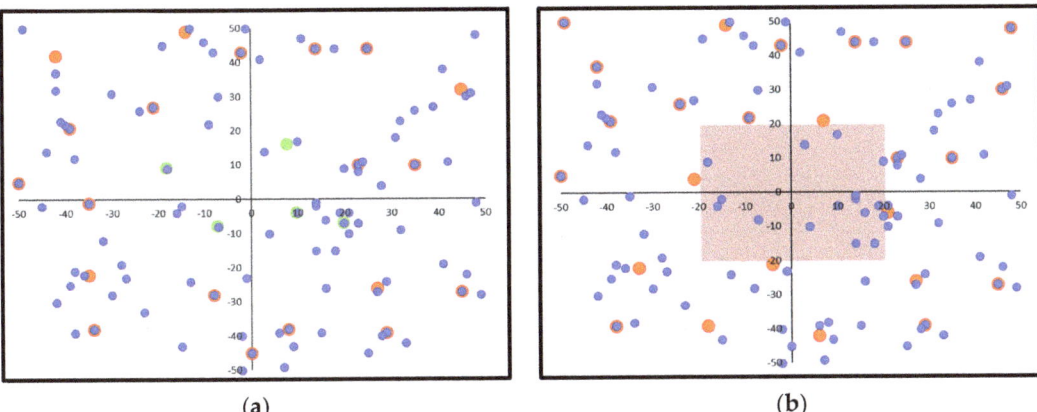

Figure 15. Indicative deployment optimization results of a 25-station network: (**a**) without infeasible zones; (**b**) with infeasible zones.

5. Discussion

The EV charging station deployment within a spatial area is a combinatorial problem that aims to place the stations in such a pattern that facilitates different objectives, the most important being the cost of station deployment and the necessary traveling distance of EV users to reach a station. The problem includes several numerical and practical considerations that need to be considered to enhance its practical applicability in the real world. In this paper, a simple yet effective optimization model has been developed and tested in several cases to provide a rather complete view of its capability, applicability, and scalability in tackling practical problems.

Focusing on the main analysis and result features, the model considers the two cost functions described above with certain weight coefficients and can be quite easily implemented in a spreadsheet mode. Based on that, several aspects of the problem can be modeled and tested through an optimization software. Among them, some interesting findings include the following: The model can provide an optimal solution for both the number and placement of charging stations based on the demand level and dispersion as well as the cost of station construction and user movement to the nearest station. The optimal station positioning outcome is slightly affected by the area size and the number of demand spots. In fact, the level of solution detail (both in terms of the fitness value and the required computation time) may be controlled by the grid discretization of the analysis area. As anticipated, the more coercive discretization leads to a declining efficiency in terms of fitness value. In the case of very aggregate discretization, the problem may resemble the facility-to-location structure. The model responds expectedly to variations in the charging demand levels and assigns stations close to or right at high-demand spots.

The case studies that were analyzed in the previous section aimed at highlighting the practical implications and summarizing the usefulness of this research. The consideration of several characteristics that simulate the actual EV charging problem and environment (in terms of both optimization goals and constraints) and the exploration of methods for modeling these characteristics provide a tool for efficiently addressing the real-world challenges associated with this problem. In fact, it appears that many of the real-world characteristics can be modeled quite efficiently by exploring suitable ways to describe them with mathematics and developing the respective optimization problem.

The charging station allocation represents the first stage of the whole optimization process. The second part that integrates the process is the EV distribution to stations. Unlike the former part, which is rather of a stationary nature, the second one is highly dynamic as the demand distribution typically changes over time while the actual routes from EV demand spots to any nearby station as well as the traffic conditions need to be taken into account.

Supplementary Materials: The following supporting information can be downloaded at: https://www.mdpi.com/article/10.3390/app13084867/s1, Table S1: Dataset used for the case study development and analysis.

Author Contributions: Conceptualization, V.L. and A.C.; methodology, V.L. and A.C.; software, V.L. and A.C.; validation, V.L. and A.C.; formal analysis, V.L. and A.C.; investigation, V.L. and A.C.; resources, V.L. and A.C.; data curation, V.L.; writing—original draft preparation, V.L. and A.C.; writing—review and editing, V.L. and A.C.; visualization, V.L. and A.C.; supervision, A.C.; funding acquisition, V.L. All authors have read and agreed to the published version of the manuscript.

Funding: The present work was financially supported by the "Andreas Mentzelopoulos Foundation".

Institutional Review Board Statement: Not applicable.

Informed Consent Statement: Not applicable.

Data Availability Statement: Data is contained within the Supplementary Materials.

Acknowledgments: The author Vasiliki Lazari expresses her deep gratitude to the "Andreas Mentzelopoulos Foundation" for the financial support of her doctoral study.

Conflicts of Interest: The authors declare no conflict of interest. The funders had no role in the design of the study; in the collection, analyses, or interpretation of data; in the writing of the manuscript; or in the decision to publish the results.

References

1. Shafiei, M.; Ghasemi-Marzbali, A. Fast-Charging Station for Electric Vehicles, Challenges and Issues: A Comprehensive Review. *J. Energy Storage* **2022**, *49*, 104136. [CrossRef]
2. Mastoi, M.S.; Zhuang, S.; Munir, H.M.; Haris, M.; Hassan, M.; Usman, M.; Bukhari, S.S.H.; Ro, J.S. An In-Depth Analysis of Electric Vehicle Charging Station Infrastructure, Policy Implications, and Future Trends. *Energy Rep.* **2022**, *8*, 11504–11529. [CrossRef]
3. Kong, W.; Luo, Y.; Feng, G.; Li, K.; Peng, H. Optimal Location Planning Method of Fast Charging Station for Electric Vehicles Considering Operators, Drivers, Vehicles, Traffic Flow and Power Grid. *Energy* **2019**, *186*, 115826. [CrossRef]
4. Meng, J.; Yue, M.; Diallo, D. Nonlinear Extension of Battery Constrained Predictive Charging Control with Transmission of Jacobian Matrix. *Int. J. Electr. Power Energy Syst.* **2023**, *146*, 108762. [CrossRef]
5. Pal, A.; Bhattacharya, A.; Member, S.; Chakraborty, A.K.; Member, S. Planning of EV Charging Station with Distribution. *IEEE Trans. Ind. Appl.* **2023**. [CrossRef]
6. Lam, A.Y.S.; Leung, Y.W.; Chu, X. Electric Vehicle Charging Station Placement: Formulation, Complexity, and Solutions. *IEEE Trans. Smart Grid* **2014**, *5*, 2846–2856. [CrossRef]
7. Kizhakkan, A.R.; Rathore, A.K.; Awasthi, A. Review of Electric Vehicle Charging Station Location Planning. In Proceedings of the 2019 IEEE Transportation Electrification Conference (ITEC-India), Bengaluru, India, 4–9 December 2019. [CrossRef]
8. Ahmad, F.; Iqbal, A.; Ashraf, I.; Marzband, M.; Khan, I. Optimal Location of Electric Vehicle Charging Station and Its Impact on Distribution Network: A Review. *Energy Rep.* **2022**, *8*, 2314–2333. [CrossRef]
9. He, J.; Yang, H.; Tang, T.Q.; Huang, H.J. An Optimal Charging Station Location Model with the Consideration of Electric Vehicle's Driving Range. *Transp. Res. Part C Emerg. Technol.* **2018**, *86*, 641–654. [CrossRef]
10. Efthymiou, D.; Chrysostomou, K.; Morfoulaki, M.; Aifantopoulou, G. Electric Vehicles Charging Infrastructure Location: A Genetic Algorithm Approach. *Eur. Transp. Res. Rev.* **2017**, *9*, 27. [CrossRef]
11. Tao, Y.; Huang, M.; Yang, L. Data-Driven Optimized Layout of Battery Electric Vehicle Charging Infrastructure. *Energy* **2018**, *150*, 735–744. [CrossRef]
12. Liu, Q.; Liu, J.; Liu, D. Intelligent Multi-Objective Public Charging Station Location with Sustainable Objectives. *Sustainability* **2018**, *10*, 3760. [CrossRef]
13. Tian, Z.; Hou, W.; Gu, X.; Gu, F.; Yao, B. The Location Optimization of Electric Vehicle Charging Stations Considering Charging Behavior. *Simulation* **2018**, *94*, 625–636. [CrossRef]

14. He, S.Y.; Kuo, Y.H.; Wu, D. Incorporating Institutional and Spatial Factors in the Selection of the Optimal Locations of Public Electric Vehicle Charging Facilities: A Case Study of Beijing, China. *Transp. Res. Part C Emerg. Technol.* **2016**, *67*, 131–148. [CrossRef]
15. Niccolai, A.; Bettini, L.; Zich, R. Optimization of Electric Vehicles Charging Station Deployment by Means of Evolutionary Algorithms. *Int. J. Intell. Syst.* **2021**, *36*, 5359–5383. [CrossRef]
16. Anjos, M.F.; Gendron, B.; Joyce-Moniz, M. Increasing Electric Vehicle Adoption through the Optimal Deployment of Fast-Charging Stations for Local and Long-Distance Travel. *Eur. J. Oper. Res.* **2020**, *285*, 263–278. [CrossRef]
17. Erbaş, M.; Kabak, M.; Özceylan, E.; Çetinkaya, C. Optimal Siting of Electric Vehicle Charging Stations: A GIS-Based Fuzzy Multi-Criteria Decision Analysis. *Energy* **2018**, *163*, 1017–1031. [CrossRef]
18. Kong, C.; Jovanovic, R.; Bayram, I.S.; Devetsikiotis, M. A Hierarchical Optimization Model for a Network of Electric Vehicle Charging Stations. *Energies* **2017**, *10*, 675. [CrossRef]
19. Mortimer, B.J.; Hecht, C.; Goldbeck, R.; Sauer, D.U.; De Doncker, R.W. Electric Vehicle Public Charging Infrastructure Planning Using Real-World Charging Data. *World Electr. Veh. J.* **2022**, *13*, 94. [CrossRef]
20. Xiao, D.; An, S.; Cai, H.; Wang, J.; Cai, H. An Optimization Model for Electric Vehicle Charging Infrastructure Planning Considering Queuing Behavior with Finite Queue Length. *J. Energy Storage* **2020**, *29*, 101317. [CrossRef]
21. Huang, Y.; Kockelman, K.M. Electric Vehicle Charging Station Locations: Elastic Demand, Station Congestion, and Network Equilibrium. *Transp. Res. Part D Transp. Environ.* **2020**, *78*, 102179. [CrossRef]
22. Zhang, Y.; Wang, Y.; Li, F.; Wu, B.; Chiang, Y.Y.; Zhang, X. Efficient Deployment of Electric Vehicle Charging Infrastructure: Simultaneous Optimization of Charging Station Placement and Charging Pile Assignment. *IEEE Trans. Intell. Transp. Syst.* **2021**, *22*, 6654–6659. [CrossRef]
23. Yi, T.; Cheng, X.; Zheng, H.; Liu, J. Research on Location and Capacity Optimization Method for Electric Vehicle Charging Stations Considering User's Comprehensive Satisfaction. *Energies* **2019**, *12*, 1915. [CrossRef]
24. Wang, Y.; Liu, D.; Wu, Y.; Xue, H.; Mi, Y. Locating and sizing of charging station based on neighborhood mutation immune clonal selection algorithm. *Electr. Power Syst. Res.* **2023**, *215*, 109013. [CrossRef]
25. Genevois, M.E. Locating Electric Vehicle Charging Stations in Istanbul with AHP Based Mathematical Modelling. *Int. J. Transp. Syst.* **2018**, *3*, 1–10.
26. Luo, L.; Gu, W.; Zhou, S.; Huang, H.; Gao, S.; Han, J.; Wu, Z.; Dou, X. Optimal Planning of Electric Vehicle Charging Stations Comprising Multi-Types of Charging Facilities. *Appl. Energy* **2018**, *226*, 1087–1099. [CrossRef]
27. Zhang, Y.; Zhang, Q.; Farnoosh, A.; Chen, S.; Li, Y. GIS-Based Multi-Objective Particle Swarm Optimization of Charging Stations for Electric Vehicles. *Energy* **2019**, *169*, 844–853. [CrossRef]
28. Bai, X.; Chin, K.S.; Zhou, Z. A Bi-Objective Model for Location Planning of Electric Vehicle Charging Stations with GPS Trajectory Data. *Comput. Ind. Eng.* **2019**, *128*, 591–604. [CrossRef]
29. Akbari, M.; Brenna, M.; Longo, M. Optimal Locating of Electric Vehicle Charging Stations by Application of Genetic Algorithm. *Sustainability* **2018**, *10*, 1076. [CrossRef]

Disclaimer/Publisher's Note: The statements, opinions and data contained in all publications are solely those of the individual author(s) and contributor(s) and not of MDPI and/or the editor(s). MDPI and/or the editor(s) disclaim responsibility for any injury to people or property resulting from any ideas, methods, instructions or products referred to in the content.

Article

Semantic-Based Multi-Objective Optimization for QoS and Energy Efficiency in IoT, Fog, and Cloud ERP Using Dynamic Cooperative NSGA-II

Hamza Reffad [1] and Adel Alti [1,2,*]

[1] LRSD, Faculty of Sciences, Computer Science Department, University Ferhat Abbas Sétif 1, Sétif 19000, Algeria; hamza.reffad@univ-setif.dz
[2] Department of Management Information Systems, College of Business & Economics, Qassim University, Buraidah 51452, Saudi Arabia
* Correspondence: a.alti@qu.edu.sa; Tel.: +966-(0)55-990-1163

Citation: Reffad, H.; Alti, A. Semantic-Based Multi-Objective Optimization for QoS and Energy Efficiency in IoT, Fog, and Cloud ERP Using Dynamic Cooperative NSGA-II. *Appl. Sci.* **2023**, *13*, 5218. https://doi.org/10.3390/app13085218

Academic Editor: Vincent A. Cicirello

Received: 1 April 2023
Revised: 18 April 2023
Accepted: 19 April 2023
Published: 21 April 2023

Copyright: © 2023 by the authors. Licensee MDPI, Basel, Switzerland. This article is an open access article distributed under the terms and conditions of the Creative Commons Attribution (CC BY) license (https://creativecommons.org/licenses/by/4.0/).

Abstract: Regarding enterprise service management, optimizing business processes must achieve a balance between several service quality factors such as speed, flexibility, and cost. Recent advances in industrial wireless technology and the Internet of Things (IoT) have brought about a paradigm shift in smart applications, such as manufacturing, predictive maintenance, smart logistics, and energy networks. This has been assisted by smart devices and intelligent machines that aim to leverage flexible smart Enterprise Resource Planning (ERP) regarding all the needs of the company. Many emerging research approaches are still in progress with the view to composing IoT and Cloud services for meeting the expectation of companies. Many of these approaches use ontologies and metaheuristics to optimize service quality of composite IoT and Cloud services. These approaches lack responsiveness to changing customer needs as well as changes in the power capacity of IoT devices. This means that optimization approaches need an effective adaptive strategy that replaces one or more services with another at runtime, which improves system performance and reduces energy consumption. The idea is to have a system that optimizes the selection and composition of services to meet both service quality and energy saving by constantly reacting to context changes. In this paper, we present a semantic dynamic cooperative service selection and composition approach while maximizing customer non-functional needs and quickly selecting the relevant service drive with energy saving. Particularly, we introduce a new QoS energy violation degree with a cooperative energy-saving mechanism to ensure application durability while different IoT devices are run-out of energy. We conduct experiments on a real business process of the company SETIF IRIS using different cooperative strategies. Experimental results showed that the smart ERP system with the proposed approach achieved optimized ERP performance in terms of average service quality and average energy consumption ratio equal to 0.985 and 0.057, respectively, in all simulated configurations compared to ring and maser/slave methods.

Keywords: ERP; services composition; QoS optimization; context-aware; IoT-Fog cloud service ontology; multi-agent strategies; improved NSGA-II algorithm

1. Introduction

The Internet of Things (IoT) is a technology recently introduced in the industrial field to provide many services facilities, including data collection for service monitoring and analysis, control, and maintenance. Thus, this technology is revolutionizing the ways factories and industrial organizations improve their tasks. With the technological development in the IoT industrial world, emphasis has been placed on designing advanced industrial applications, including manufacturing, predictive maintenance, smart logistics, and energy networks.

For Enterprise Resource Planning (ERP) [1], the Industrial Internet of Things (IIoT) is crucial as it can provide support for widespread business tasks, including smart tracking items, a smart Cloud ERP system using warehouse robotics and drone deliveries, and so on. With the fast advances of Cloud computing and IIoT, we have witnessed many paradigms to assure the quality and safety of many business tasks. Those solutions provide essential advantages for significant quality enhancement of business outputs and services. Moreover, Cloud and IIoT can provide smart and flexible ERP deliveries, which have become a great shift for Small and Medium Enterprises (SMEs). Nevertheless, the difficulty lies in mastering the evolution of customer needs as well as the considerable growth in the number of IoT devices that include heterogeneous smart sensors. These devices significantly increase the amount of data exposed to processing. Therefore, it is crucial for ERP when managing an increased number of IoT devices, including CPU load, latency, bandwidth, and energy.

On the other hand, a great deal of network latency and congestion is managed through Fog computing. This is a new computing model, which is placed between the Cloud and IoT devices, and is a very important solution for facilitating the deployment of relevant IoT services. Furthermore, it is essential to know that the selection of relevant services is not a standardized solution. It can vary according to specific enterprise needs [2]. To reach the goal, more advanced and intelligent techniques might be used via the exploitation of additional semantic information to effectively select and compose services among a large number of service candidates.

To this end, researchers shift out to support intelligent QoS-aware service selection and composition approaches as a solution that could make a significant contribution to deal with a wide range of IoT, Fog, and Cloud services. Furthermore, selecting and composing services can be achieved in three main ways: (1) ontologies [3–9], (2) metaheuristic [10–21], and (3) IoT-Fog enabled services [22,23]. Each one of them has its advantages as well as its drawbacks. Recently, Polyakov et al. [24] proposed an approach to estimate the QoS of composite services based on intuitionistic fuzzy sets. The importance of the proposed approach is due to the fact that it can be applied to all types of service systems. What can be deduced from techniques mentioned is that none of them can achieve an optimum consensus between QoS, energy, and execution time. These techniques are not efficient and flexible enough to support service failures, and none of them can achieve an optimum consensus between service quality, service speed, service cost, continuous changes of customer's needs, limited device energy, etc. In fact, they consume a large amount of processing time to intelligently find optimal composite services. Furthermore, existing works lack a parallel strategy that effectively finds relevant services, customizes ERP business processes, and improves the quality of the solution. A hybrid dynamic multi-agent system that combines both genetic and cooperative agents can solve the optimization of the service composition problem in an acceptable amount of time. A multi-agent system can discover and select relevant services by launching many intelligent agents to potential sites based on the ontology of the IoT-Fog-Cloud semantic service model in order to optimize when composing business processes both in terms of QoS and energy. In addition, smart and mobile devices with limited energy for the deployment of services should not be neglected. The most striking case is the exponential increase of heterogeneous IoT-based services and the continuous evolution of various business requirements based on both users' requirements and preferences. So, it is difficult to obtain ideal customization results of ERP systems and realize an effective service selection and composition method for ERP applications. There is thus a requirement to explore the relative importance of a semantic-based constraint violation strategy to ensure energy saving in IoT devices.

We focus on semantic web technology, smart devices, and, especially, the fragmented system process of enterprise departments using cooperative intelligent agents, which are widely used despite the fact that the advantages of these technologies can help us to build an effective and efficient business process management model.

1.1. Research Gaps

Despite the significance of the IIoT for the flexibility of smart ERP, many such technologies frequently do not include both QoS and energy attributes. This is because both customer needs and the energy capacity of IoT devices are frequently not considered in existing smart ERP management approaches. As business tasks are influenced by their quality energy contexts, it is important to manage such services and energy contexts during the selection and composition of appropriate services at the runtime.

1.2. Motivations and Contributions

This paper presents an efficient and effective cooperative method for smart ERP IoT-Fog-Cloud service selection and composition. It extends the CCS-2S component of the global framework architecture introduced in [25] to enhance the precision and optimize the execution time of IoT with Fog and Cloud services composition in order to offer useful and optimal business processes depending on the user's QoS constraints and contextual preferences. In particular, we introduce a new QoS energy violation with a leverage energy-efficient cooperative strategy. Since there exist different types of cooperative strategies for service composition, we propose a QoS-Energy aware multi-agent strategy, which leverages relevant services in the selection phase and optimizes energy consumption. The proposed study is very important because it is carried out in one of the new research fields, especially in light of the improvement of distributed business processes in Algerian companies, since semantic-based context-aware strategies can help to filter business services that violate both energy and customer constraints. An evaluation was undertaken of the proposed method using several kinds of agents through the IoT business process of the IRIS company in Setif city in Algeria. The study achieved promising ERP performance in terms of QoS and saving energy. The contributions to this dynamic and adaptive smart ERP are outlined as follows:

- We extend the previous model of CxQSCloudSERP ontology [25] with a variety of industrial IoT concepts such as robotics, drones, I-IoT, etc. for describing rich information of ERP services to ensure shifting and business agility. It enables the classification of services by category, role, QoS, and energy consumption.
- We propose a Semantic Context-aware Agent-based Concrete Composite Service with three stages (SCwA-CCS-3S) for smart ERP to accelerate the service composition task and offer customers a flexible personalized smart ERP.
- We propose an improved SCwA-CCS-3S, which provides a three-stage composition and optimization process. First, we introduced a novel dominance relationship with a new QoS energy violation degree through NSGA-II (Non-Dominated Sorting Genetic Algorithm II) [17]. Intelligent cooperative agents aim to quickly discover available composite services that meet customer's constraints. We then selected a composite service (concrete Business Process) among the set of all generated composite services, according to customer's QoS preferences. Finally, we adapted a composite service for managing unpredicted events, such as changes of customers' QoS needs, critical sensor events, and low latency that have stopped working.
- The proposed approach targeted the problem of agility and dynamicity of ERP in IRIS Setif Company of Algeria. It is based on different cooperative multi-agent strategies to compose business services quickly and inexpensively.

1.3. Organization of the Paper

The organization of this paper is as follows. Section 2 presents a motivating case study. Section 3 provides a review of different service composition solutions. An extended CxQSCloudSERP ontology model is introduced in Section 4. Section 5 describes the problem description. The proposed framework is detailed in Section 6. Section 7 presents experiments and the interpretation of results. Finally, the last section concludes this work.

2. Motivating Case Study: IRIS SETIF Company

In order to create a clear scientific vision of the research problem, this section analyses the IoT services and fog computing to illustrate their utility and effectiveness in distributed business processes using a real-life scenario. This scenario is based on the production and transport business processes of IRIS Company of SETIF city, Algeria. This company manufactures electronic products and provides wheels and mobile IT services to clients. These services are accessible through the IRIS Web portal. The IRIS services are hosted on a Cloud provider using the SaaS service type to provide additional support for services that are not in-house. This is illustrated in Figure 1, which covers the Cloud and the user application layers. A centralized Cloud ERP proxy selects and composes quality Cloud services based on IRIS company's needs and execution context. The Cloud ERP proxy, however, is in charge of satisfying the company's needs and preferences, deploying and maintaining ERP business processes. When the number of requests becomes enormous, it slows down the execution speed of the Cloud ERP proxy with considerable latency and can thus influence the QoS.

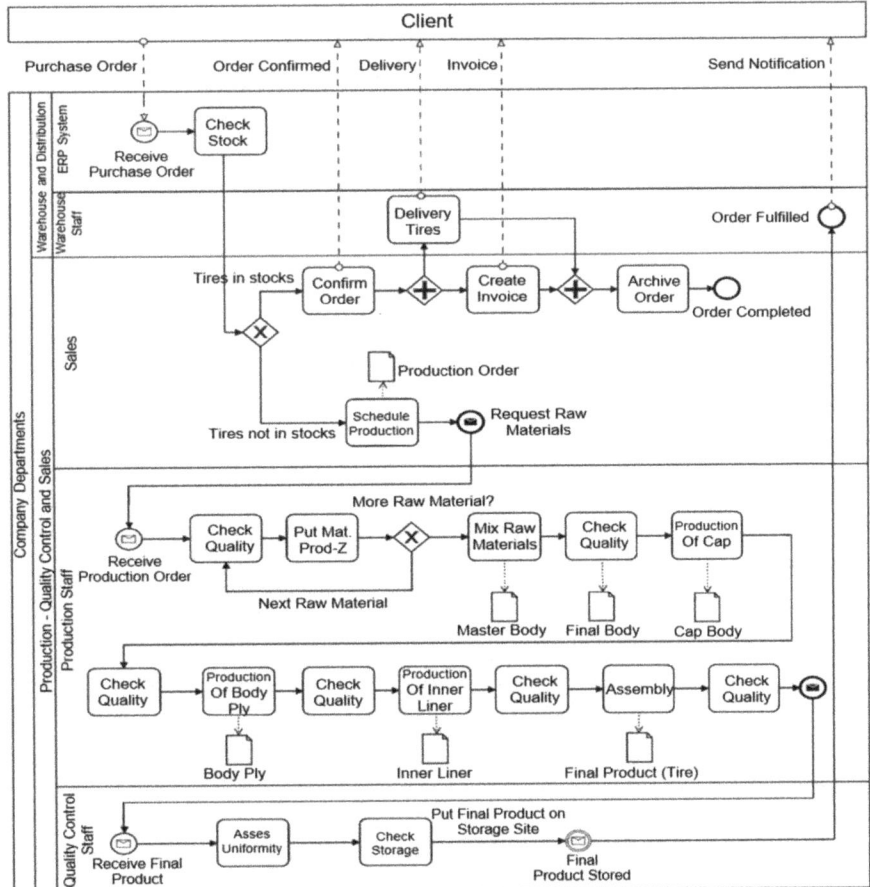

Figure 1. Cloud-based ERP production business process for IRIS.

Besides, IRIS is facing a continuous increase in business and suffering from resource limitations that can lead to production delays and significant economic losses to control raw materials and products in many different sites with different storage conditions of its growth cycle. Thus, its current traditional Cloud ERP software has become inefficient and lacks flexibility and efficiency in production and transport management services. To overcome these constraints, it is necessary to modernize current traditional Cloud ERP services and introduce modern technologies, such as industrial IoT services, in this company to improve production and adopt effective transport. Many of benefits of industrial IoT that have been identified include: (1) robots to optimize the transport of raw materials and products on different locations, (2) smart sites equipped with sensors and actuators to ensure remote control of suitable storage conditions for growing products, (3) production time and energy cost optimization to effectively transport the appropriate quantity of raw materials, and (4) synchronization and monitoring of different robots with ERP services. We aim to dynamically integrate industrial IoT with existing production and transport management processes. All the challenges faced by integrating industrial IoT with existing Cloud ERP are as follows:

- How could we integrate different mobile robotics and monitoring services to ERP?
- How to choose reliable IoT services to be used in Cloud business processes to get better QoS at optimized cost?
- How can we manage the increase in IoT services while increasing the amount of data?
- How would manage data among IoT services and Cloud ERP affect latency and QoS?

As services can be heterogeneous, there will be different data formats. Furthermore, different types of protocols might be used. Thus, it will be difficult to integrate such robotics and compose them with existing business processes. To handle these problems, a semantic ontology model is proposed. The ontology undertakes this problem by hiding a large amount of heterogeneous IoT data and their different communication protocols. However, the selection and integration of quality IoT services are still difficult and differ according to the context of ERP application, time constraints, and evolution of the QoS needs and preferences of the company (non-functional needs and preferences). For example, an IoT service that detects incorrect data in real-time can cause a service failure or synchronization problem detection. Thus, it increases the complexity of IoT integration to the business processes. If the complexity of the ERP business processes augments using the semantic model, the most efficient way to resolve it is to combine IoT-Fog-Cloud services with the semantic web and multi-agent approach to make the business processes dynamic and optimal for every QoS and context change. Furthermore, Cloud-based ERP requires a high execution time to integrate and select quality services from a large candidate service and further requires both high IoT data processing and transmission time. A slight execution time and transmission delay mismatches the SLA agreement. To handle such a scenario, fog computing provides better QoS for production and transports IoT Cloud-based business processes. Finally, the estimation of the QoS of different services should meet comprehensive model normalization for appropriate prediction and management of QoS in ERP.

Figure 2 shows the use of an intelligent agents-based ontology to discover and select relevant services according to customer's needs, as well as fog components to improve QoS and transmission delay and reduce Cloud access by delocalizing business services at decentralized fogs. The proposed research work needs to provide:

1. A description of IoT services with richer semantic information for seamless integration with ERP business processes. Moreover, they need to integrate mobility-related features to represent mobile devices and monitor context changes in their locations.
2. The management of business processes efficiently and the optimization of QoS needs and preferences (i.e., the parallel discovery of services and effectively optimizing business processes based on context-aware multi-objective QoS attributes using multi-agent strategies).
3. Context monitoring (e.g., customer's needs, enterprise, and services) and the management of various adaptation tasks on multiple business process parts using a dynamic parallel strategy (e.g., easy replacement of a service with another equivalent, fast integration of new composite IoT service, and continuous discovery of available enhanced services to ensure the quality of ERP).

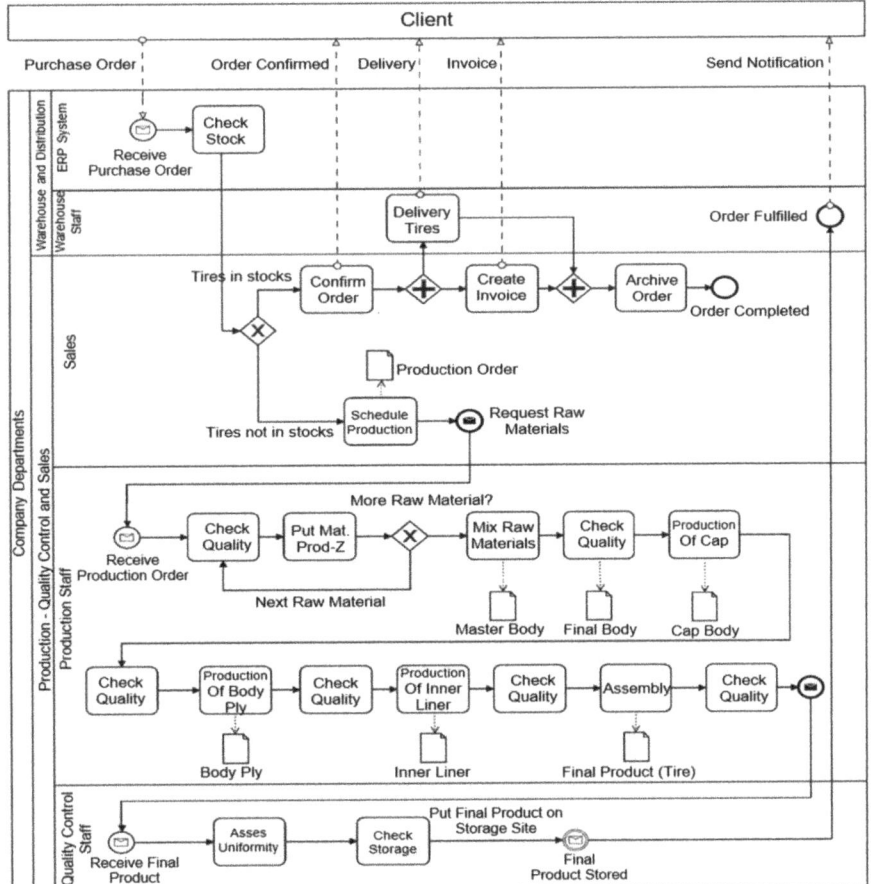

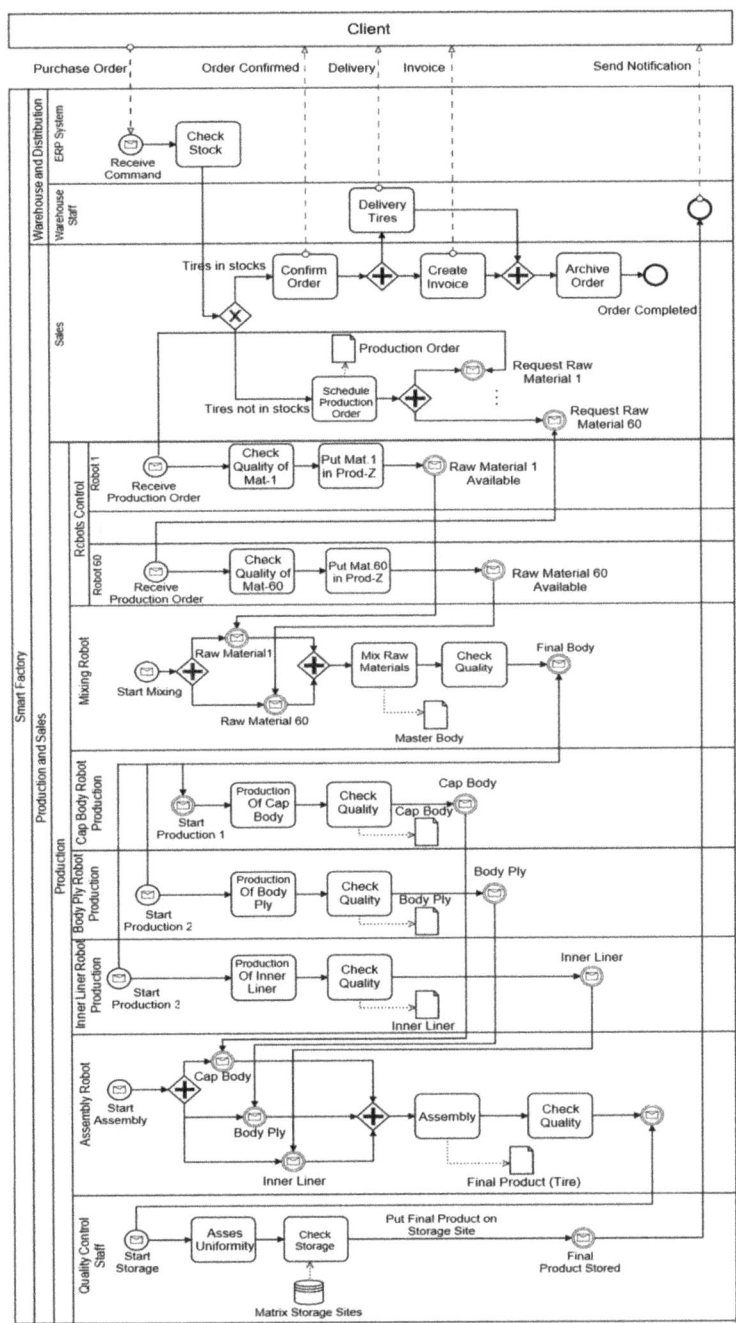

Figure 2. Proposed IoT Cloud-based ERP production business process for IRIS.

3. Related Work

In this section, we will review recent ontology models and optimization approaches related to IoT and Cloud ERPs. Most recent multi-objective optimization approaches can ensure Cloud ERP services description in general and integrate IoT services in particular. However, related works suffer from various limitations, including the energy of IoT device management and control in the context of business processes. Our work is based on semantic multi-objective optimization for QoS and energy efficiency for IoT-Fog-Cloud ERP using NSGA-II multi-agent cooperation.

3.1. Ontology Models for Representation of IoT and Cloud ERP Services

Our work deals with a semantic dynamic approach that produces and optimizes context-aware IoT business processes for industry 4.0 based on an ontology model and multi-agent strategies. Many IoT and Cloud ERP ontology models have been proposed in recent decades [3–9,25,26]. They have been used for different business domains but do not have enough concepts to manage context-aware constraints, describe all business changes, and manage quality semantically equivalent IoT services. However, they all share the following issues: (1) low coverage of business domain-independent semantic service information that describes the Cloud, Fog, and IoT services; (2) lack of consideration of flow sharing (e.g., data flow sharing, event flow sharing), sensor mobility, and usage constraints (e.g., service dependency, energy consumption, and time-space service accessibility), which helps to reduce the impact of context changes on service composition. We distinguish three main categories: IoT-based ontology model, Cloud ERP services ontology model, and combine both ontology models.

Andročec et al. [3] proposed an ontology model for digitalizing different business activities, including IoT services and Cloud ERP APIs. They aim to integrate IoT with various Cloud ERP APIs that can be used to add automated functionalities to ERP systems. They connect various resources with IoT devices and integrate them with Cloud ERP APIs in one module. However, this model neither includes the semantic concepts of contextual management constraints and preferences nor the execution context. Moreover, this work does not consider the classification of Cloud ERP APIs and neglects the semantic grouping of IoT services by functionality, category, role, and QoS, which is considered very important in services selection.

In [4], the authors present a new ontology model called OntoSLAM based on autonomous robots for solving simultaneous localization and mapping (SLAM). It described all aspects related to autonomous robots as well as temporal–spatial movement regions. They integrated it into the Robot Operating System (ROS) to increase the application flexibility. The ontology model relies on many ontology models, including ISRO ontology [5], POS ontology [6], CORA ontology [7], and KnowRob ontology [8], which leads to standard robotics metadata. However, the authors wasted most of their work in the classification of robots and physical things, and hence it neglected significant contextual and QoS constraints for efficiently managing robotic services.

Authors in [9] used semantic Web and ontologies (OWL-S and SAWSDL) for describing OPC UA industrial applications to ensure semantic service annotation and interoperability. The approach builds a common ontology that integrates the production abstract services with concrete IoT-based services. Concrete services will be used on the shop floor. This common ontology makes the approach more interoperable in all levels of automation and infers the potential information. It provides relevant information accurately that helps to describe business processes. Unfortunately, it suffers from contextual information and custom conditions at each stage of the production business process.

3.2. Service Cloud Composition and Optimization

The first research domain is related to techniques for the dynamic composition of Cloud services. Three main categories of approaches are distinguished: constraints-based [26,27], semantic-based [28,29], and metaheuristic-based [30,31]. Rosenberg et al. [26] explored constraint specification language to describe rich constraints. The authors conclude that service composition is related to explicit user constraints. However, this approach lacks semantic descriptions of heterogeneous services provided by different service providers. This heterogeneity raises the problem of the right decision making during the services selection [32]. In [28], Alti et al. automated the generation of quality composite service through context-aware service ontologies and semantic forward chaining. They grouped equivalent services by category, context, and QoS. However, this work lacks efficient and consistent quality Cloud services cooperation. Therefore, in such cases, reducing research time requires an intelligent cooperative strategy to obtain timely and quality solutions while the constraints are dealt at the right time.

From the above discussions, constraints with semantic models do not provide priority to the development and implementation of optimized management ERP systems as they are not aware of the importance of time in decision making. Other researchers recommend metaheuristic-based approaches as iterative methods to solve NP-hard optimization problems. Usually, two optimization approaches are used, which are outlined below.

3.2.1. Mono-Objective Approaches

Many QoS (time, cost, availability, etc.) are provided by Cloud services. The optimal composite Cloud service considered multiple QoS based on mono-objective techniques. It calculates the fitness function of each solution based on certain utilities to choose the optimal solution among feasible ones [33]. The mono-objective function based on the weighted sum [34] is used for optimization of multiple QoS with real cloud services using taboo research [10], GA [11], PSO [12], GWO [13], ACO [14], ICA [15], CHHO [16], and SMO [35]. However, results showed that the mono-objective function is effective with respect to the power loss reduction, but penalizes some QoS attributes.

3.2.2. Multi-Objective Approaches

Yao et al. [17] presented NSGA-II to enhance diversity preservation and convergence. Here, parent and offspring populations were generated and performed non-dominated sorting, and crowding operators were there to find out valuable non-dominated front. The Pareto-optimal front is reached with minimal iterations [36]. Wada et al. [19] modeled the multi-objective optimization problem with E^3 for ensuring service quality as well as effectiveness by employing multiple SLAs. Sadeghiram et al. [21] developed a distributed knowledge-based repair model to increase the efficiency and QoS satisfiability of NSGA-II. It considered QoS constraints for composition process and distributed nature of services, but failed to include industrial IoT and energy model. Peng et al. [37] proposed an improved multi-objective algorithm to seek a representative set of solutions with a dual strategy to adjust the usage of different creation operators using neighbor search in order to achieve a fine-grained search. In our work, we have included both QoS and energy aspects to enhance system performance.

3.3. IoT and Service-Based ERP

In [22], authors developed a CloudERP platform on which the client company customizes an entire ERP system to match their needs. CloudERP provides corporate customers with personalized ERP composite services from multiple providers. This work applied a genetic algorithm with a "rough set theory" and extracted web services from a web service platform. This purpose proposal stands out from the rest due to several elements, including the adaptive way that it reacts to changes in the customer context (customer-evolved QoS constraints and semantic preferences) and the Cloud service context (QoS) at any time.

Recently, in [23], Bolu et al. developed a task-planning approach for efficient management of IoT-based smart warehouses. This approach defines a heuristic model to select order tasks that are assigned to robots. It used a mathematical model to describe the spatial and temporal priority level of tasks, allowing multi-robots to perform tasks in less completion time. However, we noticed that scheduling a large number of tasks increased costs. Therefore, they can achieve effective results with other techniques such as early services selection.

3.4. Discussions

Table 1 presents a comparison between existing ontologies and optimization techniques for ERP semantic description and services composition in terms of (1) reusability, (2) extensibility, (3) dynamicity, (4) adaptability, (5) usage constraints (time-space constraints, services dependencies, energy), and (6) information coverage of thing diversity, thing mobility, service diversity, and events sharing.

Table 1. Review of existing ontologies, service composition, and IoT Cloud ERPs.

Related Works		Criteria	Reusability	Extensibility	Dynamicity	Adaptability	Usage Constraints				Information Modeling					
							Time Constraints	Spatial Constraints	QoS	Energy	Things Diversity	Things Mobility	Services Diversity	Service Dependency	Event Sharing Flow	Data Sharing Flow
	Ontologies	Andročec et al. [3]	✓	✓	×	×	×	×	×	×	×	×	✓	×	×	×
		OntoSLAM [4]	✓	✓	✓	×	✓	✓	×	×	✓	✓	×	×	×	×
		ISRO [5]	✓	✓	✓	×	✓	✓	×	×	✓	✓	×	×	×	×
		POS [6]	✓	✓	✓	×	✓	✓	×	×	✓	✓	×	×	×	×
		CORA [7]	×	✓	×	×	✓	✓	×	✓	×	×	×	×	×	×
		KnowRob [8]	✓	✓	×	×	✓	✓	×	✓	✓	✓	×	×	×	×
		SAWSDL [9]	✓	✓	✓	×	✓	✓	✓	×	✓	✓	✓	✓	×	✓
		SSN [7]	✓	✓	×	×	×	×	×	×	✓	✓	✓	×	×	×
Optimization approaches	Scalarizations	Latency [32]	×	×	✓	×	✓	✓	✓	×	×	×	×	×	×	×
		Taboo [10]	×	×	✓	×	✓	✓	✓	×	×	×	×	×	×	×
		GA [12]	×	×	✓	×	✓	✓	✓	×	×	×	×	×	×	×
		PSO [12]	×	×	✓	×	✓	✓	✓	×	×	×	×	×	×	×
		ACO [14]	×	×	✓	×	✓	✓	✓	×	×	×	×	×	×	×
		ICA [15]	×	×	✓	×	✓	✓	✓	×	×	×	×	×	×	×
		CHHO [16]	×	×	✓	×	✓	✓	×	×	×	×	×	×	×	×
	Multi-Object.	NSGA-II [17]	×	×	✓	×	✓	✓	✓	×	×	×	×	×	×	×
		SPEA2 [18]	×	×	✓	×	✓	✓	✓	×	×	×	×	×	×	×
		E3-MOGA [19]	×	×	✓	×	✓	✓	✓	×	×	×	×	×	×	×
		MOMS-GA [20]	×	×	✓	×	✓	✓	✓	×	×	×	×	×	×	×
		DWSC [21]	×	×	✓	×	✓	✓	✓	×	×	×	×	×	×	×
		QoS-Aware [33]	×	×	✓	×	✓	✓	✓	×	×	×	×	×	×	×
		SMO-MSF [35]	×	×	✓	×	✓	✓	✓	×	×	×	×	×	×	×
		Diff-Evolution [37]	×	×	✓	×	✓	✓	✓	×	×	×	×	×	×	×
IoT Service based ERP		CloudERP [22]	✓	✓	✓	×	✓	✓	✓	×	✓	✓	✓	×	×	×
		Bolu et al. [23]	✓	×	✓	×	✓	✓	×	×	✓	✓	✓	×	×	×
		Fuzzy-Estimation [24]	×	×	✓	×	✓	✓	✓	✓	✓	✓	✓	×	×	×
Proposed Approach			✓	✓	✓	✓	✓	✓	✓	✓	✓	✓	✓	✓	✓	✓

These criteria were chosen for several reasons. Reusability refers to how existing concepts can be reused across multiple applications, as well as the choice of extensibility criteria that determine whether or not the model is extensible for domain-specific business models. The choice of dynamicity refers to how existing models monitor context changes and adapt business processes dynamically. The choice of adaptability refers to how existing techniques adapt business processes and improves their quality. Usage constraints refers to how existing models describe spatiotemporal constraints. Limited resources and device constraints are parameters needed in managing sensor mobility in terms of energy consumption. We have identified the common limitations of the studied approaches, particularly for integrating hybrid heterogeneous services providers (IoT, fog, and Cloud) with a heuristic-based optimization strategy to dynamically compose rich services for ERP in smart factories.

Most existing works have neglected both energy and QoS optimization when monitoring one or more products to ensure their safety. We need to integrate mobility-related features to represent mobile devices and monitor changes in their locations and coverage area. Moreover, when examining service optimization approaches, we notice that most works in [10–22,34,36] use bio-inspired and meta-heuristics approaches, enabling the optimization of QoS criteria for composite Cloud service regardless of their semantics, such as category, role, energy, spatial temporal constraints, etc. Furthermore, other related works [3–9,23,25,26,38,39] integrate IoT and Cloud services-related semantic features to handle service heterogeneity but without optimizing QoS and energy for IoT-Fog-Cloud ERP. We notice that no approach covers semantic service description for IoT, fog, and Cloud ERP with QoS and energy optimization under sensor mobility and context changes.

Current works have neglected the management and control of energy of IoT devices when composing one or more services and do not propose an energy-aware strategy. In this work, efficient selection of multiple local solutions with a multi-agent approach based on heuristics that considers QoS and energy is defined. It is used to select IoT, Fog, and Cloud services in intelligent and dynamic ways.

4. CxQSIoT-Fog-CloudSERP: An Extension of CxQSCloudSERP Ontology Model

We extend Context-aware Quality Cloud ERP (CxQSCloudSERP) [25] to the limit of its centralized composition of ERP services, even though Cloud ERP services are associated with great IoT and fog services such as warehouse robotics and drone delivery services. The extension of CxQSIoT-Fog-CloudSERP aims to dynamically adapt smart ERP systems to usage contexts of IoT, Fog, and Cloud. The proposed ontology is built based on three semantic services categories, IoT, Fog, and Cloud services, which are used to describe business processes. It classifies context-aware IoT, fog, and Cloud services based on business category, business role, QoS, location, and time.

4.1. Abbreviations

Table 2 summarizes the different abbreviations used in this paper.

Table 2. Abbreviations used in this work.

Abbreviations	Description
ACO	Ant Colony Optimization
BP	Business Process
CCS	Concrete Composite Service
CCS-2S	Concrete Composition Service with 2 Stages
CHHO	Chaotic Harris Hawks Optimization
CxQSCloudSERP	Context-aware Quality of Semantic Cloud Service ERP
ERP	Enterprise Resource Planning
GA	Genetic Algorithm
GWO	Grey Wolf Optimization
ICA	Imperialist Competitive Algorithm
IIoT	Industrial Internet of Things
IoT	Internet of things
IoT-BP	IoT-based business process
MOGA	Multiple Objective Genetic Algorithm
MOMS-GA	Multi-Objective Multi-State Genetic Algorithm
NSGA-II	Non-dominated Sorting Genetic Algorithm
OWL-S	Semantic Markup for Web Services
SCwA-IFCCS-3S	Semantic Context-aWare Agents on IoT-Fog Concrete Composition Service with three Stages
SMA	Small and Medium Enterprises
PSO	Particle Swarm Optimization
QoS	Quality of Service
ROS	Robot Operating System
SAWSDL	Semantic Annotations for WSDL and XML Schema
SCwA-CCS	Semantic-Based Context-aWare Agents for the Composition of Cloud Services
SCwA-CCS-3S	Semantic-Based Context-aWare Agents for Composition of Services
SLA	Service Level Agreement
SMO	Sequential Minimal Optimization
SPEA2	Strength Pareto Evolutionary Algorithm
SWRL	Semantic Web Rule Language
VCS	Virtual Composite Service'
VS	Virtual Service

4.2. An Overview of CxQSCloudSERP

CxQSCloudSERP defines Cloud ERP in terms of virtual and concrete services-based business processes [25]. Figure 3 shows the CxQSCloudSERP ontology model of the context-aware services responsible for ERP management. The ontology is designed using the following main classes:

- **Customer:** Manages requirements and preferences of customers. It is specified with GUI and stored as ontology individuals. The customer's requirements are classified into functional needs and QoS constraints. Each functional needs to have a specific language based on logical operators (AND, OR, and Not). It is employed to infer virtual BP. The QoS constraint is split into regular intervals for each QoS attribute based on three semantic values:
    - High: quality values in $[qmin_i^{High}, qmax_i^{High}]$.
    - Medium: quality values in $[qmin_i^{Medium}, qmax_i^{Medium}[$.
    - Low: quality values in $[qmin_i^{Low}, qmax_i^{Low}[$.

    The constraint (C_i) of the quality attribute q_i is defined as follows:

    $$C_i = qmin_i^{semantic\_value} \quad (1)$$

    In CxQSCloudSERP, preferences are defined as an ordered list of QoS attributes that are assigned with semantic values. Based on this semantic value, automatically generate the weight w_i by using Equation (2):

    $$w_i = \frac{e^{P_i}}{\sum_{j=1}^{n} e^{P_j}} \quad (2)$$

    The computed priority P_i of the QoS q_i is:

    $$P_i = explicitPriority_i + constraintRank_i \times n \quad (3)$$

    where:
    - n: Number of QoS attributes.
    - *ExplicitPriority*: A priority level of QoS. It is defined by a value between 1 and n. The *explicitPriority* is null when the customer does not specify any preference.
    - *ConstraintRank*: The importance of QoS constraint. There are four priority levels: high, medium, low, and none.

- **Service:** The ontology presents unified business concrete services, such as buy, sale, finance, etc. Each service uses resources and they are linked to other services that provide or require these resources.
- **Business Process (BP):** This BP includes a set of business tasks (*parallel, iterative, choice, repeat*) delivering the same features with varied QoS.
- **Context:** Circumstances are defined as execution contexts. Each execution context responds to a functional need that has a specific language. Execution contexts provide an efficient solution to filter different services of a business process. In our case, we divide the context into three categories: service context, environment context, and customer context.
- **QoS:** A variety of quality attributes are defined in our ontology during ERP specification, design, and execution. QoS attributes are designed and built in such a way that every service is evaluated in perfect conditions and in a real environment.

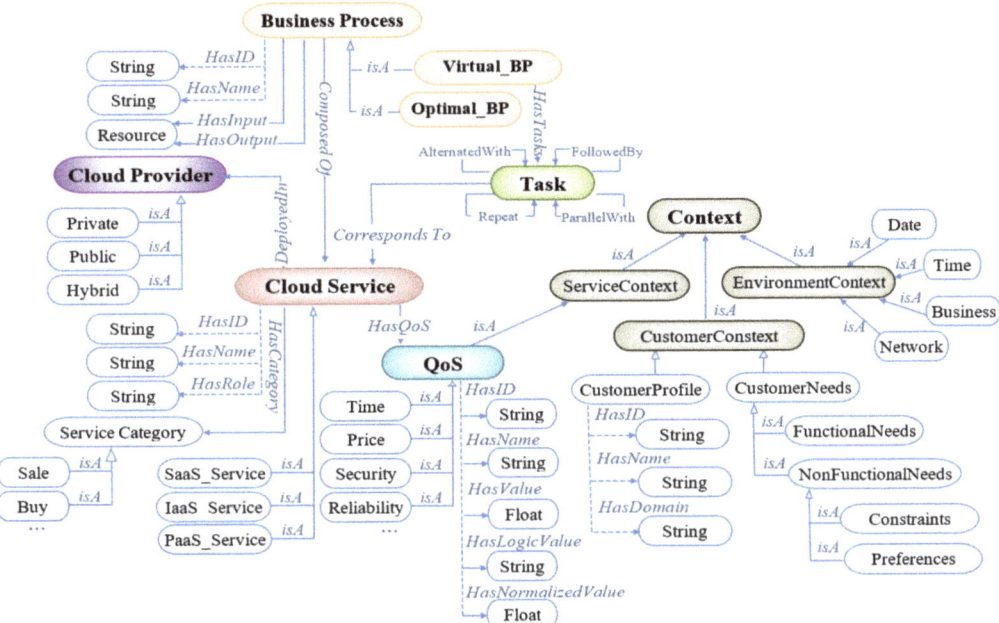

Figure 3. CxQSCloudSERP Ontology model.

4.3. CxQSIoT-Fog-CloudSERP

One of the main benefits of extending CxQSCloudSERP is that the business process can be made more flexible and interoperable through richer semantic service information of IoT field and Fog capabilities. It therefore becomes necessary to define entities related to IoT-based, Fog, and Cloud business processes, as illustrated in Figure 4.

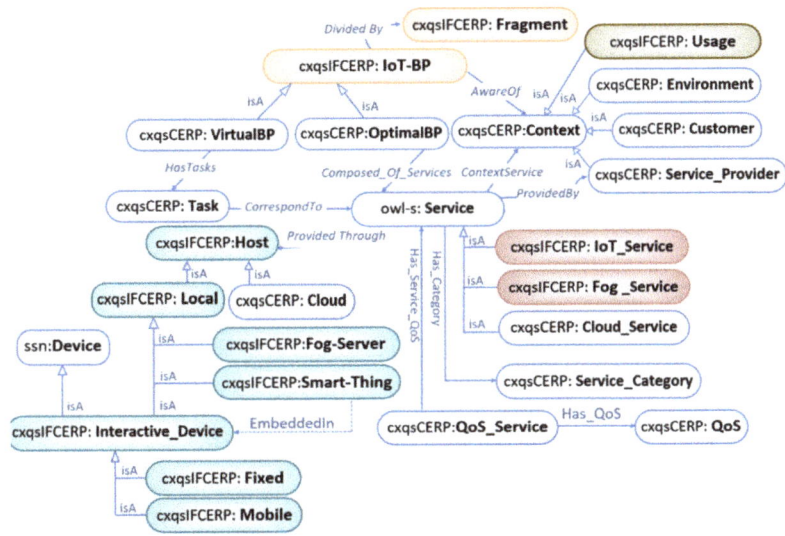

Figure 4. CxQSIoT-Fog-CloudSERP ontology model.

As illustrated in Figure 5, the service is deployed on top of the smart devices (sensors or actuators) or Fog servers as local hosts or the Cloud. Quality services are selected and composed to build intelligent, quality ERP business processes. However, business processes are mostly triggered by execution context and events.

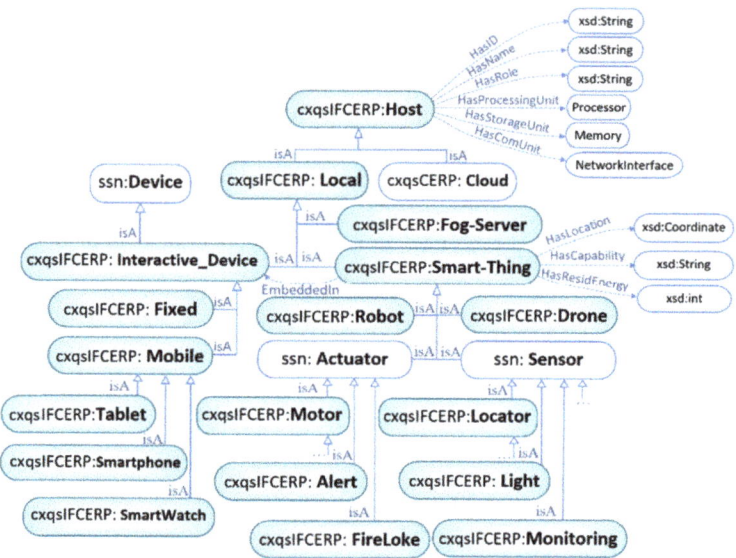

Figure 5. A semantic model of different hosts.

4.3.1. Modeling IoT, Fog, and Cloud Services

Figure 6 shows the extended ontology of CxQSCloudSERP with IoT sensors/actuators and Fog computing concepts.

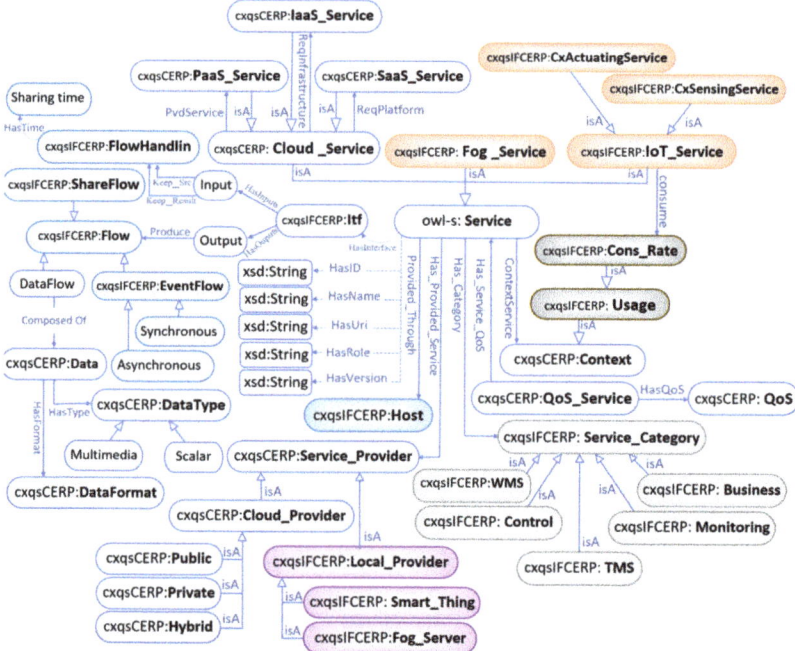

Figure 6. A semantic model of context-aware intelligent services.

In this model, a *Service* is divided into three types: context-aware IoT services (CxIService), Fog, and Cloud services which have many inputs and outputs. Inputs/Outputs can be further divided into two types, data flows defined by a "specific data type and data format", and an event flow handled by the action of a service. The service is described by an identifier, name, Uri, role, and category, and it is deployed on various hosts (IoT device, Fog server, or Cloud).

A service is defined by QoS properties such as reliability, availability, and security and a set of context quality properties such as energy. *IoT-Service* may be either *CxIEventListenerService* and *CxIActuatingService* to define the data collector and event listener and service's actions, respectively.

4.3.2. Modeling IoT-Based Business Process

Figure 7 presents a semantic model of context-aware intelligent services. In this model, an IoT-based business process (*IoT-BP*) may be modeled to provide automation facilities using sensors and actuators. Several types of sensors and actuators with different sensing and actuating capabilities may be included in IoT-based business processes. We define a new concept of *IoT-BP* as an extension of business processes as shown in Figure 7. At a minimum, IoT-BP consists of tasks that may be either *Cloud Service*, *IoT Service*, or *Fog Service*. A customer (enterprise) has some needs and preferences, such as high availability, low latency, better automation, and good flexibility. To do this, the Service provider and the customer agree on high-quality constraints and resource energy saving that define the services provided to the customer under rigorous conditions within the service and business process provided. After that, quality IoT-based services are selected, integrated into a business process, and deployed in IoT devices or Fog while reducing energy consumption.

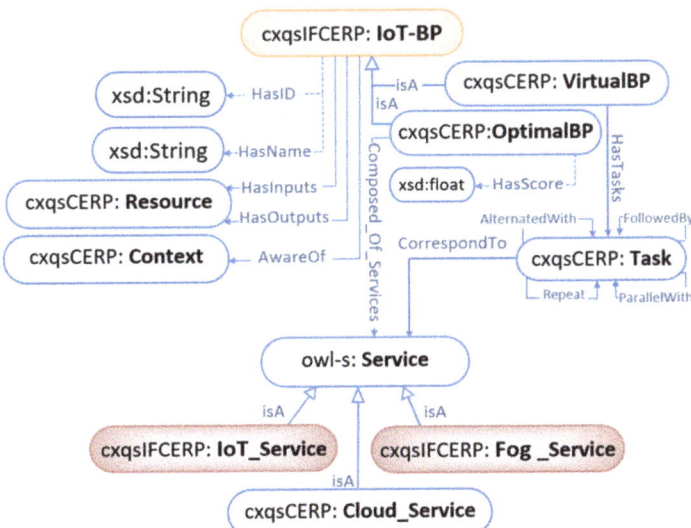

Figure 7. A semantic model of context-aware IoT business process.

4.3.3. Modeling Context

We extend the *Context* concept of CxQSCloudSERP with usage context to manage and save energy by better selecting the most appropriate services when producing and adapting concrete business processes. As illustrated in Figure 8, we introduce new concepts to *Usage context*: *service dependencies*, *service accessibility*, and *energy consumption*. We define accessibility through time constraints, space constraints, and space-time constraints. High levels of accessibility are a result of excellent services, which considers reliable and high service qualities. Energy consumption describes the average energy rate consumed when deploying services on top of IoT devices, Fog servers, or Cloud.

Figure 8. Usage context for smart and Cloud services.

5. Problem Formulation

Our work aims to come up with an intelligent multi-agent cooperative strategy to generate and manage IoT business processes. It aims to improve system performance and optimize energy when using limited smart devices. The proposed approach must describe the required and consumed energy of IoT services during the decision-making process. That will help the system to measure the energy consumption at the run-time and thus react rapidly to these changes.

5.1. Description of Physical Servers

- We consider three levels of hierarchical infrastructure $\mathcal{I}$, consisting of $\mathcal{N}$ physical nodes. The IoT layer (*level 0*) regroups a set of $\mathcal{T}$ smart devices, sensors, actuators, and mobile devices. The fog computing layer (*level 1*) is a set of $\mathcal{F}$ Fog servers and getaways.
- The Cloud layer (*level 2*) consists of a set of $\mathcal{C}$ large-scale data centers.
- The physical infrastructure consists of a set of nodes (or servers) $\mathcal{N}$. Each node $\mathcal{N} \in [1..\mathcal{T}]$ has the following features:
  - Each node N has a current workload $W_N^{current}$.
  - $Capacity_N$ denotes the energy capacity of the node N.
  - If N is a fixed node and continuously powered then its $Capacity_N = \infty$.

5.2. Description of Business Process

- We consider VBP a virtual business process consisting of m tasks $\mathcal{T} = \{t_1, t_2, \ldots, t_m\}$ described using the proposed ontology.
- Each concrete composite service CCS has m services S. Each CCS is composition of atomic or composite services from all providers (IoT, Fog, and Cloud) that we have to select and compose according to a customer's constraints and preferences.
- Each service S_i has a set of QoS attributes qos_i^j.
- Each service S_i is defined by consumed energy S_i^{Energy}.

5.3. Problem Description

One problem is finding out an optimal composite service of ERP on IoT-Fog-Cloud infrastructure for minimizing energy consumption while optimizing customer QoS (i.e., response time, reliability, and security). To solve the problem, we use a three stages approach. First, we generate the virtual composite service (VCS), which includes all needed functional business tasks. Secondly, a multi-agent intelligence research algorithm gives all the possible composite service chains providing VCS. The optimal composite service solution is based on multi-objective QoS and energy-saving functions. Finally, we adapt the current composite service for incoming new customer's needs and/or context changes, including device power consumption. The problem is formalized as follows:

- The required workload $W_N^{required}$ is the total energy to be consumed while S_i is invoked on node N.

$$W_N^{required} = W_N^{current} + S_i^{Energy} \quad (4)$$

- The consumption rate R_N^i of service S_i on node N is defined as follows:

$$R_N^i = \begin{cases} \frac{Capacity_N}{Capacity_N - W_N^{required}} & if \quad Capacity_N > W_N^{required} \\ \infty & otherwise \end{cases} \quad (5)$$

where f_{energy}^i is considered a negative attribute that will be minimized and the service S_i on a node, N is considered infeasible when $R_N^i = \infty$.

- The energy function of the CCS: $f_{energy}(CCS) = agg\left(f_{energy}^i\right)$, is calculated according to Table 3 where i varies from 1 to m.

$$f_{energy}^i = S_i^{Energy} + R_N^i \quad (6)$$

- Each CCS is evaluated by combining the normalized QoS values of IoT, Fog, and Cloud services via Table 3. The total QoS are normalized by the following equations:

$$q_j'(CCS) = \begin{cases} \frac{agg(q_j) - agg\left(q_j^{min}\right)}{agg\left(q_j^{max}\right) - agg\left(q_j^{min}\right)} & If \quad agg\left(q_j^{max}\right) \neq agg\left(q_j^{min}\right) \\ 1 & Otherwise \end{cases} \quad (7)$$

$$q_j'(CCS) = \begin{cases} \frac{agg\left(q_j^{max}\right) - agg(q_j)}{agg\left(q_j^{max}\right) - agg\left(q_j^{min}\right)} & If \quad agg\left(q_j^{max}\right) \neq agg\left(q_j^{min}\right) \\ 1 & Otherwise \end{cases} \quad (8)$$

$agg\left(q_j^{max}\right)$: The highest combined score of the jth QoS criterion of CCS,

$agg\left(q_j^{min}\right)$: The lowest combined score of the jth QoS criterion of CCS.

$agg(q_j)$: The combined score of the jth QoS criterion of CCS.

Table 3. QoS and Energy Process Metrics.

QoS Dimensions	Iterative Flow	Parallel Flow	Choice Flow	Repeat Flow
Time	$\sum_{j=1}^{n} Time_j$	$Max_{j=1}^{n} Time_j$	$Max_{j=1}^{n} Time_j$	$Time_j * k$
Security	$Min_{j=1}^{n} Sec_j$	$Min_{j=1}^{n} Sec_j$	$Min_{j=1}^{n} Sec_j$	Sec_k
Availability	$\prod_{j=1}^{n} Av_j$	$\prod_{j=1}^{n} Av_j$	$Min_{j=1}^{n} Av_j$	$\left(Av_j\right)^k$
Cost	$\sum_{j=1}^{n} Cost_j$	$\sum_{j=1}^{n} Cost_j$	$\sum_{j=1}^{n} Cost_j$	$Cost_j \times k$
Energy	$\sum_{j=1}^{n} Energy_j$	$\sum_{j=1}^{n} Energy_j$	$\sum_{j=1}^{n} Energy_j$	$Energy_j \times k$

We define multi-objective functions (f_{energy}, f_{QoS}) to minimize the total energy and maximize the total quality of service of the CSS as follows:

$$Find\ CCS\ with \begin{cases} \min f_{energy}(CCS) \\ \max f_{QoS}(CCS) \end{cases} \quad (9)$$

where $f_{energy}(CCS)$ and $f_{QoS}(CCS)$ are the energy aggregated function of the CCS and QoS aggregated function respectively.

And $f_{QoS}(CCS) = \sum_{j=1}^{nb} w_i \times qos_{ccs}^{j}$ where: the dimension of QoS.

Equation (10) defines the final score by combining the energy and QoS of each CCS:

$$score(CCS) = \sum_{j=1}^{nb} w_i \times qos_{ccs}^{j} + w_{energy} \times energy_{ccs} \quad (10)$$

6. The Framework

To develop an efficient service selection and composition approach that can provide minimum energy costs and optimal QoS for customers, we propose a new adaptive and dynamic system based on NSGA-II for multi-objective optimization and cooperative agents with more capacities like intelligence, scalability, and efficiency. Our goal is to ensure coordination between different agents of the IoT-Fog-Cloud ERP system. Such a method discovers relevant services based on customer needs while selecting services with high QoS and optimal energy cost, and provides a continuously optimized composite service based on context changes and/or customer preference changes. To achieve this goal, we use CxQSIoTFogCloudSERP ontology, the multi-agent approach, and multi-objective optimization NSGA-II. Using both QoS and energy saving and modified NSGA-II multi-agents improves the solution accuracy, minimizing the execution time and energy consumption, and it is effectively used in remote controlling smart services ERP appliances, monitoring the state of the products with an exact angle measurement.

We present a new multi-objective function based on both QoS and energy consumption for selecting the optimal personalized composite service and saving energy with satisfactory accuracy and minimal execution time. To exploit this new multi-objective function, we propose two modes: dynamic and adaptive. The dynamic mode consists of three service management layers, IoT, Fog, and Cloud. The adaptive mode allows adaption of some business process parts according to different context changes and customer preferences to integrate IoT and Fog services and compose them with Cloud services. Such dynamic and adaptive modes are believed to efficient and economical, to be (potentially) utilized in several virtual services and their corresponding concrete services. They are stored, respectively, in distributed virtual service registries and concrete service registries of smart interactive devices, Fog servers, and Cloud. We leveraged the advantages of ontology-based agents by forming a well-hierarchical service management model to reduce the discovery time of appropriate services and to target potential service providers.

The work presented here focuses on a typical scenario, production, and transport process in a wheel production domain, as described in Section 2, and on different agents to generate an optimized smart ERP business process, as illustrated in Figures 1 and 2. In this work, we have used the multi-agent platform Jade [40] for cooperative IoT, Fog, and Cloud services composition. The CxQSIoTFogCloudSERP ontology is implemented using Protégé [41] to match customer needs to virtual services using inference rules.

6.1. General Architecture

Our system architecture of the SCwA- CCS (Semantic-Based Context-aWare Agents for the Composition of IoT with Fog and Cloud Services) is a semantic multi-agent architecture that is composed of the following agents (as illustrated in Figure 9): Customer, Semantic Context Broker, Context Supervisor, Knowledge, and SCwA-CCS Agent.

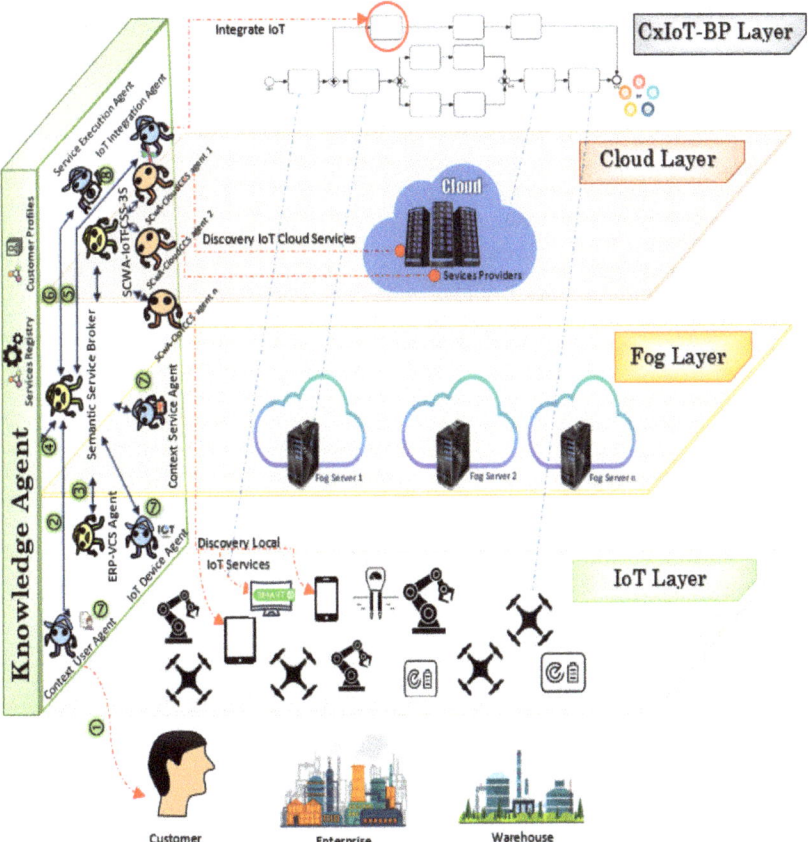

Figure 9. An overview of the proposed approach.

- **Customer**: The enterprise that wants a personalized IoT-ERP, which communicates with the system through GUI developed in Java.
- **Semantic Context Broker**: Acts as a semantic mediator between a customer and different service providers, with six kinds of agents interacting and collaborating to realize and manage concrete composition processes. The context broker is composed of the following four main agents:
    a. **SCwA-CCS Agent:** Allows the customer to specify their requirements and receive their responses. This agent calls the following agents:
        - *Virtual Business Process Planner*: Agent generates virtual business process. It is implemented using JESS as an inference engine and SWRL as an inference rule.

- *SCwA-IFCCS-3S*: Agent discovers IoT, fog, and Cloud services in terms of customer preferences and energy constraints.
- *ERP–CCS-3S Agent*: Finds the optimal concrete composite service.
- *Context Supervisor Agent*: Monitors context changes (*customer context, service context, IoT devices context, environment context, and usage context*).

  b. **ERP–CCS-3S Agent:** Generates an optimal business process in the form of a Concrete Composite Service (CCS) based on QoS and energy constraints, costumer's preferences, and description of IoT, Fog, and Cloud services.

  c. **SCwA-IFCCS-3S Agent:** Discovers local and Cloud services with respect to a customer's needs and energy constraints, and generates possible composite services. It calls *ERP-ICS* and *ERP-FCS Agent* for discovering IoT Fog services, respectively, and calls *ERP-CCS Agent* for discovering Cloud services.

- **Context Supervisor Agent:** Monitors customer's context in real-time by reading the data from the sensors. It also enables service context monitoring (network connection, Cloud server load rate, available service provider, QoS, etc.) and the IoT context (availability, energy capacity, etc.) to provide the desired services for the customer.
- **Knowledge Agent:** Manages Cloud services and customer profiles that are defined by the user. It also ensures continuous updates of the ontology model.
- **Service Executor Agent:** Calls composite services and execution context management. It also takes over the management of IoT, Fog, and Cloud services.
- **The Services and Client Repository:** Made up of five sets of data descriptions: data from the sensor network, definition of IoT services for all smart interactive devices, definition of Fog service providers, and Cloud service providers, and a description of customer profiles using our ontology.

6.2. Functional Model

Initially, a customer sends their request, which consists of a customized ERP business process in terms of functional and QoS needs to a *Semantic Context Broker* Agent who plays the role of a mediator. Here, the *Semantic Context Broker* Agent is a core agent that calls the *SCwA-IFCCS-3S* Agent that is responsible for launching several *ERP-ICS* and *ERP-FCS* agents, which can migrate to IoT devices and Fog servers to look for the requested services. These agents search for local IoT and Fog services with the help of a knowledge-agent able to make partial or full services compositions of requested ERP business processes.

The *ERP–CCS-3S* Agent, from all the possible composite services, selects the best solution (single composite service or a set of composite services) that optimizes service qualities and energy cost. The *ERP–CCS-3S* Agent uses a customer's preferences to select an optimal composite service. Once the optimal composite service has been selected and the ERP business process has been deployed, the deployment results are sent back to the customer. If the *SCwA-IFCCS Agent* does not find any local services that match the customer's needs, it transmits the customer's query to the *ERP-CCS Agent* to discover and compose services involved in the business process on Cloud servers.

The Context Supervisor Agent is responsible for handling the problems encountered during the execution of the various services (such as changes of user's needs and/or customer preferences) and incoming new customer's needs in order to continuously provide an optimal composite service.

- If it is a simple service update (replacing Cloud service by another IoT service), a *SCwA-IFCCS-3S* Agent will replace it with another equivalent IoT/Fog service based on the Knowledge agent.
- If it is about more than one Cloud service, he calls *ERP-ICS* and *ERP-FCS* agents. It moves towards IoT devices and Fog servers, and then sends discovered services to ensure adaptation of business processes. The discovered services will be transmitted to the *ERP–CCS-3S* Agent. The *ERP–CCS-3S* Agent composes these services and selects the optimal composite service.

6.3. Detailed Algorithms

The proposed SCwA-IFCCS-3S agent consists in dynamic and adaptive modes. The flowchart of *SCwA-IFCCS-3S* is illustrated in Figure 10. The algorithm for establishing an optimal composite service for IoT, Fog, and Cloud environments changes over time due to context changes (processing load, change of customer's needs, low latency, insufficient energy of devices, and other factors). As a result, it is critical to adapt the actual composite service while taking into account a variety of management elements. The algorithm consists mainly of three main stages: *virtual business process generating, parallel multi-objective concrete composite service optimization,* and *adaptation of concrete composite service*. The stages involved in the implementation of SCwA-IFCCS-3S is given below:

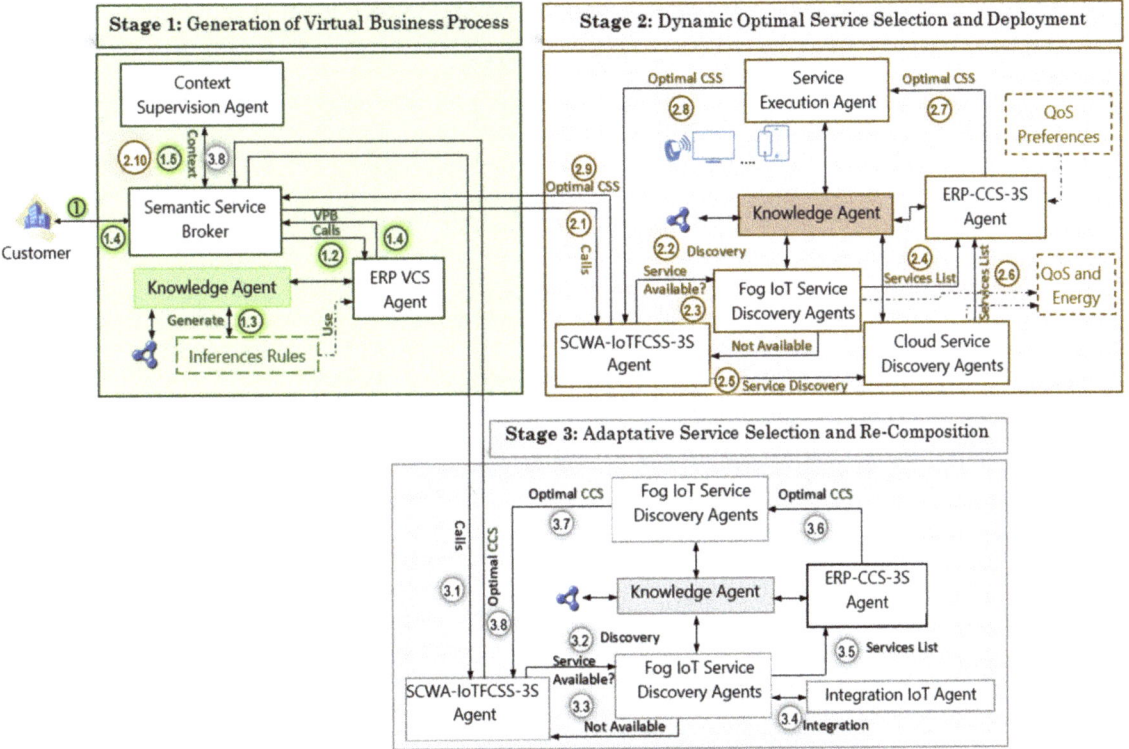

Figure 10. SCwA-IFCCS-3S functional model.

Stage 1. Generate virtual business process: In this stage, the research before the algorithm is used to systematically generate virtual IoT-BP according to the client's functional needs. Virtual services (e.g., tasks) are generated and optimized based on semantic relationships between VSs [28] and the reputation of each virtual service [42]. We automate this process to ensure dynamic changes in functional needs of the customer and his context.

Stage 2. Find local optimal solutions: We focus on dynamic multi-objective service selection and composition problems, we target each service provider, and the discovery agent closes in towards it. Each service provider, whether IoT, Fog, or Cloud, is modeled by ERP-ICS, ERP-FCS, and ERP-CCS agents, respectively, that cooperate to share local services. Therefore, each service provider has a set of services S_i which is a subset of the set S. The service provider agents create a set of NSGA-II agents to select the best Concrete Composite Service (CCS). It initializes them by a subset S_i and the energy cost and aggregated QoS to its composite service. Simultaneous use of penalties for energy consumption and penalty for QoS violation may hamper the reliability of service selection as well. The NSGA-II agent finishes when finding the local optimal composite service for a set of service providers, thus *ERP-ICS* and *ERP-FCS* end by sending a set of optimal composite services to ERP–CCS-3S Agent. The collaborative mechanism between different agents via messages allows us to find the best optimal composite service.

- **Step 1.** Each individual (composite service) is encoded by an array of n concrete services, called chromosome.
- **Step 2. Evaluation:** Each individual is evaluated using the aggregated fitness function.
- **Step 3. Selection:** To pick the optimal individual, tournament selection is used. In the proposed approach, the relationship of non-domination has been adapted by adding a constraint violation criterion of selection to respect the customer's constraints and energy consumption. The new CSS violation degree (deg_{v_CCS}) is evaluated to penalize IoT, Fog, and Cloud services that do not agree with both QoS and energy constraints. This degree is calculated as follows:

$$\deg_{v\_CCS} = w_{Qos} \times \deg_{v\_ccs}^{QoS} + w_{energy} \times \deg_{v\_ccs}^{energy} \quad (11)$$

$$\text{With } \deg_{v\_ccs}^{qos} = \begin{cases} C_{qos} - qos & \text{If } C_{qos} > qos \\ 0 & \text{Otherwise} \end{cases} \text{ and,}$$
$$\deg_{v\_ccs}^{energy} = \begin{cases} C_{energy} - energy & \text{If } C_{energy} > energy \\ 0 & \text{Otherwise} \end{cases} \quad (12)$$

  - Here w_{Qos} and w_{energy} denote weights of the QoS and energy, respectively, which is increased when the use of IoT services is greater than the Fog and Cloud services;
  - *qos*: Aggregated value of QoS.
  - *energy*: Aggregated value of energy.
  - C_{Qos}: The QoS constraint.
  - C_{energy}: The energy constraint.
  - For every CSS and its new dominance relation, we select the CSS based on the lowest dominance value.

- **Step 4. Merge and evaluate local optimal business process.** Merging and comparing the fitness value of local BPs of each agent with the other BPs of another agent. Based on largest fitness value, the global composite service is selected.

Stage 3. Adaptive Strategy for re-composition of services: Generally, the proposed approach is designed to be able to perform in two modes: Dynamic or Adaptive cases when context changes are present or not. We compute the adaptation cost in terms of energy and QoS difference between the current optimal CCS_t and new CCS_{t+1}, with desired services to be added or deleted in order to manage context changes. More precisely, the list of services to be added is S_{add} and the list of services to be deleted is S_{del}. The next CCS_{t+1} contains $\overline{CCS_t \cup S_{add}) \cap S_{del}}$. So, the multi-objective function of Equation (13) will include the third objective, which is the adaptation cost of newly added services and/or the cost of removed existing services in terms of energy and QoS. The new multi-objective function is defined in Equation (13).

$$\text{Find } CCS \text{ with } \begin{cases} \min f_{energy}(CCS_{t+1}) \\ \max f_{QoS}(CCS_{t+1}) \\ \min f_{cost}(CCS_{t+1} - CCS_t) \end{cases} \quad (13)$$

where $CCS_{t+1} - CCS_t$ is the difference between the current optimal *CCS* and new *CCS* according to the same VBP, since the existing optimal VBP does not change.

The detailed process is described as Algorithm 1 and illustrated in Figure 10.

Algorithm 1: Parallel NSGAII strategy for dynamic Cloud services composition

Inputs: agent_pop_size, max_itr, populations
Outputs: P_OP, Pareto optimal solutions
Begin
  1 : Generate initial agent populations as a set of CCS;
  2 : P_OP ← ∅
  3 : Iteration ← 0;
  4 : **while** (Iteration ≤ max_itr) **do**
  5 :   Fork (agent using agent_pop **in** populations)
  6 :   Evaluate the objective functions (QoS and Energy) for each individual;
  7 :   Local Pareto optimal ← Select best individuals by agent_pop;
  8 :   P_OP ← P_OP ∪ Local Pareto optimal
  9 :   s ← 0
10 :   **while** (s ≤ agent_pop/2) **do**
11 :   (P_1,P_2) ← select two individuals parents of current agent_pop;
12 :   (C1,C2) ← Crossing both parents (P_1,P_2) to obtain two children;
13 :   agent_pop.add(C_1); agent_pop.add(C_2);
14 :   s ← s + 1;
15 :   **end**
16 :   s ← 0
17 :   **while** (s ≤ agent_pop/2) **do**
18 :   new_ind ← mutate children(ind);
19 :   agent_pop.add(new_ind);
20 :   s ← s + 1;
21 :   **end**
22 :   **end**
23 :   P_OP ← Selection of optimal solution from all local Pareto-Optimal;
24 :   Update new global Pareto optimal solutions for all agents;
25 :   Iteration++;
26 : **end**
End
Return $P_{OPTIMAL}$;

The pseudo-code of the semantic context broker algorithm in adaptive intelligent strategy is illustrated in Algorithm 2. At first, the semantic context broker keeps up with the context changes. It then calls the *SCwA-IFCCS* Agent, which in turn calls *ERP-ICS* and *ERP-FCS* agents for discovering services. It ends with sending the services list to the *ERP–CCS-3S* Agent. The *ERP–CCS-3S* Agent selects the optimal composite service as the optimal fragment according to the customer's preferences, and then recomposes it with the current composite service, subsequently sending it to the service deployer. Finally, the optimal composite service fragment is deployed and results are returned to customer.

Considering an existing production *CCS* while monitoring services as a fragment of BP (see Figure 2) and the new monitoring IoT-BP generated by the adaptive multi-agent and NSGA-II algorithm (Figure 11), the adaptation cost is the cost of adapting existing BP to IoT-BP while considering only such a fragment. In this case, we need to add three services: new sensed temperature service (S1), receive temperature service (S1), and control temperature service (S3) of final products, and remove service monitoring (S'3). The adaptation cost will then equal the energy and time cost of deploying S1 and S2.

Algorithm 2: Adaptive composite service algorithm based on Multi-agent and NSGAII
Inputs: IoT-Business-Process, context, preferences, energy evolution
Outputs: Optimal adaptive IoT-Business-Process
Begin
    1 : Acquisition of context features
    2 : Calculate new constraints parameters
    3 : Recalculate the energy capacity of IoT-Devices
    4 : **For each** service s_{ij} in IoT-Devices **Do**
    5 :   **If** energy capacity of device $d_{ij} \leq energy$ of service s_{ij} **Then**
    6 :   Delete service s_{ij} from the available services
    7 :   **Foreach** service s_{ik} in $\mathcal{N}$ **Do**
    8 :   select services s_{ik} from $\mathcal{N}$ with minimal $f_{cost}(CCS_{t+1} - CCS_t)$
    9 :   End
   10 : **Endif**
   11 : **If** number of services = 1 **Then**
   12 :   Use ontology to find next equivalent service
   11 : **Else**
   12 :   Find the optimal IoT-BP with new changes according to the algorithm 1;
   13 : **Endif**
End
Return Optimal IoT-BP;

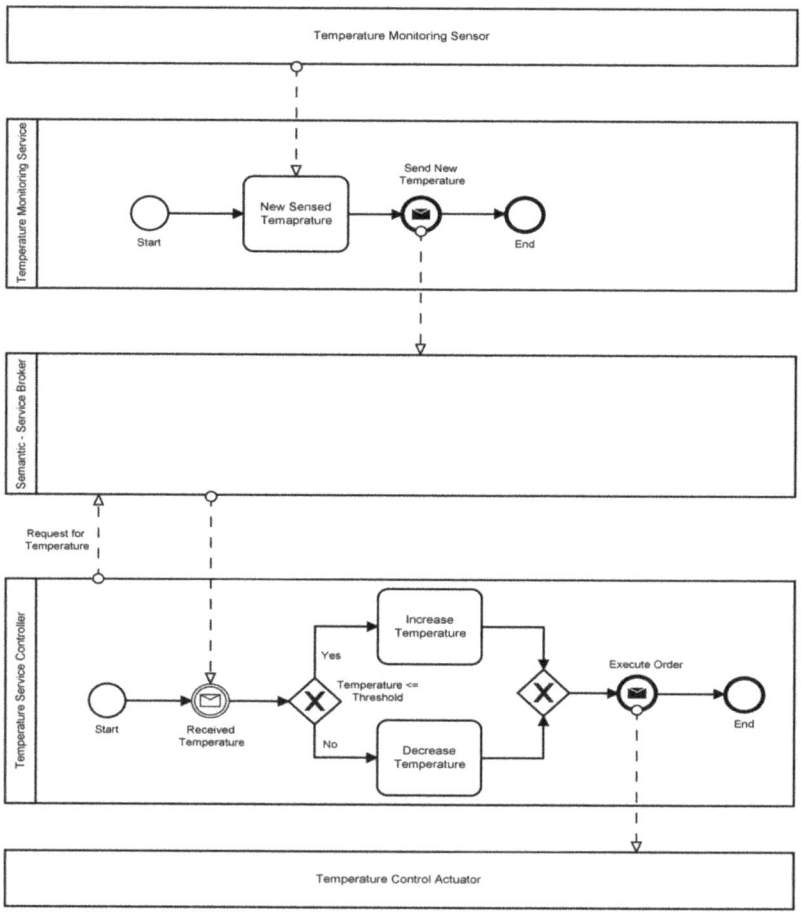

Figure 11. Dynamic semantic-based multi-agent flowchart for services composition.

7. Result and Discussion

This proposed work is implemented in Eclipse and compared with different cooperative strategies (Master/Slave, Ring, and Hybrid). The aim is to determine which strategy performed best in terms of QoS and energy for dynamic composition and adaptation of composite services. All experiments are conducted on a personal laptop that has the following configurations: 6th Generation Intel® CoreTM i5-8250U processor, 128 SSD with 12GB of RAM with Eclipse, and Matlab 2020 running on Windows 10 OS.

7.1. Dataset

To explore the strategy of parallel dynamic IoT Fog and Cloud services composition for ERP via intelligent agents, a random dataset has been used to record the QoS attributes and devices configuration that are depicted in Figure 11. The dataset comprises 10 different virtual services performed by 100 different concrete IoT, Fog, and Cloud services. We consider 333 IoT services, and 667 Fog/Cloud services. Each service contains QoS values for five attributes. Each QoS is scored by the weighted sum of their normalized values, and the energy of each service is scored by its normalized value. QoS values are generated randomly to ensure that the results of our experiments are not biased by a specific dataset. The comparisons are carried out between different strategies (NSGA-II Standard, NSGA-II Ring, NSGA-II Master/Slave, and hybrid strategy) based on Equation (10) of each evaluated solution. It should be noted that each experiment is evaluated through the following parameters: Max-Generation, Population-Size, Mutation-Ratio fixed to 0.3, and Agent-Number.

The effectiveness and energy saving of our proposal has been validated and evaluated through multiple experiments on random datasets. To show how our validation works, we detail its validation procedure as follows:

1. The user specifies their preferences in terms of QoS and required functionalities.
2. All performance metrics are evaluated and recorded by different concrete configurations on three multi-agent models: ring, master/slave, and hybrid using simulations of the different number of services.
3. These simulations include results obtained from four performance metrics (violation QoS degree, normalized average execution time, average energy consumption rate, and cost) with different multi-agent models through several simulated configurations.

7.2. The Accuracy Solution's Comparison

The contribution to smart ERP is evaluated within the NSGA-II algorithm without agents and the NSGA-II algorithm using different multi-agent strategies (Ring, Master/Slave, and Hybrid). These strategies are described as follows:

- **Master/Slave strategy:** The first strategy is focused on determining optimized global solutions from local best solutions. This strategy involves the services' local discovery and solution local evaluation using slaves, while the master agent chooses the best global solution from the selected best local solutions.
- **Ring strategy:** The second strategy explores some sub-populations in parallel to find the best solution. Each agent shares its optimal solution with its neighbor.
- **Hybrid Strategy:** To explore the advantages of both previous strategies and ensure further improvement, each slave sends its optimal solution to the master agent. Each slave shares his optimal solution with his neighbor. Finally, the master agent returns the global solution.

All three strategies are involved in determining quality solutions in an acceptable execution time. The results of Equation (10) are crucial in determining the most efficient solution in terms of QoS and energy. The obtained results on NSGA-II with/without agents are tabulated in Tables 4 and 5. It is quite remarkable that the inclusion of the multi-agent approach performed by both QoS and energy violation degree reached about 97% accuracy, which is perfect compared to the strategy without agents.

Table 4. NSGA-II without Agents in different Population Sizes and Generation Sizes.

Agents Number	Number of Generations									
	50	100	150	200	250	300	400	500	700	1000
20	0.799	0.774	0.761	0.746	0.792	0.905	0.766	0.771	0.859	0.779
40	0.774	0.811	0.849	0.834	0.845	0.820	0.851	0.880	0.847	0.782
60	0.881	0.934	0.876	0.895	0.874	0.872	0.910	0.869	0.849	0.905
80	0.931	0.875	0.915	0.91	0.977	0.941	0.912	0.941	0.977	0.905
100	0.808	0.909	0.939	0.939	0.972	0.925	0.977	0.888	0.946	0.974

Table 5. NSGA-II using different strategies Agents (Ring, Master/Slave, and Hybrid).

	Agents Number	Number of Generations									
		50	100	150	200	250	300	400	500	700	1000
Ring	2	0.78	0.884	0.895	0.911	0.921	0.943	0.971	0.973	0.975	0.976
	4	0.913	0.94	0.939	0.971	0.967	0.97	0.973	0.972	0.975	0.975
	6	0.912	0.941	0.949	0.969	0.971	0.972	0.972	0.973	0.975	0.977
M/S	2	0.796	0.821	0.811	0.856	0.861	0.849	0.851	0.859	0.964	0.975
	4	0.931	0.957	0.968	0.969	0.972	0.974	0.976	0.976	0.976	0.977
	6	0.937	0.945	0.966	0.973	0.975	0.977	0.977	0.977	0.977	0.977
Hybrid	2	0.856	0.895	0.932	0.868	0.948	0.975	0.977	0.977	0.974	0.977
	4	0.946	0.957	0.977	0.972	0.977	0.977	0.976	0.977	0.977	0.977
	6	0.961	0.945	0.977	0.977	0.975	0.977	0.977	0.977	0.977	0.977

Figure 12 depicts the scores of different strategies. It was evaluated using Equation (10) to measure how accurately the solution was made by each strategy, with a higher score indicating better performance. It appears that in general, the hybrid strategy has the highest scores across most of the cases, indicating that is the most accurate model. However, it is important to note that the results of the hybrid strategy vary depending on the number of generations. Some solutions are better determined by different strategies.

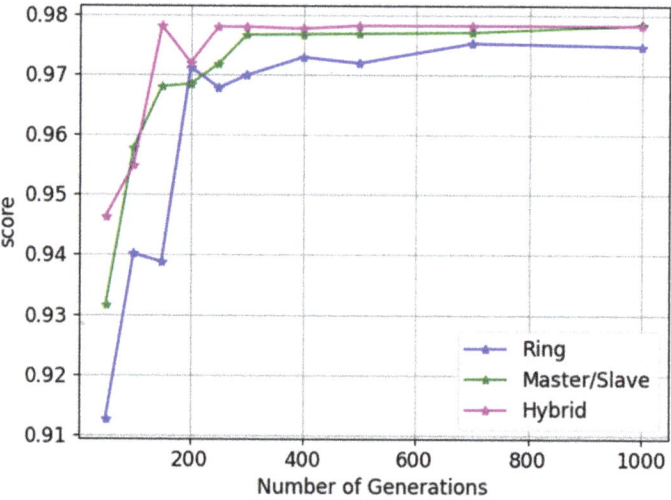

Figure 12. Scores of different strategy agents (Ring, Master/Slave, and Hybrid).

7.3. Performance Comparison

Figure 13 shows the execution time comparison for different strategies and different numbers of agents. Execution time is the measure of how the strategy performs the service dataset, where a minimum execution time means a perfect performance. From the results, it appears that the hybrid strategy has the minimum execution time. A practical increase of agents increased the volume of exchanged messages.

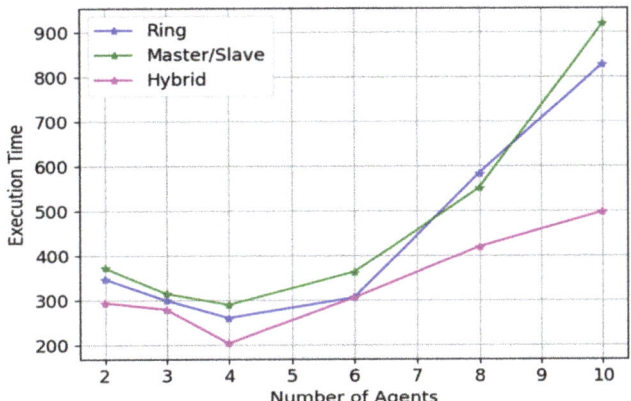

Figure 13. Execution time of different strategy agents (Ring, Master/Slave, and Hybrid).

7.4. The Energy Consumption in Dynamic and Adaptive Cases

For the dynamic case, we compared the proposed semantic-based parallel NSGA-II on three different multi-agent strategies (Ring, Master/Slave, and Hybrid) in terms of energy consumption and QoS. Figure 14 shows the results obtained by the two objective functions f_{energy} and f_{QoS} using the three agents' strategies. From the obtained results, we notice that the proposed dynamic agent-based NSGAII with hybrid strategy is better than other strategies in terms of optimized energy consumption and QoS scores.

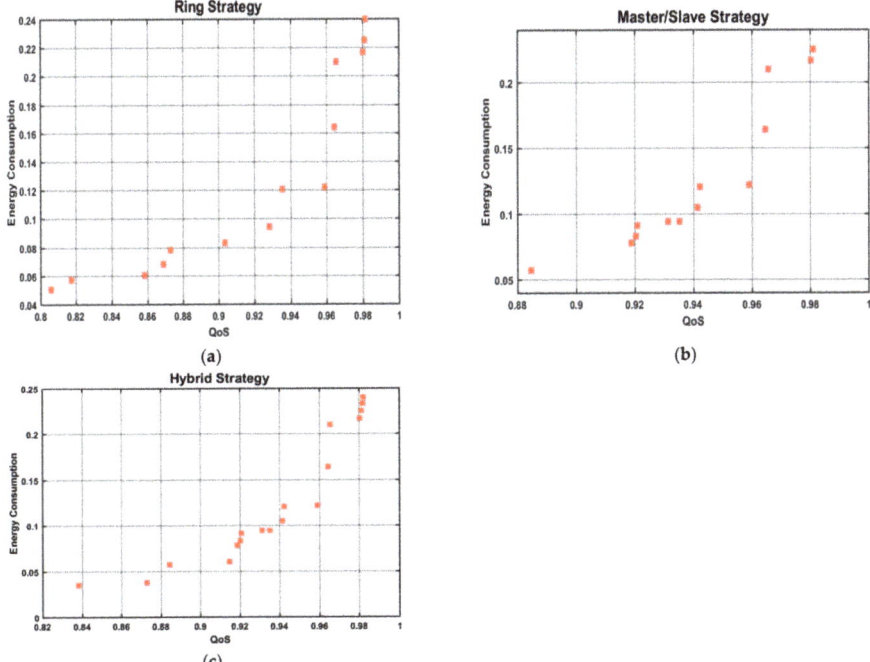

Figure 14. Pareto Optimal for Qos score and energy consumption in different strategies agents (**a**) Ring; (**b**) Master/Slave; (**c**) Hybrid.

For the adaptive case, we use three objective functions f_{energy} f_{QoS} and f_{cost} on incremental changes of workload, energy device capacity, and changes in customer needs/preferences. The strategy adopted is hybrid because the results in Figures 12–14 show that it is the best strategy compared with Ring and Master/Slave strategies. From the obtained results of Table 6, we can observe that the adaptation cost is correlated with the number of rejected services (infeasible services). This is due to the number of deleted services from the current CCS that should be replaced with other services having enough energy capacity. On the other hand, when the needs/preferences of the customer change, the new CCS replaces the majority or the totality of its services, so the adaptation cost increment changes considerably.

Table 6. Different adaptations of CCS with context changes.

CCS	f_{energy}		f_{QoS}		f_{cost}	Number of Unfeasible
	Min	Max	Min	Max		
CCS_t	0.0473	0.2481	0.8386	0.9827	0	6
CCS_{t+1}	0.0521	0.2672	0.8290	0.9844	0.22	29
CCS_{t+2}	0.0677	0.2581	0.8306	0.9879	0.37	41
CCS_{t+3}	0.0572	0.2509	0.8424	0.9883	1	73
CCS_{t+4}	0.0653	0.2618	0.8532	0.9811	1	112

Table 6 shows the different adaptations of CCS with context changes. The CCS_{t+1} is the composite service when the global workload of nodes changes. The CCS_{t+2} is the composite service adapted to new device's capacity after a lapse of execution time. CCS_{t+3} is the adapted composite service to the new needs of the customer. CCS_{t+4} is the new composite service when the needs of the customer change and the capacity of many devices are incapable of executing their elementary service (infeasible services). In the case of CCS_{t+3} and CCS_{t+4}, f_{cost} equal to 1 because all services of the current composite service are replaced.

7.5. Discussion

The above results show that the hybrid strategy is faster and more precise than Master/Slave and Ring strategy. This is due to the benefit of the advantages of both strategies while avoiding some of their disadvantages. In the Master-Slave strategy, it is hard to predict the behavior of the agents or the waiting time by the master to get the Pareto from his slave. In the ring strategy, the failure of one agent will break the chain and does not invoke the other Agent, which means breaking the whole topology. In the adaptive mode, practical increase of replaced services increased the cost function. In addition, simultaneous inclusion of agent-based approaches along with the ontology approaches and inheritance context domain adaptation techniques can be a promising remedy to the changes in the energy consumption problem.

8. Conclusions

This paper presents an adaptive and dynamic solution in IoT, fog, and Cloud environments. Semantic multi-agent intelligence features have been exploited to quickly compose IoT-Fog-Cloud services concerned by QoS and energy constraints, which help us to obtain a personalized ERP and highlight advantages regarding related works. The approach consists of guiding the dynamic services composition process by generating IoT virtual business processes that meet the customer's functional needs based on the ontology model and inference rules. A multi-agent approach using a three-stage algorithm has been proposed for the optimization of the IoT business process (IoT-BP) in terms of QoS and energy. Simultaneous inclusion of agents with QoS constraints' violation and energy device capacity can be a promising remedy to changes in customer needs and preferences. Another promising factor of this proposed article is the adapted IoT-BP to context changes and energy degradation on IoT devices. The Fog is a layer used in our work to select local services for reducing communications to the Cloud and performing global IoT-BP. An executable prototype is developed using the Protégé and Jade tools, subsequently displaying the overall optimal smart business process. Finally, we conclude that the proposed approach gives satisfactory performance on a random dataset involving multiple context change scenarios. It shows remarkable effectiveness through an intelligent adaptation mechanism for energy saving on IoT devices. The works presented use the deterministic measurement of QoS. In future work, an indeterministic measurement of QoS would be considered, as it has significant potential to improve the accuracy of smart factories, smart transportation systems, and smart production systems.

Author Contributions: Conceptualization, H.R. and A.A.; methodology, H.R. and A.A.; software, H.R. validation, H.R. and A.A.; formal analysis, H.R. and A.A.; investigation, H.R. and A.A.; resources, H.R. and A.A.; data curation, H.R. and A.A.; writing—original draft preparation, H.R.; writing—review and editing, A.A.; visualization, H.R.; supervision, A.A.; project administration, H.R. and A.A. All authors have read and agreed to the published version of the manuscript.

Funding: This research received no external funding.

Institutional Review Board Statement: Not applicable.

Informed Consent Statement: Not applicable.

Data Availability Statement: Data is available on request due to restrictions.

Acknowledgments: The researchers would like to thank the Deanship of Scientific Research, Qassim University for funding the publication of this project.

Conflicts of Interest: The authors declare no conflict of interest.

References

1. Olsen, K.A.; Sætre, P. ERP for SMEs–is proprietary software an alternative? *Bus. Process Manag. J.* **2017**, *13*, 379–389. [CrossRef]
2. Yousefpour, A.; Fung, C.; Nguyen, T.; Kadiyala, K.; Jalali, F.; Niakanlahiji, A.; Jue, J.P. All one needs to know about fog computing and related edge computing paradigms: A complete survey. *J. Syst. Archit.* **2019**, *98*, 289–330. [CrossRef]
3. Andročec, D.; Picek, R. Cloud ERP API Ontology. In Proceedings of the IEEE International Conference on Electrical, Computer and Energy Technologies (ICECET), Prague, Czech Republic, 20 July 2022; pp. 1–5.
4. Cornejo-Lupa, M.A.; Cardinale, Y.; Ticona-Herrera, R.; Barrios-Aranibar, D.; Andrade, M.; Diaz-Amado, J. OntoSLAM: An Ontology for Representing Location and Simultaneous Mapping Information for Autonomous Robots. *Robotics* **2021**, *10*, 125. [CrossRef]
5. Cornejo-Lupa, M.A.; Ticona-Herrera, R.P.; Cardinale, Y.; Barrios-Aranibar, D. A survey of ontologies for simultaneous localization and mapping in mobile robots. *ACM Comput. Surv. (CSUR)* **2020**, *53*, 1–26. [CrossRef]
6. Fiorini, S.R.; Carbonera, J.L.; Gonçalves, P.; Jorge, V.A.; Rey, V.F.; Heidegger, T.; Prestes, E. Extensions to the core ontology for robotics and automation. *Robot. Comput.-Integr. Manuf.* **2015**, *33*, 3–11. [CrossRef]
7. Olszewska, J.I.; Barreto, M.; Bermejo-Alonso, J.; Carbonera, J.; Chibani, A.; Fiorini, S.; Li, H. Ontology for autonomous robotics. In Proceedings of the 26th IEEE International Symposium on Robot and Human Interactive Communication (RO-MAN), Lisbon, Portugal, 28–31 August 2017; pp. 189–194.
8. Beetz, M.; Beßler, D.; Haidu, A.; Pomarlan, M.; Bozcuoğlu, A.K.; Bartels, G. Know rob 2.0—A 2nd generation knowledge processing framework for cognition-enabled robotic agents. In Proceedings of the 2018 IEEE International Conference on Robotics and Automation (ICRA), Brisbane, Australia, 20–25 May 2018; pp. 512–519.
9. Katti, B.; Plociennik, C.; Schweitzer, M. A jumpstart framework for semantically enhanced opc-ua. *KI Künstl. Intell.* **2019**, *33*, 131–140. [CrossRef]
10. Pop, C.B.; Vlad, M.; Chifu, V.R.; Salomie, I.; Dinsoreanu, M. A tabu search optimization approach for semantic web service composition. In Proceedings of the 10th International Symposium on Parallel and Distributed Computing, Cluj-Napoca, Romania, 6–8 July 2011; pp. 274–277.
11. Sasikaladevi, N.; Arockiam, L. Genetic approach for service selection problem in composite web service. *Int. J. Comput. Appl.* **2012**, *44*, 22–29. [CrossRef]
12. Long, J.; Gui, W. An environment-aware particle swarm optimization algorithm for services composition. In Proceedings of the International Conference on Computational Intelligence and Software Engineering, Wuhan, China, 11–13 December 2009; pp. 1–4.
13. Bouzary, H.; Frank Chen, F. A hybrid grey wolf optimizer algorithm with evolutionary operators for optimal QoS-aware service composition and optimal selection in cloud manufacturing. *Int. J. Adv. Manuf. Technol.* **2019**, *101*, 2771–2784. [CrossRef]
14. Yu, Q.; Chen, L.; Li, B. Ant colony optimization applied to web service compositions in cloud computing. *Comput. Electr. Eng.* **2015**, *41*, 18–27. [CrossRef]

15. Jula, A.; Othman, Z.; Sundararajan, E. A hybrid imperialist competitive-gravitational attraction search algorithm to optimize cloud service composition. In Proceedings of the 2013 IEEE Workshop on Memetic Computing (MC), Singapore, 16–19 April 2013; pp. 37–43.
16. Li, C.; Li, J.; Chen, H. A meta-heuristic-based approach for QoS-aware service composition. *IEEE Access* **2020**, *8*, 69579–69592. [CrossRef]
17. Yao, Y.; Chen, H. Qos-aware service composition using NSGA-II. In Proceedings of the 2nd International Conference on Interaction Sciences: Information Technology, Culture and Human, Seoul, Republic of Korea, 24–26 November 2009; pp. 358–366.
18. Li, L.; Cheng, P.; Ou, L.; Zhang, Z. Applying multi-objective evolutionary algorithms to QoS-aware web service composition. In Proceedings of the International Conference on Advanced Data Mining and Applications, Chongqing, China, 19–21 November 2010; Springer: Berlin/Heidelberg, Germany, 2010; pp. 270–281.
19. Wada, H.; Champrasert, P.; Suzuki, J.; Oba, K. Multi-objective optimization of SLA-aware service composition. In Proceedings of the 2008 IEEE Congress on Services-Part I, Honolulu, HW, USA, 6–11 July 2008; pp. 368–375.
20. Taboada, H.A.; Espiritu, J.F.; Coit, D.W. MOMS-GA: A multi-objective multi-state genetic algorithm for system reliability optimization design problems. *IEEE Trans. Reliab.* **2008**, *57*, 182–191. [CrossRef]
21. Sadeghiram, S.; Ma, H.; Chen, G. A Novel Repair-Based Multi-objective Algorithm for QoS-Constrained Distributed Data-Intensive Web Service Composition. In Proceedings of the International Conference on Web Information Systems Engineering, Amsterdam, The Netherlands, 20–24 October 2020; Springer: Cham, Switzerland, 2020; pp. 489–502.
22. Chen, C.S.; Liang, W.Y.; Hsu, H.Y. A cloud computing platform for ERP applications. *Appl. Soft Comput.* **2015**, *27*, 127–136. [CrossRef]
23. Bolu, A.; Korçak, Ö. Adaptive task planning for multi-robot smart warehouse. *IEEE Access* **2021**, *9*, 27346–27358. [CrossRef]
24. Poryazov, S.; Andonov, V.; Saranova, E.; Atanassov, K. Two Approaches to the Traffic Quality Intuitionistic Fuzzy Estimation of Service Compositions. *Mathematics* **2022**, *10*, 4439. [CrossRef]
25. Reffad, H.; Alti, A. New approach for optimal semantic-based context-aware cloud service composition for ERP. *New Gener. Comput.* **2018**, *36*, 307–347. [CrossRef]
26. Rosenberg, F.; Leitner, P.; Michlmayr, A.; Celikovic, P.; Dustdar, S. Towards composition as a service-a quality of service driven approach. In Proceedings of the IEEE 25th International Conference on Data Engineering, Shanghai, China, 29 March–2 April 2009; pp. 1733–1740.
27. Deng, S.; Huang, L.; Wu, H.; Wu, Z. Constraints-driven service composition in mobile cloud computing. In Proceedings of the 2016 IEEE International Conference on Web Services (ICWS), San Francisco, CA, USA, 27 June–2 July 2016; pp. 228–235.
28. Alti, A.; Laborie, S.; Phillipe, R. Dynamic semantic-based adaptation of multimedia documents. *Trans. Emerg. Telecommun. Technol.* **2014**, *25*, 239–258. [CrossRef]
29. Reffad, H.; Alti, A.; Roose, P. Cloud-based Semantic Platform for Dynamic Management of Context-aware mobile ERP applications. In Proceedings of the 8th International Conference on Management of Digital Ecosystems, Biarritz, France, 1–4 November 2016; pp. 181–188.
30. Asghari, S.; Navimipour, N.J. Review and comparison of meta-heuristic algorithms for service composition in cloud computing. *Majlesi J. Multimed. Process.* **2015**, *4*, 239–258.
31. Safaei, A.; Nassiri, R.; Rahmani, A.M. Enterprise service composition models in IoT context: Solutions comparison. *J. Supercomput.* **2022**, *78*, 2015–2042. [CrossRef]
32. Chang, H.; Liu, H.; Leung, Y.W.; Chu, X. Minimum latency server selection for heterogeneous cloud services. In Proceedings of the 2014 IEEE Global Communications Conference, Austin, TX, USA, 8–12 December 2014; pp. 2276–2282.
33. Strunk, A. QoS-aware service composition: A survey. In Proceedings of the 2010 Eighth IEEE European Conference on Web Services, Ayia Napa, Cyprus, 1–3 December 2010; IEEE: Piscataway, NJ, USA, 2010; pp. 67–74.
34. Rowley, H.V.; Peters, G.M.; Lundie, S.; Moore, S.J. Aggregating sustainability indicators: Beyond the weighted sum. *J. Environ. Manag.* **2012**, *111*, 24–33. [CrossRef]
35. Tarawneh, H.; Alhadid, I.; Khwaldeh, S.; Afaneh, S. An intelligent cloud service composition optimization using spider monkey and multistage forward search algorithms. *Symmetry* **2022**, *14*, 82. [CrossRef]
36. Huang, X.; Lei, X.; Jiang, Y. Comparison of three multi-objective optimization algorithms for hydrological model. In Proceedings of the International Symposium on Intelligence Computation and Applications, Wuhan, China, 27–28 October 2012; Springer: Berlin/Heidelberg, Germany, 2012; pp. 209–216.
37. Peng, S.; Guo, T. Multi-Objective Service Composition Using Enhanced Multi-Objective Differential Evolution Algorithm. *Comput. Intell. Neurosci.* **2023**, *2023*, 8184367. [CrossRef]
38. Tiwary, M.; Kumar, S.; Agrawal, P.K.; Puthal, D.; Rodrigues, J.J.; Sahoo, K.S.; Sahoo, B. Introducing network multi-tenancy for cloud-based enterprise resource planning: An IoT application. In Proceedings of the 2018 IEEE 27th International Symposium on Industrial Electronics (ISIE), Banja Luka, Bosnia and Herzegovina, 1–3 November 2018; pp. 1263–1269.
39. Tavana, M.; Hajipour, V.; Oveisi, S. IoT-based enterprise resource planning: Challenges, open issues, applications, architecture, and future research directions. *Internet Things* **2020**, *11*, 100262. [CrossRef]
40. Bellifemine, F.; Caire, G.; Poggi, A.; Rimassa, G. JADE: A software framework for developing multi-agent applications. Lessons learned. *Inf. Softw. Technol.* **2008**, *50*, 10–21. [CrossRef]

41. Musen, M.A. The protégé project: A look back and a look forward. *AI Matters* **2015**, *1*, 4–12. [CrossRef]
42. Tari, K.; Amirat, Y.; Chibani, A.; Yachir, A.; Mellouk, A. Context-aware dynamic service composition in ubiquitous environment. In Proceedings of the 2010 IEEE International Conference on Communications, Cape Town, South Africa, 23–27 May 2010; pp. 1–6.

Disclaimer/Publisher's Note: The statements, opinions and data contained in all publications are solely those of the individual author(s) and contributor(s) and not of MDPI and/or the editor(s). MDPI and/or the editor(s) disclaim responsibility for any injury to people or property resulting from any ideas, methods, instructions or products referred to in the content.

MDPI AG
Grosspeteranlage 5
4052 Basel
Switzerland
Tel.: +41 61 683 77 34

Applied Sciences Editorial Office
E-mail: applsci@mdpi.com
www.mdpi.com/journal/applsci

Disclaimer/Publisher's Note: The statements, opinions and data contained in all publications are solely those of the individual author(s) and contributor(s) and not of MDPI and/or the editor(s). MDPI and/or the editor(s) disclaim responsibility for any injury to people or property resulting from any ideas, methods, instructions or products referred to in the content.